C0-CCC-286

THE GENUS *COELOMOMYCES*

Contributors

Charles E. Bland
H. C. Chapman
John N. Couch
Brian A. Federici
Marshall Laird
Richard A. Nolan
Howard C. Whisler

THE GENUS *COELOMOMYCES*

Edited by

John N. Couch
Department of Biology
University of North Carolina
Chapel Hill, North Carolina

Charles E. Bland
Department of Biology
East Carolina University
Greenville, North Carolina

1985

ACADEMIC PRESS, INC.
Harcourt Brace Jovanovich, Publishers
ORLANDO SAN DIEGO NEW YORK
AUSTIN LONDON MONTREAL SYDNEY
TOKYO TORONTO

ACADEMIC PRESS, INC.
Orlando, Florida 32887

United Kingdom Edition published by
ACADEMIC PRESS INC. (LONDON) LTD.
24–28 Oval Road, London NW1 7DX

Library of Congress Cataloging in Publication Data
Main entry under title:

The Genus Coelomomyces.

1. Coelomomyces. I. Couch, John Nathaniel, Date.
II. Bland, Charles E.
QK621.C63G46 1985 589.2'5 85-1298
ISBN 0–12–192650–8 (alk. paper)

PRINTED IN THE UNITED STATES OF AMERICA

85 86 87 88 9 8 7 6 5 4 3 2 1

This work is dedicated to
Else Ruprecht Couch,
wife of John N. Couch

Contents

Contributors

Numbers in parentheses indicate the pages on which the authors' contributions begin.

CHARLES E. BLAND (1, 23, 81, 349), Department of Biology, East Carolina University, Greenville, North Carolina 27834

H. C. CHAPMAN (361), Gulf Coast Mosquito Research, Agricultural Research Service, U.S. Department of Agriculture, Lake Charles, Louisiana 70616

JOHN N. COUCH (1, 23, 81), Department of Biology, University of North Carolina, Chapel Hill, North Carolina 27514

BRIAN A. FEDERICI (299), Division of Biological Control, Department of Entomology, University of California, Riverside, California 92521

MARSHALL LAIRD* (369), Research Unit on Vector Pathology, Memorial University of Newfoundland, St. John's, Newfoundland, Canada A1C 5S7

RICHARD A. NOLAN (321), Department of Biology, Memorial University of Newfoundland, St. John's, Newfoundland, Canada A1B 3X9

HOWARD C. WHISLER (9), Department of Botany, University of Washington, Seattle, Washington 98105

* Present address: Whangaripo Valley Road, Rural Delivery No. 2, Wellsford, Northland, New Zealand.

Preface

Since their discovery by Keilin in 1921, fungi of the genus *Coelomomyces* have intrigued and mystified mycologists, entomologists, and all others who have come in contact with them. The primary reasons for this have been both biological and practical. In this regard, biological interest has centered around the unique host–parasite relationship between species of *Coelomomyces* and their obligate mosquito or chironomid primary hosts. One major outcome of this has been the elucidation of a life cycle that is unique among all known fungi, i.e., a pattern involving alternation of sporophytic and gametophytic phases between obligate mosquito/chironomid hosts and copepod hosts, respectively. Interest in this relationship continues today and is an incentive for study in several laboratories around the world.

A practical reason for interest in species of *Coelomomyces* involves the potential use of these organisms as biological control agents for mosquitoes, vectors for some of the worst diseases of humans. Because of their narrow host range and high degree of lethality, members of this group have long been considered primary candidates for biocontrol. In fact, interest in using *Coelomomyces* spp. in this capacity has been so keen that research on members of this genus has at times been given top priority status by the World Health Organization. At present, although complications have arisen in the development of a mass-produced inoculum for infecting large populations of mosquitoes, interest in using *Coelomomyces* spp. in biocontrol continues. We hope that the information presented herein will serve as an incentive for new and rapid developments in the culture and future successful utilization of *Coelomomyces* spp. as agents for the biological control of mosquitoes.

Some thoughts and comments relative to the origin of this volume are in order. In this regard, it was Keilin who in 1960, acting on "request[s] from workers in different parts of the world for some additional information as references on these parasites," invited M. Laird to prepare a review of the genus *Coelomomyces* for the journal *Parasitology*. Laird, responding favorably to this request, indicated to Keilin that in preparing the review he would solicit the assistance of a mycologist, J. N. Couch. Since 1961, when an agreement was reached between Laird and Couch to produce a review of the genus *Coelomomyces,* this work has been in preparation. During this time, Laird, among other things, has been involved in describing several new and interesting species of *Coelomomyces* and in conducting highly significant field studies concerning the use of *Coelomomyces stegomyiae* in controlling mosquitoes in the Tokelau Islands. Couch, on the other hand, developed the first highly successful means for *in vivo* laboratory culture of *Coelomomyces* spp. and established his laboratory as a repository where new forms of *Coelomomyces* from around the world were identified and studied. It is predominantly through the efforts of Laird and Couch that this volume has been made possible. Also of tremendous help in this study have been the large number of individuals who have deposited specimens of *Coelomomyces* in Couch's laboratory. From these, over 25 new species and varieties are described herein.

Finally, on behalf of John Couch and myself, sincere appreciation is extended to all who made this work possible, especially those who contributed time and effort in preparing chapters for this volume. Their contributions have been commendable, and any errors of omission or otherwise should be attributed to ourselves rather than to them. We hope that the information presented here will serve as a catalyst for new and expanded studies of one of the most interesting organisms known in nature. For ourselves, we are confident that our efforts have raised more questions than we have answered; questions not of strictly biological interest but also of practical interest!

On a personal note, I, as a former student of J. N. Couch and as a colleague of his for the past 15 years, have always found it a privilege and an honor to work with him. This is especially true in now helping to complete what has been a dream of his for some time.

Charles E. Bland

THE GENUS *COELOMOMYCES*

1 Introduction

CHARLES E. BLAND

Department of Biology
East Carolina University
Greenville, North Carolina

JOHN N. COUCH

Department of Biology
University of North Carolina
Chapel Hill, North Carolina

The genus *Coelomomyces* was established by Keilin (1921) to include a fungus found growing in the coelom of a single larva of the yellow fever mosquito, *Stegomyia scutellaris* Walker,[1] collected in Malaya by W. A. Lamborn. The type species, *C. stegomyiae* Keilin, was described to have thick-walled, yellowish, oval bodies that, along with a coenocytic, irregular mycelium, completely filled the body cavity of the larva. Although working with a preserved specimen, through careful observation Keilin was able to discern that the oval bodies were sporangia, characterized by a double-layered wall composed of a thin, hyaline inner layer and an outer layer that was thick, yellowish, pitted, and with a ''fine line'' (dehiscence slit) running from pole to pole. Keilin also described stages in spore formation and, by compressing sporangia between cover glass and slide, was able to observe release of spores through the dehiscence slit. Although Keilin concluded that the systematic position of *Coelomomyces* could not be determined until living material had been examined, he spec-

[1] A likely misidentification, probably *Aedes albopictus* (Skuse) (Laird, 1956a).

ulated that it bore some resemblance to members of the "Chytridineae," mentioning in particular the genera *Catenaria* and *Physoderma*.

Since the original description of *Coelomomyces*, interest in members of this genus has grown not only because of their unique nature as organisms but also because of their potential as agents for the biological control of mosquitoes. To illustrate this increased interest, during the period 1921–1960 there were only approximately 24 published references involving species of *Coelomomyces*. However, from 1960 to the present there have been over 135 publications dealing with members of this genus. While most of the more recent references are fully documented in the present work, there is the possibility that some of the earlier, historically significant papers may have been omitted or mentioned only briefly. Because of this, an annotated chronology of published references to the genus *Coelomomyces* for the period 1921–1960 follows.

Annotated Chronology of Published References to *Coelomomyces* spp.: 1921–1960

Keilin, 1921. Described *Coelomomyces stegomyiae* Keilin gen. nov., sp. nov. from *Stegomyia scutellaris* Walker (see note, p. 1) collected in Malaya.

Kobayashi, 1921. Reported infection of a larva of *Anopheles sinensis* from Okayama, Japan, by an unidentified species of *Coelomomyces*.

Bogoyavlensky, 1922. Described a *Coelomomyces*-like organism, *Zographia notonectae*, from a specimen of *Notonecta* sp. collected near Moscow, U.S.S.R. (see Keilin, 1927).

Ekstein, 1922. Described an apparent infection of *Aedes cinereus* and *Culiseta vexans* from Strasbourg, France, by an unidentified species of *Coelomomyces*.

Thompson, 1924. Reported an unidentified species of *Coelomomyces* from Malaya.

Keilin, 1927. Reduced *Zographia* Bogoyavlensky (1922) to synonomy with *Coelomomyces* Keilin (1921) and renamed *Zographia notonectae* as *Coelomomyces notonectae* (see Chapter 4, Section III).

Răsín, 1928. Reported *Coelomomyces chironomi* parasitic in larval chironomids collected near Prague, Czechoslovakia.

Răsín, 1929. Described hyphal structure and sporangial structure and development in *C. chironomi*.

Manalang, 1930. Described as "coccidiosis" an infection of larval *Anopheles tessellatus* and *Anopheles philippinensis* from the Philippines by what was probably an unidentified species of *Coelomomyces*.

Gibbins, 1932. Reported an unidentified species of *Coelomomyces* parasitizing larvae of *Anopheles funestus* and *Anopheles gambiae* from Uganda, Africa.

Feng, 1933. Reported an apparently unidentified species of *Coelomomyces* parasitizing larvae of *Anopheles hyrcanus* var. *sinensis* and *Culex tritaeniorhynchus* from Woosung, China.

Iyengar, 1935. Described *Coelomomyces anophelesicus* and *Coelomomyces indicus* from larval anophelines collected in India. Concurred with Keilin (1921) in placing the genus *Coelomomyces* in the Chytridiales.

Walker, 1938. Recognized four sporangial types from larvae of *Anopheles funestus* and *Anopheles gambiae* collected in Sierra Leone, Africa, and described the commonest type ("3") as *C. africanus*. Successfully induced infection when laboratory-bred larvae of *Anopheles gambiae* were placed in a concrete tank with water, soil, vegetation from a pool in which naturally infected larvae had been collected. First to mention potential of *Coelomomyces* spp. as a biological control of mosquitoes.

Haddow, 1942. Reported *Coelomomyces africanus* on larvae and adult females of *Anopheles funestus* and *Anopheles gambiae* from Kenya, Africa.

de Meillon and Muspratt, 1943. First to describe sporulation and zoospore flagellation in an unidentified species of *Coelomomyces* on *Aedes* sp. from Zambia, Africa.

Couch, 1945. Recognized the blastocladiaceous affinities of members of the genus *Coelomomyces* and erected the family Coelomomycetaceae in the order Blastocladiales. Described five new species (*C. dodgei, C. psorophorae, C. quadrangulatus, C. pentangulatus,* and *C. uranotaeniae*) from Georgia, United States, the first report of members of this genus in the Western Hemisphere.

Muspratt, 1946a. Reported three species of *Coelomomyces* on mosquito larvae collected in Livingstone, Zambia: "type a" (*C. indicus*) on several species of *Anopheles*, "type b" (described here as *C. orbicularis* sp. nov.) on *Anopheles gambiae* and *Anopheles squamosus*, and "type c" (*C. stegomyiae* var. *stegomyiae*) on *Aedes scatophagoides*. Observed presence of thin- and thick-walled sporangia and described sporulation of each.

Muspratt, 1946b. Obtained infection of *Anopheles gambiae* in a concrete trough by adding soil and resting sporangia from a positive infection site to the water in the trough. There was only one instance of infection involving 15 of 100 larvae.

Couch and Dodge, 1947. Subdivided *C. dodgei* Couch (1945) into *C. dodgei, C. punctatus,* and *C. lativittatus,* and described four new species and one new variety from Georgia, United States (*C. bisymmetricus, C. sculptosporus, C. cribrosus, C. keilini,* and *C. quadrangulatus* var. *irregularis*), and one new variety from Malaya (*C. quadrangulatus* var. *lamborni*). Provided also the first detailed study of sporulation in resting sporangia.

Van Thiel, 1954. Described *Coelomomyces walkeri* from two females of *Anopheles tesselatus* collected in Indonesia and claimed it to be identical with "type 1" of Walker (1938). Described "type 4" of Walker (1938) as *C. ascariformis* and considered it identical to the "coccidium" of Manalang (1930) (see Chapter 4, Section III).

Laird, 1956a. Described *C. solomonis* (see Chapter 4, Section III) on *Anopheles punctulatus* from Guadalcanal and *C. cairnsensis* on *Anopheles farauti* from Australia. Reported also the occurrence of *C. stegomyiae* on *Aedes scutellaris scutellaris* from Rennell Island and *C. indicus* on *Aedomyia catasticta* from Australia.

Laird, 1956b. Described *C. tasmaniensis* (see *C. psorophorae* var. *tasmaniensis*) on *Aedes australis* from Tasmania.

Lacour and Rageau, 1957. Reported an unidentified species of *Coelomomyces* on *Culex pipiens fatigans* from New Caledonia.

Garnham and Lewis, 1959. A highly questionable report of an unidentified species of *Coelomomyces* on *Simulium metallicum* from British Honduras.

Laird, 1959a. Described *C. quadrangulatus* var. *parvus* on *Culex taeniorhynchus siamensis* from Singapore and reported also the occurrence there of *C. cribrosus* on *Culex taeniorhynchus siamensis* (see *C. couchii*) and *C. stegomyiae* on *Aedes albopictus* and *Aedes aegypti*.

Laird, 1959b. Described *C. finlayae* and *C. macleayae* on *Aedes notoscriptus* from Australia and *C. stegomyiae* var. *rotumae* (see *C. stegomyiae* var. *stegomyiae*) on *Aedes* sp. from Rotuma Island. Reported *C. stegomyiae* var. *stegomyiae* on *Armigeres obturbans* from Singapore, *C. indicus* on *Anopheles subpictus* from Cambodia, and *C. cribrosus* (see *C. couchii*) on *Culex fraudatrix* from British North Borneo.

Shemanchuk, 1959. Reported *C. psorophorae* var. *psorophorae* on *Culiseta inornata* from Canada.

Since publication of the aforementioned early papers on *Coelomomyces* spp., many new species have been described and much has been learned of the biology of members of this genus. In fact, each of the unknowns cited by Couch (1945) for the life history of *Coelomomyces* spp. has now been answered, at least in part. While space will not permit a discussion here of all contributions providing insight into these unknowns, a brief mention of the major contributions is appropriate. The following, therefore, lists the major "unknowns" given by Couch (1945) and provides details of the primary publications contributing to an understanding of each:

1. "Whether the zoospores from the resting sporangia are the agents of infection." Although it was assumed in this statement that zoospores as "agents of infection" would potentially reinfect mosquito larvae, it has now been shown that such zoospores are indeed "agents of infection," but of copepods (Whisler, 1975; Zebold *et al.*, 1979; Wong and Pillai, 1980) or ostracods (Weiser, 1977), not mosquito larvae.

2. "How infection takes place." In infection involving the gametophytic phase, Zebold *et al.* (1979) and Wong and Pillai (1980) demonstrated clearly that infection of copepods occurs externally via pentration of the host's cuticle by a germ tube arising from the encysted zoospore. For infection of an ostracod, however, Weiser (1977) reported infection through the gut. In infection of mosquito larvae during the sporophytic phase, Travland (1979) and Zebold *et al.* (1979) showed infection to occur externally via penetration of the epicuticle.

3. "Whether infection is congenital." Since infection of adult female mosquitoes has been shown to always involve complete destruction of the ovaries (Walker, 1938; Coz, 1973; Zharov, 1973), the possibility that infection is congenital is minimal.

4. "What parts of the insect's body are attacked." Although the early studies of Keilin (1921) and Iyengar (1935) indicated an association between fungal hyphae and the host's fat body, the recent fine structural studies of Powell (1976) demonstrate clearly this relationship.

5. "Whether one species of fungus will parasitize more than one species of insect." The numerous collections of several species of *Coelomomyces* from a variety of hosts indicate clearly that at least some species have a wide host range. In this regard, the works of Romney (1971), Federici *et al.* (1975), and Wong and Pillai (1978) explore this question in some detail and indicate that degree of host specificity may vary from species to species.

6. "Whether there is an alternation of hosts." The studies of Whisler (1975), Federici and Roberts (1976), and Weiser (1977) indicate that in all forms studied thus far, copepods or ostracods (depending on species of *Coelomomyces*) serve with mosquitoes or chironomids as alternate hosts in the life cycle.

7. "Whether there is an alternation of generations." The recent work of Whisler (1975) confirms the existence of an alternation between discrete sporophytic and gametophytic phases in the life cycle of at least one species of *Coelomomyces,* with a similar phenomenon undoubtedly occurring in most if not all other forms.

8. "Whether any of the fungi can be cultivated on artifical media." The recent limited success of Shapiro and Roberts (1976) and Castillo and Roberts (1980) in culturing *C. psorophorae* and *C. punctatus,* respectively, provide evidence that *in vitro* culture of species of *Coelomomyces* is indeed possible.

9. "Whether any of these parasitic species can be used for artifical biological control." Although the usefulness of *Coelomomyces* spp. in the biological control of mosquitoes may be questioned on the basis of their complex life cycle and difficulty of culture, both the work of Umphlett (1970), demonstrating suppression of a natural mosquito population by an infection involving *C. punctatus,* and the pioneer work of Laird (1967), involving introduction and establishment of *C. stegomyiae* among two populations of mosquitoes in the Tokelau Islands, indicate that species of *Coelomomyces* may already be acting to control natural mosquito populations and may also have potential as an introduced biological control.

In addition to the works just mentioned that provided insight into the "unknowns" listed by Couch (1945), other studies have contributed significantly to our understanding of the basic biology of *Coelomomyces* spp. and are cited in the following chapters. However, in spite of all these studies many questions remain. Some of the more obvious include the following:

1. What is the function of the thin-walled sporangia occurring in many species? Do they serve as zoosporangia, such as found in some species of *Allomyces,* and produce zoospores that perpetuate the sporophytic phase in appropriate hosts? Why are forms with thin-walled sporangia rare in North America, but apparently common elsewhere?
2. Are there forms of *Coelomomyces* having life cycles comparable to those occurring in some species of *Allomyces* in which there is suppression of the gametophytic phase? If there are such forms, what are the major stages in their life cycle and on what hosts do they occur?
3. Are there hosts other than those known that may be susceptible to infection by species of *Coelomomyces*? What factors govern host specificity?
4. Are there genera of fungi similar to *Coelomomyces* that should be included in the Coelomomycetaceae?
5. Do all species exhibit relatively similar patterns of structure and development? Although this appears likely, few forms have been studied in detail.
6. What physiological and biochemical relationships exist between parasite and host, and how do these relate to habitat? In this regard, what are the factors governing distribution?
7. Can species of *Coelomomyces* be maintained in *in vitro* culture so as to allow completion of all life cycle phases? If so, can either phase be cultured in bulk (i.e., mass produced)?
8. What genetic factors are involved in speciation and mating?
9. Can this fungus be effectively used as an agent for the biological control of mosquitoes or possibly other disease-transmitting insects? If thin-walled sporangia are shown to produce zoospores infective to mosquito larvae, might they be mass cultured for use in control? What about the possibility of genetic manipulation of strains to produce forms especially virulent against given species of mosquitoes? Could zygotes be mass produced and held for later use in infection in natural mosquito populations?

Through continued study it is hoped that answers to many of these questions will be provided during the coming years. It is of course recognized also that from such study many new and exciting questions and discoveries will arise concerning members of the genus *Coelomomyces*.

REFERENCES

Bogoyavlensky, N. (1922). *Zoografia notonectae*, n.g., n. sp. *Arch. Soc. Russe Protistol.* **1**, 113.

Castillo, J. M., and Roberts, D. W. (1980). In vitro studies of *Coelomomyces punctatus* from *Anopheles quadrimaculatus* and *Cyclops vernalis*. *J. Invertebr. Pathol.* **35**, 144–157.

Couch, J. N. (1945). Revision of the genus *Coelomomyces* parasitic in insect larvae. *J. Elisha Mitchell Sci. Soc.* **61**, 124–136.

Couch, N. J., and Dodge, H. R. (1947). Further observations on *Coelomomyces*, parasitic on mosquito larvae. *J. Elisha Mitchell Sci. Soc.* **63**, 69–79.

Coz, J. (1973). Contribution à l'étude du parasitisme des *Anopheles* Ouest-Africain. Mermithidae et *Coelomomyces*. *Cah. ORSTOM, Ser. Entomol. Med. Parasitol.* **11**(4), 237–241.

de Meillon, B., and Muspratt, J. (1943). Germination of the sporangia of *Coelomomyces* Keilin. *Nature (London)* **152**, 507.

Ekstein, F. (1922). Beiträge zur Kentniss der Stechmückenparasiten. *Zentralbl. Bakteriol., Parasitenkd., Infektionskr. Hyg., Abt. 1: Orig.* **88**, 128.

Federici, B. A., and Roberts, D. W. (1976). Experimental laboratory infection of mosquito larvae with fungi of the genus *Coelomomyces*. II. Experimental with *Coelomomyces punctatus* in *Anopheles quadrimaculatus*. *J. Invertebr. Pathol.* **27**, 333–341.

Federici, B. A., Smedley, G., and Van Leuken, W. (1975). Mosquito host range tests with *Coelomomyces punctatus*. *Ann. Entomol. Soc. Am.* **68**, 669–670.

Feng, L. (1933). Some parasites of mosquitoes and flies found in China. *Lingnan Sci. J.* **12**, 23.

Garnham, P. C. C., and Lewis, D. J. (1959). Parasites of British Honduras with special reference to leishmaniasis. *Trans. R. Soc. Trop. Med. Hyg.* **53**, 12–40.

Gibbins, E. G. (1932). Natural malaria infection of house-frequenting *Anopheles* mosquitoes in Uganda. *Ann. Trop. Med. Parasitol.* **26**, 239.

Haddow, A. J. (1942). The mosquito fauna and climate of native huts at Kisumu, Keneya. *Bull. Entomol. Res.* **33**, 91–142.

Iyengar, M. O. T. (1935). Two new fungi of the genus *Coelomomyces* parasitic in larvae of *Anopheles*. *Parasitology* **27**, 440–449.

Keilin, D. (1921). On a new type of fungi: *Coelomomyces stegomyiae*, n.g., n. sp., parasitic in the body-cavity of the larva of *Stegomyia scutellaris* Walker (Diptera, Nematocera, Culicidae). *Parasitology* **13**, 225–234.

Keilin, D. (1927). On *Coelomomyces stegomyiae* and *Zografia notonectae*, fungi parasitic in insects. *Parasitology* **19**, 365–367.

Kobayashi, H. (1921). A fungus parasitic on mosquito larvae. *Dobutsugaku Zasshi* **33**, 475.

Lacour, M., and Rageau, J. (1957). Equête épidémiologique et entomologuique sur la filariose de Bancroft en Nouvelle Calédonie et Dépendances. *South Pac. Comm., Tech. Pap.* **110**, 1–24.

Laird, M. (1956a). Studies of mosquitoes and fresh water ecology in the South Pacific. *Bull.—R. Soc. N. Z.* **6**, 1–213.

Laird, M. (1956b). A new species of *Coelomomyces* (fungi) from Tasmanian mosquito larvae. *J. Parasitol.* **42**, 53–55.

Laird, M. (1959a). Parasites of Singapore mosquitoes, with particular reference to the significance of larval epibionts as an index of habit pollution. *Ecology* **40**, 206–221.

Laird, M. (1959b). Fungal parasites of mosquito larvae from the Oriental and Australian regions, with a key to the genus *Coelomomyces* (Blastocladiales: Coelomomycetaceae). *Can. J. Zool.* **37**, 781–791.

Laird, M. (1967). A coral island experiment: A new approach to mosquito control. *WHO Chron.* **21,** 18–26.

Manalang, C. (1930). Coccidiosis in *Anopheles* mosquitoes. *Philipp. J. Sci.* **42,** 279.

Muspratt, J. (1946a). On *Coelomomyces* fungi causing high mortality of *Anopheles gambiae* larvae in Rhodesia. *Ann. Trop. Med. Parasitol.* **40,** 10–17.

Muspratt, J. (1946b). Experimental infection of the larvae of *Anopheles gambiae* (Dipt., Culicidae) with a Coelomomyces fungus. *Nature (London)* **158,** 202.

Powell, M. J. (1976). Ultrastructural changes in the cell surface of *Coelomomyces punctatus* infecting mosquito larvae. *Can. J. Bot.* **54,** 1419–1437.

Răsín, K. (1928). *Coelomomyces chironomi* n. sp. *Vestn. Sjezdu Čes. Prirodozp.* **3,** 146.

Răsín, K. (1929). *Coelomomyces chironomi* n. sp. houba crzopasici v dutině tělní larev chironoma. *Biol. Spisy Vys. Sk. Zverolek.* **8,** 1–13.

Romney, S. V. (1971). Intergeneric transmission of *Coelomomyces* infection in the laboratory. *Utah Mosq. Abat. Assoc. Proc.* **24,** 18–19.

Shapiro, M., and Roberts, D. W. (1976). Growth of *Coelomomyces psorophorae* mycelium in vitro. *J. Invertebr. Pathol.* **27,** 399–402.

Shemanchuk, J. A. (1959). Note on *Coelomomyces psorophorae* Couch, a fungus parasitic on mosquito larvae. *Can. Entomol.* **91,** 743–744.

Thompson, A. (1924). Annual report of the mycologist for 1943. *Malay. Agric. J.* **12,** 246–251.

Travland, L. B. (1979). Structures of the motile cells of *Coelomomyces psorophorae* and function of the zygote in encystment on a host. *Can. J. Bot.* **57,** 1021–1035.

Umphlett, C. J. (1970). Infection levels of *Coelomomyces punctatus,* an aquatic fungus parasite, in a natural population of the common malaria mosquito, *Anopheles quadrimaculatus. J. Invertebr. Pathol.* **15**(3), 299–305.

Van Thiel, P. H. (1954). Trematode, gregarine and fungus parasites of *Anopheles* mosquitoes. *J. Parasitol.* **40**(3), 271–279.

Walker, A. J. (1938). Fungal infections of mosquitoes, especially of *Anopheles costalis. Ann. Trop. Med. Parasitol.* **32,** 231–244.

Weiser, J. (1977). The crustacean intermediary host of the fungus Coelomomyces chironomi. *Ceska Mykol.* **31**(2), 81–90.

Whisler, H. C. (1975). Life History of *Coelomomyces psorophorae. Proc. Natl. Acad. Sci. U.S.A.* **72,** 963–966.

Wong, T. L., and Pillai, J. S. (1978). *Coelomomyces opifexi* Pillai & Smith (Coelomomycetaceae: Blastocladiales). IV. Host range and relative susceptibility of *Aedes australis* and *Opifex fuscus* larvae. *N. Z. J. Zool.* **5,** 807–810.

Wong, T. L., and Pillai, J. S. (1980). *Coelomomyces opifexi* Coelomomycetaceae: Blastocladiales VI. Observations on the mode of entry into *Aedes australis* larvae. *N. Z. J. Zool.* **7,** 135–140.

Zebold, S. L., Whisler, H. C., Shemanchuk, J. A., and Travland, L. B. (1979). Host specificity and penetration in the mosquito pathogen *Coelomomyces psorophorae. Can. J. Bot.* **57,** 2766–2770.

Zharov, A. A. (1973). Detection of the parasitic fungus *Coelomomyces psorophorae* in *Aedes vexans* mosquitoes in the Astraken Oblast USSR. *Med. Parazitol. Parazit. Bolezni* **42**(4), 485–487.

2 Life History of Species of *Coelomomyces*

HOWARD C. WHISLER

Department of Botany
University of Washington
Seattle, Washington

I. INTRODUCTION

The early studies on *Coelomomyces* (1921–1960) were received with keen interest by mycologists and parasitologists alike. Here was a water mold that was a specific, obligate pathogen of mosquitoes: insects that continue to be vectors of man's worst infectious diseases. Even as Keilin (1921) and Couch (1945a) described the first species in the genus, questions were raised concerning the basic biology and life history of the fungus. Walker (1938) promptly undertook both *in vivo* and *in vitro* culture trials and encountered the erratic and frustrating results that subsequent workers experienced when they attempted to bring *Coelomomyces* into the laboratory (Walker, 1938; Muspratt, 1946; Laird, 1959; Madelin, 1968). It was, however, through these laboratory infection trials that it was finally possible to discover the missing actors in the scene. In such studies, Couch (1968) was able to maintain, in the laboratory, larvae of *Anopheles quadrimaculatus* infected with *Coelomomyces punctatus*. Soon thereafter, Pillai and Woo (1973) reported maintenance of laboratory infections of *Aedes australis* with *C. opifexi*, Whisler *et al.* (1974) of

Culiseta inornata with *Coelomomyces psorophorae,* and Federici and Roberts (1975) infections of *Aedes taeniorhynchus* with *C. psorophorae.* The latter workers speculated that the observed delay that existed between the addition of resting sporangia and the appearance of infected mosquitoes could be explained by an unknown phase in the life cycle of the fungus.

In 1974, Whisler, Zebold, and Shemanchuk provided a solution to the domestication and life cycle mysteries of *Coelomomyces.* Successful laboratory infections of *Culiseta inornata* by *Coelomomyces psorophorae* required the presence of the copepod *Cyclops vernalis.* Subsequent studies revealed that the copepod was a required, alternate host through which the fungus needed to pass before it could reinfect the mosquito host. The copepod was also discovered to be the site of gamete production and sexual fusion of the fungal parasite (Whisler *et al.,* 1975). The general life cycle pattern observed appeared homologous to that of the saprophytic water mold *Euallomyces,* as described by Emerson (1941) and Wilson (1952), and a theoretical life history for *Coelomomyces* was advanced (Whisler *et al.,* 1975) that indicated a unique coincidence of alternation of hosts with alternation of ploidal generations, with the sporophyte (diplophase) in the mosquito and the gametophyte (haplophase) in the copepod host. This proposed life history assumed nuclear fusion in the zygote prior to penetration of the mosquito, as subsequently established by Travland (1979a), and meiosis in the resting sporangia, as recently confirmed by Whisler *et al.* (1983).

II. LIFE CYCLE OF *COELOMOMYCES PSOROPHORAE*

The life history of *Coelomomyces psorophorae* in *Culiseta inornata* and *Cyclops vernalis* is summarized in Fig. 1. A brief synopsis of this cycle[1] will be used to introduce what appears to be the basic life history plan for all members of the genus *Coelomomyces.* This review will be followed by a discussion of specific variations on this theme, as exemplified by other species in the genus, and finally a consideration of how the life history of this organism relates to other fungi and parasites.

The fungus is most apparent in its mosquito host, whose body becomes packed with yellow-brown, ovoid, resting sporangia. Mature resting sporangia (Fig. 1A) from the infected larva sporulate readily when placed under appropriate environmental conditions. Sporulation appears to be a continuous process involving meiosis, differentiation of a discharge plug,

[1] See Chapter 3 for further details.

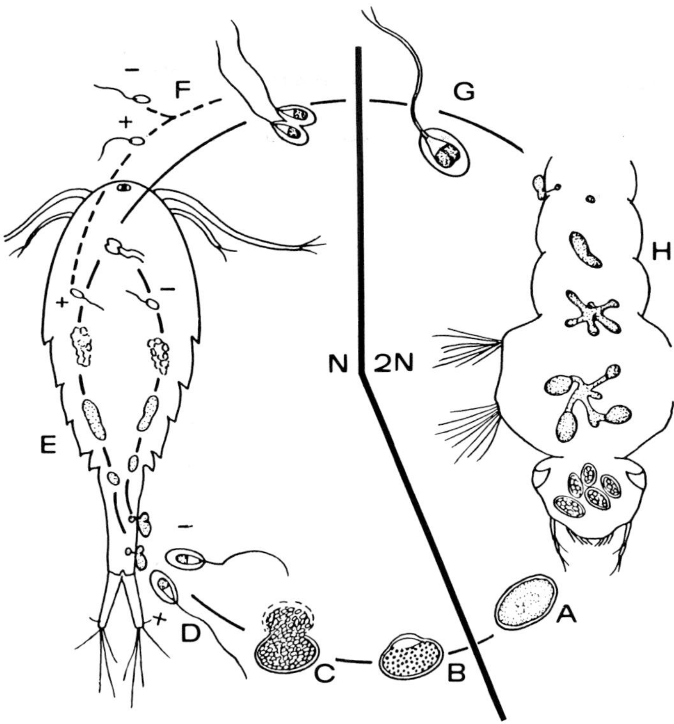

Fig. 1. Life cycle of *Coelomomyces psorophorae*. See text for description. (A) Resting sporangium. (B) "Go stage" of resting sporangium. (C) Release of meiospores. (D) Meiospores. (E) Development of gametophytic phase in the copepod *Cyclops vernalis* and release of gametes. (F) Fusion of compatible gametes. (G) Zygote. (H) Infection of larva of the mosquito *Culiseta inornata* and development of the sporophytic phase.

and opening of a preformed discharge crack in the sporangial wall (Fig. 1B). The cytoplasm of the resting sporangia differentiates into a number of posteriorly uniflagellate meiospores. The nuclei of these spores possess a plus or minus mating type, which presumably segregated at the preceding meiotic division. In *C. psorophorae,* the resting sporangia may remain for extended periods in this cracked, "go" phase (Couch, 1968; Whisler *et al.,* 1972). Meiospore release is triggered by reduction in the redox potential in the surrounding medium. Discharge is apparently dependent upon the dissolution of the discharge plug that occludes the discharge crack (Fig. 1C).

The free-swimming meiospore (Fig. 1D) terminates its flagellate phase and encysts when it encounters the alternate host, *Cyclops vernalis.* Many details of penetration and early phases of mycelial development are lacking, but the process appears to be similar to that seen in infection of the mosquito by the zygote. The relatively reduced and little-branched mycelium becomes obvious in the hemocoel of the animal (Fig. 1E), approximately 5 days after infection (at 20°C in well-fed animals).

When mature, the gametophyte or haplophase initiates gametogenesis approximately 11 hr after it is triggered by onset of darkness. This process is accompanied by apparent paralysis and then death of the host animal.

After nearly a week's cryptic existence in the copepod, the fungus engineers a brief but dramatic exit. The gametes and zygotes are seen to swarm in all parts of the copepod. The action becomes more and more frantic as the gametes from various gametangia join the scintillating mass. Eventually, the gametes and zygotes burst through the cuticle of the dead animal, frequently through tears in the intersegmental membranes of the metasome. If gametangia of opposite mating types occur in the same copepod host, the complementary isogametes will fuse to form active, motile zygotes. Gametes may also fuse with a gamete of the opposite mating type if it is encountered outside of the copepod (Fig. 1F). Cytological studies (Travland, 1979a) indicate that the zygote has a laterally tandem body structure, with the cellular organelles maintaining their individual, specialized spore organization, except for limited fusion of the nuclei and nuclear caps (Fig. 1G). The centrioles are closely paired and the two appressed flagella work as one.

When the zygote encounters a susceptible animal, it attaches and encysts. This process includes host recognition, cessation of swimming, flagellar retraction, completion of nuclear fusion, dedifferentiation of cellular organelles, and deposition of a cyst wall. This is followed by the formation of a lateral, budlike appressorium. A thin-walled penetration tube grows from the appressorium, through the cuticle, and into an epidermal cell of the mosquito. The protoplast is then injected into this cell coincident with the appearance of vacuoles in the external zygotic cyst (Travland, 1979b).

The fungus subsequently grows into the hemocoel (Fig. 1H). Within the hemolymph it multiplies as hyphal bodies, which eventually differentiate into a mycelium that may develop in all major blood cavities. At maturity, terminal ovoid segments of the mycelium differentiate into the thick, multiwalled resting sporangia. Significant mycelial development and formation of resting sporangia appear to prevent molting of the larva to the next instar. Late infections of fourth instar larvae, however, may lead to resting sporangia being produced in adults. This variant on the standard cycle is obviously important for dispersal to distant or unique habitats (i.e., tree holes).

The overall life cycle is temporally controlled by a variety of environmental signals. The sporulation of resting sporangia is initiated by light and appropriate temperature, and meiospore release is synchronized by a drop in the available oxygen. In the copepod, gametogenesis is triggered by onset of darkness, which may well time fungal development to that of the mosquito host. Thus, in both host transitions, there is a massive synchronous release of motile spores, which obviously increases the possibility of attaining the appropriate inoculum required for successful infection. In the case of gametogenesis, it would also increase the likelihood of a gamete meeting the appropriate mate.

Alternation of hosts or heteroecism is one of the notable features of this life history. This character is well known in animal parasitology, (e.g., *Plasmodium, Trypanosoma, Schistosoma),* but on the botanical side this pattern is restricted to the rust fungi (e.g., stem rust of wheat, white pine blister rust).

In contrast, alternation of ploidal generations is less obvious in animals, particularly the metazoa. In the Prostista and algae, it is a well established pattern in the Foraminifera, Myxomycetes, and brown and green algae. In the fungi, it may be experimentally maintained in some ascomycetous yeasts and is found in the Chytridiomycetes, especially the order Blastocladiales, including the genus *Coelomomyces*. The water mold *Allomyces* is perhaps the best known member of the order displaying this type of life history.

Coincidence of alternation of generations and hosts appears to be unique to *Coelomomyces*. The obligate parasite *Physoderma,* a relative in the Blastocladiales, may have alternation of generations, but both phases are on the same host (Sparrow, 1960; Olson and Lange, 1978). Perhaps the closest analogy would be to the rust fungi, where the dikaryophase predominates on one host and the haplophase on the other.

III. LIFE HISTORY OF SPECIES OF *COELOMOMYCES* OTHER THAN *COELOMOMYCES PSOROPHORAE*

Studies with other host–parasite combinations suggest that the basic life history diagramed in Fig. 1 is widespread in the genus, but each newly investigated species of *Coelomomyces* presents a number of interesting variations on the general format.

Thus, *C. punctatus* and *C. dodgei* have been shown to have "color-coded" gametangia, with one mating type being bright orange and the other light amber (Federici, 1977). *Cyclops vernalis* has also proved to be the alternate host for both *C. dodgei* (Federici and Chapman, 1977) and *C.*

punctatus (Federici and Roberts, 1976). This fact, plus the dramatic color difference in the mating types, facilitated the interesting hybridization studies reported in Chapter 5 of this volume.

Coelomomyces opifexi is one of the more completely studied members of the genus (Pillai, 1969, 1971; Pillai and Smith, 1968; Pillai and O'Loughlin, 1972; Pillai and Woo, 1973; Pillai *et al.,* 1976; Wong and Pillai, 1980). The coastal habitat of the two host mosquitoes, *Opifex fuscus* and *Aedes australis,* was of particular interest, as was the report by Pillai of a different alternate host for this fungus. Although a copepod, it proved to be *Tigriopus* sp. from the suborder Harpacticoida, rather than the Cyclopoida. In a series of well-designed and simple infection trials, Pillai *et al.* were able to quickly establish that the presence of *Tigriopus* sp. was essential for the infection of *Aedes australis* in laboratory culture. Other trials indicated the fungus from the mosquito infected the copepod, and transmission from copepod to copepod or mosquito to mosquito did not occur.

An even greater difference in the crustacen host was first suggested by Weiser (1976), who presented observations that implicated ostracods (crustaceans from an entirely different subclass than copepods) as the alternate host for *C. chironomi.* In their original description of *C. chironomi,* Weiser and Vávra (1964) noted an association of ostracods with the infected chironomids. Latter collections by Weiser revealed the presence of flagellated spores in the bodies of the ostracods.

These observations on *C. chironomi* alerted us to possible ostracod involvement in the life cycle of *C. utahensis*. This species was discovered by Romney *et al.* (1971), parasitizing representatives of three different genera of mosquitoes, *Culex, Culiseta,* and *Aedes,* occurring in the temporary rock-pools found in Arches and Canyonlands National Parks in southern Utah. These pools contain an abundant fauna of crustacea, including copepods. Experimental infection studies in the laboratory disclosed, however, that a small ostracod, *Potamocypris smaragdina,* was the alternate host for this mosquito pathogen. This particular host–parasite combination has proven to be particularly suitable for laboratory study, as all the living components may be stored in a dry condition. The ostracods are easily reared in quantity, and their eggs may be held dry for over 2 years. The mosquito host *Aedes atropalpus epactius* is also easy to rear, and its eggs may be stored until needed.

Another variation in the life history of some species of *Coelomomyces* relates to the production of more than one type of sporangia. In 1935, Iyengar noted that anopheline larvae were parasitized with a species of *Coelomomyces* that produced two types of sporangia—some yellow-brown and thick-walled with ridgelike ornamentations, and others that were smooth and colorless. This particular fungus, *C. indicus,* is a pathogen of medically significant anophelines, such as *An. gambiae,* and is under active study in a number of different laboratories. What the role of the two different sporangia may be is not yet clear. We do know that the thin-walled sporangia sporulate more readily than the thick-walled sporangia and that the basic structure of the sporangial wall is similar in both types (thick and thin) (Madelin and Beckett, 1972).

Another species of *Coelomomyces* that produces both thick- and thin-walled sporangia has been more recently described by Dubitskij *et al.* (1973) from the southern U.S.S.R. *Coelomomyces iliensis* shows some resemblance to *C. indicus* but is a parasite of culicine mosquitoes, particularly *Culex modestus*. In an extended series of observations Dubitskij (1978), Dubitskij *et al.* (1972, 1973, 1975), Dubitskij and Nam (1978), Nam (1978, 1981), and Nam and Dubitskij (1981) report a number of interesting variations in the life cycle of this species of *Coelomomyces*. They recognize four varieties in the species,[2] *C. iliensis* var. *"iliensis," C. iliensis* var. *"oriental," C. iliensis* var. *"culicis,"* and *C. iliensis* var. *"ormorii."* Five copepods are implicated as the alternate hosts for *C. iliensis* var. *iliensis* (Table I), with *Acanthocyclops viridis* and *A. longuidoides* being most important. *Cyclops strenuus* and *Mesocyclops* sp. are the alternate hosts for *C. iliensis* var. *orientalis,* while the crustacean hosts for the two other varieties are as yet unreported. As in the case of *C. indicus,* newly formed thick-walled resting sporangia resist germination, but the thin-walled sporangia sporulate readily. One of the most intriguing aspects of this research is the suggestion that there are two different life cycle patterns in *C. iliensis* var. *iliensis*. Two different thin-walled sporangia are reported to exist in the company of the thick-walled sporangia. The zoospores from one type of thin-walled sporangia are described as round and relatively small (1.4 × 1.4–3.0 μm), while the zoospores from the second category (author's "planetosporangia") are ovoid and larger (4.6 × 3.5 μm). The latter spore type is reported to infect the mosquito host directly, while the zoospores from the thick-walled sporangia and the first type of thin-walled sporangia infect the crustacean host and follow the basic life cycle pattern. The proposed existence of both short and full cycles in the development of *C. iliensis* var. *iliensis* is a very interesting possibility. Nam (1981) ties the development of observed field epizootics to the relative abundance of the two thin-walled sporangial types, with the short-cycling sporangia increasing with the onset of the epizootic. As full experimental details are not yet available, the existence of both short and full

[2] The varieties recognized by these authors have not been validated. See Chapter 4 for descriptions of varieties of this species that are recognized in this treatise.

TABLE I. Species of *Coelomomyces* with Known Alternate Hosts

Species of *Coelomomyces*	Crustacean host	Dipteran host	Reference
	COPEPODA		
	Cyclopoida		
C. psorophorae	*Cyclops vernalis*	*Culiseta inornata*	Whisler *et al.*, 1975
C. psorophorae	*Mesocyclops* sp.[a]	*Aedes vexans*	Nam, 1981
C. punctatus	*Cyclops vernalis*	*Anopheles quadrimaculatus*	Federici and Roberts, 1976
C. dodgei	*Cyclops vernalis*	*Anopheles crucians* *Anopheles quadrimaculatus*	Federici and Chapman, 1977
C. iliensis var. "*iliensis*"	*Acanthocyclops viridis*[a] *Acanthocyclops longuidoides*[a] *Microcyclops varicans*[a] *Eucyclops serrulatus*[a]	*Culex modestus*	Nam, 1981
C. iliensis var. "*orientalis*"	*Cyclops strenuus*[a] *Mesocyclops* sp.	*Culex orientalis*	Nam, 1981
	Harpacticoida		
C. opifexi	*Tigriopus angulatus*[b]	*Aedes australis* *Opifex fuscus*	Pillai *et al.*, 1976
C. psorophorae	*Harpacticus* sp.[a]	*Aedes togoi*	Nam, 1981
Coelomomyces sp.	*Elaphoidella taroi*[b]	*Aedes polynesiensis*	Toohey *et al.*, 1982
	OSTRACODA		
C. chironomi	*Heterocypris incongruens*[a]	*Chironomus plumosus*	Weiser, 1976
C. utahensis	*Potamocypris smaragdina*	*Aedes atropalpus epactius* *Culex tarsalis* *Culiseta inornata*	Unpublished observation of author

[a] Experimental infection data not available to author at time of preparation.
[b] See also Yeatman (1983).

cycles in *C. iliensis* var. *iliensis* must be treated as an intriguing hypothesis. Future research with this complex species will undoubtedly be followed with special interest.

IV. *COELOMOMYCES* AS A MEMBER OF THE BLASTOCLADIALES

Although we may have some confidence that we now understand the basic life history of several species of *Coelomomyces,* one should also anticipate the discovery of significant variations on the full-cycle format, such as those described for *C. iliensis*. This expectation is also justified by comparisons to the life cycle patterns found in other members of the Blastocladiales. Emerson (1941) has identified three different developmental cycles in the genus *Allomyces*. The basic life cycle of *Coelomomyces* (Fig. 1) corresponds to the full-cycle, *Euallomyces* type. The two other life cycle types in *Allomyces* involve drastic reduction in the gametophyte or haplophase. In the *Brachyallomyces* type, the haplophase is completely short-circuited and absent; in the *Cystogenes* type the gametophyte is reduced to a gametogenic cyst (see also Sparrow, 1960).

Both the "Brachy" and "Cyst" types are also common in other members of the Blastocladiales, including *Blastocladiella* (Couch and Whiffen, 1942) and *Catenaria* (Couch, 1945b). If either of these life cycle types were to occur in *Coelomomyces,* one would expect the crustacean host to be excluded from the host–parasite association. If such isolates exist in nature (Nam, 1981) or if they can be experimentally induced, they would also show significant differences during the course of epizootics of the mosquito host, a point of obvious interest in determining the value of *Coelomomyces* to biological control of mosquitoes.

In retrospect, one may wonder why it took so long to understand the life story of *Coelomomyces*. The early literature contains numerous hints as to the development of the fungus, but it took over 60 years to finally decipher the life cycle of this fungus. In defense of those of us who were attempting to solve this puzzle and were cognizant of the "Couch questions" (Chapter 1), it should be recalled that prior to 1974:

1. Heteroecism in fungi was not previously known or expected outside of the rust fungi.
2. The subgenus *Euallomyces* (Emerson, 1941) is the only well established example of isomorphic alternation of generations in the Blastocladiales, and fungi in general. Reports of full cycles in *Blastocladiella variabilis* (Harder and Sörgel, 1938; Stüben, 1939) have never been

confirmed, despite serious efforts by various workers to do so. Few other genera in the order Blastocladiales have been shown to have an independent, vegetative gametophyte, and isomorphic alternation of generations cannot be considered to be typical of the order.

3. Finally, the gametophyte in the living copepod is readily visible to the experienced observer only at maturity, after several days of cryptic development. It then enters into an abrupt gametogenesis, which is accompanied by a synchronous death of the infected host population and the resulting disappearance of the animals from the water column. In this situation the observer is at a clear disadvantage in chancing upon an infected alternate host.

V. FUTURE DIRECTIONS AND POSSIBLE HAZARDS

We are now gaining (Table I) a general understanding of life cycle patterns in the Coelomomycetaceae. Although some of these reports are based on limited observations of field material, as opposed to controlled and replicated tests in the laboratory, the information does point to the main types of host animals that may be parasitized by *Coelomomyces*. In examining a new strain of *Coelomomyces*, it is evident that copepods, especially cyclopoids, should receive attention, but ostracods should obviously not be ignored, particularly, as noted by Hall and Anthony (1979), in unique habitats such as the axils of bromeliads that support infected dipterans and abundant ostracods. Laboratory domestication of all probable hosts and repetitive infection trials appear to be essential to verify the identity of a potential crustacean or dipteran host. Most investigators working with species of *Coelomomyces* also remark on the necessity of rearing healthy and vigorous animals to obtain significant infection levels. Unless one can be sure of the fitness of the host, interpretation of infection trials can be a difficult task.

Another problem in host studies with species of *Coelomomyces* was brought forth recently during preliminary field trials with *C. psorophorae* in Alberta, Canada. In this, field-collected copepods included significant numbers of *Cyclops navus*. When these animals were exposed to the fungus some infection occurred, but at gametogenesis, the gametes and zygotes were unable to escape from the host. A casual observation of these copepods filled with swarming gametes might easily lead to a misinterpretation as to their role in maintaining the fungal life cycle. Further, field collections frequently include a number of copepods with a variety of other fungal parasites. Many of these, such as *Catenaria, Lagenidium,* and *Callimastix*, produce motile zoospores and can add to misinterpretation by the unwary investigator.

Of these, *Callimastix* is worthy of particular attention. In 1912, Weissenberg described a parasite of *Cyclops* that was characterized by the presence of multiflagellate cells in the hemocoel of the copepod host. Subsequent studies by Vávra and Joyon (1966) revealed that the motile cells were spores that had differentiated from a multinucleate plasmodium. Both *Callimastix* and *Coelomomyces* display a number of similarities in their morphology and development, most particularly the presence of a unique paracrystalline body in their planonts, which is rarely seen in other water molds.

However, the multiflagellate character of *Callimastix* (Manier and Loubès, 1978) distinguishes it from *Coelomomyces*, and further study is needed to clarify its relation to both *Coelomomyces* and the rumen fungus *Neocallimastix*. In this regard, Orpin (1975) isolated and cultured these anaerobic fungi and presented evidence that they are also related to the Chytridiomycetes. Life cycle studies of these organisms are needed and should yield new insights on these little-known molds and their potential relation to the genus *Coelomomyces*.

REFERENCES

Couch, J. N. (1945a). Revision of the genus *Coelomomyces*, parasitic in insect larvae. *J. Elisha Mitchell Sci. Soc.* **61,** 124–136.

Couch, J. N. (1945b). Observations on the genus *Catenaria*. *Mycologia* **37,** 163–193.

Couch, J. N. (1968). Sporangial germination of *Coelomomyces punctatus* and the conditions favoring the infection of *Anopheles quadrimaculatus* under laboratory conditions. *Proc. Jt. U.S.-Jpn. Semin. Microb. Control Insect Pests, 1967,* pp. 93–105.

Couch, J. N., and Whiffen, A. J. (1942). Observations on the genus *Blastocladiella*. *Am. J. Bot.* **29,** 582–591.

Dubitskij, A. M. (1978). "Biological Control of Bloodsucking Flies in the USSR." Alma Ata "Nauka" of the Kazakh SSR.

Dubitskij, A. M., and Nam, E. A. (1978). The possibility of asexual reproduction of *Coelomomyces iliensis* in larvae of *Culex modestus*. *Izv. Akad. Nauk* AS *Kaz. SSR, fer. Zool.*, Dep. VINITI, No. 3157-78, Dep., 1-7.

Dubitskij, A. M., Dzerzhinskii, V. A., and Deshevykh, N. D. (1972). Artificial infection of bloodsucking mosquito larvae by fungi of the genus *Coelomomyces*. *Izv. Akad. Nauk Kaz. SSR, Ser. Biol.* **4,** 36–38.

Dubitskij, A. M., Dzerzhinskii, V. A., and Danebekov, A. E. (1973). A new species of the pathogenic fungus of the *Coelomomyces* genus isolated of bloodsucking mosquito larvae. *Microl. Fitopatol.* **7,** 136–139.

Dubitskij, A. M., Deshevykh, N. D., and Dzerzhinskii, V. A. (1975). On the factors activating RS of an entomopathogenic fungus *Coelomomyces iliensis* Dubit., Dzerzh. et Daneb., under laboratory conditions. *Microl. Fitopatol.* **9,** 7–10.

Emerson, R. (1941). An experimental study of the life cycles and taxonomy of *Allomyces*. *Lloydia* **4,** 77–144.

Federici, B. A. (1977). Differential pigmentation in the sexual phase of *Coelomomyces dodgei*. *Nature (London)* **267,** 514–515.

Federici, B. A., and Chapman, H. C. (1977). *Coelomomyces dodgei:* Establishment of an *In vivo* laboratory culture. *J. Invertebr. Pathol.* **30,** 288–297.

Federici, B. A., and Roberts, D. W. (1975). Experimental laboratory infection of mosquito larvae of the genus *Coelomomyces*. I. Experiments with *Coelomomyces psorophorae* in *Aedes taeniorhynchus* and *Coelomomyces psorophorae* in *Culiseta inornata*. *J. Invertebr. Pathol.* **26,** 21–27.

Federici, B. A., and Roberts, D. W. (1976). Experimental laboratory infection of mosquito larvae with fungi of the genus *Coelomomyces*. II. Experiments with *Coelomomyces punctatus* in *Anopheles quadrimaculatus*. *J. Invertebr. Pathol.* **27,** 333–341.

Hall, D. W., and Anthony, D. W. (1979). *Coelomomyces* sp. (Phycomycetes: Blastocladiales) from the mosquito *Wyeomyia vanduzeei* (Diptera: Culicidae). *J. Med. Entomol.* **16,** 84.

Harder, R., and Sörgel, G. (1938). Über einen neuen planoisogamen Phycomyceten mit Generationswechsel und seine phylogenetische Bedeutung. *Nachr. Ges. Wiss. Goettingen* **3,** 119–127.

Iyengar, M. O. T. (1935). Two new fungi of the genus *Coelomomyces* parasitic in larvae of *Anopheles*. *Parasitology* **27,** 440–449.

Keilin, D. (1921). On a new type of fungus: *Coelomomyces stegomyiae* N.G., N. sp. parasitic in the body-cavity of the larva of *Stegomyia scutellaris* Walker (Diptera, Nematocera, Culicidae). *Parasitology* **13,** 225–234.

Laird, M. (1959). Parasites of Singapore mosquitoes, with particular reference to the significance of larval epibionts as an index of habitat pollution. *Ecology* **40,** 206–221.

Madelin, M. F. (1968). Studies on the infection by *Coelomomyces indicus* of *Anopheles gambiae*. *J. Elisha Mitchell Sci. Soc.* **84,** 115–124.

Madelin, M. F., and Beckett, A. (1972). The production of planonts by the thinwalled sporangia of the fungus *Coelomomyces indicus,* a parasite of mosquitoes. *J. Gen. Microbiol.* **72,** 185–200.

Manier, J.-F., and Loubès, C. (1978). *Callimastix cyclopis* Weissenberg, 1912 (Phycomycète, Blastocladiale) parasite d'un *Microcyclops* Claus, 1893 (Copépod, Cyclopoide) Récolté au Tchad. Particularités ultrastructurales. *Protistologia* **14,** 493–501.

Muspratt, J. (1946). Experimental infection of the larvae of *Anopheles gambiae* (Dipt., Culicidae) with a *Coelomomyces* fungus. *Nature (London)* **158,** 202.

Nam, E. A. (1978). Life cycle of the fungus *Coelomomyces iliensis,* a pathogen of mosquito larvae. *Izv. Akad. Nauk Kaz. SSR, Ser. Biol.* **6,** 24–29.

Nam, E. A. (1981). Biology of the entomopathogenic fungus *Coelomomyces iliensis* Dubit., Dzersh. et Daneb. and the possibilities of utilizing it as a biological control measure against bloodsucking mosquitoes. Author's abstract of dissertation for the degree of Candidate of Biological Sciences 1981. Alma Ata Inst. Zool., Sci. Kazakh SSR.

Nam, E. A., and Dubitskij, A. M. (1981). Systematics of *Coelomomyces* fungi that affect mosquitoes of the genus *Culex*. *Izv. Akad. Nauk Kaz. SSR, Ser. Biol.* **2,** 78–84.

Olson, L. W., and Lange, L. (1978). The meiospore of *Physoderma maydis*. The causal agent of *Physoderma* disease of maize. *Protoplasma* **97,** 275–290.

Orpin, C. G. (1975). Studies on the rumen flagellate *Neocallimastix frontalis*. *J. Gen. Microbiol.* **91,** 249–262.

Pillai, J. S. (1969). A *Coelomomyces* infection of *Aedes australis* in New Zealand. *J. Invertebr. Pathol.* **14,** 93–95.

Pillai, J. S. (1971). *Coelomomyces opifexi* (Pillai and Smith). Coelomomycetaceae. Blastocladiales. I. Its distribution and the ecology of infection pools in New Zealand. *Hydrobiologia* **38,** 425–436.

Pillai, J. S., and O'Loughlin, I. H. (1972). *Coelomomyces opifexi* (Pillai and Smith). Coelomomycetaceae: Blastocladiales. II. Experiments in sporangial germination. *Hydrobiologia* **40,** 77–86.

Pillai, J. S., and Smith, M. B. (1968). Fungal pathogens of mosquitoes in New Zealand. I. *Coelomomyces opifexi* sp. n., on the mosquito *Opifex fuscus* Hutton. *J. Invertebr. Pathol.* **11,** 316–320.

Pillai, J. S., and Woo, A. (1973). *Coelomomyces opifexi* (Pillai and Smith) Coelomomycetaceae: Blastocladiales. III. The laboratory infection of *Aedes australis* (Erickson) larvae. *Hydrobiologia* **41,** 169–181.

Pillai, J. S., Wong, T. L., and Dodgshun, T. J. (1976). Copepods as essential hosts for the development of a *Coelomomyces* parasitizing mosquito larvae. *J. Med. Entomol.* **13,** 49–50.

Romey, S. V., Boreham, M. M., and Nielsen, L. T. (1971). Intergeneric transmission of *Coelomomyces* infections in the laboratory. *Utah Mosq. Abat. Assoc. Proc.* **24,** 18–19.

Sparrow, F. K. (1960). "Aquatic Phycomycetes," 2nd ed. Univ. of Michigan Press, Ann Arbor.

Stüben, H. (1939). Über Entwicklungsgeschichte und Ernährungsphysiologie eines neuen niederen Phycomyceten mit Generationswechsel. *Planta* **30,** 353–383.

Toohey, M. K., Prakash, G., Goettel, M. S., and Pillai, J. S. (1982). *Elaphoidella taroi:* The intermediate copepod host in Fiji for the mosquito pathogen fungus *Coelomomyces. J. Invertebr. Pathol.* **40,** 378–382.

Travland, L. B. (1979a). Structures of the motile cells of *Coelomomyces psorophorae* and function of the zygote in encystment on a host. *Can J. Bot.* **57,** 1021–1035.

Travland, L. B. (1979b). Initiation of infection of mosquito larvae (*Culiseta inornata*) by *Coelomomyces psorophorae. J. Invertebr. Pathol.* **33,** 95–105.

Vávra, J., and Joyon, L. (1966). Étude sur le morphologie, le cycle évolutif et la position systèmatique de *Callimastix cyclopis* Weissenberg 1912. *Protistologia* **2,** 5–16.

Walker, A. J. (1938). Fungal infections of mosquitoes, especially of *Anopheles costalis. Ann. Trop. Med. Parasitol.* **32,** 231–244.

Weiser, J. (1976). The intermediary host for the fungus *Coelomomyces chironomi. J. Invertebr. Pathol.* **28,** 273–274.

Weiser, J., and Vávra, J. (1964). Zur Verbreitung *der Coelomomyces* Pilze in europäischen Insekten. *Z. Tropenmed. Parasitol.* **15,** 38–42.

Weissenberg, R. (1912). *Callimastix cyclopis* n.g., n. sp., ein geisselträgendes Protozoon aus dem Serum von *Cyclops. Sitzungsber. Ges. Naturforsch. Freunde Berlin* **5,** 299–305.

Whisler, H. C., Shemanchuk, J. A., and Travland, L. B. (1972). Germination of the resistant sporangia of *Coelomomyces psorophorae. J. Invertebr. Pathol.* **19,** 139–147.

Whisler, H. C., Zebold, S. L., and Shemanchuk, J. A. (1974). Alternate host for mosquito parasite *Coelomomyces. Nature (London)* **251,** 715–716.

Whisler, H. C., Zebold, S. L., and Shemanchuk, J. A. (1975). Life history of *Coelomomyces psorophorae. Proc. Natl. Acad. Sci. U.S.A.* **72,** 693–696.

Whisler, H. C., Wilson, C. M., Travland, L. B., Olson, L. W., Borkhardt, B., Aldrich, J., Therrien, C. D., and Zebold, S. L. (1983). Meiosis in *Coelomomyces*. *J. Exp. Mycol.* **7,** 319–327.

Wilson, C. M. (1952). Meiosis in *Allomyces*. *Bull. Torrey Bot. Club* **79,** 139–160.

Wong, T. L., and Pillai, J. S. (1980). *Coelomomyces opifexi* Pillai and Smith (Coelomomycetaceae, Blastocladiales). Observations on the mode of entry into *Aedes australis* larvae. *N. Z. J. Zool.* **7,** 135–139.

Yeatman, H. C. (1983). Copepods from microhabitats in Fiji, Western Samoa, and Tonga. *Micronesica* **19,** 57–90.

3 Structure and Development

CHARLES E. BLAND

Department of Biology
East Carolina University
Greenville, North Carolina

JOHN N. COUCH

Department of Biology
University of North Carolina
Chapel Hill, North Carolina

I. INTRODUCTION

The following account of structure and development in *Coelomomyces* spp. begins with the zoospore and follows stages in the life cycle described previously (Fig. 1 in Chapter 2). While efforts have been made to present complete details relative to species of *Coelomomyces,* references to other taxa are intentionally minimal and are included only to provide the reader with a basis for comparison. For additional references to related literature, one should consult the citations given for specific aspects of structure and development.

II. THE ZOOSPORE

The term zoospore is here used to refer to the structurally identical planonts produced by both the resting sporangia (meiosporangia producing meiozoospores) and, when present, the thin-walled sporangia of species of *Coelomomyces*.

Although zoospores of an unidentified species of *Coelomomyces* were first seen by de Meillon and Muspratt (1943) and described to be posteriorly uniflagellate, it remained for Couch (1945) and Couch and Dodge (1947) to observe the small nuclear cap present in such spores and to recognize their typically blastocladiaceous nature. Subsequent studies by Umphlett (1961) and Couch (1968) revealed further the structure of zoospores as seen with light microscopy. The following description is based primarily on their observations.

The typically ovoid (rarely spherical) zoospore (Figs. 2–9) measures 3.0–3.5 × 4.8–5.0 μm and has a posteriorly attached and directed, whiplash flagellum (Figs. 7 and 8) (approximately 16.0 μm long). The spore body, bound only by a thin membrane, contains a prominent, ovoid, central nucleus (Fig. 2), which is connected at its pointed end to an extension of the flagellum that continues into the spore body (Figs. 2 and 5); the flagellar extension was thought by Umphlett (1961) to be a rhizoplast (rootlet). A nuclear cap (Figs. 2 and 3) is situated atop the broad end of the nucleus. Lateral to the nucleus and along one side of the spore is a distinctive, disc-shaped side body[1] (Figs. 2–6) and associated lipid globules (4–8 per spore), which are arranged in a plate (Figs. 6 and 8). Small granules may be seen adjacent to the nucleus or scattered throughout the cytoplasm. Although zoospore organelles can be easily seen either in stained preparations or via phase-contrast microscopy, conventional, bright-field microscopy of living cells will reveal only flagellar attachment and possibly the nucleus and associated nuclear cap.

Fine-structural studies of zoospores have been limited to planonts arising from resting sporangia of *C. punctatus* (Martin, 1971) and *C. psorophorae* (Whisler *et al.*, 1972; Travland, 1979a) and from thin-walled sporangia of *C. indicus* (Madelin and Beckett, 1972). However, from these studies it is possible to discern clearly the typically blastocladiaceous but unique structure of the zoospore of *Coelomomyces* spp.

The zoospore (Figs. 10 and 23), bound by a simple plasmalemma, is propelled by a typical, eukaryotic flagellum (Figs. 11 and 16) (i.e., an axoneme composed of a 9 + 2 arrangement of microtubules surrounded

[1] The structure referred to as a side body by Umphlett (1961) probably represents the large mitochondrion seen via electron microscopy to occupy this position.

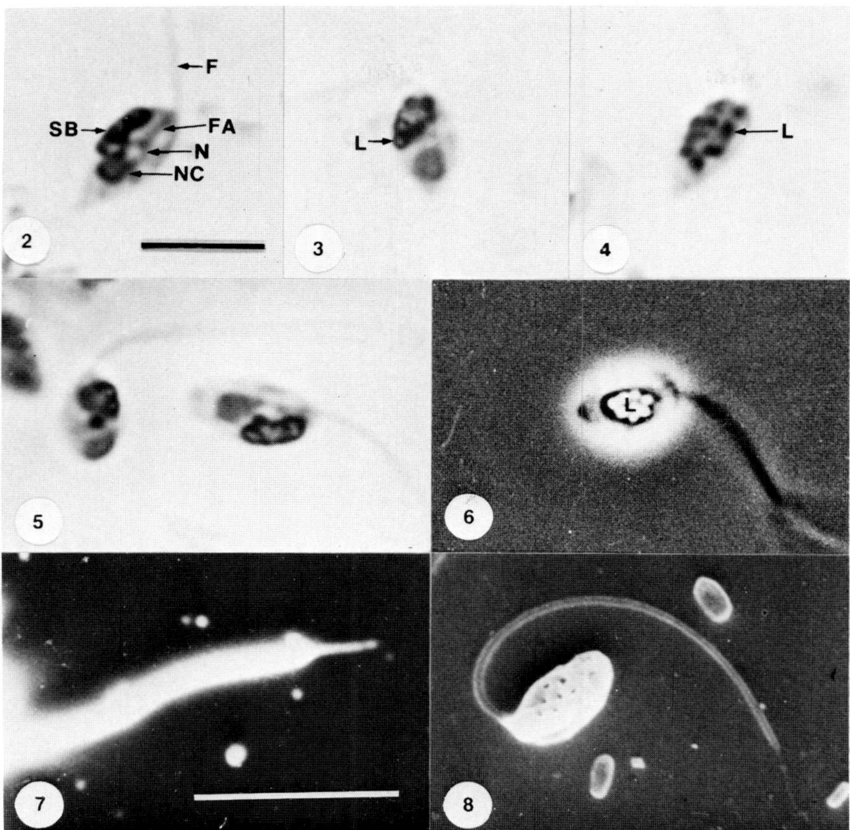

Figs. 2–5. Meiozoospores of *C. indicus* stained with 1% aqueous crystal violet. Flagellum (F), side-body complex (SB), nuclear cap (NC), nucleus (N), flagellar axoneme extension (FA), lipid droplets (L). Bar = 0.5 μm.

Fig. 6. Meiozoospore of *C. punctatus* viewed by phase-contrast microscopy. Note prominent plate of lipid droplets. Scale as for Fig. 2.

Fig. 7. Flagellar tip of *C. punctatus* viewed by dark-field microscopy to illustrate "whip lash." Bar = 0.5 μm.

Fig. 8. Scanning electron micrograph of meiospore of *C. dodgei*. Note pits in spore body where lipid droplets were leached out during processing. Scale as for Fig. 7. [From B. A. Federici, unpublished.]

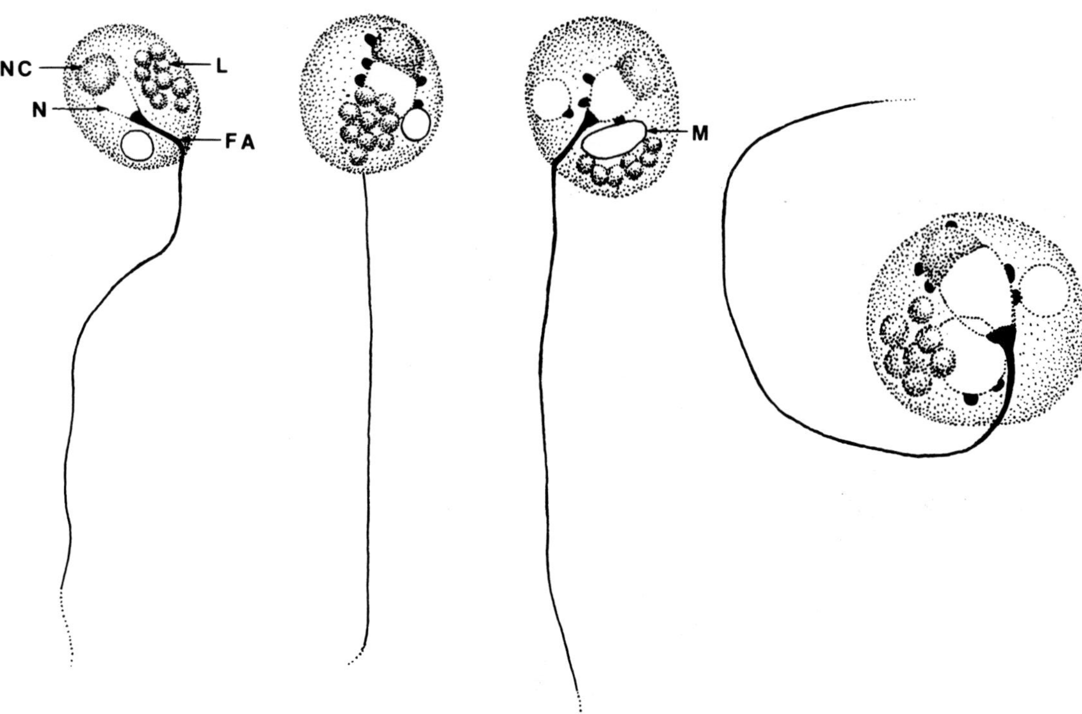

Fig. 9. Interpretative drawings of meiozoospores of *C. punctatus*. Nuclear cap (NC), nucleus (N), lipid (L), flagellar axoneme (FA), mitochondrion (M). [Redrawn from Umphlett (1961).]

by a flagellar sheath that is continuous with the plasmalemma of the spore). On entering the spore body, the axoneme is attached to the plasmalemma at its juncture with the flagellar sheath via a series of electron-dense props (Fig. 15). The axoneme, rather than terminating at this point in a proximal kinetosome, extends for a distance of 0.6–0.8 μm toward the beaked portion of the central, conical nucleus (Figs. 10, 12, 15, and 16). On reaching the nuclear beak, the axoneme does end in a typical kinetosome (Figs. 13 and 17), which, however, is not separated from the axoneme proper via a basal plate (Fig. 18). A ring of electron-dense, amorphous material surrounds the proximal portion of the kinetosome (Figs. 17 and 18) and extends a short distance along the beaked portion of the nucleus. Microtubules, in groups of two, three, or four (Travland, 1979a), originate from the amorphous material surrounding the kinetosome and extend along the nuclear beak and nuclear cap (Figs. 14 and 15), excluding the area of close proximity between nucleus and adjacent mitochondrion (Fig. 14) (Travland, 1979a). The nuclear cap sits apically on the nucleus and is composed of numerous ribosomes surrounded by a double-membraned envelope (Fig. 10), a common envelope being shared between the nuclear cap and the nucleus. Adjacent to and along one side of the nucleus and nuclear cap are, in sequence toward the exterior of the spore, a large mitochondrion, approximately 4–10 lipid droplets, a microbody, and a backing membrane (Figs. 14 and 19). Together, these components constitute the side-body complex as defined by Cantino and Truesdell (1970) for *Blastocladiella emersonii*. The mitochondrion, which may or may not surround the beaked portion of the nucleus (Travland, 1979a) and adjacent flagellar kinetosome, is connected to the kinetosome by fine fibrils (Figs. 17 and 18). The lipid droplets are sandwiched in a layer between the mitochondrion and the microbody. The backing membrane, which separates the microbody from the plasmalemma of the spore, is continuous with segments of smooth endoplasmic reticulum. Scattered throughout the cytoplasm are gamma-like bodies (Fig. 22) (Travland, 1979a), profiles of smooth endoplasmic reticulum, adhesion vesicles (Fig. 22) (Travland, 1979a), and vacuoles of various sizes with electron-transparent contents. Present also in the cytoplasm is an unusual paracrystalline body (Figs. 20 and 21) [see Travland (1979a) for a discussion of its possible nature and Lange and Olson (1979) for references to other fungi having a similar inclusion], appearing either as parallel rods or as a hexagonal pattern of electron-dense, circular profiles, depending on the plane of sectioning.

According to Travland (1979a), unique and/or interesting features of the zoospores of *Coelomomyces* spp. include a lack of striate, flagellar rootlets, the absence of a second or vestigial kinetosome, the lack of a termi-

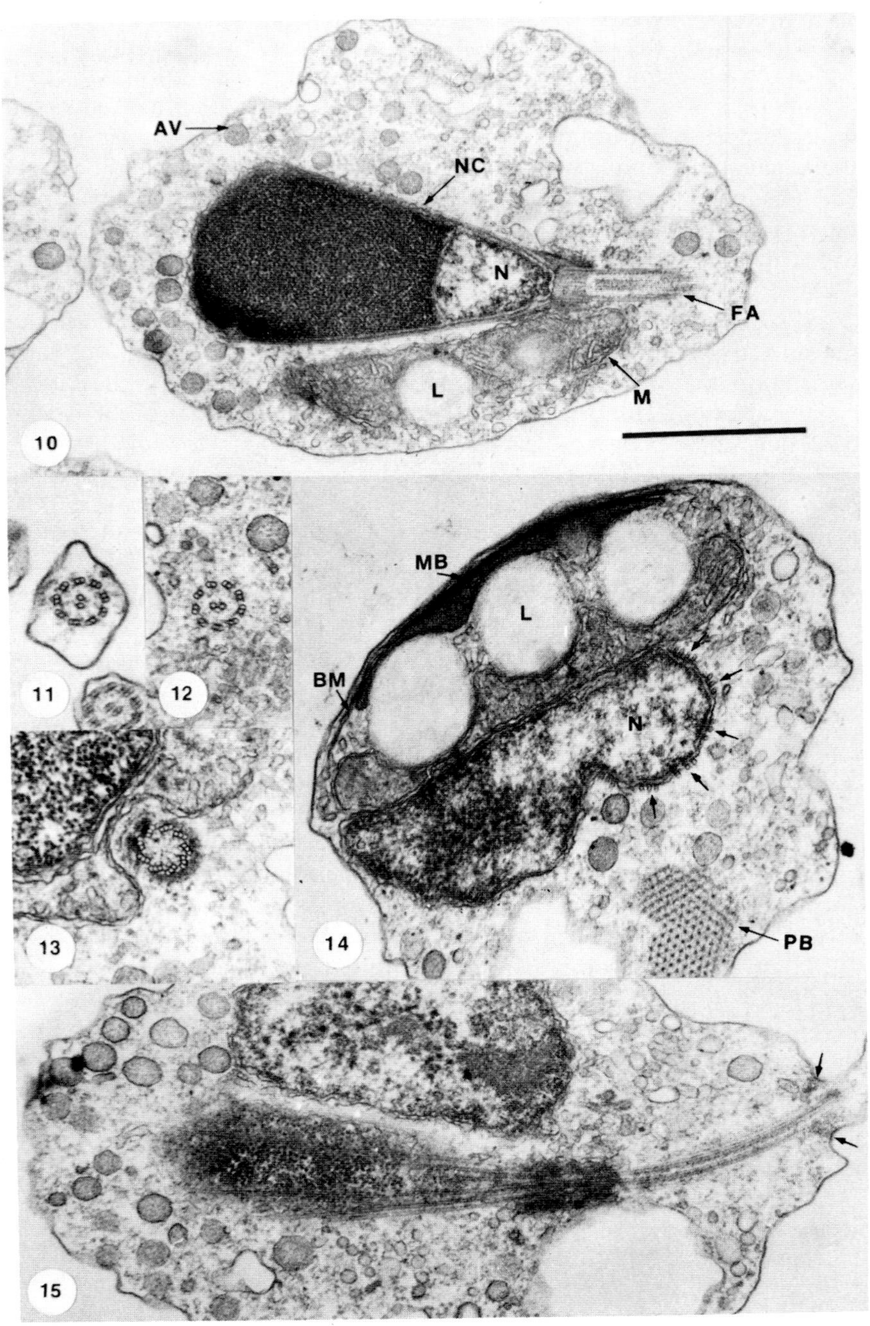
AV
NC
N
FA
L
M
10
MB
L
BM
N
11
12
13
14
PB
15

nal plate between kinetosome and flagellar axoneme, the irregular arrangement of the ascending microtubules surrounding the nuclear beak, the presence of an extended cytoplasmic axoneme, and the presence of a paracrystalline body. Based on comparisons between zoospore structure in parasitic versus saprophytic members of the Blastocladiales, Travland (1979a) postulates that the features cited above for zoospores of *Coelomomyces* spp. may be consistent with its parasitic mode of nutrition. Lange and Olson (1979), in a comprehensive survey of structural relationship among zoospores of uniflagellate phycomycetes, recognized similarities between zoospores of species of the genera *Coelomomyces, Blastocladiella, Catenaria, Allomyces,* and *Physoderma* and designated such zoospores as "type 4." Although the zoospore of *Coelomomyces* spp. was included in this group, it was noted that zoospores of members of this genus are distinct in having a centrally located nucleus and lacking a striated rootlet.

III. INFECTION OF ALTERNATE HOSTS AND DEVELOPMENT OF THE GAMETOPHYTE

Infection of alternate hosts by zoospores of *Coelomomyces* spp. has been reported to occur either through the gut, in infection of the ostracod *Heterocypris* by *C. chironomi* (Weiser, 1977), or through the exoskeleton, in infection of the copepods *Cyclops vernalis* by *C. psorophorae* (Zebold *et al.,* 1979) and *C. dodgei* (Federici, 1980) and by *C. opifexi* (*C. psorophorae* var. *tasmaniensis*) (Wong and Pillai, 1980).

In infection involving *Heterocypris* sp., Weiser (1977) reports infection to occur via zoospores, from ingested resting sporangia, that "cross" the gut wall and enter the body cavity. Weiser observed further that following penetration by 30–40 zoospores, development of the fungus is first evident as oval hyphae occurring throughout the coelom and appressed to the gut and fat body. Such hyphae reportedly grow into large oval masses

Figs. 10–15. Electron micrographs of meiozoospores of *C. punctatus.* Fig. 10. Longitudinal section. Fig. 11. Cross-section of flagellum. Fig. 12. Cross-section of flagellar axoneme inside spore body. Fig. 13. Cross-section of flagellar basal body. Fig. 14. Cross-section through meiozoospore at the upper level of the upper conical portion of the nucleus. Microtubule triplets (arrows) are spaced equidistant about the nuclear periphery. Fig. 15. Tangential section through a meiozoospore showing microtubular nuclear "basket" and flagellar props (arrows). Adhesion vesicles (AV), lipid (L), mitochondrion (M), nucleus (N), nuclear cap (NC), paracrystalline body (PB), flagellar axoneme (FA), microbody (MB), backing membrane (BM). Bar = 0.5 μm. [From Martin (1971).]

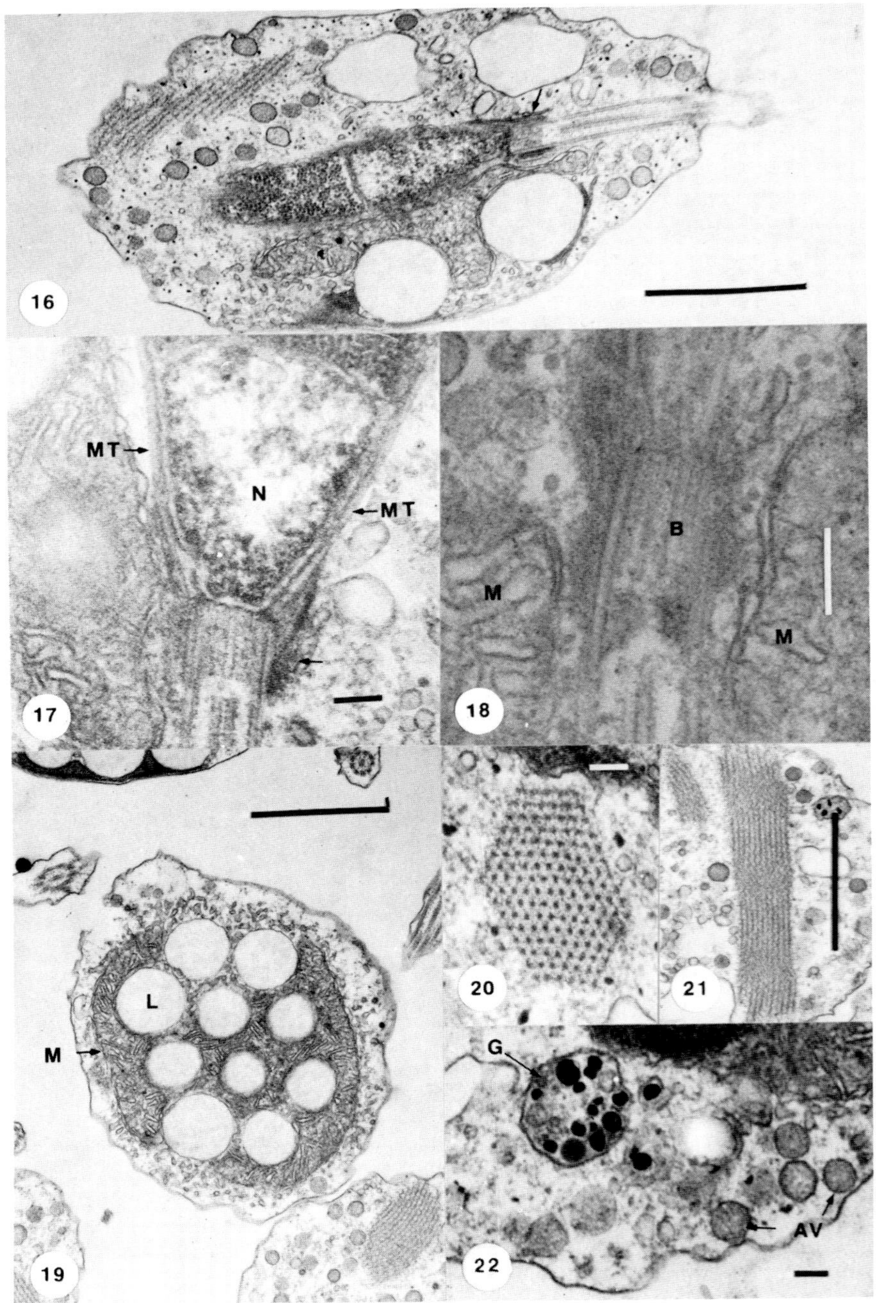
16
MT
N
MT
17
B
M
M
18
L
M
19
20
21
G
AV
22

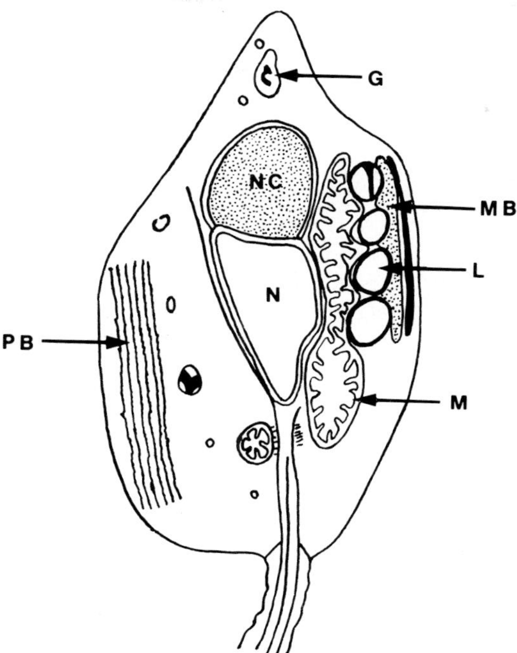

Fig. 23. Meiozoospore of *C. psorophorae.* Paracrystalline body (PB), gamma-like body (G), mitochondrion (M), lipid (L), microbody (MB), nuclear cap (NC), nucleus (N). [Redrawn from Travland (1979a).]

Figs. 16–17. Electron micrographs of meiozoospores of *C. punctatus.* Note electron-dense material located along the proximal portion of the basal body (arrows). Microtubules (MT) radiate out from the basal body to surround the nucleus (N). Fig. 16. Bar = 1 μm. Fig. 17. Bar = 0.1 μm. [From Martin (1971).]

Fig. 18. Electron micrograph of meiozoospore of *C. psorophorae* showing electron-dense deposit located between the basal body (B) and adjacent mitochondrion (M). Bar = 0.1 μm. [From Travland (1979a).]

Fig. 19. Electron micrograph of a tangential section through the mitochondrion (M)–lipid (L) complex of a meiozoospore of *C. punctatus.* Bar = 1 μm. [From Martin (1971).]

Figs. 20 and 21. Cross-section (Fig. 20) and longitudinal section (Fig. 21) through the paracrystalline body of meiozoospore of *C. punctatus.* Fig. 20. Bar = 0.1 μm. Fig. 21. Bar = 1 μm. [From Martin (1971).]

Fig. 22. Gamma-like bodies (G) and adhesion vesicles (AV) in meiozoospores of *C. psorophorae.* Bar = 0.1 μm. [From Martin (1971).]

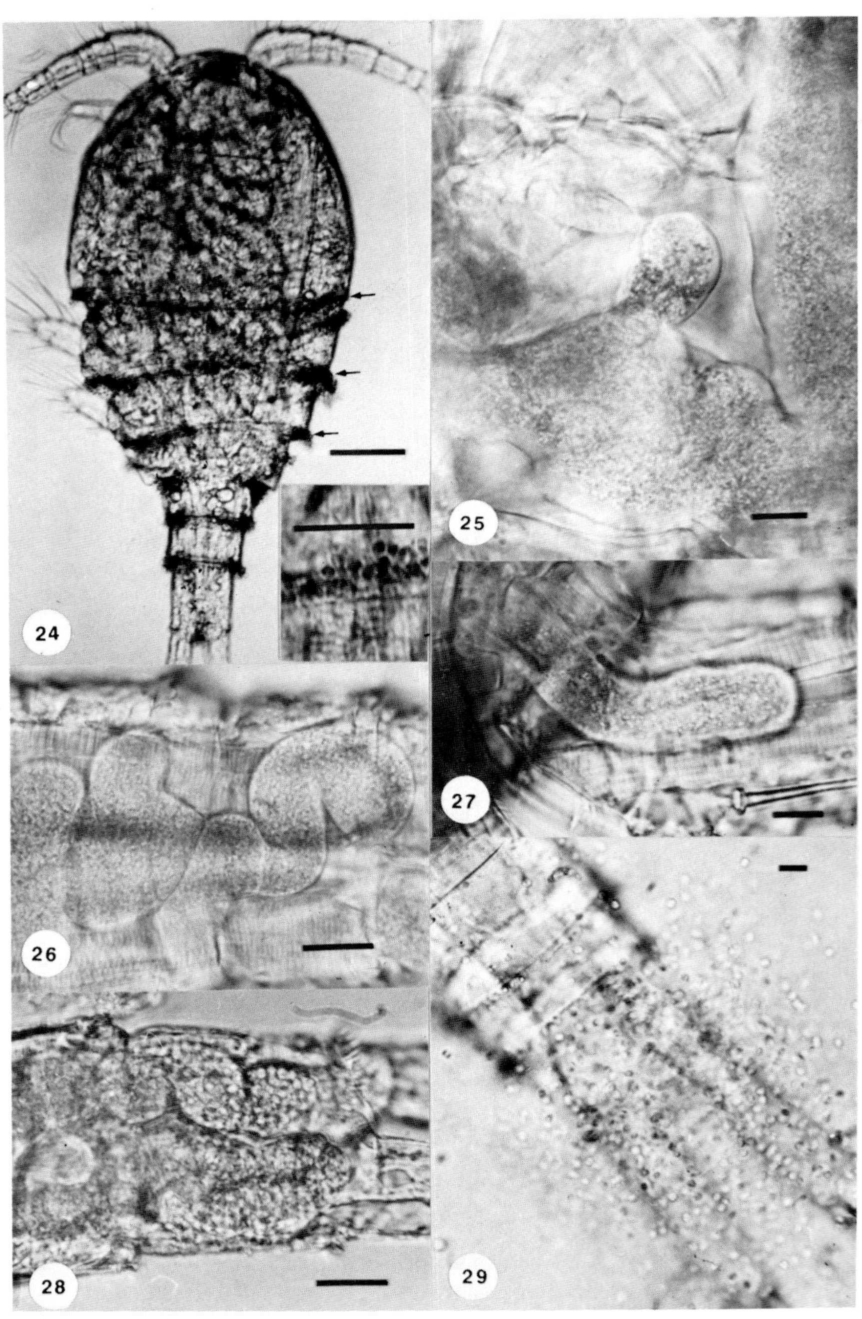
24
25
26
27
28
29

ranging from 15 to 150 μm or more in diameter. Following repeated nuclear divisions, the thallus divides into several hundred, uninucleate "secondary zoospores" (gametes) of different mating types. Fusion of gametes results in formation of a biflagellate zygote (Weiser, 1977).

During infection of copepods by *Coelomomyces* spp., Zebold *et al.* (1979) and Wong and Pillai (1980) report zoospore attachment and encystment to occur primarily in bands along the intersegmental regions of the dorsal thorax, with limited attachment and encystment occurring on other body areas, usually involving crevices or folds (Figs. 24 and 25). Although little is known of the penetration process of zoospores on copepod hosts, Zebold *et al.* (1979) report formation of appressoria by germinating zoospore cysts and a mode of entry probably similar to that occurring during infection of mosquite hosts by germinating zygote cysts (Zebold *et al.*, 1979; Travland, 1979b). The earliest stage of gametophyte development observed (about day 4 following infection) is apparent as dichotomously branched hyphae (Figs. 25–27) growing throughout most of the coelom (Federici and Chapman, 1977). Growth continues over the next 24–72 hr until hyphae have penetrated into most body cavities and occupy a major portion of the coelom (Figs. 30; and 155, facing p. 80). Infected copepods may contain hyphae of one or both of the mating types necessary for sexual reproduction to occur (Figs. 155–160, facing p. 80) (see Chapter 2). In some species of *Coelomomyces* [*C. psorophorae* (Whisler *et al.*, 1975), *C. opifexi* (Pillai *et al.*, 1976), *C. chironomi* (Weiser, 1976)], hyphae of both mating types are colorless to light amber and impossible to distinguish from each other. In other species [*C. punctatus* and *C. dodgei* (Federici, 1977; Federici and Chapman, 1977; J. N. Couch, unpublished)], the male is bright orange[2] and the female is pale to light amber, thereby making them easy to distinguish in copepods infected with one or both types (Figs. 155–160). Following proliferation of one or both types of gametophytic hyphae throughout the coelom, gametogenesis, apparently

[2] The orange pigmentation evident in the gametophytic phase of *C. dodgei* was shown by Federici and Thompson (1979) to be β-carotene, the first known instance of this isomer as the dominant carotene associated with a gametophyte in species of the Blastocladiales.

Figs. 24–29. Gametophytic phase of *C. psorophorae* (Fig. 24) and *C. punctatus* (Figs. 25–29) on alternate host, *Cyclops vernalis*. Fig. 24. A fifth-stage copepodid showing adhesion of meiozoospore cysts in the intersegmental areas (arrows). Inset at higher magnification (X160) to show individual cysts on the urosome. Cysts stained with 0.01% aqueous methylene blue to enhance visibility. Bar = 0.1 mm. [From Zebold *et al.* (1979).] Figs. 25–27. Gametophytic thallus in coelom of copepod. Bar = 10 μm. Fig. 28. Gametophytic thallus with partially cleaved gametes. Bar = 10 μm. Fig. 29. Gamete release. Bar = 10 μm.

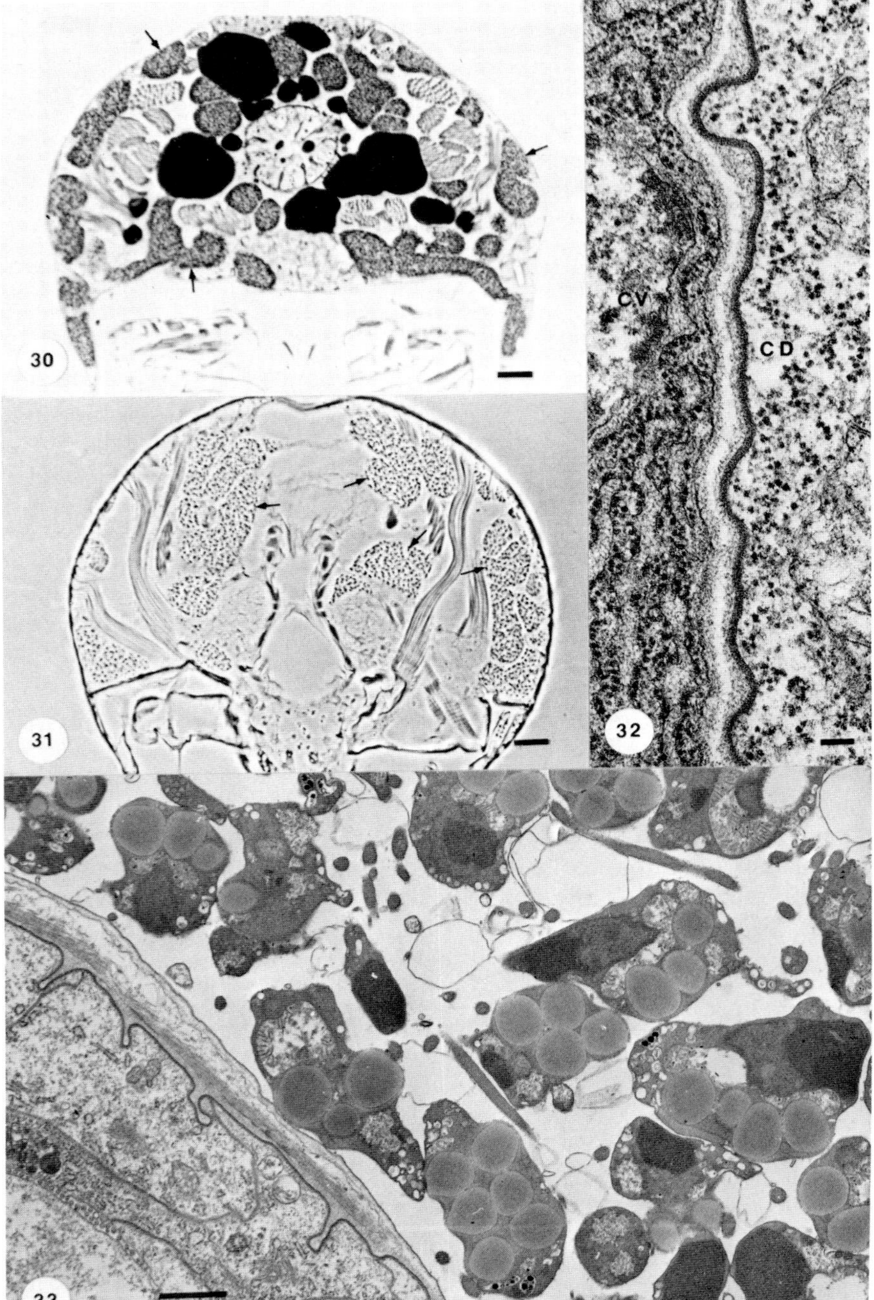
30
31
32
33
CV
CD

simultaneous, is initiated (Figs. 28–33, 158, and 159, facing p. 80); the hyphae (i.e., the entire gametophyte) thereby become gametangia (Federici and Chapman, 1977). During gametogenesis, light microscopy reveals the gametangial cytoplasm to become increasingly granular, a condition attributable to organization of cellular organelles and inclusions in accordance with their arrangement in fully formed gametes. In this regard, fine-structural studies of gametogenesis in *C. dodgei* (Figs. 34–39) (Lucarotti and Federici, 1984a) have shown that with an 8:16-hr dark:light regime at 8 hr after initiation of the last dark period (6–10 days after infection), the centriole associated with each nucleus in the multinucleate gametophytic thallus has elongated to form a kinetosome with associated flagellar props; by 9 hr, a flagellar vesicle, probably formed via coalescence of several smaller vesicles, is evident distal to the kinetosome; by 10 hr, the flagellar axoneme has elongated into the flagellar vesicle, the membranes of which fuse with the plasmalemma and give rise to the flagellar sheath; by 13 hr, cleavage is largely complete and there is evidence of side-body formation, i.e., fusion of mitochondria and aggregation of lipid droplets and associated microbodies; by 15 hr, the nuclear cap begins formation atop a now conical nucleus; and by 17 hr, gametes are completely formed. The fully formed, uniflagellate gametes (Fig. 33), described in the next section, are contained within a distinct hyphal sheath (Fig. 32) that is similar in structure to the granular hyphal coat occurring on hyphae of the sporophyte and persisting occasionally as a sheath around mature sporangia. Fully formed gametes "wait" until the appropriate time window appears, generally around 0.5–3 hr after cleavage, before swarming is initiated (B. A. Federici, unpublished). Increasing activity of the gametes results in rupture of gametangia and release of gametes into the coelomic cavity of the copepod. If the copepod is still living at this stage (often the case), death occurs shortly. Escape of gametes and/or newly formed zygotes (possible only when both mating types

---

Fig. 30. Photomicrograph of cross-section of copepod, *C. vernalis*, infected with *C. punctatus*. Arrows indicate segments of the gametophytic thallus of the fungus. Bar = 10 μm. [From Federici and Roberts (1976).]

Fig. 31. Photomicrograph of cross-section of copepod, *C. vernalis*, infected with gametophyte of *C. dodgei*. Fully cleared gametes are discernible in the gametophytic thallus (arrows). Bar = 10 μm. [From Federici and Chapman (1977).]

Fig. 32. Electron micrograph showing host–parasite interface between *C. vernalis* (CV) infected with the gametophyte of *C. dodgei* (CD). Note the fibrillar coating on the plasmalemma of the hypha of *C. dodgei*. Bar = 100 nm. [From B. A. Federici (unpublished).]

Fig. 33. Electron micrograph showing fully cleaved gametes of *C. dodgei* in coelom of *C. vernalis*. Host tissue is evident in the lower left corner of the micrograph. Bar = 1 μm. [From B. A. Federici (unpublished).]

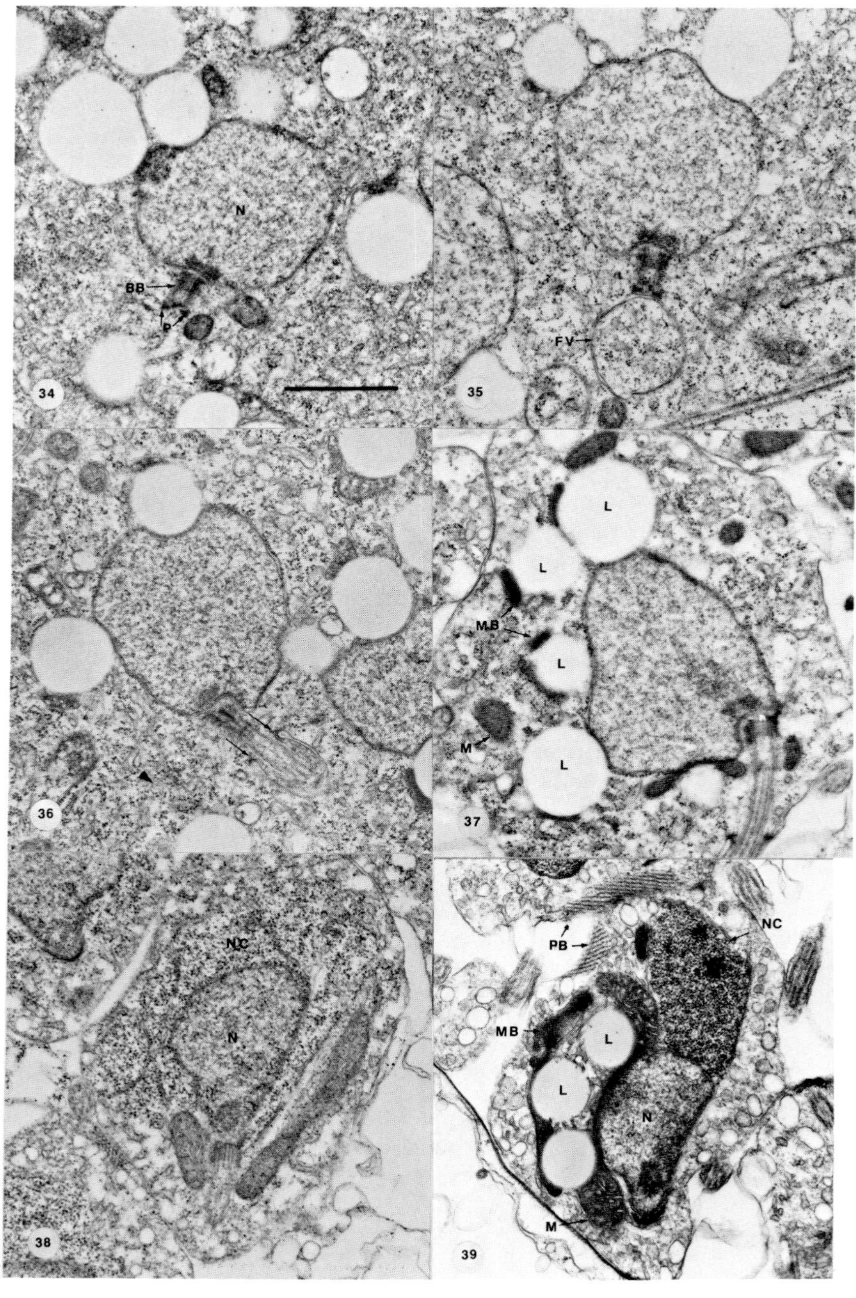
N
BB
P
34
FV
35
36
L
L
MB
L
M
L
37
NC
N
38
PB
NC
MB
L
L
N
M
39

are present in a single copepod) from the carcass of the copepod (Figs. 29; and 160, facing p. 80) happens within minutes and occurs primarily in the vicinity of the dorsal intersegmental abdominal membranes (Federici and Chapman, 1977).

IV. THE GAMETE

Although gamete structure has been observed in relatively few species of *Coelomomyces,* it appears from the studies of Whisler *et al.* (1975), Weiser (1977), Travland (1979a), Lucarotti and Federici (1984b), and J. N. Couch (unpublished) that the uniflagellate isogametes (Figs. 40 and 41) of *Coelomomyces* spp. are structurally almost identical to zoospores (described previously). Of similar size and shape, zoospores and gametes are similar also in cytoplasmic organization (Figs. 42 and 43), including the nucleus and nuclear cap, flagellar apparatus and associated structures, side-body complex, and most cytoplasmic inclusions (Travland, 1979a). As described by Travland (1979a) for *C. psorophorae,* minor structural differences between zoospores and gametes include the apparently larger size of the lipid globules in the zoospore (thought to possibly be due to different techniques of fixation), the failure of the zoospore mitochondrion to always encircle the kinetosome (a feature felt to be consistent in gametes), and the absence of a paracrystalline body in gametes. For the latter feature, however, recent observations by Lucarotti and Federici (1984b) have shown that a paracrystalline body is present in gametes of *C. dodgei* (Fig. 39). Also, in the mature gametes of *C. dodgei,* the mitochondrion does not completely surround the kinetosome, except for a short period late in gametogenesis at 14 hr.

Figs. 34–39. Transmission electron micrographs illustrating gametogenesis in *Coelomomyces dodgei.* Bar = 1 μm. [Figures 35 and 36 from Lucarotti and Federici (1984a); Figs. 34, 37–39 from C. J. Lucarotti and B. A. Federici (unpublished).] Fig. 34. At 8 hr after the initiation of the last dark period (time of the onset of the light period), the centriole has elongated to form the basal body (BB). Nucleus (N). Fig. 35. At 9 hr. The secondary flagellar vesicle (FV) at the base of the basal body has been formed by the fusion of smaller primary flagellar vesicles. Fig. 36. At 10 hr. The flagellum elongates (arrows), pushing into the secondary flagellar vesicle, whose membranes will form the flagellar membrane. Fig. 37. At 13 hr. At this time, cleavage is largely complete and the mitochondria (M) have begun to fuse and become more closely associated with lipid bodies (L) and microbodies (MB) to initiate the formation of the microbody–lipid complex. Fig. 38. At 15 hr. Initiation of formation of the nuclear cap (NC). Note the more conical shape of the nucleus (N). Fig. 39. At 17 hr. Mature gamete just prior to release from the gametangium. Lipid (L), nucleus (N), microbody (MB) mitochondrion (M), paracrystalline body (PB), nuclear cap (NC).

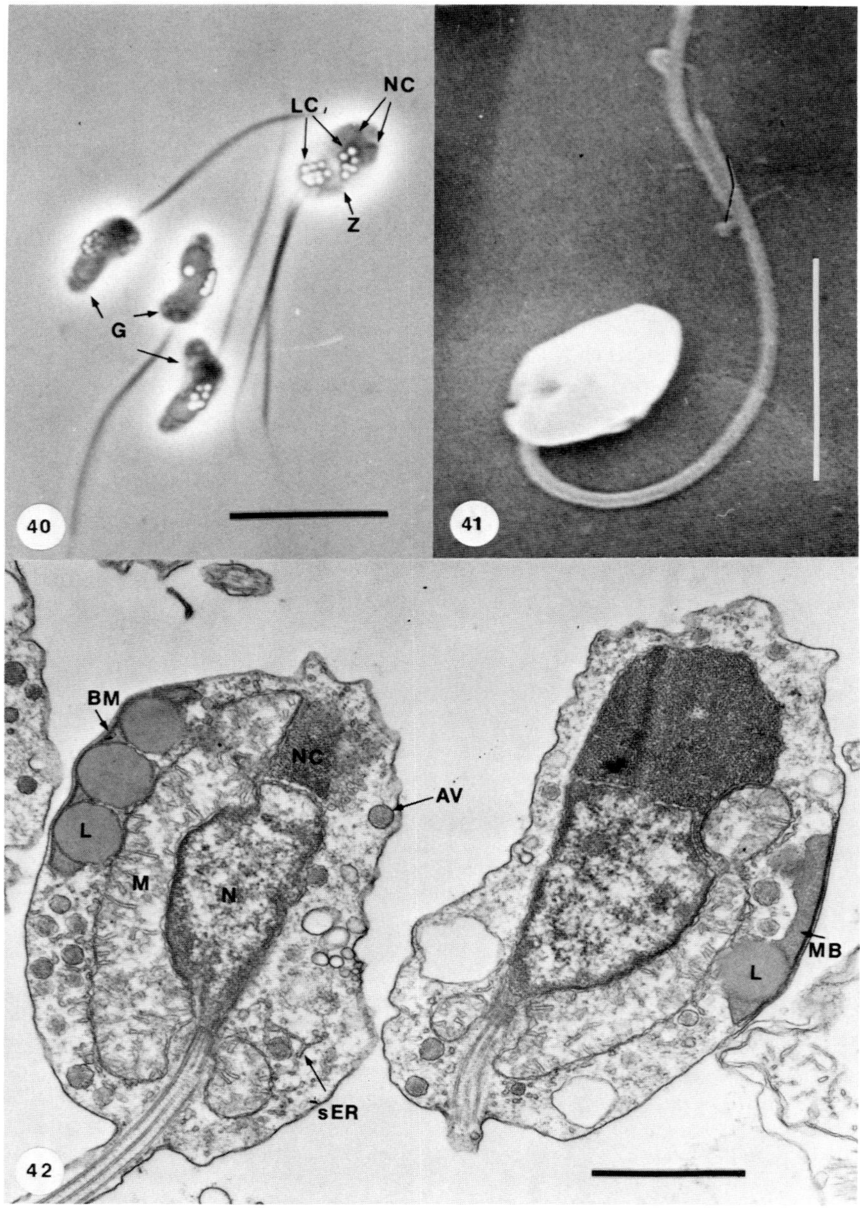

Fig. 40. Photomicrograph of three gametes (G) and one zygote (Z) of *C. psorophorae.* Note the two lipid complexes (LC) and two nuclear caps (NC) evident in the zygote. Phase-contrast microscopy. Bar = 10 μm. [From Whisler *et al.* (1975).]

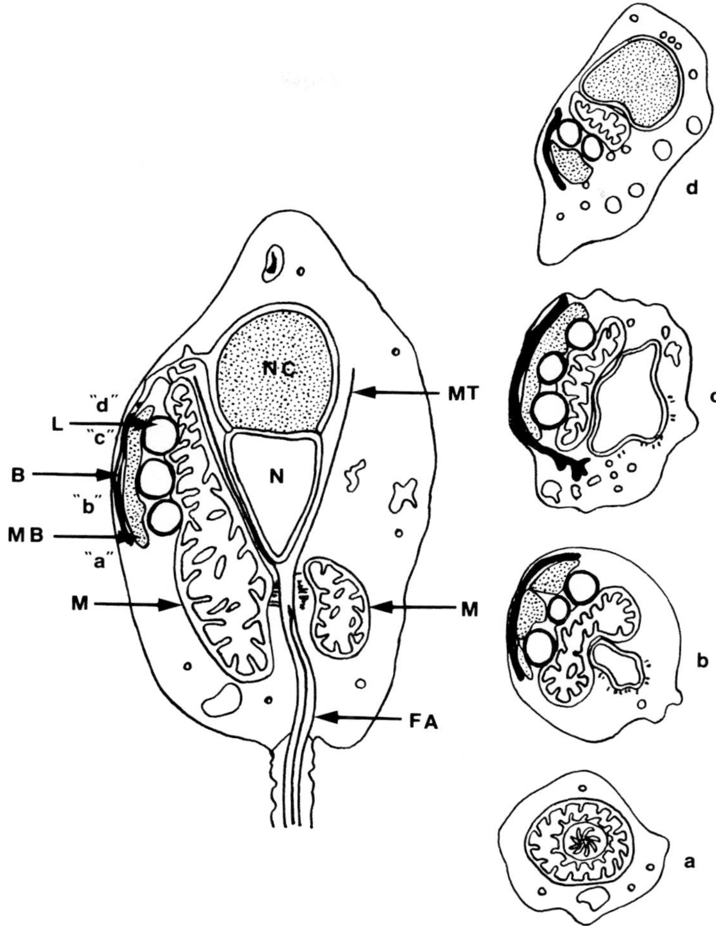

Fig. 43. Interpretative drawings from transmission electron micrographs of gametes of *C. psorophorae.* Cross-sections a–d are from the corresponding levels of the longitudinal section indicated. Flagellar axoneme (FA), mitochondrion (M), backing membrane (B), lipid (L), nuclear cap (NC), nucleus (N), microtubules (MT), microbody (MB). [Redrawn from Travland (1979a).]

Fig. 41. Scanning electron micrograph of gamete of *C. dodgei.* Scale as in Fig. 40. [From B. A. Federici (unpublished).]

Fig. 42. Transmission electron micrograph showing longitudinal sections through two gametes of *C. psorophorae.* Mitochondrion (M), nucleus (N), nuclear cap (NC), lipid (L), microbody (MB), backing membrane (BM), vacuoles (V), smooth endoplasmic reticulum (sER), adhesion vesicle (AV). Bar = 1 μm. [From Travland (1979a).]

As noted previously, while both gametangia and resulting gametes are light amber to colorless in some species of *Coelomomyces* [*C. psorophorae* (Whisler *et al.*, 1975), *C. opifexi* (Pillai *et al.*, 1976), *C. chironomi* (Weiser, 1976)], in both *C. punctatus* and *C. dodgei*, gametangia and gametes of opposite mating types are differentially pigmented, light amber or bright orange (Federici, 1977; Federici and Chapman, 1977; J. N. Couch, unpublished). Based on comparisons between the differentially pigmented gametangia and gametes occurring in species of *Coelomomyces* and a similar condition occurring in other blastocladialean species (*Allomyces arbuscula* and *Allomyces javanicus:* Emerson and Fox, 1940; Emerson, 1941; *Blastocladiella variabilis*, Harder and Sörgel, 1938), Federici (1979) designated the orange-pigmented gametes as male and the light amber gametes as female. Structurally, gametes of both mating types appear to be similar.

V. THE ZYGOTE

Although the structural phenomena involved in zygote formation via fusion of isogametes have not been studied for any species of *Coelomomyces*, the work of Travland (1979a) involving *C. psorophorae* provides a detailed account of the structure of the mature zygote of this species (Figs. 44 and 45); a similar structure probably occurs in other species of *Coelomomyces*. During syngamy, while some of the structures contributed by the isogametes merge completely, others fuse only partially, and some remain discrete. The flagella, in remaining distinct, result in the typically biflagellate appearance of the zygote. Proximally, the flagellar axonemes continue into the zygote, as was the case in the gametes. The mitochondria from both gametes fuse and completely surround one of the axonemes, while enclosing the other axoneme in a C-shaped segment. Lateral to the mitochondrion, the side-body components become associated as a single backing membrane, a single microbody, and a plate of lipid droplets totaling the number contributed by both gametes. Nuclear fusion is only partial and occurs as a bridge between adjacent nuclei; nuclear integrity is evident in swimming zygotes after 12.5 hr and even in recently encysted zygotes. The microtubular basket surrounding the nuclei remains intact, except that microtubules appear to be excluded in the nuclear bridge area and in the area of nuclear–mitochondrion association. Both nuclear caps persist in their apical association with the nuclei. Also, profiles of smooth endoplasmic reticulum (ER), gamma-like bodies, large clear vacuoles, and adhesion vesicles continue to be present throughout the zygote cytoplasm.

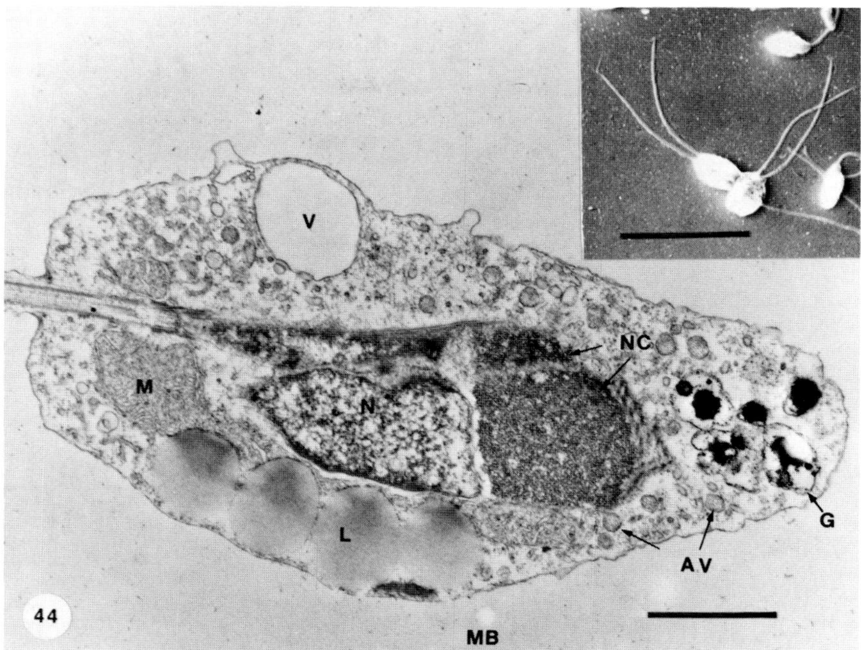

Fig. 44. Transmission electron micrograph (bar = 1 μm) of a longitudinal section of a zygote of *C. psorophorae* and a scanning electron micrograph (inset, bar = 10 μm) of zygotes of *C. dodgei.* The section shown is part of a series in which the other basal body was seen also. The single side-body complex [mitochondrion (M), lipid (L), microbody (MB)] is seen flanking the bridged nuclear lobes (N) and nuclear cap (NC). Vacuole (V), gamma-like bodies (G), adhesion vesicles (AV). Note the clearly biflagellate nature of the zygotes illustrated in the scanning electron micrograph inset. [Transmission electron micrograph from Travland (1979a). Scanning electron micrograph inset from B. A. Federici (unpublished).]

VI. INFECTION OF MOSQUITO LARVAE AND DEVELOPMENT OF THE SPOROPHYTE

Infection of mosquito larvae by zygotes of *Coelomomyces* spp. was first shown by Zebold *et al.* (1979) to occur by direct penetration of the cuticle and not through the digestive tract. In this process, motile zygotes attach to the host cuticle (Figs. 46–48), encyst and secrete a thin wall (Fig. 49), germinate via development of an appressorium that is closely appressed to the host cuticle, and produce a narrow penetration tube that grows through the larval cuticle (Zebold *et al.,* 1979; Travland, 1979b). Details of these processes follow.

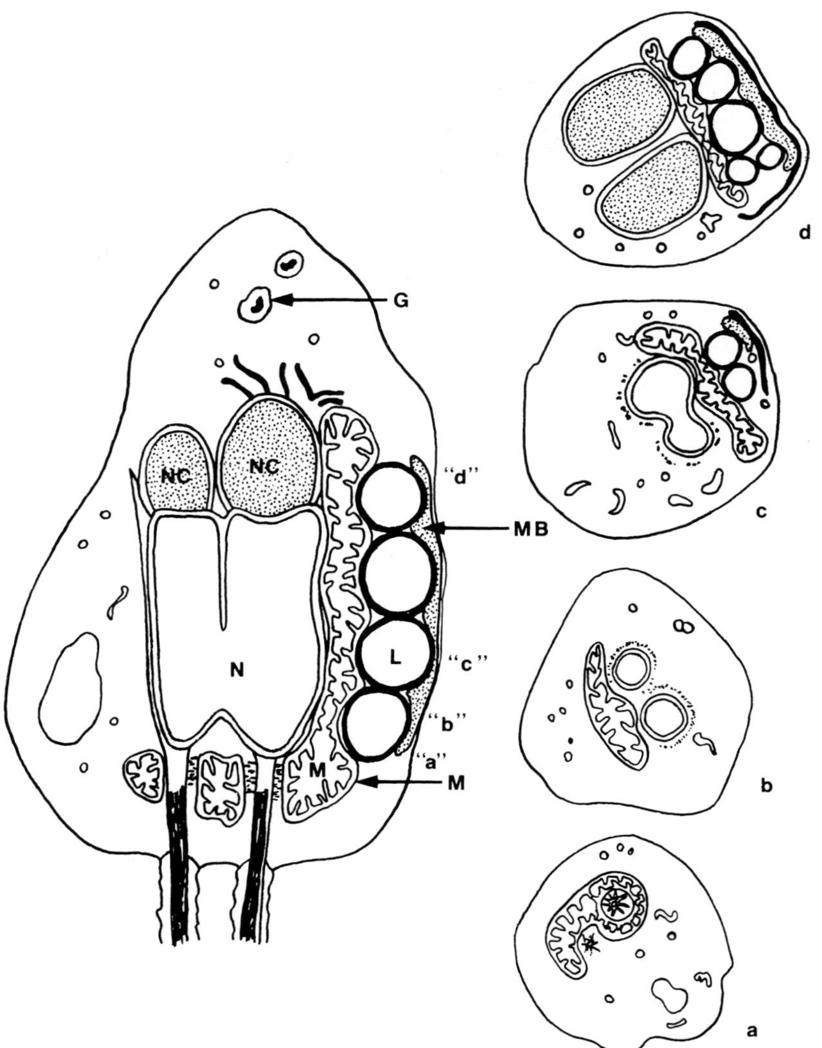

Fig. 45. Interpretative drawings from transmission electron micrographs of zygotes of *C. psorophorae*. Cross-sections a–d are from the levels indicated beside the longitudinal section. Gamma-like body (G), lipid (L), microbody (MB), mitochondrion (M), nucleus (N), nuclear cap (NC). [Redrawn from Travland (1979a).]

Fig. 46–49. Attachment to and encystment of zygote on cuticle of larva of *Culiseta inornata*. [From Travland (1979a).] Fig. 46. Early stages of encystment before wall formation. Note accumulation of adhesion vesicles (arrows) in the region of contact with the host cuticle (C). Bar = 1 μm. Fig. 47. Adhesion vesicle (AV) appressed to the plasma membrane (PM) of the zygote. Bar = 0.1 μm. Fig. 48. Gamma-like bodies in encysting zygote. Bar = 0.25 μm. Fig. 49. Five-minute-old cyst with a discrete wall, vesiculating gamma-like bodies (G), and highly irregular mitochondrion (M). Note fla-

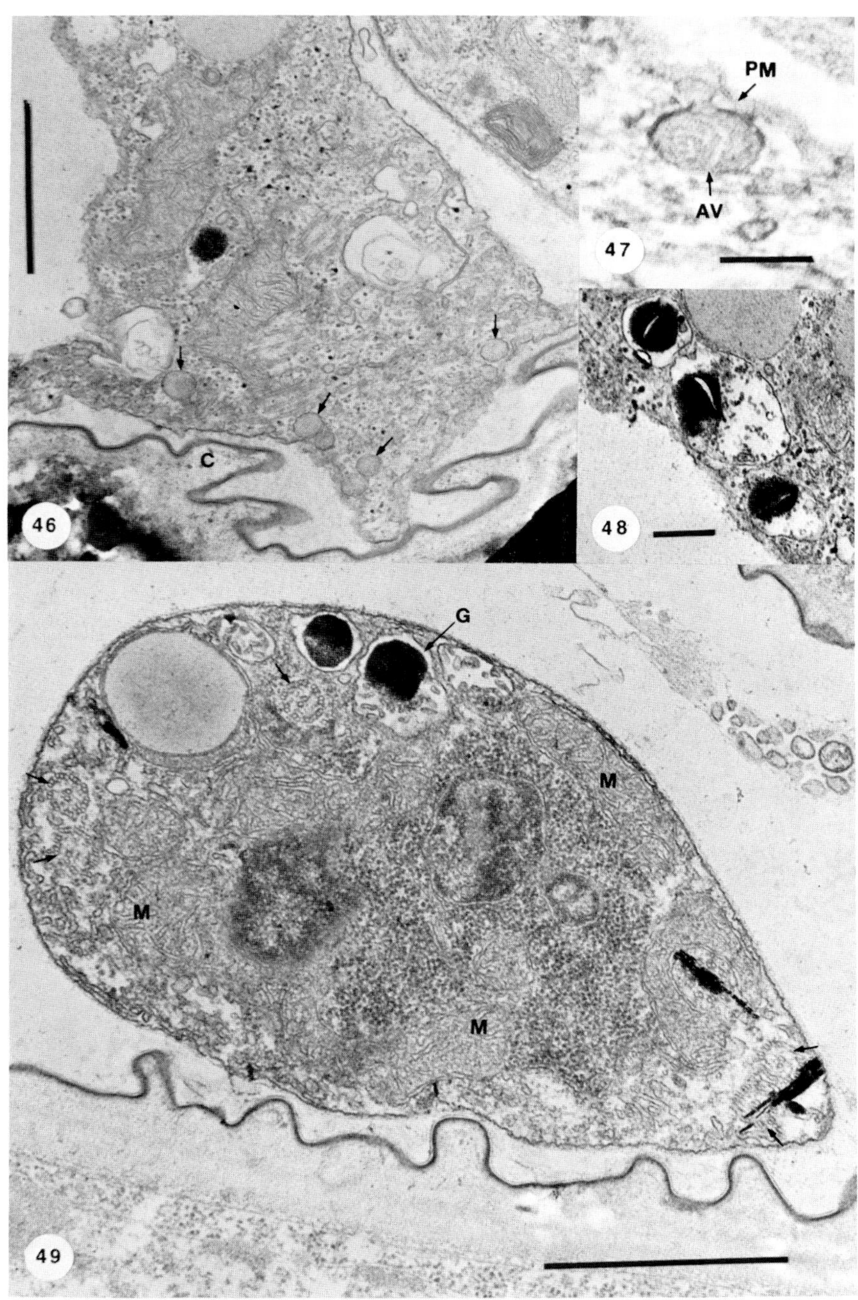

gellar cross-sections (arrow) evident in cyst cytoplasm, indicating flagellum was retracted and coiled inside the encysting zygote. Bar = 1 μm.

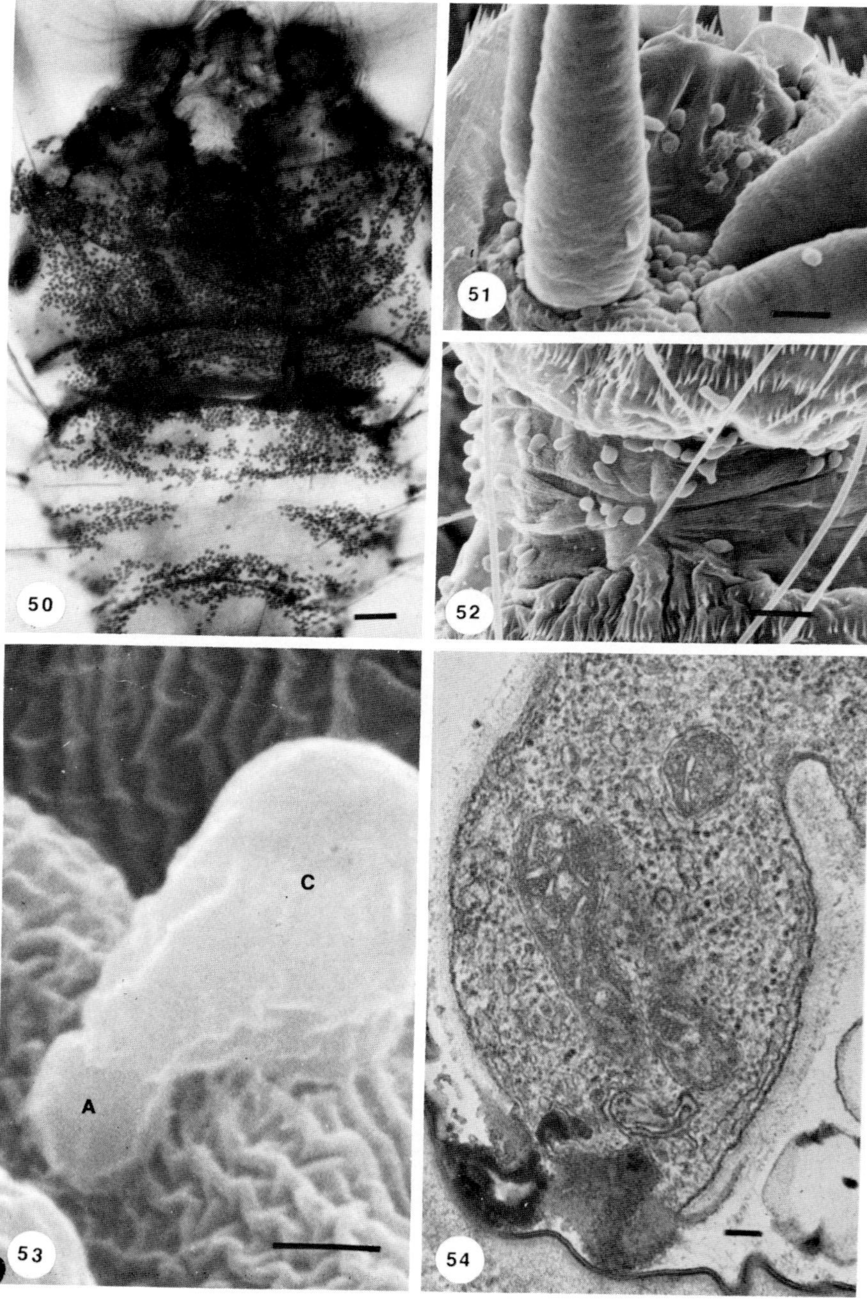
50
51
52
53
C
A
54

In attaching to hosts, zygotes show preferential sites of adhesion (Figs. 50–52) on the larval cuticle, most attachments occurring at the bases of the anal gills, in the intersegmental folds, around the anus, and on the head (Zebold *et al.*, 1979; Travland, 1979b; Wong and Pillai, 1980). Although infrequent, other sites of attachment may include the siphon, the anal gills, and the setae. In the earliest stage of attachment observed thus far (Travland, 1979b), it is evident that adhesion of zygote to cuticle is accomplished by a "gluelike," fibrillar substance secreted by the zygote (Fig. 46). Production and secretion of the adhesive substance is apparently mediated by adhesion vesicles (Fig. 47), which concentrate at the site of contact between zygote and host but are gone within 5 min after attachment (Travland, 1979b).

Stages in zygote encystment and germination and growth of the penetration tube into the host were described by Travland (1979a,b). The following summarizes her observations on these processes.

During early attachment, the zygote has an irregular profile and is characterized by one to several pseudopodia in the region of contact with the host cuticle (Fig. 46). Internally, zygote cytoplasm shows signs of reorganization leading to encystment. There are faint indications that formation of a cyst wall has been initiated. The flagellum has been retracted and the axoneme appears coiled inside the zygote (Fig. 49). The gamma-like bodies now appear as an electron-dense matrix surrounded by an electron-transparent peripheral zone (Figs. 48 and 49). Finally, the nucleus–side-body–nuclear cap association is reoriented following flagellar retraction.

In later stages of encystment, the zygote becomes ovoid and is surrounded by a thin but definite wall. Wall formation is concomitant with vesiculation of the gamma-like bodies (Fig. 49), a relationship thought by Travland (1979a) to indicate participation of the bodies in wall synthesis. As encystment continues, there is gradual disorganization of the nucleus–

Fig. 50. Photomicrograph of first instar larva of *Culiseta inornata,* showing stained (0.01% aqueous methylene blue) zygote cysts of *C. psorophorae* attached to the head capsule and arranged in distinct bands on intersegmental areas of thorax and abdomen. Bar = 50 μm. [From Zebold *et al.* (1979).]

Figs. 51 and 52. Scanning electron micrographs showing zygote cysts of *C. psorophorae* attached in the anal gill region (Fig. 51) and intersegmental region (Fig. 52) of a larva of *C. inornata.* Bar = 10 μm. [From Travland (1979b).]

Fig. 53. Scanning electron micrograph of a zygote cyst (C) and appressorium (A) of *C. psorophorae* on *Culiseta inornata.* Bar = μm. [From Zebold *et al.* (1979).]

Fig. 54. Transmission electron micrograph of a 3-hr-old appressorium of *C. psorophorae* that has not yet penetrated the cuticle of a larva of *C. psorophorae.* Note the accumulation of dense material external to the plasma membrane. Bar = 0.1 μm. [From Travland (1979b).]

nuclear cap and side-body complex components such that within 30 min after encystment the nuclear cap has dispersed, there are numerous profiles of rough ER, and the single large mitochondrion of the side-body complex has divided into numerous, smaller mitochondria. The microtubular basket encasing the nucleus is still evident, however. At approximately 1 hr after encystment there is complete fusion of the previously, partly fused, paired zygote nuclei. This is accompanied by disappearance of the microtubular nuclear basket. The fully encysted zygote is now ready for germination.

Germination of the encysted zygote (Figs. 53 and 54) typically occurs at the pointed end of the ovoid cyst. A short hyphal neck grows from this site and expands immediately into a bulbous appressorium, which is in contact with the host's cuticle, then an accumulation of electron-dense material often present between appressoria and host cuticle (Fig. 54). For infection (Fig. 55), penetration of the cuticle is accomplished by a narrow (0.3 μm in diameter) germ tube which grows out from the appressorium. Host response to the invading germ tube is generally apparent due to melanization of the cuticle in the vicinity of the germ tube (see Travland, 1979b, for a discussion of this phenomenon). With continued growth of the germ tube (Figs. 55–58) (sometimes a distance of up to several micrometers), there is concurrent alignment of organelles within the cyst and the appearance of intranuclear microtubules and a distal cytoplasmic vacuole. Infection continues with injection of cyst cytoplasm through the germ tube and into a host epidermal cell (Travland, 1979b) or hemocoel (Wong and Pillai, 1980). The injected cytoplasm appears as a spherical hyphal body (hyphagen; Martin, 1969) containing typical cell organelles and surrounded by two plasma membranes (Fig. 58), the innermost that of the fungus and the outermost probably of host origin (Travland, 1979b). Further vegetative development of the fungus is initiated from the injected fungal cell.

Following injection of the zygote cyst cytoplasm into the intracellular infecting cell, distribution of infection throughout the coelomic cavity of larvae is accomplished via spherical hyphal bodies or hyphagens (Martin, 1969). Such hyphagens (Figs. 59–61; see also Fig. 162, facing p. 81) may float free in the hemolymph or adhere to various tissues in the head, thorax, or abdomen prior to proliferating vegetative hyphae. Although these structures have long been recognized as early stages in infection (Iyengar, 1935; Couch, 1945; Umphlett, 1962; Couch and Umphlett, 1963; Muspratt, 1963; Madelin, 1968; Martin, 1969), the mechanism of their proliferation has only recently been elucidated. In this regard, Wong and Pillai (1980) report that during infection of *Aedes australis* larvae with *C. opifexi* (*C. psorophorae* var. *tasmaniensis*, see Chapter 4) the injected

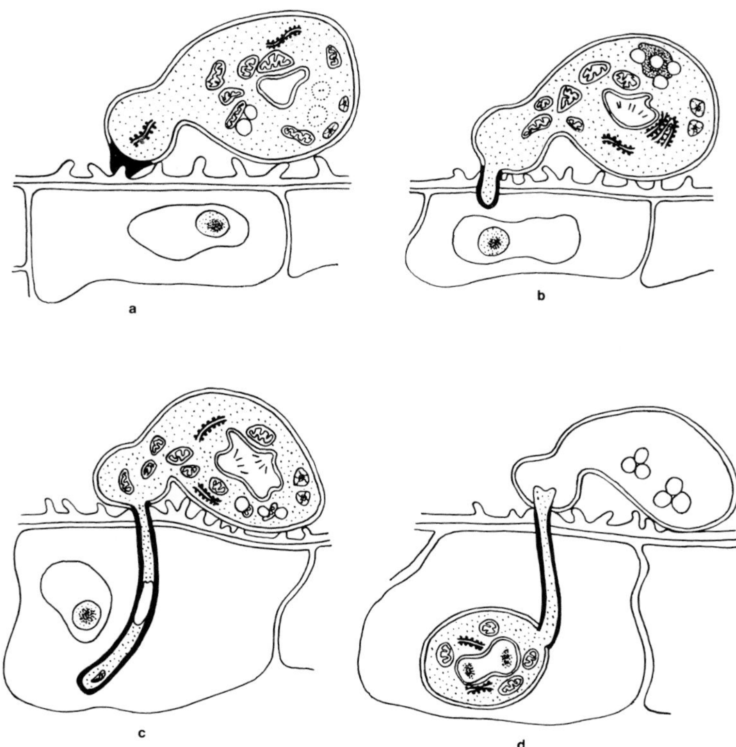

Fig. 55. Interpretative drawings showing infection of larva of *Culiseta inornata* with *C. psorophoraea*. a. Zygote cyst attachment to host cuticle and deposition of electron-dense material between appressorium tip and cuticle. b. Penetration of cuticle by germ tube. c. Elongation of germ tube and nuclear division in the zygote cyst. d. Formation of hyphagen by discharge of zygote cyst cytoplasm through the germ tube and into a host epidermal cell. [Redrawn from L. B. Travland (unpublished).]

spore plasm (in the form of a hyphal body or hyphagen) either may be released to circulate in the hemolymph or may remain attached to the epidermis. The presence of many hyphagens during early stages of infection and their persistence through later stages may indicate that continuous and multiple infection of larvae occurs in nature (Muspratt, 1946; Wong and Pillai, 1980) or that there is continual fragmentation of hyphae and/or differential growth and development of hyphagens during infection (Martin, 1969; Wong and Pillai, 1980).

Although not necessarily valid for all species of *Coelomomyces*, the observations of Martin (1969) on development and structure of *C. punctatus* show that hyphagens in this species vary in shape from spherical to

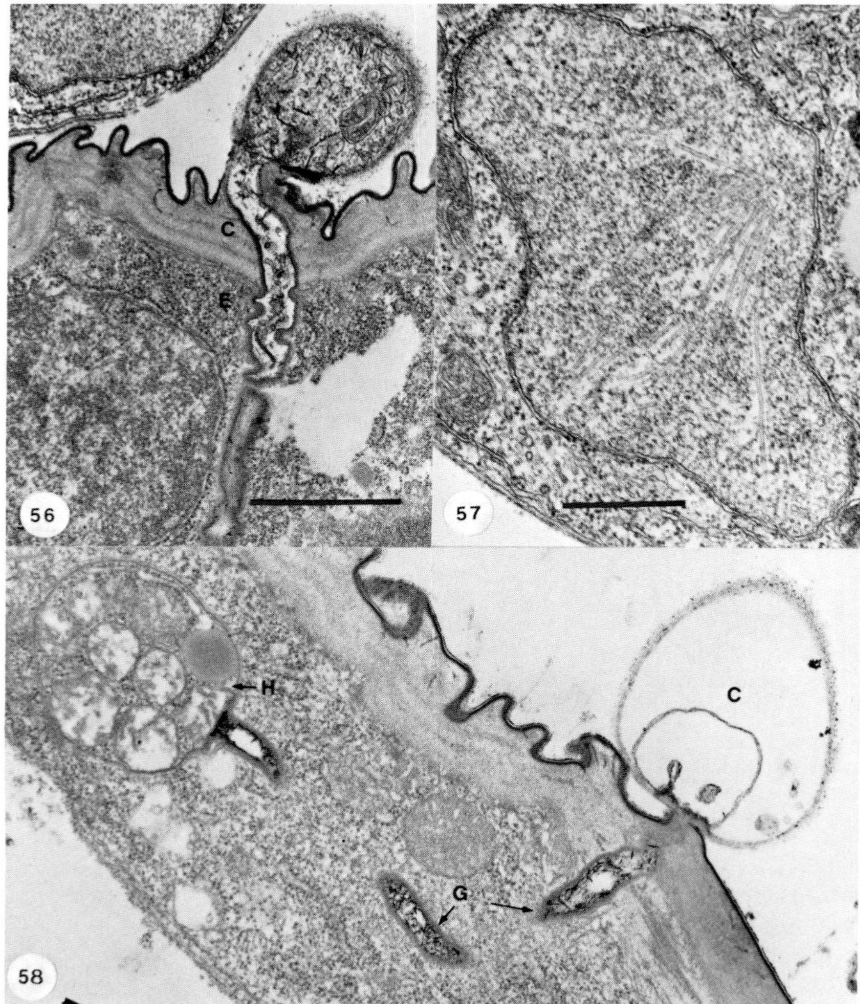

Figs. 56–58. Transmission electron micrographs of *Coelomomyces psorophorae* on larva of *Culiseta inornata.* [From Travland (1979b).] Fig. 56. Penetration of cuticle (C) and epidermis (E) of larva by germ tube from zygote cyst. Bar = 10 μm. Fig. 57. A nucleus in a zygote cyst that has germinated. Note intranuclear microtubules. Bar = 5 μm. Fig. 58. At 8 hr after infection. Note degenerating zygote cyst (C) and germ tube (G), and the newly formed hyphagen (H). Bar = 10 μm.

ovoid to rarely irregular and measure from 5.4 to 27.0 μm in diameter, larger sizes typically occurring in later instars. Distinction of young hyphagens may be made in that a cluster of one or more, centric to encentric, lipid globules is generally present in each. However, before or during the proliferation of hyphae, such clusters disperse and the cytoplasm of the hyphagen assumes the granular appearance characteristic of hyphae of *Coelomomyces* spp. Seen with light microscopy (Fig. 60; also see Fig. 67) to have a dark outer border (Martin, 1969), hyphagens have been shown by electron microscopy (Figs. 62–66) to be surrounded by a plasmalemma coated by an external, densely granular to fibrillar polysaccharide layer (Powell, 1976). Numerous simple, branched, or contorted cytoplasmic protuberances cover the hyphagens (Figs. 62–64) (Martin, 1969; Powell, 1976) and, no doubt, contribute to the dark appearance of its outer "membrane" seen in light microscopy. In addition to the cytoplasmic protuberances, the surface of hyphagens adjacent to the fat body of the host has occasional involutions of the plasmalemma that are sheathed by cisternae of endoplasmic reticulum (Fig. 64). As suggested by Martin (1969) and Powell (1976), such involutions and protuberances probably provide increased area for absorption of nutrients from the host, a factor of potential significance to a fungal parasite in which the vegetative (trophic) stage does not penetrate host cells (Powell, 1976). During advanced stages of infection when sporangia are present, cytoplasmic protuberances are apparently absent from hyphagens (Figs. 65 and 66). Instead, vesicles of uniform size are scattered around the smooth plasmalemma of these structures (Powell, 1976). Based on similarities in extracellular coat, thickness of plasmalemma, and ribosome-like particulate contents of the vesicles and protuberances, Powell (1976) suggests that the vesicles originate from the detachment of protuberances from hyphagens. [For a further discussion of the nature and possible function of the extracellular coat, protuberances, and surrounding vesicles of hyphagens, see Powell (1976).] Internally, hyphagens are multinucleate and contain other typical, fungal cell organelles (Martin, 1969; Powell, 1976).

Following production of hyphagens and their distribution throughout the coelomic cavity, the next stage in infection involves proliferation of vegetative hyphae from hyphagens (Figs. 67–75). Although there have been few studies of this process, the keen observations of Martin (1969) provide insight into its mechanism in *C. punctatus*. While hyphagen size has little effect on the time of hyphal initiation, size does apparently affect the number of hyphae arising from a given hyphagen. In this regard, hyphagens measuring 10.0–17.3 μm typically proliferate one or more hyphae of roughly the same diameter as the hyphagen (Fig. 76), while larger hyphagens usually produce several hyphae of smaller diameter than the

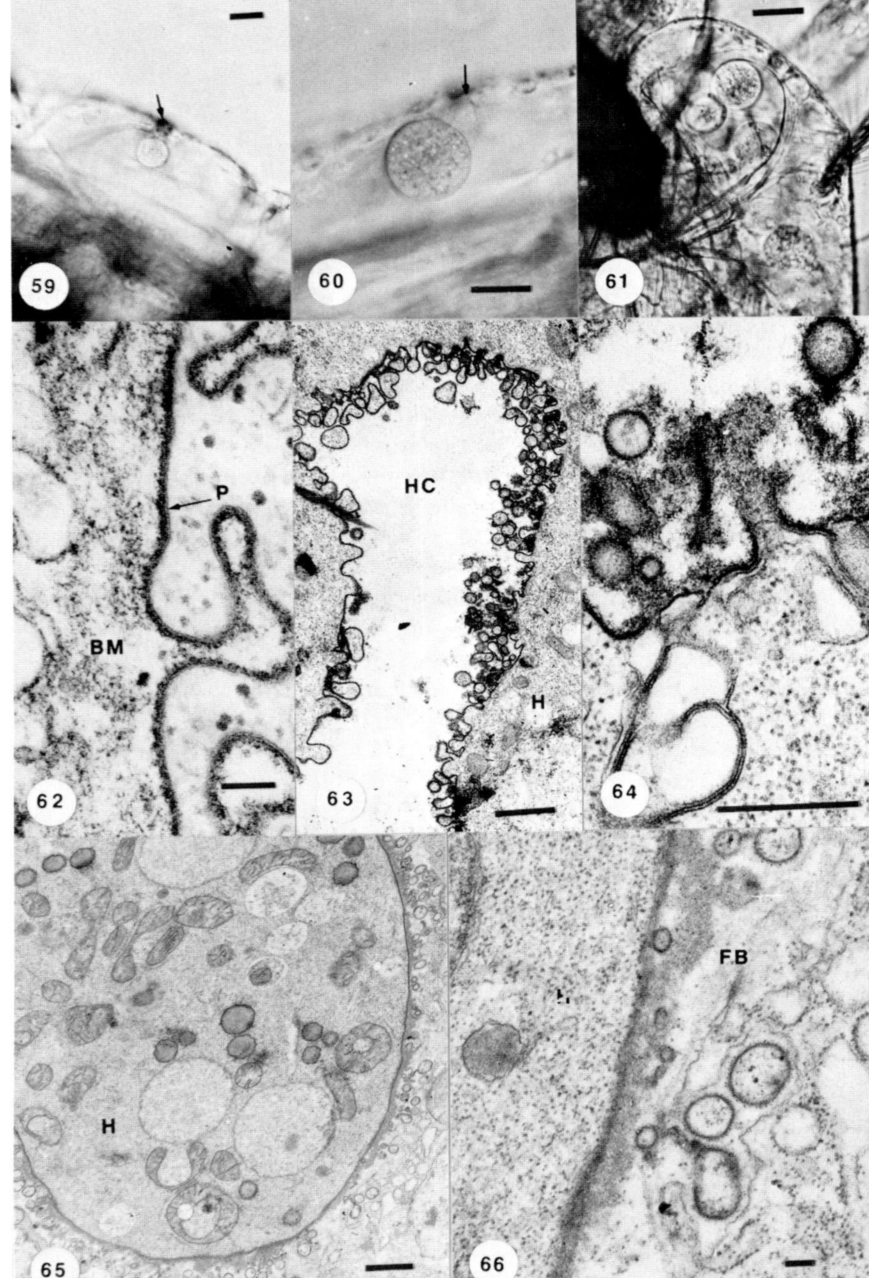
59
60
61
P
BM
62
HC
H
63
64
H
65
FB
66

hyphagen (Fig. 77). Occasionally, hyphagen "centers" as large as 25.0–35.3 μm have been seen proliferating 5–18 hyphae up to 85 μm in length. In agreement with Martin's observations, Wong and Pillai (1980) report also the formation of hyphae ("thallic protrusions") from hyphagens.

First seen by Keilin (1921) in *C. stegomyiae* and referred to as "a branched, anastomosing mycelium," the vegetative hyphae of *Coelomomyces* spp. have long been structures of speculation. Of special interest has been the presence or absence of a wall around the hyphae. Due to their apparent lack of a wall, Bogoyavlensky (1922) referred to the nonreproductive portion of the thallus of *C. notonectae* as a plasmodium. Later, Iyengar (1935) described the hyphae of *C. indicus* to be "extremely thin-walled." Following several attempts to disclose the real nature of the hyphal covering of *Coelomomyces* spp., Couch (1945), Umphlett (1962), Couch and Umphlett (1963), and Martin (1969) concluded that the hyphae are bounded by only a plasmalemma. Powell (1976), using improved techniques of fixation and electron microscopy, revealed new and interesting features of the hyphal covering of *C. punctatus*. Although this covering is similar to that described previously for hyphagens, a summary of her findings relative to hyphae will be presented following a brief account of hyphal development.

Following their proliferation from hyphagens, the coenocytic, vegetative hyphae of *Coelomomyces* spp. grow throughout the coelomic cavity of parasitized larvae, often in close proximity to the fat body of the host (Figs. 78 and 79; also see Figs. 163 and 164, facing p. 81). Depending on severity of infection and stage of development of the parasite, hyphae may be abundant, scarce, or absent in the final stages of infection when all have been converted to sporangia. Varying from species to species and

---

Figs. 59 and 60. Early stage of infection showing hyphagen of *C. punctatus* immediately inside cuticle host, *Anopheles quadrimaculatus*. Note melanigation (arrow) of host cuticle of probable site of entry. Bar = 10 μm.

Fig. 61. Several hyphagens of *C. punctatus* in coelom of *Anopheles quadrimaculatus*. Bar = 10 μm.

Figs. 62–66. Transmission electron micrographs of hyphagens of *C. punctatus* in coelom of larvae of *Anopheles quadrimaculatus*. [From Powell (1976). Fig. 62. Surface of a hyphagen showing protuberances with exterior granular coat on plasmalemma of the fungus (P). Basement membrane of host (BM). Bar = 0.1 μm. Fig. 63. Portion of two irregularly shaped hyphagens in the abdominal coelom of a lightly infected larva. Note the numerous fingerlike-to-lobed projections. Hemocoel of host (HC), hyphagen (H). Bar = 1 μm. Fig. 64. Enlargement of the surface of a hyphagen. The involution contains whorls of membrane-like material. Bar = 0.5 μm. Fig. 65. Hyphagen with a smooth cell surface in a heavily infected second instar larva. Hyphagen (H). Bar = 1 μm. Fig. 66. Enlargement from Fig. 65 showing thick extracellular layer in the region of the fat body (FB). Hyphagen (H). Bar = 0.1 μm.

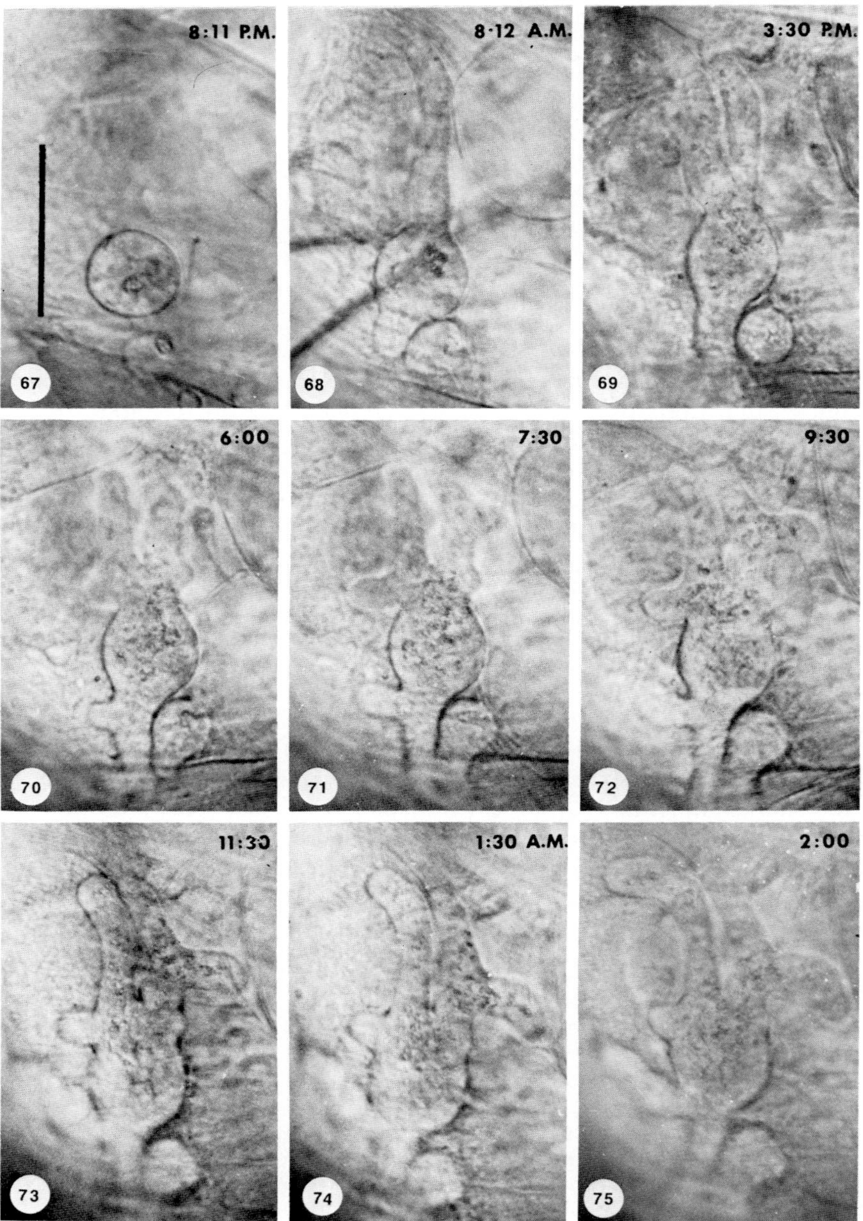

Figs. 67–75. Proliferation of hyphae from hyphagens of *C. punctatus* in living larva of *Anopheles quadrimaculatus*. The time of each micrograph is indicated in the upper right-hand corner. Bar = 50 μm. [From Martin (1969).]

usually showing a relationship to sporangial size, hyphae range in diameter from 2.6–6.2 μm in small forms such as *C. pentangulatus* (Umphlett, 1962) to 10–20 μm in some of the larger forms. Similarly, hyphae may vary in length, from only slightly greater than their diameter up to 200 μm (Martin, 1969). While hyphal branching has been described as highly irregular (Martin, 1969) and to occur at a variety of angles (Umphlett, 1962), branches most often arise at approximately right angles (Fig. 77) along an elongate parent hypha (Martin, 1969). New hyphal branches may grow linearly and give rise to additional branches, or they may differentiate into rounded knobs, presumably resting sporangia or hyphagens (Martin, 1969). Although anastomosis of hyphae has been described for some species of *Coelomomyces* (Keilin, 1921; Bogoyavlensky, 1922; Couch, 1945), anastomoses were not observed by Umphlett (1962) or Martin (1969) and, in most instances, probably represent branching rather than actual anastomoses. The "rhizoids" reported by Iyengar (1935) and Umphlett (1962) to occur on hyphae of two species of *Coelomomyces* have not been observed by other investigators and probably represent empty, wrinkled, hyphal membranes (Couch and Umphlett, 1963).

Light microscopy of living hyphae (Figs. 76, 78, and 79) reveals a densely granular cytoplasm containing many central, refractive granules (Martin, 1969) and nonsynchronously dividing nuclei (Umphlett, 1962). In agreement with these observations, fine-structural studies by Martin (1969) and Powell (1976) show hyphae of *C. punctatus* to contain typical fungal cell organelles, including numerous spherical nuclei and prominent lipid globules (Fig. 80). Also seen are mitochondrial profiles ranging from ring- or plate-shaped to elongate, and abundant ER. As in hyphagens, vegetative hyphae are bordered by a hyphal envelope comprised of plasmalemma and an outer, granular to fibrillar, polysaccharide coat (Fig. 81) varying in thickness from a "superficial layer to a copious matrix" (Powell, 1976). Depending on the stage of infection, the hyphal envelope may be highly irregular (Fig. 81) and have numerous protuberances and involutions in early and mid infection, or it may be smooth and surrounded by numerous vesicles in advanced infection; the vesicles no doubt represent detached protuberances (Figs. 80 and 82). In some larvae, hyphae with both smooth and irregular envelopes may be present. This variation in morphology of the hyphal envelope probably explains the differing images of this structure described by Martin (1969) and Whisler *et al.* (1972). As postulated by Powell (1976), it seems likely that the specialized hyphal envelope of *Coelomomyces* spp. functions in absorption of nutrients from the host, primarily from the fat body. As reflected by its variable morphology, it in some way regulates the rate of this absorption. [See Powell

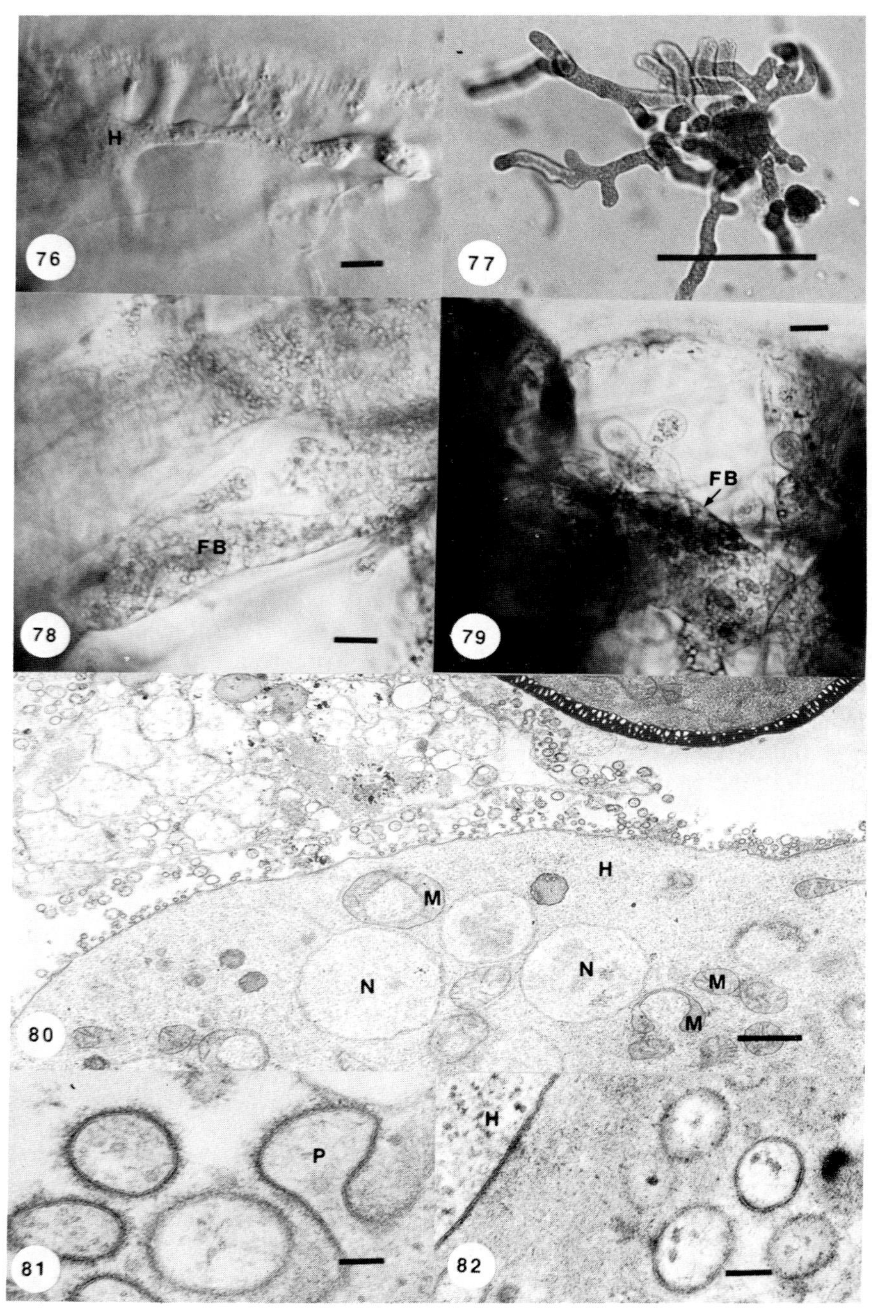
H
76
77
FB
78
FB
79
H
M
N
N
M
M
80
P
81
H
82

(1976) for a further discussion of the nature of the hyphal envelope and its significance at the host–parasite interface.]

While the mechanism triggering differentiation of vegetative hyphae into sporangia is unknown, it is apparent that, in some instances, sporangia are produced during early stages of infection when hyphae are sparse; in other instances, sporangia are not produced until a larva is completely filled with hyphae. Such differences are likely to be due to variations in host species and nutritional state of hosts. In some situations, however, vegetative hyphae are eventually converted entirely into sporangia, the thallus thus being holocarpic. Although resting sporangia and, when present, thin-walled sporangia are both produced from vegetative hyphae, only the process of resting sporangial formation has been studied and will be described here. The two processes are, however, probably quite similar.

Formation of resting sporangia typically begins by a swelling at the tip of a main hypha or at the tips of lateral branches of hyphae (Figs. 83 and 84; also see Figs. 161 and 165, facing p. 81). Arising singly or in clusters, such swellings may be distinguished as incipient sporangia in that their hyphal envelope appears thicker than on supporting hyphae (Fig. 86), an observation made by Couch and Umphlett (1963), Umphlett (1964), and Martin (1969) for several species of *Coelomomyces*. This observed thickening of the hyphal envelope is no doubt a result of the preliminary stage in sporangial wall deposition occurring at this time, a process studied by Powell (1976) and summarized subsequently. As described by Martin (1969) for *C. punctatus,* sporangial formation continues with migration of the multinucleate hyphal cytoplasm into the now rounding hyphal tip (Figs. 85–98). Next, the hyphal cytoplasm separates from that of the rounded, incipient sporangium and leaves it attached to the parent hypha via a thin sheath (Fig. 94). The forming sporangium then assumes a shape

---

Figs. 76–79. Photomicrographs of hyphae of *C. punctatus* on *Anopheles quadrimaculatus.* Fig. 76. Interference contrast optics showing hyphae growing out from hyphagen (H). Bar = 10 μm. Fig. 77. Excised and stained (lactophenol with cotton blue) hyphagen center. Bar = 50 μm. [From Martin (1969).] Figs. 78 and 79. Hyphae growing in association with fat body (FB). Bar = 10 μm.

Figs. 80–82. Transmission electron micrographs of hyphae of *C. punctatus* in larvae of *Anopheles quadrimaculatus.* [From Powell (1976).] Fig. 80. Hypha (H) from a heavily infected third instar larva. Note that the hyphal surface is smooth and is surrounded by an extracellular layer and unattached vesicles. Mitochondria (M), nuclei (N). Bar = 1 μm. Fig. 81. Detail of a protuberance (P) on a hypha. Note that the tripartite plasmalemma is covered by a granular to fibrillar coat. Bar = 0.1 μm. Fig. 82. Detail of the smooth plasmalemma and unattached vesicles of a hyphagen (H). A granular to fibrillar layer coats the vesicles, which contain ribosome-like particles. Bar = 0.1 μm.

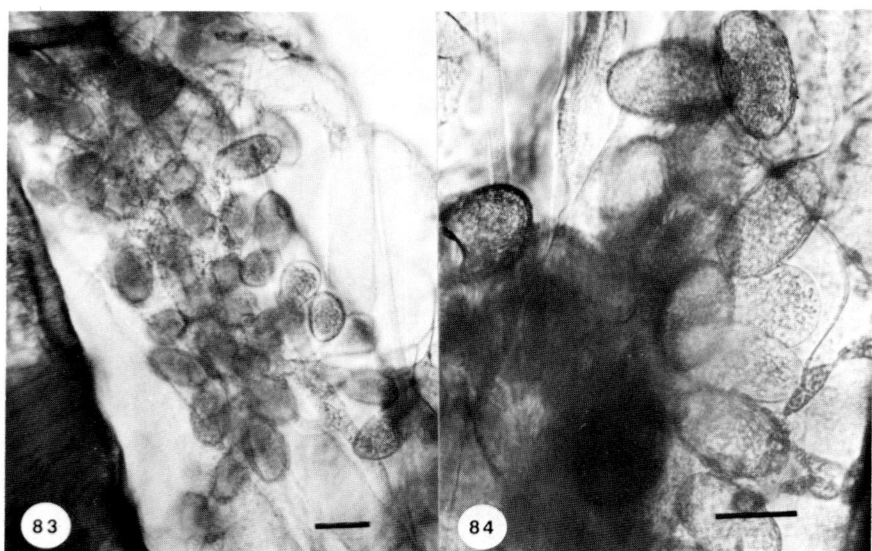

Figs. 83 and 84. *Coelomomyces punctatus* on *Anopheles quadrimaculatus.* Early stages in formation of sporangia. Photomicrographs from a living larva. Bar = 10 μm.

characteristic for sporangia of the species, with elongation occurring at a right angle to the long axis of the parent hypha (Figs. 95–100). The final stage in sporangial development involves completion of deposition of a complex, two-layered wall around the sporangial cytoplasm (Figs. 99 and 100), a process described subsequently. With light microscopy, the outer wall layer of mature sporangia appears amber to golden-brown, whereas the inner layer is colorless. The cytoplasm in such sporangia (Fig. 101) is densely granular and has central and peripheral accumulations of lipid globules (Couch, 1968; Martin, 1969). Internal structure, however, may

---

Figs. 85–101. Photomicrographs showing formation of resting sporangia of *C. punctatus* on *Anopheles quadrimaculatus.* [From Martin (1969).] Figs. 85–92. Time-sequence series showing sporangium formation from an expanded hyphal tip in the coelom of a living larva. Bar = 30 μm. Figs. 93–98. Whole mounts of excised young sporangia mounted in lactophenol with cotton blue. Figs. 93 and 94. Sporangial initials at hyphal tip. Phase-contrast microscopy. Bar = 30 μm. Figs. 95–98. Young resting sporangia still attached to parent hypha by hyphal sheath. Phase-contrast microscopy. Scale as in Fig. 93. Figs. 99 and 100. Surface views of a mature resting sporangium. Fig. 99, dorsal. Fig. 100, ventral. Note ornamentation. Scale as in Fig. 93. Fig. 101. Mid-optical section through mature resting sporangium showing peripheral and central granular-appearing accumulation of lipid droplets. Scale as in Fig. 93.

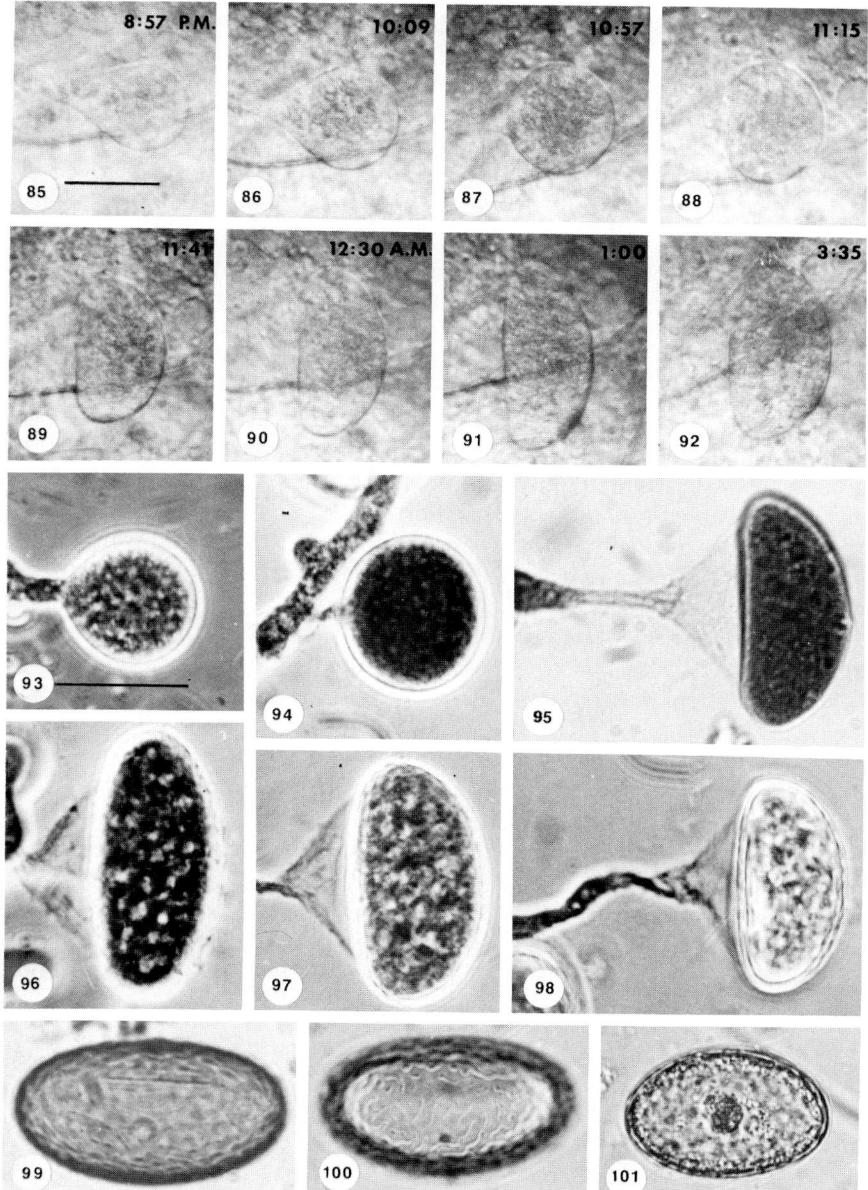
8:57 P.M.
10:09
10:57
11:15
11:41
12:30 A.M.
1:00
3:35
85
86
87
88
89
90
91
92
93
94
95
96
97
98
99
100
101

be difficult to discern in forms with thick and/or heavily sculptured sporangial walls.

Sporangial maturation in different species of *Coelomomyces* has been reported to occur either while sporangia remain attached to the parent hyphae by an enveloping hyphal sheath (Iyengar, 1935; Lum, 1963; Martin, 1969) or following their detachment from the parent hypha (Keilin, 1921; Couch, 1945; Umphlett, 1964). Martin (1969) speculated that detachment or retention of incipient sporangia may be a species characteristic. While it is possible that his phenomenon may be a function of sheath thickness in different species, it is also possible that detachment is related simply to larval activity during sporangial formation, greater larval activity resulting in increased numbers of detached sporangia. Support for this is found in the report by Martin (1969) that, although the majority of sporangia of *C. punctatus* complete development while attached to the parent hypha, the presence of free-floating sporangia in the hemolymph indicates that detachment may occur also. In other observations concerning sporangial formation, Bogoyavlensky (1922) noted a decrease in sporangial size during maturation. However, this observation is probably erroneous and likely to be a result of his observations having been made on material dissected in water, in which case young sporangia swell appreciably more than mature ones (Umphlett, 1964; Martin, 1969; Madelin and Beckett, 1972).

Although there have been no fine-structural studies involving early stages in sporangial formation, the works of Martin (1969), Whisler *et al.* (1972), and Powell (1976) provide insight into the structure of resting sporangia and into the stages in deposition of the complex, two-layered sporangial wall of *C. punctatus* and *C. psorophorae,* with similar structure and stages probably occurring in other species. In the species studied, sporangial cytoplasm (as will be shown in Fig. 108) contains an assemblage of typical fungal cell organelles such as those described previously for hyphagens and hyphae. The "unknown inclusion" observed in sporangial cytoplasm by Martin (1969) may be the same as the "gamma-like bodies" and/or "adhesion vesicles" reported by Travland (1979a) to occur in zoospores. The paracrystalline inclusion, known also to occur in zoospores (Martin, 1969; Whisler *et al.,* 1972; Travland, 1979a), is not evident in sporangial cytoplasm and becomes apparent only during late stages of sporogenesis.

While the origin(s) and means of deposition of the material comprising the primary (outer) and secondary (inner) sporangial walls of *Coelomomyces* spp. is unknown, the structural phenomena associated with primary wall synthesis have been described (Powell, 1976). From this work on *C. punctatus,* the first indication of primary wall formation around sporangial initials is the appearance of a continuous, compact, granular layer of electron-dense material between the plasmalemma and the granular/fibrillar layer of the hyphal envelope (Figs. 102 and 103). As noted previously, it is probably this thin layer of wall material that accounts for the thickened walls of sporangial initials visible via light microscopy. Continued but unequal deposition of wall material coincides with the appearance of radially arranged lacunae in the forming wall (Fig. 104). Depending on species, such lacunae may apparently vary in their pattern of formation and appearance, e.g., parallel in *C. punctatus* (Martin, 1969; Powell, 1976), but parallel distally and anastomosing proximally in *C. psorophorae* (Whisler, 1972). The lacunae no doubt constitute the "internal vertical striae" visible in sporangial walls with light microscopy and often referred to in taxonomic descriptions. Although wall lacunae are open proximally during development, in the completely formed primary wall the inner surface is closed by an electron-opaque covering (Fig. 105). Therefore, when completed, the primary wall consists of three layers or zones: a thin, outer layer, "D_1" (Whisler *et al.*, 1972) or "tectum" (Martin, 1969); a thick, lacunate layer, "D_2" Whisler *et al.*, 1972) or "columellae" (Martin, 1969); and a thin, inner layer, "D_3" (Whisler *et al.*, 1972) or "foot" (Martin, 1969). It is through uneven deposition of primary wall material that the wall sculpturing and dehiscence slit characteristic of different species of *Coelomomyces* are formed, thin-walled areas generally being subtended during development by several cisternae of endoplasmic reticulum (Figs. 106 and 107) [see Powell (1976) for the possible significance of this]. Following formation of the primary wall, a second, homogeneous, electron-translucent layer is deposited between the primary wall and the sporangial cytoplasm. Although formation of this layer was not observed by Powell (1976), the layer is evident with light microscopy and has been seen also in fine-structural studies (Figs. 109 and 110) (Martin, 1969; Whisler *et al.*, 1972). A third wall layer, described by Martin (1969) to occur inside the second layer, is not evident in the works of others (Whisler *et al.*, 1972; M. J. Powell, personal communication) and is likely to be an artifact of the fixation procedure used by Martin. The granular hyphal coat, often appearing during wall formation with vesicles embedded in it (Powell, 1976), persists throughout wall formation and probably represents the "hyphal sheath" (Martin, 1969) visible with light (Figs. 95–98) and scanning electron microscopy and responsible for anchoring sporangia to parent hyphae (Powell, 1976). As noted previously, although resting sporangial wall structure has been studied only in *C. punctatus* and *C. psorophorae*, it is likely that, while varying in thickness and sculpturing from species to species, a basically similar structural pattern exists in resting sporangia of all species. A similar structure has even been

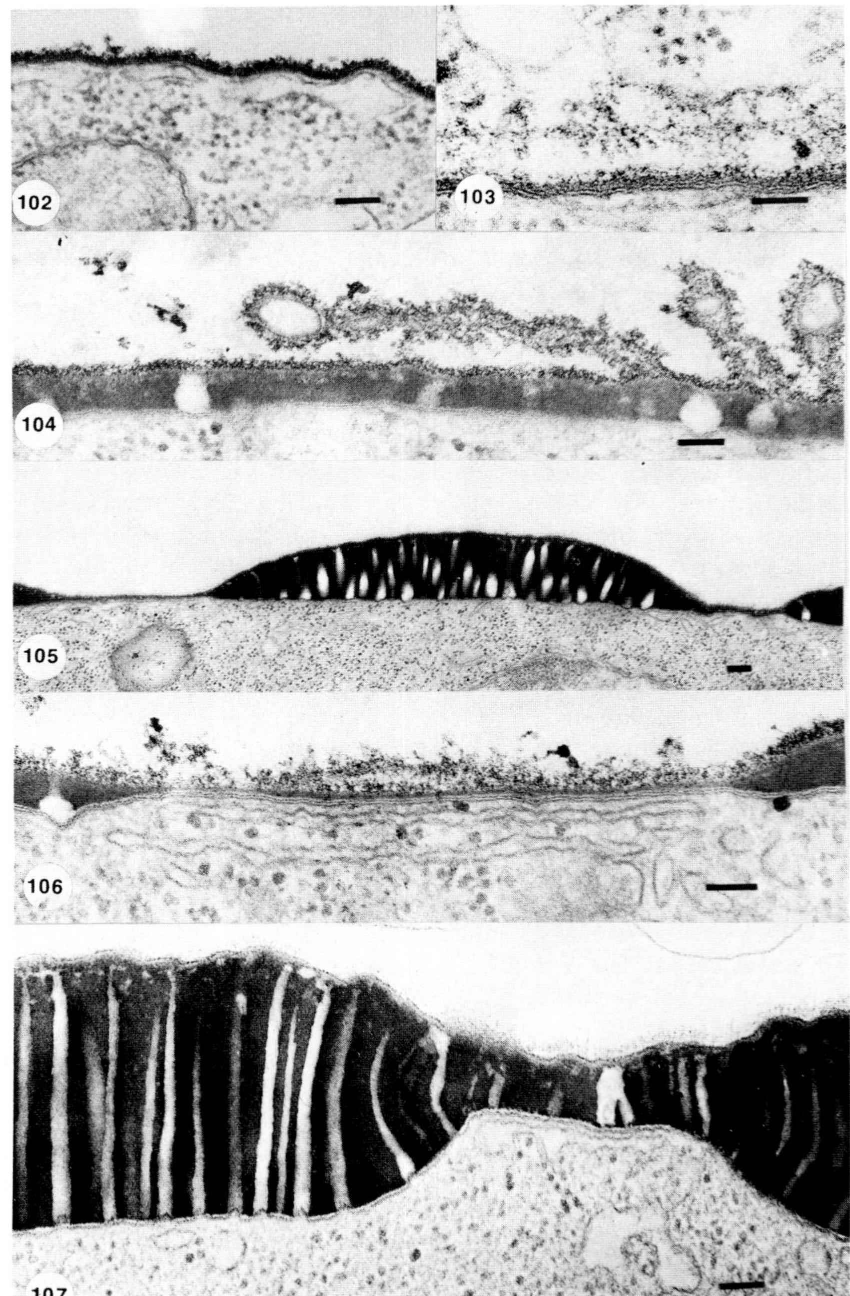
102
103
104
105
106
107

described for thin-walled sporangia of *C. indicus* (Madelin and Beckett, 1972) and probably explains the "smooth, and finely punctate" appearance described by Laird *et al.* (1975) for thin-walled sporangia of *C. omorii* (*C. iliensis*). For thin-walled sporangia, the primary wall is approximately half the thickness of the secondary wall (Fig. 147), the latter being of about the same thickness as the corresponding structure in resting sporangia (Madelin and Beckett, 1972).

VII. ZOOSPOROGENESIS AND SPORE RELEASE

Keilin (1921) first observed resting and thin-walled sporangia in a species of *Coelomomyces* (*C. stegomyiae*) and accurately suggested their function, the former representing a resistant or resting form and the latter serving for immediate reproduction. Although Keilin was unable to observe spore release in the preserved specimens he had available for study, he precisely predicted the mode of sporangial dehiscence and spore release in members of this genus. Since this early work there have been several references to dehiscence of both resting sporangia (Răsín, 1929; de Meillon and Muspratt, 1943; Muspratt, 1946; Couch, 1945, 1968; Couch and Umphlett, 1963; Lum, 1963; Whisler *et al.*, 1972) and thin-walled sporangia (Muspratt, 1946; Couch and Umphlett, 1963; Couch, 1968; Madelin and Beckett, 1972). While each of these studies documents sporangial dehiscence, few (Couch and Umphlett, 1963; Couch, 1968; Madelin and Beckett, 1972; Whisler *et al.*, 1972) provide the details of sporogenesis and/or sporangial dehiscence and spore release. Many questions remain concerning these processes, so the following summarizes

---

Figs. 102–107. Transmission electron micrographs showing stages in the formation of the outer wall of resting sporangia of *C. punctatus*. Bar = 0.1 μm. [From Powell (1976).] Fig. 102. The surface of the granular hyphal coat appears fibrillar. Notice the cisterna of endoplasmic reticulum lying parallel to the plasma membrane. Fig. 103. Appearance of a compact, continuous granular wall layer between the hyphal coat and the plasma membrane of a sporangial initial. The sporangial initial is appressed against the basement membrane of the mosquito. Fig. 104. Homogeneously osmiophilic wall layer transected by lacunae and developing beneath the hyphal coat and granular wall layer. Vesicles are still embedded in the coat material. Fig. 105. Uneven thickening of the wall. Lacunae are only in the thickened regions of the wall. Cisternae of endoplasmic reticulum subtend the thin areas of pits in the wall. Fig. 106. Detail of a pit. Two cisternae of smooth endoplasmic reticulum subtend the pit. The hyphal coat and the continuous wall layer cover the surface of the plasma membrane in the pit and extend over the thickened regions of the wall. Fig. 107. Endoplasmic reticulum cisternae are slightly inflated as the pit area of the wall thickens. Plasma membrane protrudes into the lacunae.

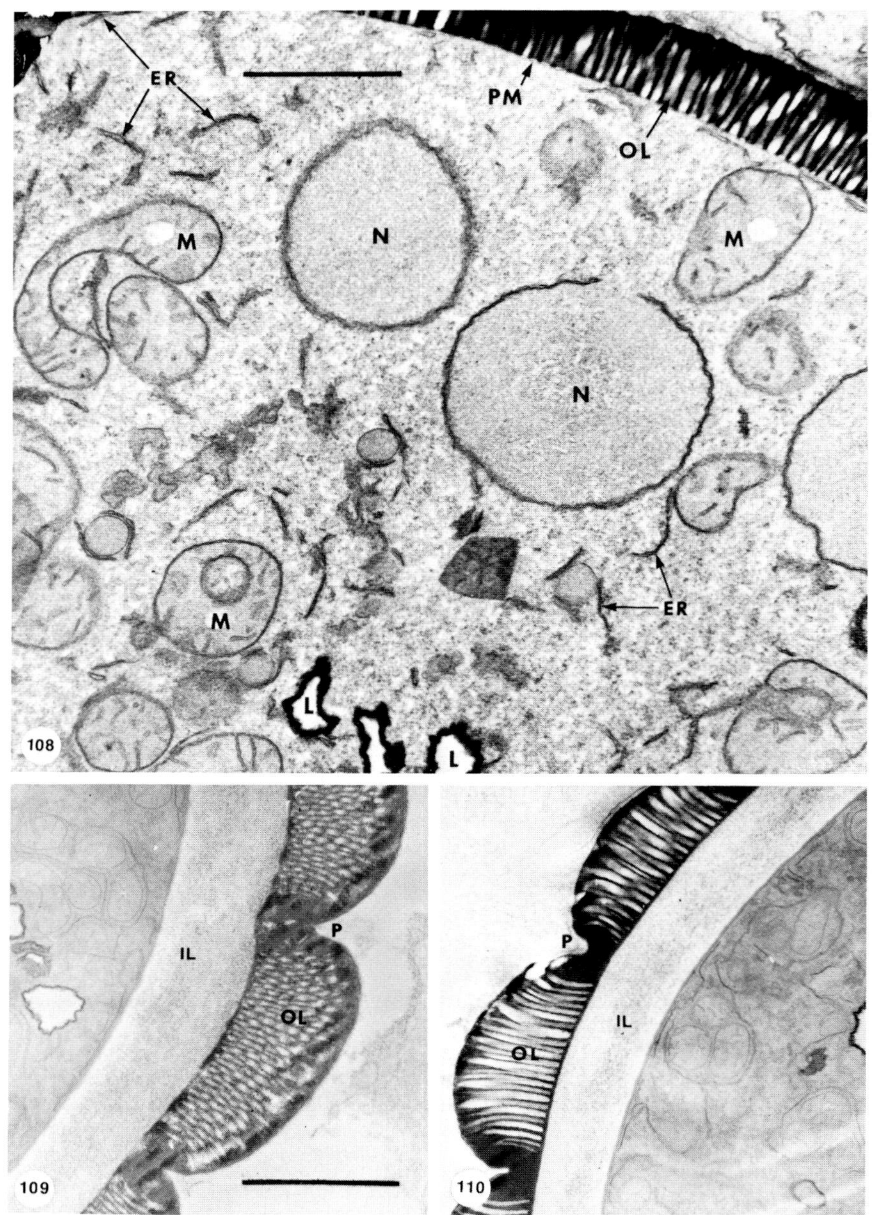
ER
PM
OL
M
N
M
N
ER
M
L
L
108
IL
P
OL
109
P
OL
IL
110

what is currently known of them. Although resting sporangia and thin-walled sporangia sporulate[3] under different conditions (Muspratt, 1946; Couch and Umphlett, 1963; Madelin and Beckett, 1972) and produce spores presumably serving different functions (see Chapter 2), the events of sporogenesis, dehiscence, and spore release appear to be practically identical in the two. For this reason, they will here be described jointly, with small points of difference being indicated. As first recognized by Whisler *et al.* (1972) and Madelin and Beckett (1972), one should note for comparison that the cytoplasmic events of sporogenesis in *Coelomomyces* spp. appear to be similar to those of *Blastocladiella emersonii* (Lessie and Lovett, 1968) and also to those of *Allomyces* spp. (Moore, 1968; Barron and Hill, 1974), all members of the order Blastocladiales.

Before discussing the events of sporogenesis and dehiscence, a consideration of the structure and appearance of mature sporangia is in order. Although resting sporangia and thin-walled sporangia may be functionally different (see Chapter 2), the internal cytoplasm of mature, viable sporangia of both forms (Figs. 111 and 112; also see Figs. 128 and 129) appears quite similar with light microscopy (Couch, 1968; Madelin and Beckett, 1972). In this regard, both are characterized by a central mass of granular material, "lipoid bodies" (Couch, 1968), from which arms of similar granules radiate to a layer of granules at the sporangial periphery (Fig. 112). Although the nature of these granules has not been confirmed by electron microscopy, it is likely from comparative studies by light and electron microscope of other developmental stages that they are in fact primarily lipid. Whether other cytoplasmic organelles are unevenly distributed at this stage is uncertain. However, Couch and Umphlett (1963) do describe nuclei as being peripherally distributed in mature sporangia of *C. indicus*.

In most studies involving sporulation (noted previously) of both thick- and thin-walled sporangia, observations have been made on sporangia dissected from infected larvae into a drop of water on a microscope slide. Such slides are normally stored in a damp chamber, from which they may be removed at intervals for study (Couch, 1968). Using this technique, it has been possible to investigate the factors involved in conditioning of

[3] The term sporulate (including the events of spore formation, dehiscence, and spore release) is used here instead of germinate, a term often misapplied to this process.

Figs. 108–110. Transmission electron micrographs of resting sporangia of *C. punctatus* that were fixed in $KMNO_4$. [From Martin (1969).] Fig. 108. Cross-section of a young sporangium in which deposition of the outer wall layer is incomplete. Bar = 1.5 μm. Figs. 109 and 110. Mature sporangia with both inner and outer wall layers formed. Endoplasmic reticulum (ER), inner wall layer (IL), lipid (L), mitochondria (M), nuclei (N), outer wall layer (OL), pits (P), plasma membrane (PM). Bar = 2 μm.

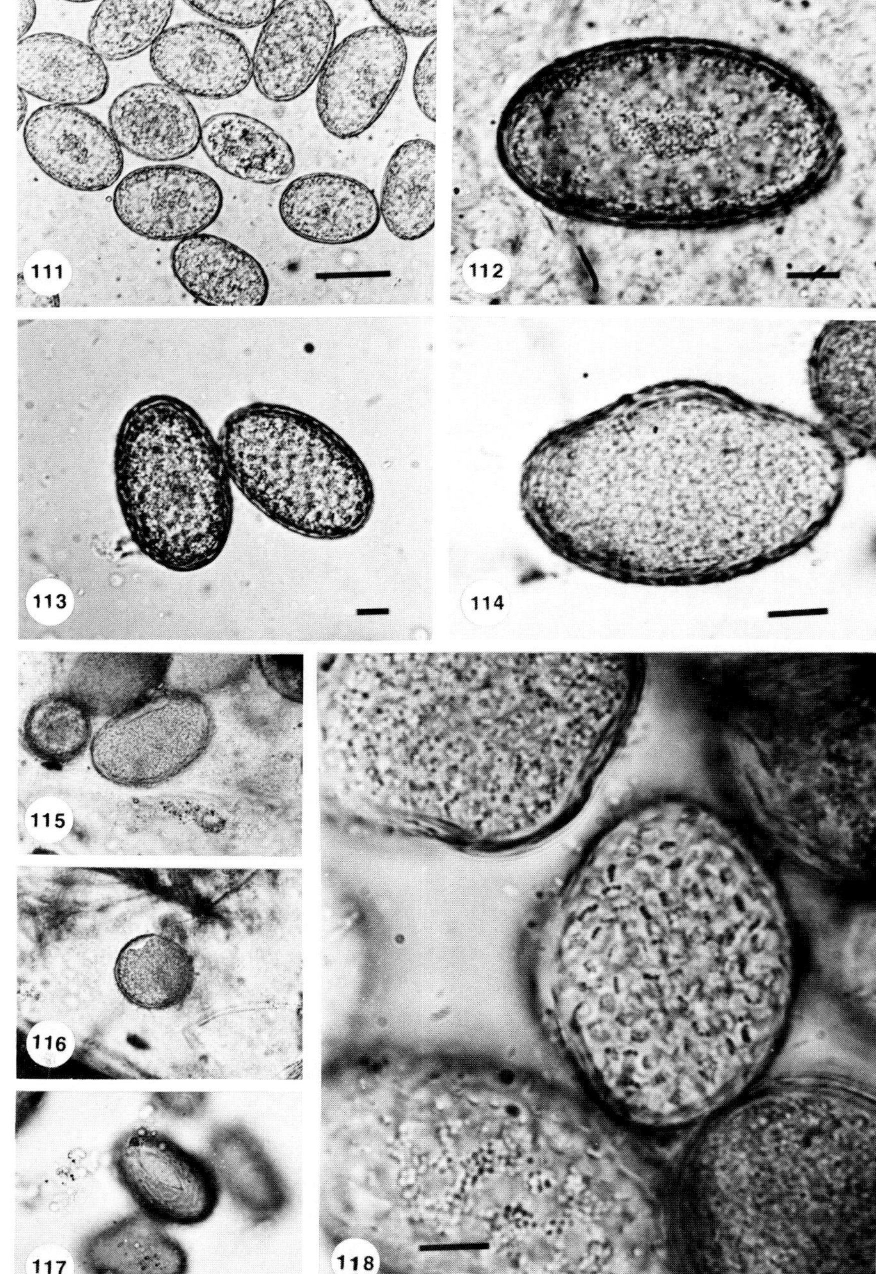

111
112
113
114
115
116
117
118

sporangia prior to sporulation as well as the actual events of sporulation, including sporogenesis, sporangial dehiscence, and spore release.

A. Light-Microscope Observations of Sporulation

Initiation of sporogenesis in resting and thin-walled sporangia is first evident as a change in sporangial cytoplasm from unevenly granular, to reticulate or alveolate (Fig. 113; also see Figs. 128 and 129), to finally uniformly granular or homogeneous (Couch, 1968; Madelin and Beckett, 1972). Subsequently, an expanding, hyaline layer of material is deposited beneath the dehiscence slit (Figs. 114–116; also see Fig. 130). The increase in thickness of this material results in opening of the dehiscence slit (Figs. 117 and 120; also see Fig. 131), an event followed by extension of the hyaline material as a plug into the opening. After formation of the plug [rarely before, in thin-walled sporangia (Madelin and Beckett, 1972)], cleavage of the cytoplasm into closely packed spore initials becomes evident. Concurrently, there is aggregation of the cytoplasmic lipid bodies as plates of 6–12 droplets in each spore initial (Fig. 118; also see Fig. 131). In side view, such clusters appear as slightly curved rows of 3–8 droplets. Resting sporangia at this stage have been referred to by Couch (1968) as being in the "go" condition and are ready to discharge their spores, an event apparently triggered by a decrease in O_2 level. Under such conditions (achieved experimentally by adding a cover glass to a slide mount or by placing the slide in an O_2-free chamber), spore release (Figs. 119–127) commences in 10–15 min (Couch, 1968; Whisler *et al.*, 1972). Thin-walled sporangia are also ready to release spores at this stage (Figs. 132–135) and will do so within 2 hr if adequate aeration is provided (Madelin and Beckett, 1972). The time necessary for conversion of sporangia from the uniformly reticulate stage to the "go" stage has been noted to require "several hours" for thin-walled sporangia of *C. indicus* (Madelin and Beckett, 1972) and approximately 6 hr for resting sporangia of *C. punctatus* (Couch, 1968).

Impending spore release is first indicated by a swelling of the hyaline plug material in the dehiscence slit and bulging of the material through the

---

Figs. 111–118. Stages in sporulation in resting sporangia of *C. punctatus*. Figs. 111 and 112. Mature, viable, resting sporangia with lipid accumulations at periphery and in the center. Fig. 111. Bar = 35 μm. Fig. 112. Bar = 10 μm. Fig. 113. Uniformly reticulate stage. Bar = 10 μm. Fig. 114. "Go" stage. Bar = 10 μm. Figs. 115 and 116. Side and end views, respectively, showing accumulation of hyaline plug material under the dehiscence slit. Bar = 35 μm. Fig. 117. Face view of sporangium in the "go" stage. Bar = 35 μm. Fig. 118. Spore initial stage with lipid plates evident. Bar = 10 μm.

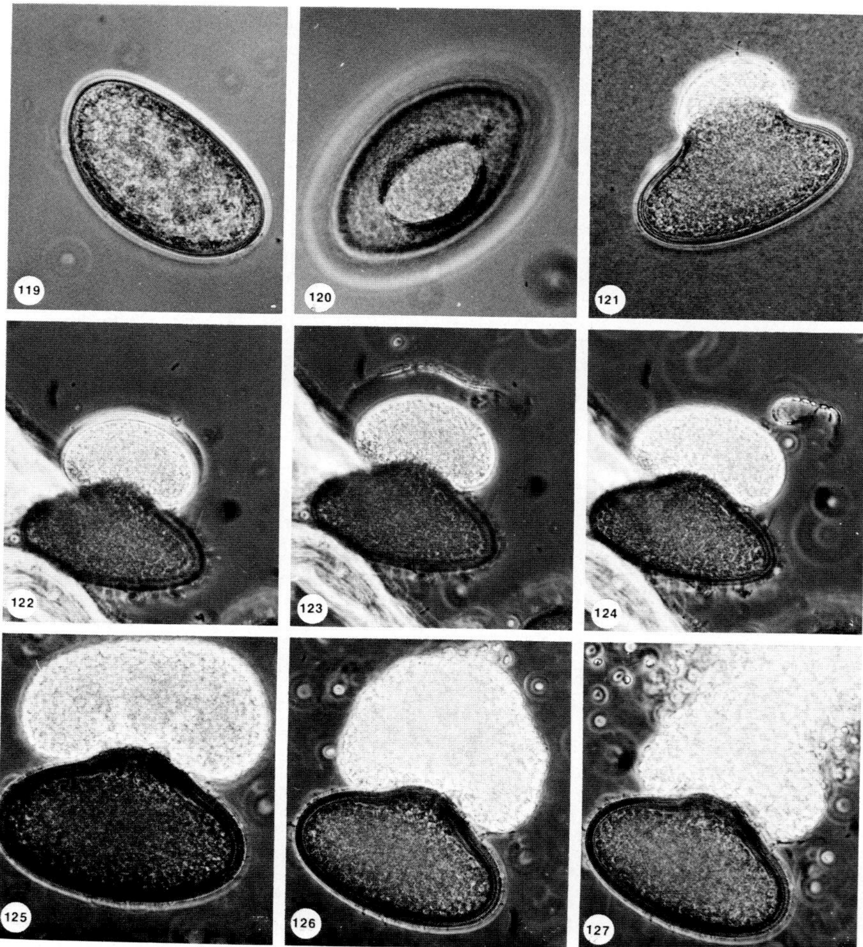

Figs. 119–127. Stages in resting sporangial dehiscence in *C. psorophorae.* Phase-contrast microscopy. [From Whisler *et al.* (1972).] Fig. 119. Mature resting sporangium. Fig. 120. "Go" stage sporangium. Figs. 121–124. Extrusion of sporogenic cytoplasm into an expanding vesicle. Figs. 125–127. Completion of cleavage and spore release (Fig. 127).

widening slit (Figs. 120 and 121; also see Fig. 136). At this stage it is apparent that the plug material is bounded peripherally by a thin, refractive covering (Fig. 121). Within minutes, the cleaved, tightly-packed spores are extruded through the dehiscence slit into a constraining vesicle formed by expansion of the plug (Figs. 121–125). With this expansion, the refractive outer layer of the plug (now vesicle) fragments and may actu-

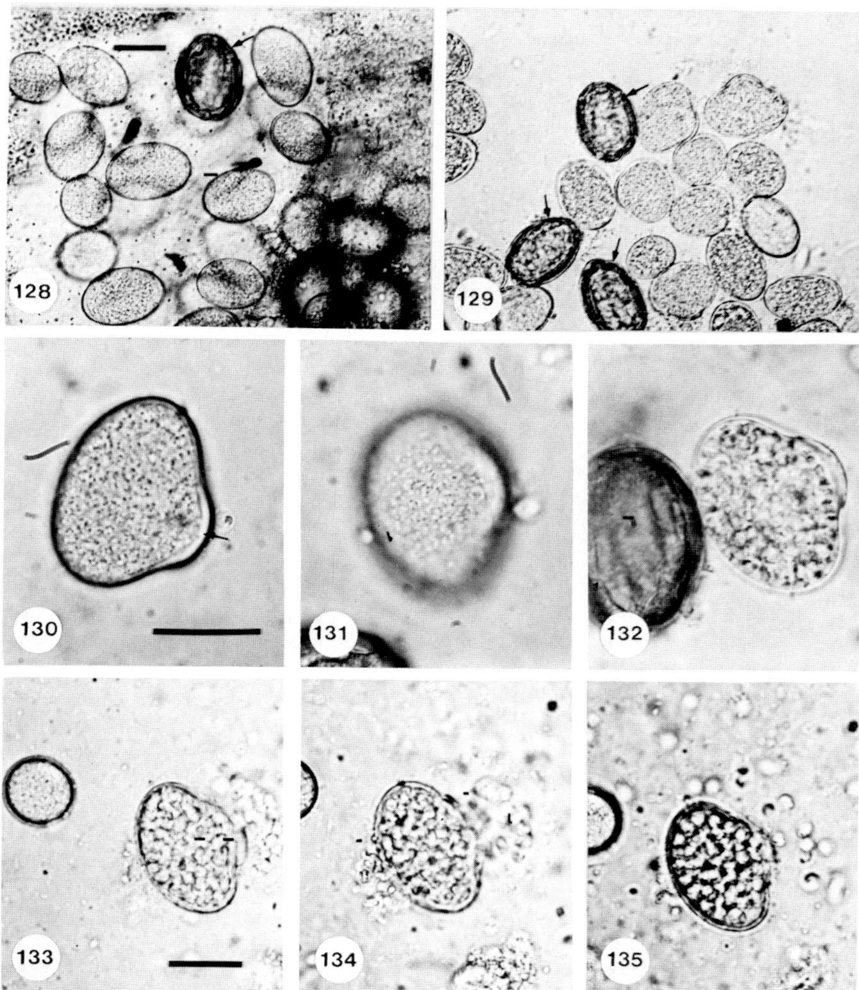

Figs. 128–135. Sporulation in thin-walled sporangia of *C. indicus*. Figs. 128 and 129. Note difference in appearance of thick-walled (arrows) and thin-walled sporangia. Bar = 10 μm. Fig. 130. Stage in early sporogenesis with finely dispersed lipid droplets and prominent gel (arrow) under the dehiscence slit. Scale as for Fig. 128. Fig. 131. Lipid plate stage. Scale as for Fig. 128. Fig. 132. Fully cleaved zoospores. Scale as for Fig. 128. Figs. 133–135. Zoospore release. Scale as for Fig. 128.

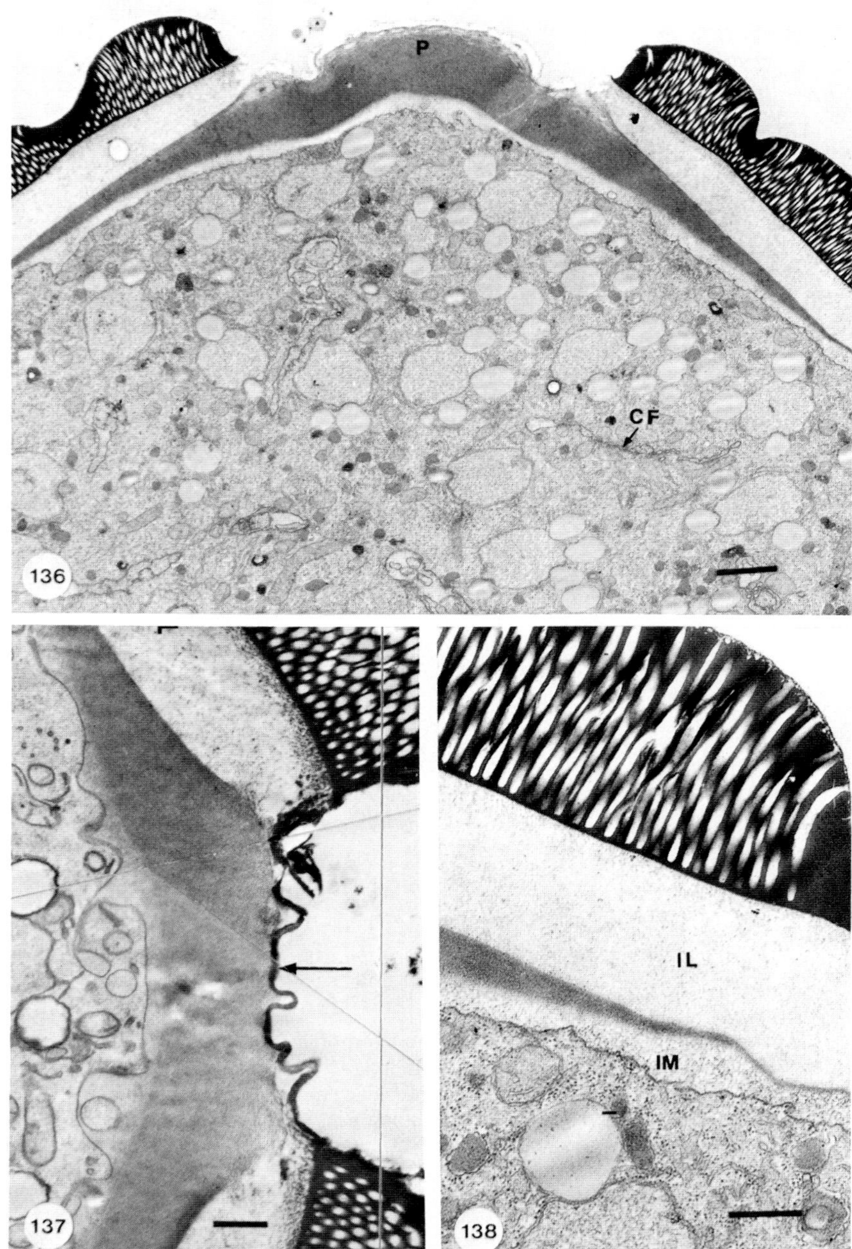
P
CF
136
137
IL
IM
138

ally break away (Figs. 123 and 124). Extrusion of the spore mass continues until the vesicle and its enclosed spores reach a size approximating that of the sporangium itself, approximately half the spore mass now being in the vesicle and the other half being retained in the sporangium. Spore motility, noted immediately as a churning of the spore mass, becomes progressively more vigorous until the vesicle ruptures, an event accompanied by the apparent forcible expulsion of a large group of spores from the vesicle (Figs. 126 and 127). The extruded spores separate from one another and swim away. The remaining spores swim actively from the vesicle and sporangium. The empty vesicle soon disappears. Release of all spores is usually accomplished within 15 min after initiation.

B. Electron-Microscope Observations of Sporulation

Although fine-structural studies of sporulation in *Coelomomyces* spp. have been hampered by difficulties in obtaining adequate fixation, the recent, unpublished observations of M. J. Powell (personal communication) combined with the observations of Madelin and Beckett (1972) on thin-walled sporangia of *C. indicus* and Martin (1970) and Whisler *et al.* (1972) on resting sporangia of *C. punctatus* and *C. psorophorae,* respectively, provide insight into this process. As already noted for light-microscope observations, fine-structural studies reveal the events of sporulation to be practically identical in resting and thin-walled sporangia. In both, however, the earliest stage of sporulation observed thus far with electron microscopy coincides approximately with the "go" stage described in light-microscope studies. Beginning with observations at this stage, formation of the plug material is first evident as an accumulation, in the area under the dehiscence slit, of an electron-transparent, homogeneous substance between the secondary wall and the sporogenic cytoplasm (Figs. 136 and 137; also see Figs. 139 and 147). As noted by Martin (1970), this substance probably extends the entire length of the dehiscence slit and, through its continued accumulation, causes rupture along the slit of the primary and secondary wall layers, thereby resulting in the "go"

Figs. 136–138. Transmission electron micrographs of "go" stage resting sporangia of *C. punctatus.* Fig. 136. Note two wall layers, open dehiscence slit, discharge plug (P), partially cleared cytoplasm. Cleavage furrows (CF). Bar = μm. [From M. J. Powell (unpublished).] Fig. 137. Open dehiscence slit showing covering of wall material (arrow) over the plug material. Bar = 0.5 μm. [From W. W. Martin, III (unpublished).] Fig. 138. Detail at edge of plug material. The plug material tapers as it extends away from the dehiscence slit and forms a thin layer between the inner wall layer (IL) and the interstitial material (IM). Bar = 0.5 μm. [From M. J. Powell (unpublished).]

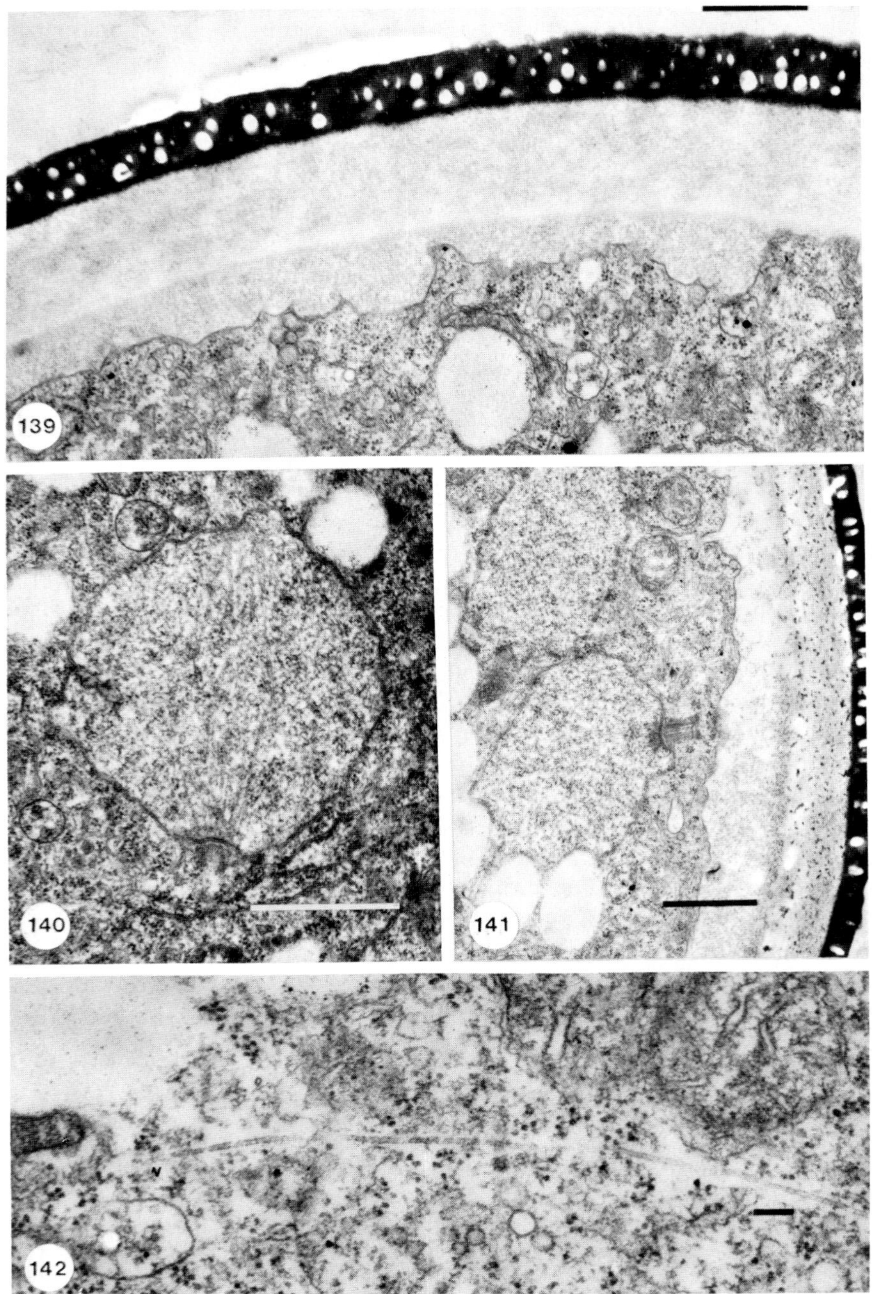
139
140
141
142
N

stage appearance of the sporangium. In cross-sectional view, the plug appears as a crescent under the dehiscence slit and surrounding wall area, the thickest portion of the crescent being under the opening in the sporangial wall. From the recent work of M. J. Powell (unpublished), it appears that a thin layer of plug material may be continuous over the entire inner edge of the inner wall layer (Fig. 138). The exposed periphery of the plug (Fig. 136) is covered by a thin layer of electron-dense material (appearing via light microscopy as a thin, refractile covering), which is thought to be a continuation of the inner portion of the primary wall [''D_3'' layer (Whisler *et al.*, 1972), ''foot'' layer (Martin, 1969)]. While the origin of the plug is unknown, it is likely that it is formed by a process of vesicular secretion similar to that producing the papilla of *Blastocladiella emersonii* (Lessie and Lovett, 1968) and other blastocladiaceous fungi, with the two structures serving essentially similar and probably homologous functions.

In agreement with the proposed life cycle of *Coelomomyces* spp. (Whisler *et al.*, 1975), recent studies by H. C. Whisler *et al.* (unpublished) have confirmed the existence of synaptonemal complexes in resting sporangia of *C. psorophorae,* thereby establishing that meiosis precedes sporulation and that spores arising from resting sporangia are meiospores. Although the nature of nuclear divisions (meiotic or mitotic) occurring in thin-walled sporangia of *Coelomomyces* spp. has not been determined, Madelin and Beckett (1972) showed such divisions to be centric (Fig. 140) and to have a persistent nuclear envelope and an intranuclear spindle. These authors showed further that the nuclear envelope in the vicinity of centrioles is slightly concave and appears more electron-dense than in other areas. Following the nuclear divisions in both types of sporangia, the nucleus-associated centrioles function as kinetosomes in proliferating flagellar axonemes (Figs. 141 and 143). Elongation of axonemes occurs inside the cleavage furrows forming also at this time. Although origin of the cleavage furrows (Figs. 146 and 148) has not been demonstrated in *Coelomomyces* spp., they probably form by coalescence of cleavage vesicles (Figs. 145 and 146) in a manner similar to that described for *Blastocladiella emersonii* (Lessie and Lovett, 1968) and *Allomyces macrogynus* (Barron and Hill, 1974). As in *B. emersonii,* cleavage furrows of *Coelomo-*

Figs. 139–142. Transmission electron micrographs of thin-walled sporangia of *C. indicus.* [From Madelin and Beckett (1972).] Fig. 139. Two-layered wall. Note accumulation of gel between sporogenic cytoplasm and the inner wall. Bar = 0.5 μm. Fig. 140. Nuclear division during early stage of sporogenesis. Note the intranuclear spindle and the centrioles. Bar = 1 μm. Fig. 141. Nuclear centriole association during late precleavage. Bar = 1 μm. Fig. 142. Cytoplasmic microtubule evident during late precleavage. Bar = 0.1 μm.

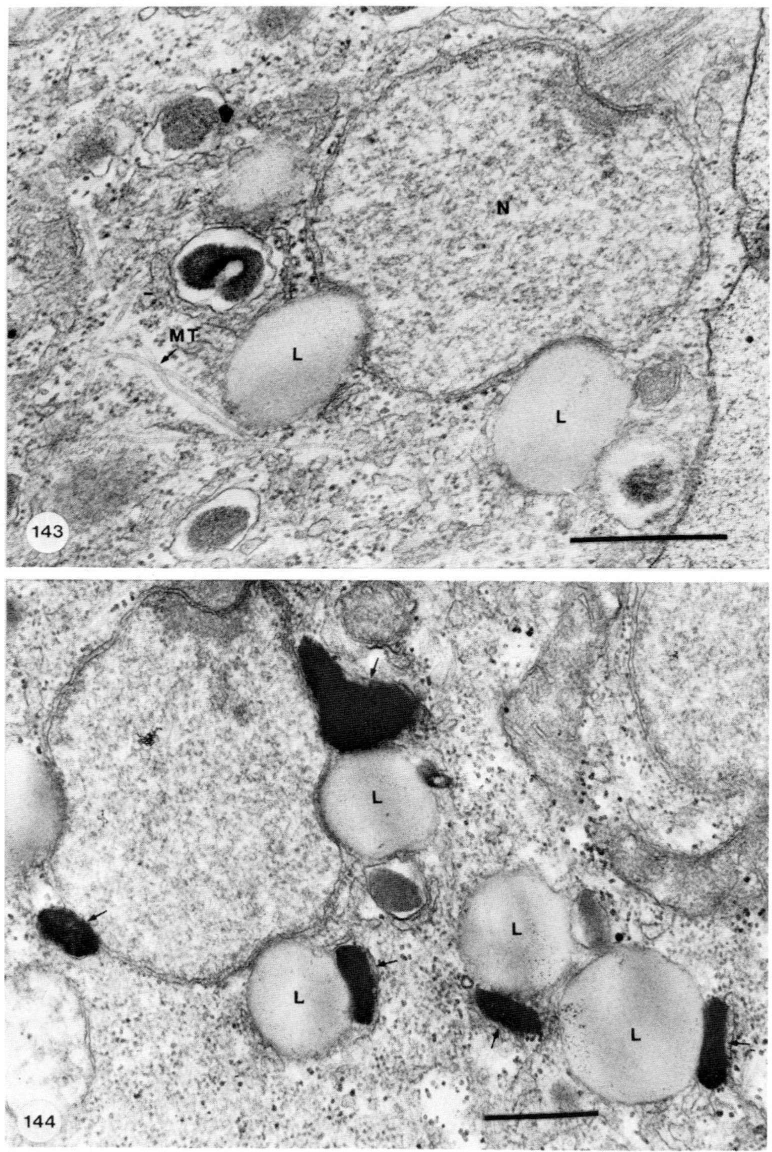

Figs. 143–144. Transmission electron micrographs of resting sporangia of *C. punctatus.* Bar = 0.5 μm. [From M. J. Powell (unpublished).] Fig. 143. Partially cleaved cytoplasm. Note that the lipid droplets (L) have associated with the nucleus (N) and that microtubules (MT) are prevalent. Fig. 144. Later stage in sporogenesis showing association of microbodies (arrows) associated with lipid droplets (L). Microbodies stained with the diaminobenzidine technique at pH 7.2.

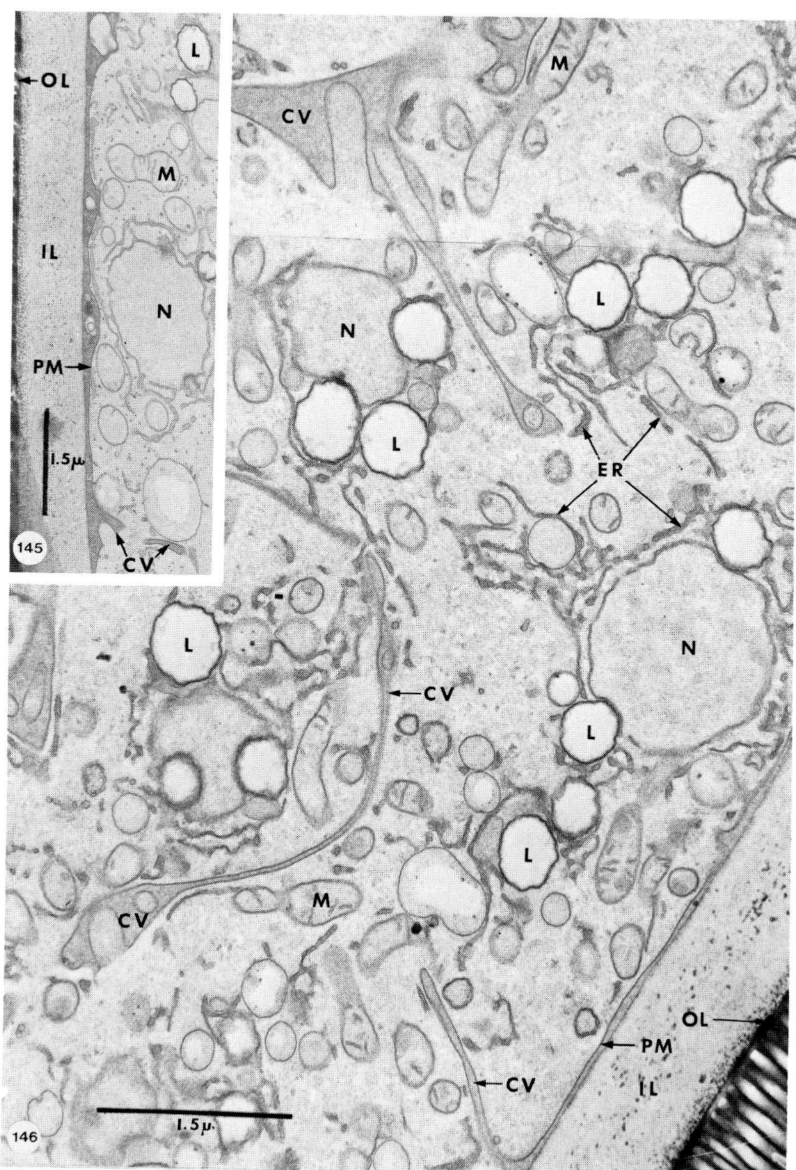

Figs. 145–146. Transmission electron micrographs showing early cleavage in resting sporangia of *C. punctatus*. Fixation in $KMNO_4$. Endoplasmic reticulum (ER). Bar = 1.5 μm. [From W. W. Martin, III (unpublished).] Fig. 145. Detail of sporogenic cytoplasm adjacent to sporangial wall. Fig. 146. Cleavage of sporogenic cytoplasm by cleavage vesicles. Note association of lipid with nuclei. Cleavage vesicles (CV), inner wall layer (IL), lipid (L), mitochondria (M), nucleus, (N), outer wall layer (OL), plasmalemma (PM).

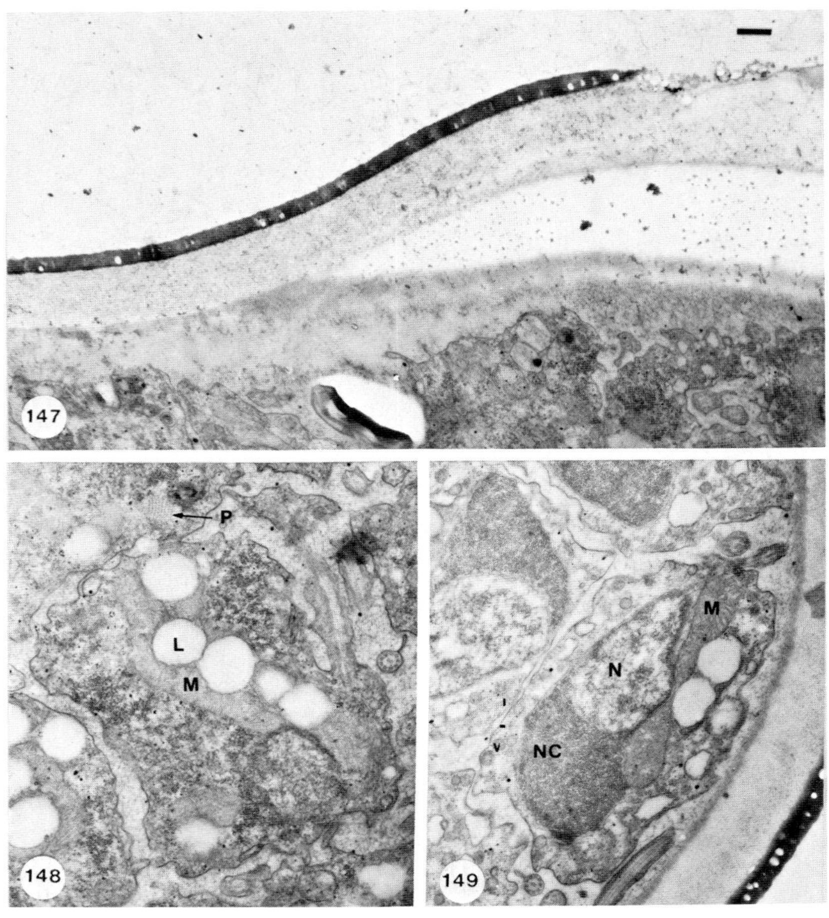

Figs. 147–149. Transmission electron micrographs of thin-walled sporangia of *C. indicus.* Bar = 0.5 μm. [From Madelin and Beckett (1972).] Fig. 147. Section through area of the dehiscence slit at "go" stage. Figs. 148 and 149. Fully cleaved zoospores. Note nucleus (N), nuclear cap (NC), mitochondrion (M), lipid (L), and paracrystalline inclusion (P).

Figs. 150–151. Cleavage of zoospores. Fig. 150. Transmission electron micrograph showing almost cleaved spores in a resting sporangium of *C. punctatus.* $KMNO_4$ fixation. Note that flagellar sheaths (S) are evident in the space between spores. Axonemal microtubules were not preserved by this manner of fixation. Endoplasmic reticulum (ER), inner wall layer (IL), lipid (L), mitochondria (M), nucleus (N), outer wall layer (OL), plasmalemma (P). Bar = 2 μm. [From W. W. Martin, III (unpublished).] Fig. 151. Transmission electron micrograph showing fully cleaved spores of *C. punctatus* inside a resting sporangium. See Figs. 2–22 for details of spore structure. Bar = 2 μm. [From M. J. N. Powell (unpublished).]

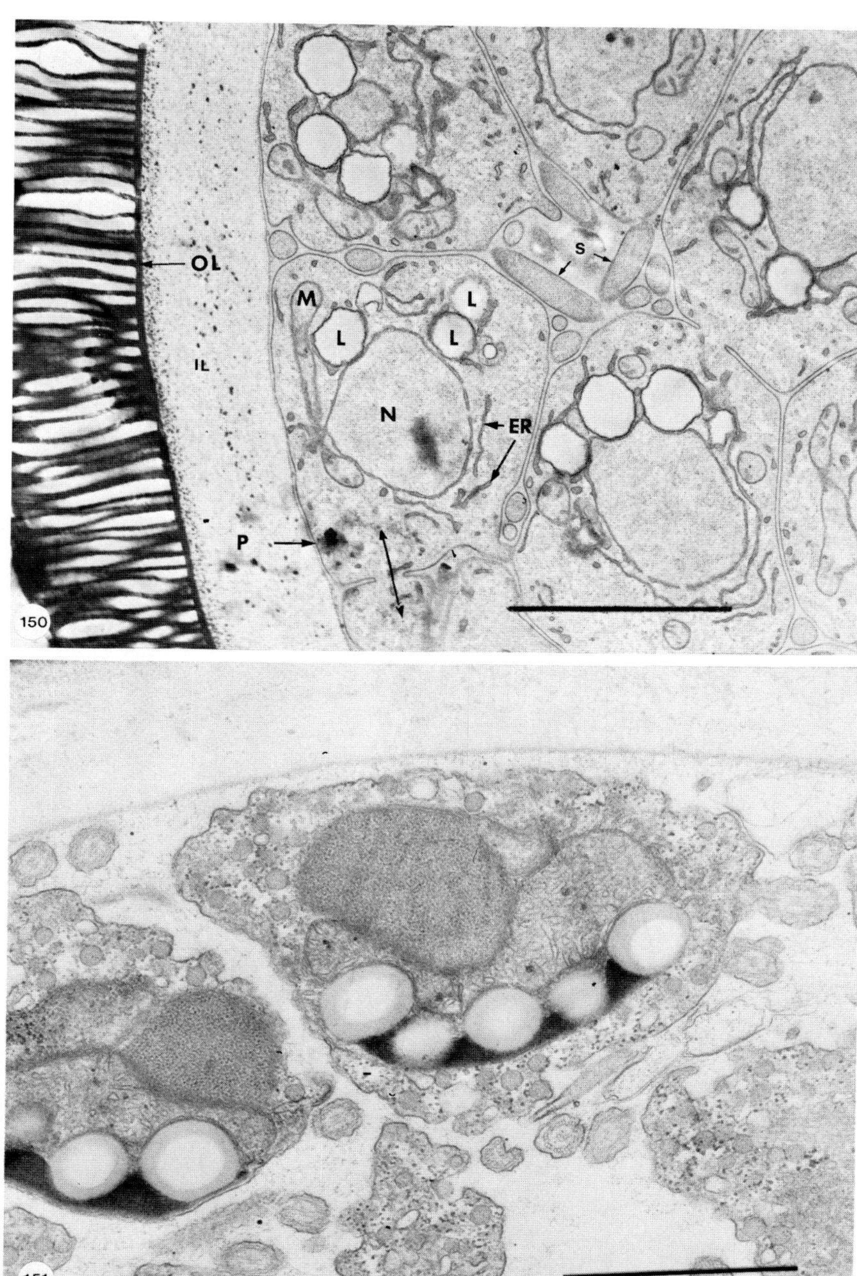
OL
IL
M
L
L
L
L
N
ER
S
P
150
151

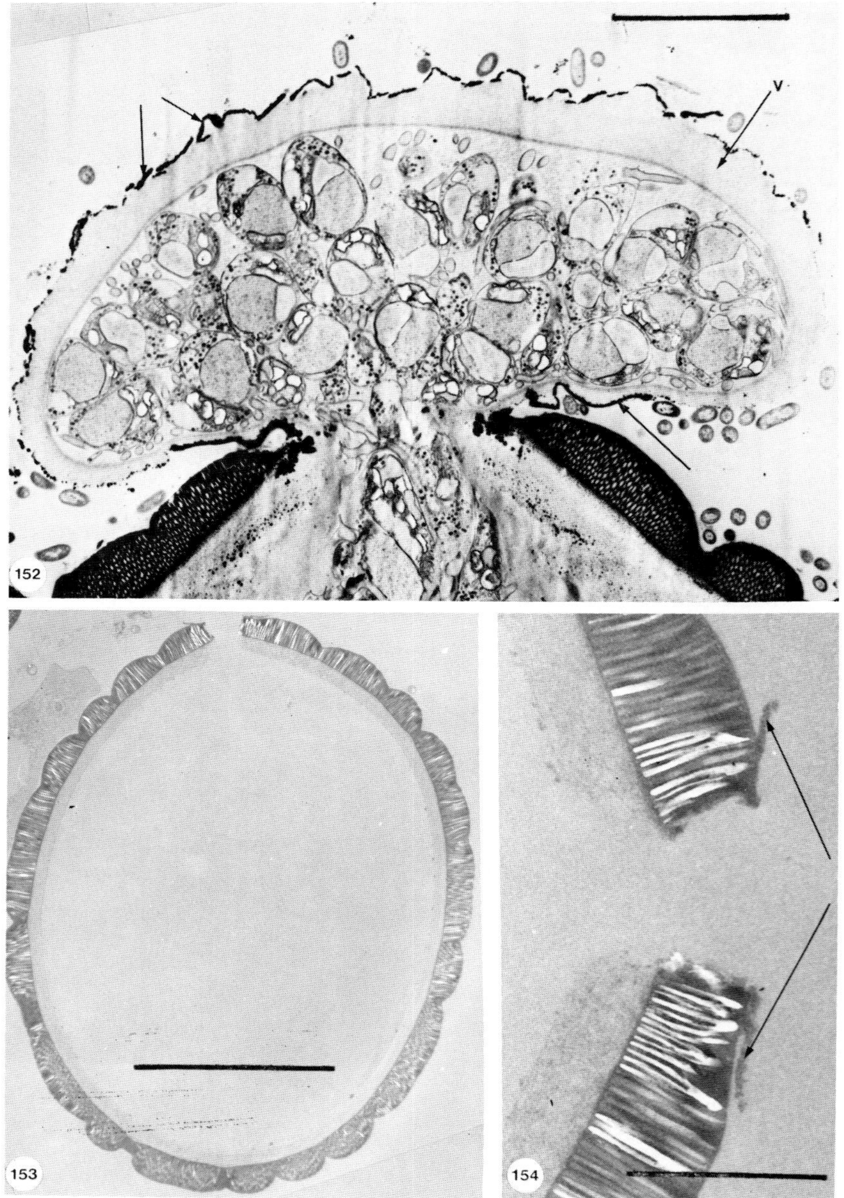

Figs. 152–154. Transmission electron micrographs of resting sporangia of *C. punctatus.* $KMNO_4$ fixation. [From W. W. Martin, III (unpublished).] Fig. 152. Sporangium caught in the act of releasing spores into the vesicle (V). Note the wall remnants

myces spp. contain a fibrillar material that is undoubtedly dumped into them by fusion of cleavage vesicles. Microbodies appear at this stage in association with lipid droplets, which are themselves often so intimately associated with nuclei as to actually indent their envelope (Figs. 143 and 144). A similar phenomenon has been noted during sporogenesis in *Blastocladiella emersonii* (Lessie and Lovett, 1968) and *Allomyces macrogynus* (Barron and Hill, 1974). Gamma-like bodies, mitochondria, free ribosomes, and rough endoplasmic reticulum are now scattered throughout the spore plasm (Fig. 146). Also, microtubules are abundant and may be seen in close association with kinetosomes (Figs. 142 and 143), presumably a stage in formation of the nuclear basket of microtubules evident in later stages of sporogenesis. Further delineation of zoospores by cleavage furrows (Figs. 150 and 151; also see Figs. 148 and 149) is accompanied by reorientation of the cell organelles of spore initials so as to conform to their arrangement in completely formed zoospores. Several events occur during this process: (1) The side-body complex is formed via migration of the lipid droplets to one side of the nucleus and their arrangement as a curved plate, fusion of mitochondria resulting in formation of a single large mitochondrion between the nucleus and the lipid plate and encircling the kinetosome, and coalesence of microbodies external to, but in close association with the lipid droplets. (2) A microtubular basket arises from the kinetosome and ensheaths the nucleus, giving it a characteristic beaked appearance. (3) The paracrystalline inclusion and gamma-like inclusions become evident. (4) The ribosomes accumulate at the nuclear apex and are ensheathed by an enveloping, membranous envelope.

Therefore, during formation of the "go-stage" sporangium, sporogenesis is completed and results in zoospores that are ready for release. Electron-microscope observations of sporangia caught in the preliminary stages of spore release reveal, as suspected from light-microscope observations, that the spore vesicle is formed by expansion and apparent swelling of the plug material (Fig. 152). The thin, electron-dense layer covering the plug is seen to fragment along the periphery of the expanding vesicle (Fig. 152). Following spore release, the vesicle is no longer evident, and all that remains is the two-layered sporangial wall (Figs. 153 and 154).

(arrows) evident around the periphery of the vesicle. Bar = 5 μm. Fig. 153. Empty sporangium showing remaining two wall layers. Bar = 10 μm. Fig. 154. Enlargement of dehiscence slit from Fig. 153. Note wall remnants at orifice (arrows). Bar = 2 μm.

REFERENCES

Barron, J. L., and Hill, E. P. (1974). Ultrastructure of zoosporogenesis in *Allomyces*. *J. Gen. Microbiol.* **80,** 319–327.

Bogoyavlensky, N. (1922). *Zoografia notonectae,* n.g., n. sp. *Arch. Soc. Russe Protistol.* **1,** 113.

Cantino, E. C., and Truesdell, L. C. (1970). Organization and fine structure of the side body and its lipid sac in the zoospore of *Blastocladiella emersonii*. *Mycologia* **62,** 548–567.

Couch, J. N. (1945). Revision of the genus *Coelomomyces* parasitic in insect larvae. *J. Elisha Mitchell Sci. Soc.* **61,** 124–136.

Couch, J. N. (1968). Sporangial germination of *Coelomomyces punctatus* and the conditions favoring the infection of *Anopheles quadrimaculatus* under laboratory conditions. *Proc. J. U.S.-Jpn. Semin. Microb. Control Insect P[illegible]sts, 1967,* pp. 93–105.

Couch, J. N., and Dodge, H. R. (1947). Further observations on *Coelomomyces* parasitic on mosquito larvae. *J. Elisha Mitchell Sci. Soc.* **63,** 69–79.

Couch, J. N., and Umphlett, C. J. (1963). *Coelomomyces* infections. *In* "Insect Pathology" (E. A. Steinhaus, ed.), Vol. 2, pp. 149–188. Academic Press, New York.

De Meillon, B., and Muspratt, J. (1943). Germination of the sporangia of *Coelomomyces* Keilin. *Nature (London)* **152,** 507.

Emerson, R. (1941). An experimental study of the life cycle and taxonomy of *Allomyces*. *Lloydia* **4,** 72–144.

Emerson, R., and Fox, D. L. (1940). γ-Carotene in the sexual phase of the aquatic fungus *Allomyces*. *Proc. R. Soc. London, Ser. B* **128,** 275–293.

Federici, B. A. (1977). Differential pigmentation in the sexual phase of *Coelomomyces dodgei*. *Nature (London)* **267,** 514–515.

Federici, B. A. (1979). Experimental hybridization of *Coelomomyces dodgei* and *Coelomomyces punctatus*. *Proc. Natl. Acad. Sci. U.S.A.* **76,** 4425–4428.

Federici, B. A. (1980). Production of the mosquito parasitic fungus *Coelomomyces dodgei* through synchronized infection and growth of the intermediate host, *Cyclops vernalis*. *Entomophaga* **25,** 209–217.

Federici, B. A., and Chapman, H. C. (1977). *Coelomomyces dodgei* establishment of an in-vivo laboratory culture. *J. Invertebr. Pathol.* **30**(3), 288–297.

Federici, B. A., and Roberts, D. W. (1976). Experimental laboratory infection of mosquito larvae with fungi of the genus *Coelomomyces*. II. Experiments with *Coelomomyces punctatus* in *Anopheles guadrimaculatus*. *J. Invertebr. Pathol.* **27,** 333–341.

Federici, B. A., and Thompson, S. N. (1979). β Carotene in the gametophytic phase of *Coelomomyces dodgei*. *Exp. Mycol.* **3,** 281–284.

Harder, R., and Sörgel, G. (1938). Über einen neuen planoisogamen Phycomyceten mit Generationswechsel und seine phylogenetische Bedeutung. *Nachr. Ges. Wiss. Goettingen* **3,** 119–127.

Iyengar, M. O. T. (1935). Two new fungi of the genus *Coelomomyces* parasitic in larvae of *Anopheles*. *Parasitology* **27,** 440–449.

Keilin, D. (1921). On a new type of fungus. *Coelomomyces stegomyiae,* n. g., n. sp., parasitic in the body cavity of the larva of *Stegomyia scutellaris* Walker (Diptera, Nematocera, Culicidae). *Parasitology* **13,** 225–234.

Laird, M., Nolan, R. A., and Mogi, M. (1975). *Coelomomyces omorii* sp. n. and *C. raffaeli* Coluzzi and Rioux var. *parvum* var. n. from mosquitoes in Japan. *J. Parasitol.* **61**(3), 539–544.

Lange, L., and Olson, L. W. (1979). The uniflagillale phycomycete zoospore. *Dan. Bot. Ark.* **33,** 1–95.

Lessie, P. E., and Lovett, J. S. (1968). Ultrastructural changes during sporangium formation and zoospore differentiation in *Blastocladiella emersonii*. *Am. J. Bot.* **55,** 220–236.

Lucarotti, C. J., and Federici, B. A. (1984a). Gametogenesis in *Coelomomyces dodgei*. Couch (Blastocladiales, Chytridiomycetes). *Protoplasma* **121,** 65–76.

Lucarotti, C. J., and Federici, B. A. (1984b). Ultrastructure of the gametes of *Coelomomyces dodgei*. Couch (Blastocladiales, Chytridiomycetes). *Protoplasma* **121,** 77–86.

Lum, P. T. M. (1963). The infection of *Aedes taeniorhynchus* (Wiedemann) and *Psorophora howardii* Coquillett by the fungus *Coelomomyces*. *J. Insect Pathol.* **5,** 157–166.

Madelin, M. F. (1968). Studies on the infection by *Coelomomyces indicus* of *Anopheles gambiae*. *J. Elisha Mitchell Sci. Soc.* **84,** 115–124.

Madelin, M. F., and Beckett, A. (1972). The production of planonts by thin walled sporangia of the fungus *Coelomomyces indicus*. A parasite of mosquitoes. *J. Gen. Microbiol.* **72,** 185–200.

Martin, W. W., III (1969). A morphological and cytological study of the development of *Coelomomyces punctatus* parasitic in *Anopheles quadrimacutatus*. *J. Elisha Mitchell Sci. Soc.* **85,** 59–72.

Martin, W. W., III (1970). A morphological and cytological study of *Coelomomyces punctatus*. Master's Thesis, University of North Carolina, Chapel Hill.

Martin, W. W., III (1971). The ultrastructure of *Coelomomyces punctatus* zoospores. *J. Elisha Mitchell Sci. Soc.* **87,** 209–221.

Moore, R. T. (1968). Fine structure of Mycota. XIII. Zoospore and nuclear cap formation in *Allomyces*. *J. Elisha Mitchell Sci. Soc.* **84,** 147–165.

Muspratt, J. (1946). On *Coelomomyces* fungi causing high mortality of *Anopheles gambiae* larvae in Rhodesia. *Ann. Trop. Med. Parasitol.* **40,** 10–17.

Muspratt, J. (1963). Destruction of the larvae of *Anopheles gambiae* by *Coelomomyces* fungus. *Bull. W.H.O.* **29,** 81–86.

Pillai, J. S., Wong, T. L., and Dodgshun, T. J. (1976). Copepods as essential hosts for the development of a *Coelomomyces* parasitizing mosquito larvae. *J. Med. Entomol.* **13,** 49–50.

Powell, M. J. (1976). Ultrastructural changes in the cell surface of *Coelomomyces punctatus* infecting mosquito larvae. *Can. J. Bot.* **54,** 1419–1437.

Răsín, K. (1929). *Coelomomyces chironomi* n. sp. houba czopasici r dutině tělní larev chironoma. *Biol. Spisy Vys. Sk. Zverolek.* **8,** 1–13.

Travland, L. B. (1979a). Structures of the motile cells of *Coelomomyces psorophorae* and function of the zygote in encystment on a host. *Can. J. Bot.* **57,** 1021–1035.

Travland, L. B. (1979b). Initiation of infection of mosquito larvae (*Culiseta inornata*) by *Coelomomyces psorophorae*. *J. Invertebr. Pathol.* **33,** 95–105.

Umphlett, C. J. (1961). Comparative studies in the genus *Coelomomyces* Keilin. Ph.D. Dissertation, University of North Carolina, Chapel Hill.

Umphlett, C. J. (1962). Morphological and cytological observations on the mycelium of *Coelomomyces*. *Mycologia* **54,** 540–554.

Umphlett, C. J. (1964). Development of the resting sporangia of two species of *Coelomomyces*. *Mycologia* **56,** 488–497.

Weiser, J. (1976). The intermediary host for the fungus *Coelomomyces chironomi*. *J. Invertebr. Pathol.* **28,** 273–274.

Weiser, J. (1977). The crustacean intermediary host of the fungus *Coelomomyces chironomi*. *Ceska Mykol.* **31**(2), 81–90.

Whisler, H. C., Shemanchuk, J. A., and Travland, L. B. (1972). Germination of the resistant sporangia of *Coelomomyces psorophorae*. *J. Invertebr. Pathol.* **19,** 139–147.

Whisler, H. C., Zebold, S. L., and Shemanchuk, J. A. (1975). Life history of *Coelomomyces psorophorae*. *Proc. Natl. Acad. Sci. U.S.A.* **72,** 963–966.

Wong, T. L., and Pillai, J. S. (1980). *Coelomomyces opifexi* Coelomomycetaceae blastocladiales 6. Observations on the mode of entry into *Aedes australis* larvae. *N. Z. J.* **7**(1), 135–140.

Zebold, S. L., Whisler, H. C., and Travland, L. B. (1979). Host specificity and penetration in the mosquito pathogen *Coelomomyces psorophorae*. *Can. J. Bot.* **57,** 2766–2770.

Figs. 155–160. Photomicrographs showing the living gametophytic phase of *Coelomomyces punctatus* on *Cyclops vernalis*. Bar = 10 μm. Fig. 155. Anterior of copepod with male (M) and female (F) thalli. Figs. 156 and 157. Young male thalli. Figs. 158 and 159. Male (M) and female (F) thalli with cleaved gametes. Fig. 160. Gametes immediately after release.

Figs. 161–165. Photomicrographs of a living larva showing light infection of *Anopheles quadrimaculatus* with *Coelomomyces punctatus*. Fig. 161. Entire larva showing concentration of resting sporangia in the thorax. Bar = 10 μm. Fig. 162. Hyphagens (arrows) visible through cuticle of the abdomen. Bar = 10 μm. Figs. 163 and 164. Hyphae (arrows) growing in association with the fat body. Bar = 10 μm. Fig. 165. Early stages in formation of resting sporangia from hyphae. Bar = 10 μm.

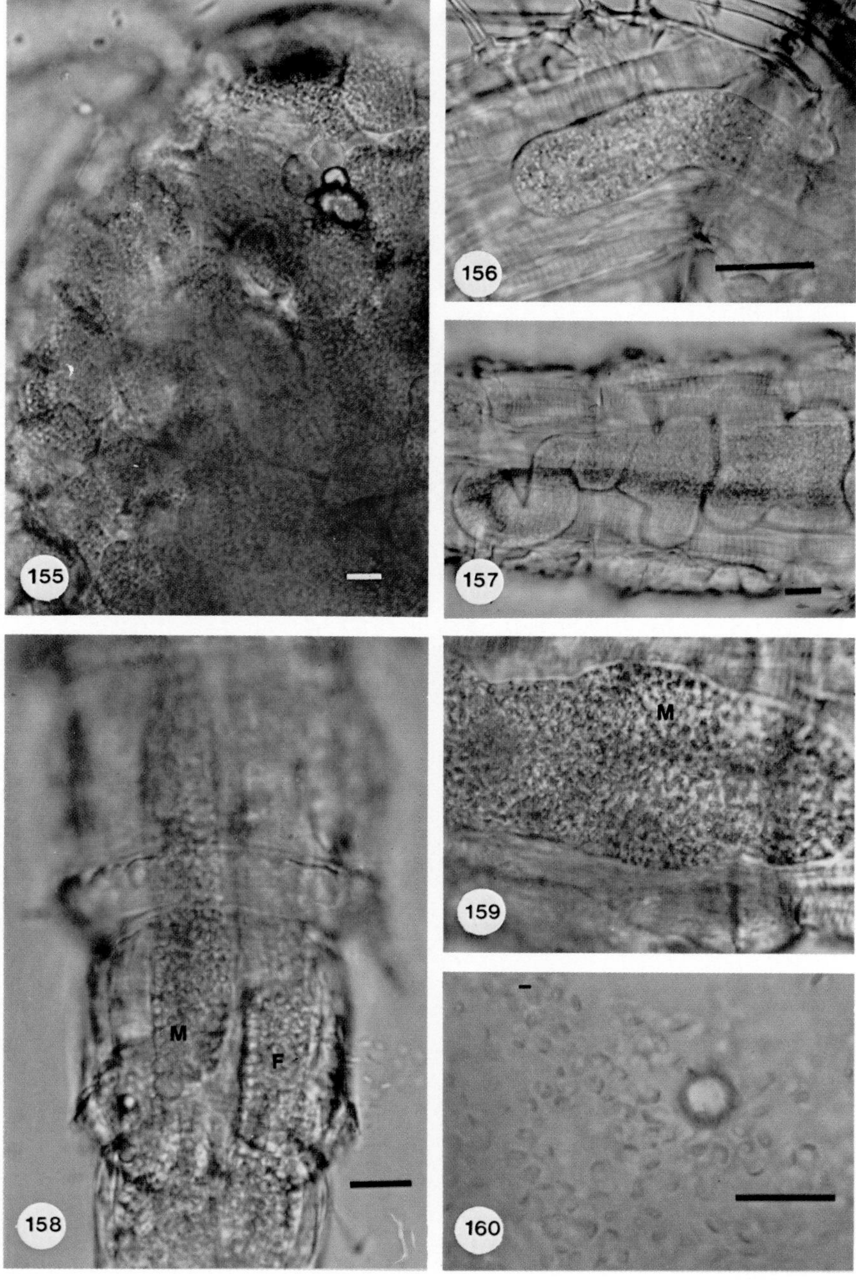
155
156
157
158
M
F
159
M
160

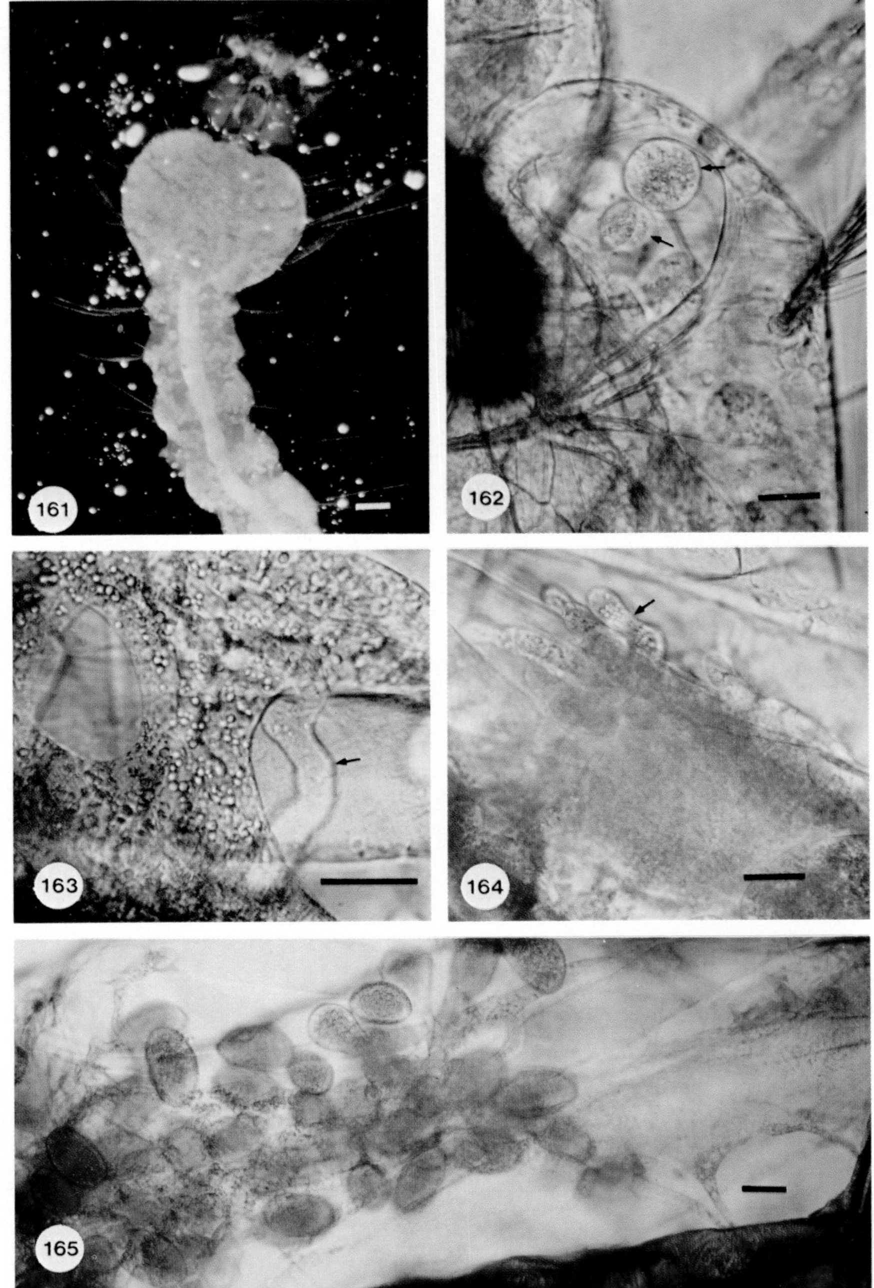

4 Taxonomy

JOHN N. COUCH

Department of Biology
University of North Carolina
Chapel Hill, North Carolina

CHARLES E. BLAND

Department of Biology
East Carolina University
Greenville, North Carolina

I. TECHNIQUES FOR COLLECTION, PRESERVATION, AND MOUNTING OF SPECIMENS

A. Collection of Specimens and Recognition of Infection

Since all known species of *Coelomomyces* are obligate parasites, collection of specimens first requires collection of hosts. In this regard, collection is somewhat simplified since hosts for all recognized species are arthropods for which all or most life forms are aquatic: larval chironomids and larval and adult mosquitoes for the sporophyte, and copepods or ostracods for the gametophyte. Detailed techniques for collection of these potential hosts are given in Peterson (1964), Service (1976), and Weiser and McCauley (1971), and are too involved to detail here. However, the dipper, by far the most commonly used tool for collecting mosquito larvae and pupae (Service, 1976), is equally suited for collecting not only these but most other aquatic hosts. In using this instrument, care should be taken to disturb surface water as little as possible since such disturbance usually results in the specimens (especially mosquito larvae and pupae) diving to the water bottom where they are inaccessible. Collection of chironomid hosts involves usually sampling of the benthos (often by coring) and subsequent screening for larvae, as described by Weiser and McCauley (1971). Removal of suspected hosts from aquatic samples may be by a variety of techniques (Peterson, 1964) but is most often achieved by their selective removal with forceps or with pipettes of sufficient bore to accomodate the desired form.

Although most species of *Coelomomyces* have been collected on aquatic hosts, many have been collected also on adult mosquitoes. Collection of these hosts may be by any of the various techniques described by Service (1976) for sampling adult mosquito populations.

Following collection of potential hosts, detection of infection may in some instances be possible with the unaided eye, since, for the sporophyte, heavily infected larvae (Figs. 166–170) generally appear rust-colored, are sluggish, and remain at or near the water surface, while, for the gametophyte, (Chapter 3, Figs. 160–165), the bright orange pigmenta-

---

Figs. 166–169. Photomicrographs of living larvae of *Anopheles quadrimaculatus* that were heavily infected with *C. punctatus*. Fig. 166. Larvae floating at water surface. Note the distinctive, golden brown coloration of the larvae. Bar = 3 mm. Fig. 167. Tightly packed resting sporangia in larval thorax. Bar = 100 μm. Figs. 168 and 169. Successive photos of a larva so filled with sporangia that the slightest pressure resulted in rupture of the larval cuticle. Bar = 1 mm.

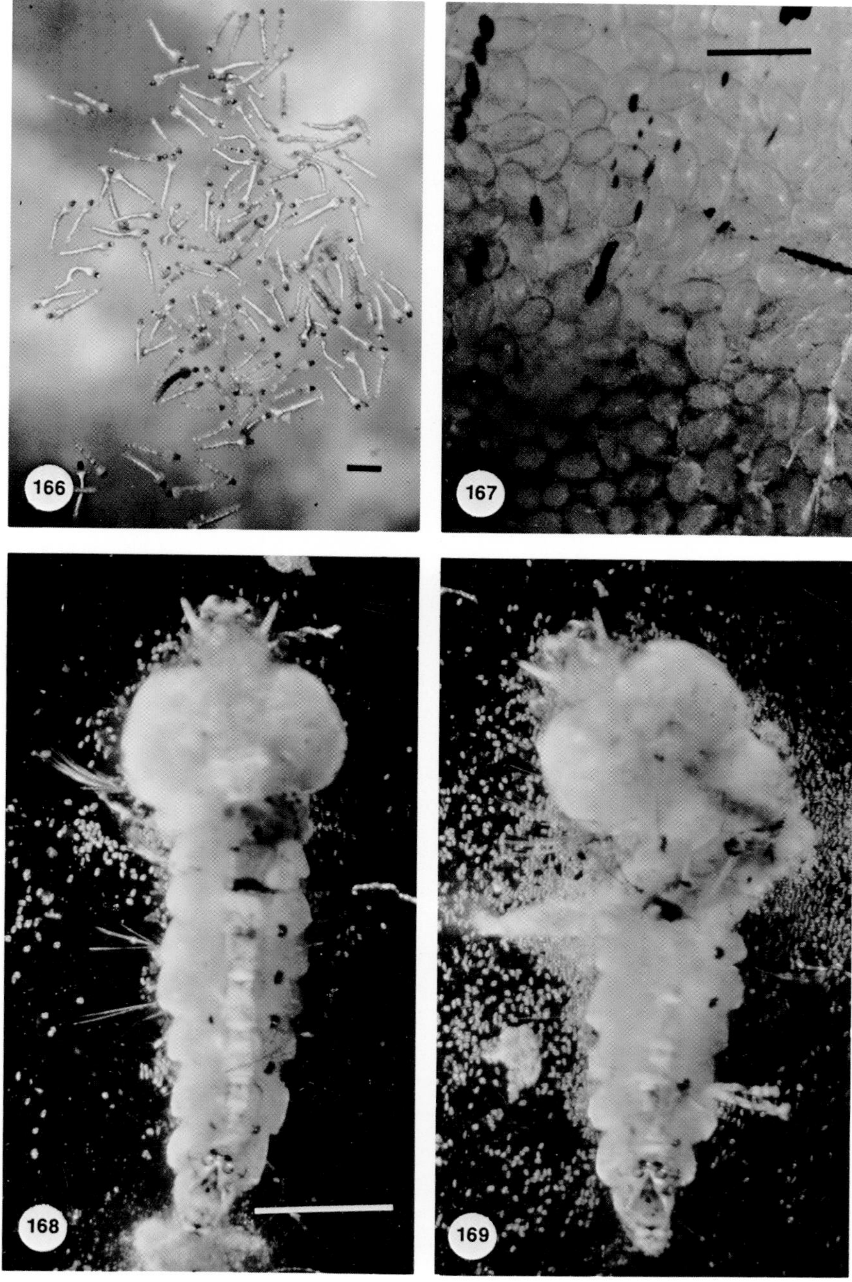

166
167
168
169

170

tion of the male makes detection of infected copepods relatively easy. In most instances, however, microscopic examination of hosts at magnifications of ×50 to ×200 is essential for infection to be recognized. For this, screening of specimens may be facilitated by their serial alignment in water on a microscope slide or in a petri dish half. In light infections (Chapter 3, Figs. 155–159) involving small numbers of sporangia or only hyphae (as for the gametophyte and in early infection for the sporophyte), closer scrutiny at higher magnification is often necessary and requires that a coverglass be placed over the specimen. Species identification requires microscopic examination of mature resting sporangia and usually necessitates dissection of infected larvae, so that surface ornamentation of liberated sporangia may be observed at magnifications of ×200 to ×400.

B. Preservation and Mounting of Specimens

Storage of living resting sporangia for several months or longer (likely to depend on species) is possible by simply placing intact infected larvae on moistened filter paper in a petri dish and storing them at 10°C. Although viability of stored sporangia declines with age, it is possible by this technique to retain specimens for study and experimental purposes for some time.

Preservation of intact infected specimens in 7% formalin results in minimum distortion of both host and parasite. Although of no apparent benefit in preserving features for gross observations and species determination, other formalin-based fixatives and preservatives may be used also. However, preservation of specimens in ethanol (70% or otherwise), or probably also alcohol based formulations, results in distortion of both hyphae and sporangia.

Semipermanent preservation of intact or dissected specimens on microscope slides is best in lactophenol (phenol crystals, 20 g; lactic acid, 16 ml; glycerol, 31 ml; distilled H_2O, 20 ml). For this, specimens from water or 7% formalin may either be mounted whole or dissected directly in the medium. To preserve specimens mounted in water and subsequently placed under a cover glass for study, 1–2 drops of lactophenol may simply be placed at the periphery of the coverglass. There, as the water evaporates from beneath the coverglass, the lactophenol will be drawn in and around the specimen via capillary action. For added permanency of such preparations, coverglasses may be ringed with lacquer (fingernail polish).

Fig. 170. Photomicrograph showing distribution of resting sporangia of *C. punctatus* in the head and thorax of a larva of *A. quadrimaculatus*. Bar = 100 μm.

Even if this is not done, mounts in lactophenol will last indefinitely if the mounting medium is replenished every 1–2 years.

Permanent mounting of specimens in Canada balsam or other hardening medium may be made following routine techniques of fixation and dehydration. In most instances, however, such procedures result in severe shrinkage and change in appearance of hyphae, thin-walled sporangia, and even resting sporangia. Observations and interpretations based on specimens so mounted must therefore be very careful and subject to revision following examination of fresh material or specimens mounted in lactophenol. In certain instances, in which permanently mounted specimens represent the only material available for study, it is possible to at least partially restore normal sporangial shape and appearance by processing specimens through steps in reverse of those necessary for mounting, i.e., xylene, graded ethanol series, H_2O. The time necessary at each step varies from specimen to specimen, but should be sufficient to allow thorough dissolution of the balsam and gradual rehydration. Once in water, specimens may be remounted in lactophenol or processed for scanning electron microscope (SEM) studies as described subsequently. Note: If sporangia have been dissected from larvae, stages after xylene may be conducted in a filter apparatus as described in Section C.

Specimens (resting sporangia) preserved by drying, or that were inadvertantly allowed to dry, may be satisfactorily restored by placing them in 7% KOH (a commonly used wetting agent for fungal spores) and subsequent mounting in lactophenol. However, since prolonged exposure to KOH may result in loss of sporangial detail, treatment should be limited to 12–24 hr.

C. Techniques for Scanning Electron Microscope Study of Sporangia

Although examination of specimens by SEM is not, and should not be, essential to identification of species of the genus *Coelomomyces,* this instrument does illustrate graphically the features of resting sporangial morphology that are of major significance in taxonomy of this group (Anthony *et al.*, 1971; Bland and Couch, 1973). Because of this, basic techniques for preparation and mounting of resting sporangia for SEM are as follows.

Sporangia suitable for use in SEM studies may be obtained from a variety of sources, including living and preserved specimens, semipermanent slides, and permanent slides. For all specimens, except those as permanent slides, larvae and/or sporangia should be dissected into distilled water so as to disperse and clean the sporangia. Specimens as permanent slides must be processed to water as described previously. From water, sporangia are collected on 13 mm/8.0 μm pore diameter polycarbonate membranes (Nuclepore Corp., Pleasanton, California) by use of 3-cc disposable syringes and suitable filter holders. The passing of 10–25 cc of distilled water through the filter apparatus ensures that sporangia are thoroughly washed. The filters are then removed, allowed to dry, and affixed by double-stick tape to specimen stubs for SEM. Following routine metal coating of specimens, observations may be made in SEMs at accelerating voltages of 5–15 kV. Note: Although critical-point drying may improve the appearance of sporangia of some species, it does not appear to be necessary in most instances. And, unless extreme care is taken, the processing of specimens necessary for critical-point drying may result in specimen loss.

II. TAXONOMIC CRITERIA IN THE GENUS *COELOMOMYCES*

A. Hyphal Characteristics

Many descriptions of *Coelomomyces* spp. include references to morphological features of vegetative hyphae such as diameter, pattern of branching, and growth form. However, for most species, host specimens were collected at stages of fungal development in which hyphae were either rudimentary or absent, thereby making their description of questionable use in comparisons or at times impossible. Even in instances in which hyphae have been observed, there is much overlap in sizes given and confusion in interpretation of characteristics from species to species. Among different species, for example, hyphal diameter has been specified to range from 3 to >100 μm, but is generally 5–15 μm; branching has been described as dichotomous to subdichotomous to right-angled, but is generally at right angles (differences in appearance likely to be due to distortion caused by mounting or by confinement within the host's coelomic cavity); and finally, growth form has been given as filamentous to compact, but probably varies within the same species during different stages of development. These factors, combined with the ease with which the delicate hyphae are distorted during preservation and/or mounting, make hyphal morphology of questionable or, at least at present, limited use in the taxonomy of *Coelomomyces* spp. Additional collections and/or culture of species of this genus will enable study of hyphae at comparable stages of development and will reveal whether variations in hyphal morphology can be used in making taxonomic decisions. However, since preliminary observations by J. N. Couch (unpublished) indicate that hyphal morphology may be distinctive for species, complete descriptions of hyphae should be included, whenever possible, in all future descriptions of new species, preferably from observations of living material.

B. Thin-Walled Sporangia

Of the relatively few instances in which thin-walled sporangia have been observed in species of *Coelomomyces* (approximately 12 species collected primarily* in Europe, Asia, and Africa), all reports indicate that they lack external wall ornamentation, have similar dimensions (ranging between 15 and 33.5 μm wide × 24–58 μm long), and are similarly shaped (mostly ellipsoid, with some references to ovoid or allantoid forms). At present, therefore, characteristics of thin-walled sporangia are of little use in taxonomy. However, in that thin-walled sporangia represent an important stage in the life cycle (see Chapter 2) of forms in which they occur, and since additional collections may reveal taxonomically significant variations in their morphology, characteristics of thin-walled sporangia (when possible) should be included in all descriptions.

C. Resting Sporangia

Morphological features of resting sporangia are by far the most significant criteria in taxonomy of *Coelomomyces* spp. Of prime importance in this regard are sporangial size, shape and symmetry, and wall ornamentation. Useful also, but to a lesser extent, are outer wall thickness and the prominence of internal wall lacunae (see Chapter 3) as internal vertical striae. A discussion of the use of these features in the delineation of taxa follows.

1. Size

Although resting sporangial dimensions overlap in many species, there are numerous instances in which species or especially varieties can be distinguished simply on the basis of sporangial size. Because of this, a list of sporangial dimensions for all taxa is given in Table II. The dimensions given include minimum and maximum measurements of sporangial width and length as viewed in midoptical section. In some species in which sporangia exhibit shapes (see Section II,C,2 below) other than the typical

* Several specimens of *Culex territans* bearing only thin-walled sporangia of an unidentified species of *Coelomomyces* have been collected in both Louisiana and Connecticut. See Section III,F for descriptions of specimens.

TABLE II. Maximum and Minimum Sporangial Dimensions for Species and Varieties of *Coelomomyces*

Organism	Dimensions (μm)
C. africanus	9–24 × 18.5–35
C. anophelesicus	34–65.5 × 33.5–54
C. arcellaneus	21–42 in diameter, face view; 18–25 × 21–42, side view
C. bisymmetricus	23–28 × 34–48
C. borealis var. *borealis*	33.5–71 × 63–126
C. borealis var. *giganteus*	76–134 × 109–182
*C. cairnsensis**	26.5–35.5 × 42–61 × 14–26 (thick)
C. canadensis	33.8–68 × 26.7–45
C. carolianus	17–26 × 32–50
C. celatus	20.5–26 × 33.5–39
C. chironomi	18–25 × 27–42
C. couchii	32–37 × 48–59
C. cribrosus	24–42 × 42–71
C. dentialatus	19–37 × 33.5–68
C. dodgei	27–42 × 37–65
C. dubitskii	26–35 × 41–54
C. elegans	39–61 × 48–71
C. fasciatus	28–39 × 33.5–54 × 22–30 (thick)
C. finlayae	18–27.5 × 31.5–47.5
C. grassei	25–40 × 42–70
C. iliensis var. *iliensis*	26–43 × 34–60 × 24–39 (thick)
C. iliensis var. *australicus*[a]	26–43 × 34–60 × 24–39 (thick)
C. iliensis var. *indus*	32–41 × 43–67 × 30–39 (thick)
C. iliensis var. *irregularis*	24–36 × 37–54
C. iliensis var. *striatus*	37–52 × 46.5–82
C. indicus	25–41 × 29–65.5
C. iyengarii	37–46.5 × 48–76
C. keilini	34–46 × 58–71
C. lacunosus var.[a]	17–33 × 25–57
C. lairdi	22–33 × 35–57
C. lativittatus	29–36 × 40–77
C. macleayae	17–37 × 35–62
C. madagascaricus[a]	15–30 × 28–48
C. musprattii	26–45 × 37–70
C. orbicularis	30–50 × 39–75
C. orbiculostriatus	32–61 × 46.5–108
C. pentangulatus	12–18 × 18–40
C. ponticulus	19–34 × 35–56
C. psorophorae var. *psorophorae*	25–80 × 56–164
C. psorophorae var. *halophilus*	26–50 × 37–90
C. psorophorae var. *tasmaniensis*	35–83 × 50–179
C. punctatus	32–41 × 42–75
C. quadrangulatus var. *quadrangulatus*	12–21 × 19–40
C. quadrangulatus var. *irregularis*	15–21 × 23–41

(*continued*)

TABLE II. (Continued)

Organism	Dimensions (μm)
C. quadrangulatus var. *lamborni*	26–31 × 36–58
C. quadrangulatus var. *parvus*	9–16 × 12–21
C. raffaelei var. *raffaelei*	18–26 × 33–49
C. raffaelei var. *parvus*	12–21 × 21–34
C. reticulatus var. *reticulatus*	27–55 × 43–80
C. reticulatus var. *parvus*	22–35 × 33.5–45
C. rugosus	26–39 × 30–62
C. sculptosporus	22–31 × 35–58
C. seriostriatus	20.5–32 × 39–59.5
C. stegomyiae vars.	15–45 × 25–100
C. sulcatus	19–24 × 30–37
C. thailandensis	24–33.5 × 38–71
C. triangulatus	11–17 × 19–35
C. tuberculatus[a]	10–12.5 × 17.5–22.5
C. tuzetae	18–30 × 27–47
C. uranotaeniae	21–30 × 29–45
C. utahensis	39–71 × 54–97
C. walkeri	30–46.5 × 45–65

[a] Measurements from specimens in permanent mounting medium.

ellipsoid, a measurement of thickness is included also. While efforts have been made to obtain all measurements from living material or from specimens preserved so as to result in minimal shrinkage, some measurements (noted by an asterisk) were of necessity from specimens mounted in balsam or other media causing shrinkage. As a result, future examination of fresh material of these species will undoubtedly yield slightly larger sporangial dimensions. It should be noted that in media inducing shrinkage, primary reduction occurs in sporangial width (sometimes resulting in complete collapse of the wall), with reductions in length being somewhat less.

In comparing reports by different individuals regarding sporangial size, one should realize that, as noted by Laird *et al.* (1975), cited dimensions may differ significantly (10% or more) as a result of the use of different optics or techniques of measurement. To minimize such discrepancies and increase accuracy, only sporangia oriented flat in the optical plane should be measured. In addition, by noting the orientation of all sporangia measured one can determine whether sporangia are symmetrical and require only measurements of width and length, or asymmetrical and require measurements of length, width, and thickness, or otherwise. When possible, all measurements should include at least 50 sporangia, and should be made either with a precisely calibrated ocular micrometer at magnifications of ×400, or via measurement of sporangial outlines drawn with the aid of a calibrated microscope drawing apparatus (camera lucida). If available, use of a measuring device such as the Zeiss Micro-Videomat employed by Laird *et al.* (1975) should provide highly accurate measurements. An additional, but time-consuming, means of obtaining precise sporangial measurements is from micrographs of known and precise magnification.

2. Shape and Symmetry

Although resting sporangia of most species of *Coelomomyces* are shaped as ellipsoids (ranging from broad to elongate depending on species) and exhibit some degree of bilateral symmetry, sporangial shape is generally consistent within species and, often in combination with sporangial size, may be of primary importance in delineating taxa (see Fig. 171 for diagrams of basic sporangial shapes). In addition to the common ellipsoid forms, a few taxa exhibit shapes ranging from spherical to allantoid in the *C. psorophorae* "complex," to the uniquely symmetrical forms found in *C. anophelesicus* (Fig. 171G), *C. arcellaneus* (Fig. 171H), *C. fasciatus* (Fig. 171I), and *C. iliensis* (Fig. 171J) varieties.* In many species, the sporangial face on the side of the dehiscense slit is slightly flattened.

In that examination of specimens by light microscopy usually presents sporangia from a variety of different perspectives, a basic understanding of sporangial symmetry is vitally important to accurate interpretation of sporangial shape, structure, and ornamentation (see below). Equally important is a standardized terminology for the major reference points found on practically all sporangia. In this regard, Fig. 172 provides recommended terminology that is applicable to most sporangia. Arbitrarily, the surface with the dehiscence slit has been designated "dorsal," the opposite surface thereby being "ventral." The tapered ends of sporangia are to be referred to as "poles," and the lateral surfaces between the dorsal and ventral surfaces, simply as "sides." By this terminology it is possible to accurately describe sporangia and to indicate in descriptions whether one is referring to dorsal (face), ventral, side, or polar (end) views (Fig. 173). Additionally, the same terminology may be used to designate the views from which descriptions of midoptical sections of sporangia are based (Fig. 173), such views generally being quite different. Although for full understanding of sporangial morphology it is often necessary to study sporangia from perspectives other than the standard ones described, it is

* Note: The occasional reference to sporangia of some species as "boat-shaped" is believed attributable to study of sporangia mounted in a medium causing collapse of the sporangial wall, therefore not a valid characteristic.

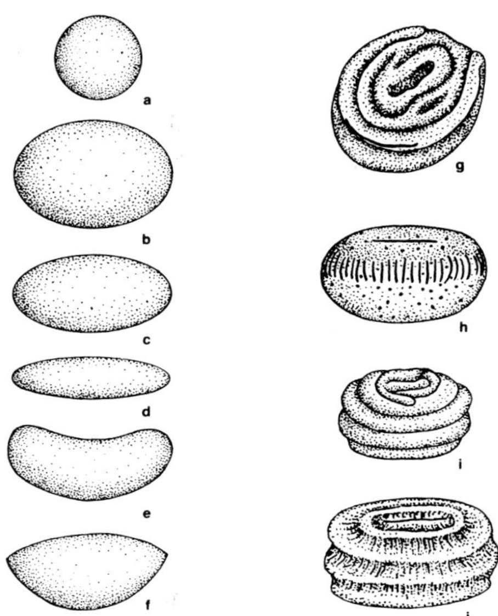

Fig. 171. Semidiagrammatic drawings illustrating variations in shape of resting sporangia. Although few forms have truly spherical sporangia (a), many have sporangia that are ellipsoid (b–d) (ranging from broadly ellipsoid to narrowly ellipsoid) or allantoid (e). A few forms such as *C. borealis* vars. and *C. stegomyiae* var. *chapmani* have resting sporangia with truly pointed poles (f). Forms with truly unique sporangial shapes include *C. anophelesicus* (g), *C. arcellaneus* (h), *C. fasciatus* (i), and *C. iliensis* (j).

essential that points of reference be known and used in all descriptions. Failure to do so will lead to frequent misinterpretation and/or misunderstanding of important diagnostic features. Note: While the basic points of reference are applicable to sporangia of practically all species, the difficult-to-discern dehiscence slit and uniquely shaped sporangia of *C. anophelesicus* make it difficult to describe sporangia of this species by recommended means.

3. Wall Ornamentation

Ornamentation of the outer wall of resting sporangia is unquestionably the most important criterion in taxonomy within the genus and is the feature by which most species are delineated. Because of this, and since it is occasionally difficult to discern and/or interpret the true nature of wall ornamentation, a summary of basic types and patterns is provided in Fig. 174. Also included are interpretations of how various forms of ornamentation appear in optical section. To facilitate identification and as a supple-

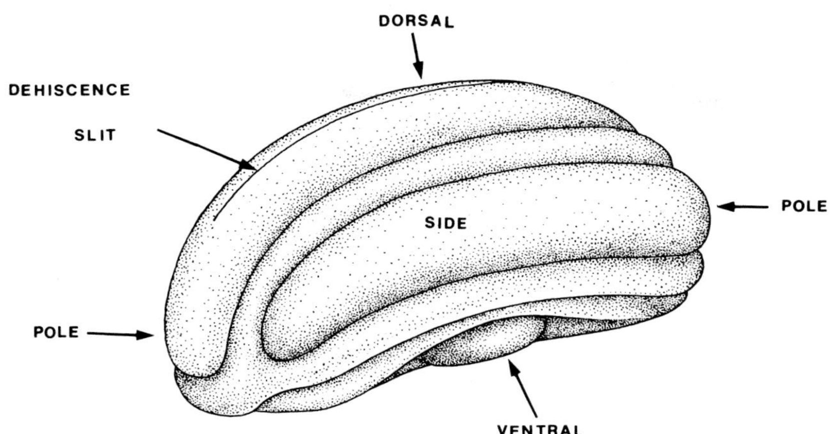

Fig. 172. Semidiagrammatic drawings of a resting sporangium of *C. bisymmetricus* giving terminology to be used for various views of sporangia.

ment to the key to the species, Table III gives taxa for each diagnostically strategic pattern of wall ornamentation depicted.

Four basic categories of wall ornamentation may be recognized among the various species. Such categories, while not necessarily reflecting relationships, do serve as a basis for interpreting and understanding the fantastic variety of wall ornamentation occurring in this group. In the first category, including forms in which wall ornamentation is lacking (Fig.

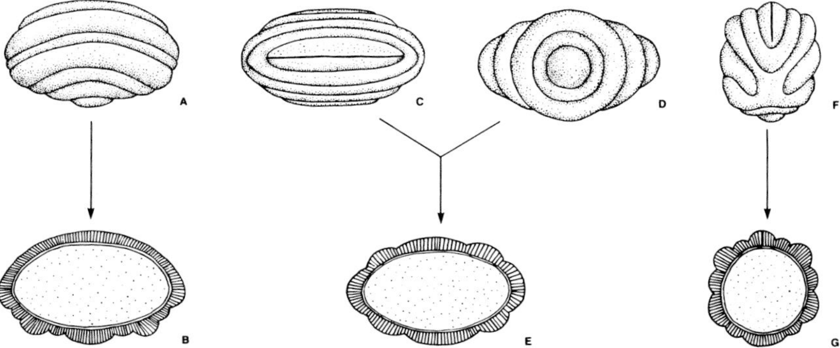

Fig. 173. Semidiagrammatic drawings of resting sporangia of *C. bisymmetricus* showing resting sporangial appearance from different views. Drawings A and B show, respectively, side view of the sporangial surface and corresponding midoptical section, side view. Drawings C and D are of dorsal (C), and ventral (D) views, with E illustrating a midoptical section from either view. Drawing F is a polar or end view, with G being a midoptical section, end view.

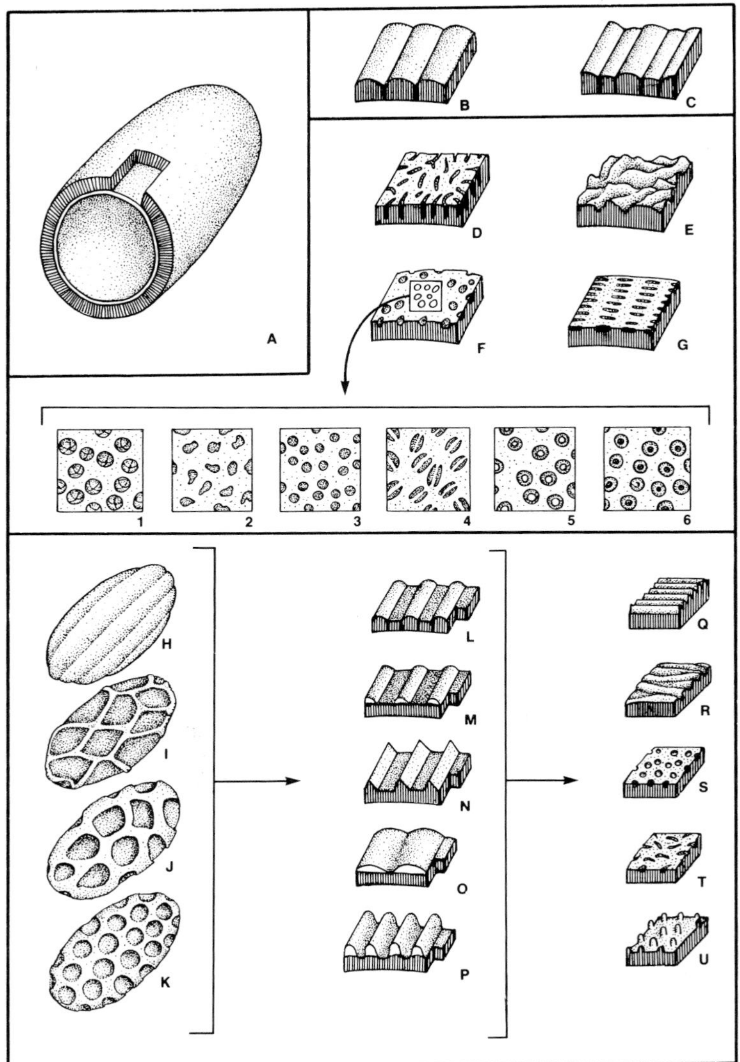

Fig. 174. Semidiagrammatic drawings illustrating variations in resting sporangial wall ornamentation in species of *Coelomomyces*. As a supplement to the key to species, these drawings may be used in conjunction with Table III to facilitate species identification. In referring to the drawings, please note that there are four basic categories of wall ornamentation, including: wall with no sculpturing (A); wall with parallel, interconnected ridges appearing as uniform bands (B) or alternating high/low bands (C); wall irregular or with punctae/striae of varying shapes and sizes (D–G); and wall with ridging that is parallel (H) or interconnected to enclose polygonal to irregular (I and J) or circular (K) depressed areas of varying ornamentation (Q–U). The ridging itself may vary from low domes with internal, vertical striae (L) or without internal vertical striae (M); or sharp ridges (N) separated by generally flat depressed areas to broad (O) or steep (P) interconnected domes. Note that drawing A shows area from which drawings B–G were taken.

TABLE III. Species Identification on the Basis of Resting Sporangial Wall Ornamentation

Wall feature(s) from Fig. 174	Possible species
A	*C. pentangulatus*
	C. tuzetae
	C. walkeri
B	*C. dodgei*
	C. fasciatus
	C. lativittatus
C	*C. bisymmetricus*
	C. sculptosporus
D	*C. dodgei*
	C. lacunosus vars.
	C. punctatus
E	*C. rugosus*
F-1	*C. canadensis*
	C. keilini
F-2	*C. lacunosus*
F-3	*C. arcellaneus*
	C. psorophorae vars.
	C. utahensis
F-4	*C. ponticulus*
F-5	*C. borealis* vars.
F-6	*C. stegomyiae* vars.
G	*C. seriostriatus*
H,L,Q	*C. iyengarii*
H,L,R	*C. iyengarii*
	C. musprattii
H,M,R	*C. carolinianus*
H,M,T	*C. triangulatus*
H,N,R	*C. uranotaeniae*
H,O or P,T	*C. dentialatus*
H,O,Q	*C. iliensis* vars.
H,O,R	*C. grassei*
	C. iliensis vars.
	C. musprattii
	C. quadrangulatus vars.
H,O,S	*C. sulcatus*
H,O,T	*C. sulcatus*
H,P,Q	*C. iyengarii*
H,P,R	*C. anophelesicus*
	C. grassei
	C. indicus
	C. iyengarii
	C. musprattii
I,N,R	*C. uranotaeniae*
I,N,S	*C. cairnsensis*
	C. finlayae
	C. macleayae
I,N,T	*C. finlayae*
	C. macleayae
I,P,S or T	*C. dentialatus*
I,P,U	*C. dentialatus*
J,L or J,M	*C. tuberculatus*
J,L,R	*C. dubitskii*
	C. indicus
J,L,S	*C. africanus*
	C. dubitskii
	C. raffaelei
J,L,T	*C. dubitskii*
J,M,R	*C. carolinianus*
J,M,S	*C. couchii*
	C. madagascaricus
	C. thilandensis
J,M,T	*C. celatus*
	C. raffaelei
J,O,S	*C. raffaelei* vars.
	C. thailandensis
J,O,T	*C. raffaelei* vars.
J,P,R	*C. indicus*
J,P,T	*C. dubitskii*
K,L,R	*C. indicus*
K,L,S	*C. orbicularis*
	C. reticulatus
K,L,T	*C. cribrosus*
K,L,U	*C. elegans*
K,M,S	*C. lairdi*
K,M,T	*C. cribrosus*
	C. orbiculostriatus vars.
K,M,U	*C. lairdi*
K,N,R	*C. chironomi*
K,P,S	*C. reticulatus* vars.
K,P,T	*C. chironomi*

174A), there are only three recognized species, *C. pentangulatus, C. tuzetae,* and *C. walkeri.* The second category includes forms with ornamentation as contiguous bands or ridges formed by elongate, parallel striae or grooves in the sporangial wall (Fig. 174B,C). The three taxa characterized by such ornamentation are *C. lativittatus, C. fasciatus,* and *C. dodgei,* the last having ornamentation as both bands and punctae. The third category includes forms with punctae, linear striae, or irregular convolutions occurring as small indentations into or projections from the sporangial wall. Such ornamentation is generally highly specific and characterizes a variety of forms (Fig. 174D–G). The fourth category of ornamentation is by far the most complex, includes the largest number of representatives, and is characterized by ornamentation as ridges separated by depressed areas. From this basic pattern, a variety of types of wall ornamentation may occur. Depending on species, ridging of different shapes, widths, heights, and sculpturing may enclose depressed areas of different shapes and ornamentation (Fig. 174H–U). In this regard, patterns of ridging vary from elongate and parallel along the length of sporangia, thereby enclosing elongate depressed areas (Fig. 174H); to anastomosing, enclosing elongate polygonal areas (Fig. 174I); to reticulate, enclosing polygonal to circular depressed areas (Fig. 174J,K). Additionally, ridging may vary from low, broad bands, appearing as low domes in edge views (Fig. 174L); to steep ridges, appearing as high domes in edge view (Fig. 174M); to sharp ridges, appearing as teeth in edge view (Fig. 174N); to truncate ridges, appearing as truncate domes in edge view (Fig. 174O,P). Ornamentation of depressed areas between ridges varies from parallel to anastomosing transverse grooves, appearing toothlike in edge view (Fig. 174Q,R); to scattered or linearly arranged punctae/striae, appearing as pits in edge view (Fig. 174S,T); to papillae, appearing also as papillae in edge view (Fig. 174U). Since combinations of these types of wall ornamentation are specific to given species of *Coelomomyces,* an understanding of their nature is essential to identification of taxa.

Another important feature of wall ornamentation is the position of the dehiscence slit relative to any sculpturing present. Of primary significance in this regard is whether the dehiscence slit occurs along a raised area of the wall (ridge) or in a depressed area. Important also is the presence or absence of ornamentation (usually as punctae or striae) adjacent to the slit. Consideration of such factors is especially important in that in some instances separation of taxa can be made on these criteria alone.

4. Wall Thickness and Internal Structure

Although sporangial wall thickness is often specified in descriptions, this criterion has never been used as a primary basis for distinguishing taxa. One reason for this is that, while there are obvious differences in wall thickness between some species, measurements for most overlap within a range of 2.5–5 μm. Another problem in using wall thickness as a taxonomic criterion lies in the fact that dimensions for wall thickness fall in a range within which accurate measurements by light microscopy are difficult. Additionally, as a result of unusual sporangial shape or improper techniques of preparation, it is often impossible to distinguish between inner and outer wall layers, resulting therefore in combined measurement of both layers in some specimens, but only one layer in others. In most situations, therefore, wall thickness has been of limited value in taxonomy. However, in certain instances involving specimens with sporangial walls having little ornamentation (i.e., the *C. psorophorae* var. *psorophorae* "complex") but exhibiting wide and consistent variations in wall thickness, this criterion may prove useful in distinguishing taxa, especially at the varietal level. To ensure accuracy, all measurements of wall thickness should be from specimens viewed in midoptical section at magnifications of ×400 or greater.

Regarding internal wall structure, it appears that the outer wall of all species of *Coelomomyces* is characterized by vertically oriented, internal lacunae or channels (see Chapter 3), which appear by light microscopy as internal vertical striae in side view, and as fine, circular punctae in end view. Although such lacunae are present in all forms, they apparently exhibit wide variation in size, as their appearance by light microscopy varies from only faintly visible in some forms (i.e., *C. africanus, C. punctatus, C. psorophorae* varieties) to very prominent (i.e., *C. iliensis* varieties, *C. iyengarii, C. musprattii*). Since this feature is consistent for taxa, it is possible in some instances to use it along with other criteria as a basis on which to distinguish species and/or varieties. When so used, references to internal vertical striae must, for the present, be comparative, as means to quantify or measure their prominence are unknown.

D. Hosts

1. Primary Host

Although it was thought initially that species of *Coelomomyces* were highly specific to given mosquito hosts and their separation of taxa could be determined by the host on which forms occurred, it is now known that most species are capable of infecting a wide range of mosquito and possibly chironomid hosts (Table IV). Because of this, and since several species of *Coelomomyces* may be infective to the same host, host specificity is of questionable or only limited use as a taxonomic criterion. However,

TABLE IV. Primary Host Index

Host	Species of *Coelomomyces*
Aedes sp.	*C. dentialatus*
	C. stegomyiae var. *stegomyiae*
	C. quadrangulatus var. *quadrangulatus*
	C. macleayae
Aedes aegypti	*C. dentialatus*
	C. macleayae
	C. stegomyiae var. *stegomyiae*
Aedes (stegomyia) alacasida	*C. macleayae*
Aedes albopictus	*C. dentialatus*
	C. macleayae
	C. stegomyiae var. *stegomyiae*
Aedes atropalpus epactius	*C. utahesis*
Aedes australis	*C. psorophorae* var. *tasmaniensis*
	C. opifexi
Aedes (Pseudokusea) australis	*C. psorophorae* var. *tasmaniensis*
	C. opifexi
Ades cantans	*C. borealis* var. *borealis*
Aedes caspius	*C. psorophorae* var. *psorophorae*
Aedes cinereus	*C. psorophorae* var. *psorophorae*
Aedes communis	*C. "borealis"* var. *borealis*
Aedes excrucians	*C. "borealis"* var. *borealis*
Aedes (stegomyia) flavopictus	*C. ponticulus*
	C. stegomyiae var. *stegomyiae*
Aedes fitchii	*C. borealis* var. *giganteus*
Aedes funestus	*C. psorophorae* var. *psorophorae*
Aedes hebrideus	*C. finlayae*
	C. stegomyiae var. *stegomyiae*
Aedes japonicus	*C. macleayae*
Aedes (finlaya); japonicus shintienensis	*C. macleayae*
Aedes mediopunctatus	*C. macleayae*
Aedes melanimon	*C. psorophorae* var. *psorophorae*
Aedes multifolium	*C. stegomyiae* var. *stegomyiae*
Aedes notoscriptus	*C. finlayae*
Aedes polynesiensis	*C. macleayae*
	C. stegomyiae var. *stegomyiae*
Aedes pseudalbopictus	*C. macleayae*
Aedes quadrispinatus	*C. stegomyiae* var. *stegomyiae*
Aedes riversi	*C. stegomyiae* var. *stegomyiae*
Aedes scatophagoides	*C. stegomyiae* var. *stegomyiae*
Aedes (stegomyia) scutellaris	*C. quadrangulatus* var. *lamborni*
	C. stegomyiae var. *stegomyiae*
Aedes sollicitans	*C. arcellaneus*
	C. psorophorae var. *halophilus*
Aedes stimulans	*C. borealis* var. *giganteus*
Aedes subalpictus	*C. stegomyiae* var. *stegomyiae*
Aedes taeniorhynchus	*C. psorophorae* var. *halophilus*

TABLE IV. (Continued)

Host	Species of *Coelomomyces*
Aedes togoi	*C. psorophorae* var. *psorophorae*
Aedes triseriatus	*C. macleayae*
Aedes variabilis	*C. stegomyiae* var. *stegomyiae*
Aedes vexans	*C. psorophorae* var. *psorophorae*
	C. quadrangulatus var. *quadrangulatus*
Aedomyia catastica	*C. indicus*
Aedomyia squamipennis	*C. reticulatus* var. *parvus*
Anopheles sp.	*C. lativittatus*
Anopheles aconitus	*C. iliensis* var. *irregularis*
	C. indicus
	C. iyengarii
Anopheles annularis	*C. anophelesicus*
	C. indicus
Anopheles annulatus	*C. walkeri*
Anopheles barbirostris	*C. africanus*
	C. indicus
Anopheles bengalensis	*C. thailandensis*
Anopheles bradleyi	*C. punctatus*
Anopheles campestris	*C. africanus*
Anopheles claviger	*C. raffaelei* var. *raffaelei*
Anopheles crucians	*C. bisymmetricus*
	C. cribrosus
	C. keilini
	C. dodgei
	C. lativittatus
	C. punctatus
	C. quadrangulatus
	C. sculptosporus
	C. triangulatus
Anopheles culicifacies	*C. indicus*
Anopheles distinctus	*C. africanus*
Anopheles earlei	*C. dodgei*
	C. lativittatus
Anopheles farauti	*C. cairnsensis*
	C. lairdi
	C. couchii
Anopheles freeborni	*C. punctatus*
Anopheles funestus	*C. africanus*
Anopheles gambiae	*C. africanus*
	C. fasciatus
	C. grassei
	C. indicus
	C. orbicularis
	C. orbiculostriatus
	C. rugosus
	C. walkeri
Anopheles georgianus	*C. quadrangulatus* var. *quadrangulatus*

(continued)

TABLE IV. (Continued)

Host	Species of *Coelomomyces*
Anopheles hyrcanus var. *nigerrimus*	*C. indicus*
Anopheles jamesi	*C. anophelesicus*
	C. fasciatus
	C. indicus
	C. iyengarii
Anopheles jeyporensis	*C. indicus*
	C. reticulatus var. *reticulatus*
Anopheles kochi	*C. reticulatus* var. *reticulatus*
Anopheles navipes	*C. couchii*
Anopheles nigerrimus	*C. iliensis* var. *indianensis*
	C. rugosus
Anopheles pallidus	*C. anophelesicus*
	C. indicus
	C. iyengarii
Anopheles pharoensis	*C. indicus*
Anopheles pretoriensis	*C. indicus*
Anopheles punctatus	*C. lativittatus*
Anopheles punctipennis	*C. carolinianus*
	C. cribrosus
	C. lativittatus
	C. punctatus
	C. quadrangulatus var. *irregularis*
	C. quadrangulatus var. *quadrangulatus*
	C. sculptosporus
	C. traingulatus
Anopheles (Cellia) punctulatus	*C. lairdi*
	C. solomonis
Anopheles quadrimaculatus	*C. punctatus*
	C. quadrangulatus var. *quadrangulatus*
Anopheles ramsayi	*C. iliensis* var. *irregularis*
	C. indicus
Anopheles rivulorum	*C. indicus*
Anopheles rufipes	*C. africanus*
	C. indicus
Anopheles sinensis	*C. raffaelei* var. *parvus*
Anopheles splendidus	*C. indicus*
Anopheles squamosus	*C. africanus*
	C. indicus
	C. grassei
	C. orbicularis
Anopheles stephensi	*C. indicus*
	C. punctatus
Anopheles subpictus	*C. anophelesicus*
	C. indicus
	C. iyengarii
Anopheles subpictus indefinitus	*C. indicus*

TABLE IV. (Continued)

Host	Species of *Coelomomyces*
Anopheles tesselatus	*C. walkeri*
Anopheles vagus	*C. anophelesicus*
	C. indicus
	C. iyengarii
	C. muspratti
Anopheles varuna	*C. anophelesicus*
	C. iliensis var. *irregularis*
	C. indicus
Anopheles walkeri	*C. quadrangulatus* var. *quadrangulatus*
	C. sculptosporus
Armigeres longipalpis	*C. macleayae*
Armigeres obturbans	*C. dentialatus*
	C. stegomyiae var. *stegomyiae*
Armigeres subalbatus	*C. stegomyiae* var. *stegomyiae*
Chironomus paraplumosus	*C. chironomi*
Chironomus plumosus	*C. chironomi*
Cricotopus	*C. tuzetae*
Cules annulirostris	*C. iliensis*
	C. iliensis var. *australicus*
	C. stegomyiae var. *stegomyiae*
Culex annulus	*C. iliensis* var. *iliensis*
Culex antennatus	*C. iliensis* var. *iliensis*
	C. muspratti
	C. tuberculatus
Culex basicinctus	*C. muspratti*
Culex bitaeniorhynchus	*C. muspratti*
Culex erraticus	*C. pentangulatus*
Culex fraudatrix	*C. muspratti*
Culex fuscocephalus	*C. muspratti*
Culex gelidus	*C. elegans*
Culex modestus	*C. dubitskii*
	C. iliensis
	C. iliensis var. *irregularis*
Culex morsitans	*C. psorophorae*
Culex orientalis	*C. iliensis*
Culex peccator	*C. pentangulatus*
Culex pipiens	*C. iliensis*
Culex pipiens modestus	*C. quadrangulatus* var. *quadrangulatus*
Culex portesi	*C. seriostriatus*
Culex quinquefasciatus	*C. muspratti*
Culex restuans	*C. psorophorae* var. *psorophorae*
Culex salinarius	*C. psorophorae* var. *psorophorae*
Culex simpsoni	*C. indicus*
	C. muspratti
Culex sp.	*C. muspratti*

(*continued*)

TABLE IV. (Continued)

Host	Species of *Coelomomyces*
Culex t. summorosus	*C. couchii*
Culex theileri	*C. musprattii*
Culex tritaeniorhynchus	*C. iliensis* var. *striatus*
	C. reticulatus var. *reticulatus*
	C. musprattii
Culex tritaeniorhynchus summorosus	*C. quadrangulatus* var. *parvus*
Culex univittatus	*C. orbicularis*
	C. musprattii
Culex vishnui	*C. orbiculostriatus*
	C. iliensis var. *striatus*
	C. sulcatus
Culiseta incidens	*C. iliensis* var. *iliensis*
	C. psorophorae var. *psorophorae*
	C. reticulatus var. *reticulatus*
	C. utahensis
Culiseta inornata	*C. psorophorae* var. *psorophorae*
Culiseta melaneura	*C. psorophorae* var. *psorophorae*
Ingramia (Ravenalites) roubaudi	*C. madagascaricus*
Opifex fuscus	*C. psorophorae* var. *tasmaniensis*
Orthocladius	*C. opifexi*
Orthocladius	*C. tuzetae*
Psectrocladius sp.	*C. canadensis*
Psorophora ciliata	*C. psorophorae* var. *psorophorae*
Psorophora howardii	*C. psorophorae* var. *psorophorae*
Rovenolats nonberuli	*C. madagascaricus*
Sabethes cyaneus	*C. lacunosus* var. *lacunosus*
	C. opifexi
Stegomyia scutellaris	*C. stegomyiae* var. *stegomyiae*
Theobaldia (Culiseta) inornata	*C. psorophorae* var. *psorophorae*
Topomyia yanbarensis	*C. stegomyiae* var. *stegomyiae*
Toxyrhynchites	*C. macleayae*
Toxyrhynchites gravelyi	*C. macleayae*
Toxyrhynchites rutilus	*C. macleayae*
Toxyrhynchites rutilus septentrionalis	*C. macleayae*
Trichoprosopon longipes	*C. lacunosus*
Tripteroides bambusa	*C. macleayae*
	C. stegomyiae var. *stegomyiae*
Uranotaenia campestris	*C. celatus*
Uranotaenia douceti	*C. lacunosus* var. *madagascaricus*
Uranotaenia sp.	*C. lacunosus* var. *madagascaricus*
	C. quadrangulatus var. *quadrangulatus*
Uranotaenia sappharina	*C. uranotaeniae*

in several instances there are indications that certain species of *Coelomomyces* may have a limited host range and infect only mosquitoes of a given taxonomic group (i.e., *C. africanus* on several species of *Anopheles* and *C. iliensis* on several species of *Culex*). Because of this and in view of the potential use of *Coelomomyces* in biological control of mosquitoes, all species descriptions and references to collections should include identification of host(s).

2. Alternate Host

The knowledge that different species of *Coelomomyces* require not always the same but often a different copepod or ostracod host for completion of their life cycle indicates that specificity of alternate host may be of potential use in delineating taxa. At present, the difficulty in determining the alternate host for a species and the limited number of known alternate hosts (see Chapter 2, Table I) make distinction of species on this basis impossible. However, since the alternate host represents a significant and little-known stage in the life cycle of *Coelomomyces* spp., and because of the potential significance of alternate host specificity to identification of taxa, efforts should be made to identify the alternate host(s) for both known and new species.

E. Habitat

Although most species of *Coelomomyces* are collected on hosts from relatively similar aquatic environments, a few forms appear to be restricted to hosts from specific and often unique habitats, including brackish water, leaf axiles, tree holes, spent nuts, discarded receptacles, etc. Because of this, information regarding the collection site can often provide a clue to the correct taxonomic placement of an unidentified specimen. In this regard, Table V lists some of the more interesting known associations between species of *Coelomomyces* and host habitats. While the species listed are most often collected from the sites indicated, this does not preclude the possibility of their being found elsewhere or that other species might occur in the same or similar habitats. Table VI gives the geographic distribution of known species of *Coelomomyces*.

III. SYSTEMATICS

Couch (1945) was the first to recognize the blastocladiaceous affinities of members of the genus *Coelomomyces* and to acknowledge their unique attributes in erecting for them the family Coelomomycetaceae in the order

TABLE V. Unusual Collection Sites for Selected Species of *Coelomomyces*[a]

Habitat	Species collected
Borrow pits (India)	*C. elegans, C. grassei*
Brackish water	*C. arcellaneus, C. psorophorae* var. *halophilus, C. psorophorae* var. *tasmaniensis*
Leaf axiles	*C. lacunosus, C. stegomyiae*
Rice fields	*C. iliensis* varieties
Rock pools (Utah)	*C. utahensis*
Treeholes, discarded receptacles, and spent nuts	*C. dentialatus, C. macleayae* *C. stegomyiae*

[a] For host citations, see species descriptions.

TABLE VI. Geographic Distribution of Species of *Coelomomyces*

Collection site	Species	Host(s)
Africa		
Angola	*C. africanus*	*Anopheles cydippis*
		Anopheles distinctus
	C. grassei	*Anopheles squamosus*
	C. indicus	*Anopheles gambiae*
Egypt	*C. indicus*	*Anopheles pharoensis*
	C. musprattii	*Culex antennatus*
		Culex theileri
Kenya	*C. africanus*	*Anopheles funestus*
	C. grassei	*Anopheles gambiae*
	C. iliensis var. *iliensis*	*Culex antennatus*
	C. indicus	*Anopheles gambiae*
	C. orbicularis	*Anopheles gambiae*
	C. rugosus	*Anopheles gambiae*
Liberia	*C. africanus*	*Anopheles funestus*
		Anopheles gambiae
Nigeria	*C. africanus*	*Anopheles gambiae*
	C. indicus	*Anopheles gambiae*
	C. orbiculostriatus	*Anopheles gambiae*
North Chad	*C. grassei*	*Anopheles gambiae*
Sierra Leone	*C. africanus*	*Anopheles gambiae*
	C. walkeri	*Anopheles funestus*
		Anopheles gambiae
South Africa	*C. fasciatus*	*Anopheles gambiae*
	C. grassei	*Anopheles gambiae*
	C. orbicularis	*Culex univittatus*
Upper Volta	*C. africanus*	*Anopheles gambiae*

TABLE VI. (Continued)

Collection site	Species	Host(s)
		Anopheles rufipes
	C. grassei	*Anopheles gambiae*
	C. orbiculostriatus	*Anopheles gambiae*
	C. walkeri	*Anopheles gambiae*
Zambia	*C. indicus*	*Anopheles funestus*
		Anopheles gambiae
		Anopheles pretoriensis
		Anopheles rivulorum
		Anopheles rufipes
		Anopheles squamosus
		Culex simpsoni
Zimbabwe	*C. africanus*	*Anopheles squamosus*
Australia		
Queensland	*C. cairnsensis*	*Anopheles farauti*
	C. finlayae	*Aedes notoscriptus*
	C. iliensis var. *iliensis*	*Culex annulirostris*
	C. iliensis var. *australicus*	*Culex annulirostris*
	C. indicus	*Aedomyia catastica*
	C. macleayae	*Aedes (Macleaya)* sp.
	C. muspratti	*Culex basicinctus*
		Culex quinquefasciatus
Borneo		
Gizo Island	*C. couchii*	*Anopheles Cellia farauti*
North Borneo	*C. couchii*	*Culex traudatrix*
		Culex traudatrix summorosus
Burma	*C. macleayae*	*Aedes aegypti*
Canada		
Belleville	*C. borealis* var. *giganteus*	*Aedes fitchii*
		Aedes stimulans
British Columbia	*C. canadensis*	*Psectrocladius*
Cayman Islands	*C. psorophorae* var. *halophilus*	*Aedes taeniorhynchus*
Corsica	*C. thailandensis*	*Anopheles claviger*
Czechoslovakia		
Radov	*C. chironomi*	*Chironomus paraplumosus*
		Chironomus plumosus
England		
Huntington	*C. borealis* var. *borealis*	*Aedes cantans*
Fiji	*C. macleayae*	*Aedes polynesiensis*
France	*C. tuzetae*	*Orthocladius lignicola*
India	*C. anophelesicus*	*Anopheles jamesi*
		Anopheles pallidus

(*continued*)

TABLE VI. (Continued)

Collection site	Species	Host(s)
		Anopheles subpictus
		Anopheles vangus
	C. fasciatus	*Anopheles jamesi*
	C. iliensis var. *indus*	*Anopheles nigerrimus*
	C. iliensis var. *irregularis*	*Anopheles aconitus*
	C. iliensis var. *irregularis*	*Anopheles ramsayi*
	C. iliensis var. *irregularis*	*Anopheles varuna*
	C. iliensis var. *striatus*	*Culex tritaeniorhynchus*
	C. iliensis var. *striatus*	*Culex vishnui*
	C. indicus	*Anopheles hyrcanus*
		Anopheles subpictus
		Anopheles aconitus
		Anopheles ramsayi
		Anopheles annularis
		Anopheles jamesi
		Anopheles culicifacies
		Anopheles jeyporensis
		Anopheles pallidus
		Anopheles splendidus
		Anopheles stephensi
		Anopheles subpictus
		Anopheles vagus
	C. iyengarii	*Anopheles vagus*
	C. musprattii	*Culex bitaeniorhynchus*
		Culex fraudatrix
		Culex fuscocephalus
	C. orbiculostriatus	*Culex vishnui*
	C. reticulatus	*Anopheles jeyporiensis*
	C. rugosus	*Anopheles nigerrimus*
	C. sulcatus	*Culex vishnui*
Italy	*C. raffaelei* var. *raffaelei*	*Anopheles claviger*
Japan	*C. ponticulas*	*Aedes (Stegomyia) flavopictus*
	C. raffaelei var. *parvus*	*Anopheles sinesis*
Java	*C. reticulatus* var. *reticulatus*	*Culex tritaeniorhynchus*
	C. walkeri	*Anopheles tesselatus*
Madagascar	*C. africanus*	*Anopheles squamosus*

TABLE VI. (Continued)

Collection site	Species	Host(s)
	C. dentialatus	*Aedes aegypti*
	C. lacunosus var. *madagascaricus*	*Uranotaenia douceti*
	C. madagascaricus	*Ingramia (Ravenalites) roubaudi*
		Rovenolats nonberuli
	C. tuberculatus	*Culex antennatus*
Malaya	*C. reticulatus* var. *reticulatus*	*Anopheles kochi*
	C. stegomyiae var. *stegomyiae*	*Stegomyia scutellaris*
Malaysia	*C. quadrangulatus* var. *parvus*	*Culex tritaeniorhynehus summorosus*
New Zealand	*C. psorophorae* var. *tasmaniensis*	*Opifex fuscus*
Panama	*C. lacunosus* var. *lacunosus*	*Trichoprosopon longipes*
Philippines	*C. indicus*	*Anopheles subpictus indefinitus*
Singapore	*C. couchii*	*Culex t. summorosus*
Solomon Islands	*C. lairdi*	*Anopheles (Cellia) punctalatus*
		Anopheles farauti
South America		
Colombia	*C. reticulatus* var. *parvus*	*Aedomyia squamipennis*
Sri Lanka (Ceylon)	*C. dentialatus*	*Aedes albopictus*
		Aedes sp.
	C. dentialatus	*Aedes aegypti*
		Armigeres obturbans
	C. elegans	*Culex gelidus*
	C. stegomyiae var. *stegomyiae*	*Aedes albopictus*
Taiwan	*C. iliensis* var. *iliensis*	*Culex annulus*
	C. macleayae	*Aedes (Stegomyia) alcasida*
		Aedes (Finlaya) japonicus shintienensis
		Tripteroides bambusa
	C. quadrangalatus	*Uranotaenia recondita*
	C. stegomyiae var. *chapmani*	*Aedes albopictus*
		Armigeres subalbatus
		Topomyia yanbarensis
Tasmania	*C. psorophorae* var. *tasmaniensis*	*Aedes (Pseudokusea) australis*

(*continued*)

TABLE VI. (Continued)

Collection site	Species	Host(s)
Thailand	*C. africanus*	*Anopheles barbirostris*
		Anopheles campestris
	C. celatus	*Uranotaenia campestris*
	C. couchii	*Anopheles navipes*
	C. indicus	*Anopheles subpictus*
		Anopheles vagus
	C. iyengarii	*Anopheles aconitus*
		Anopheles vagus
	C. macleayae	*Aedes albopictus*
		Aedes mediopunctatus
		Aedes pseudalbopictus
		Armigeres longipalpis
		Toxyrhynchites gravelyi
		Toxyrhynchites sp.
	C. orbiculostriatus	*Culex vishnui*
	C. thailandensis	*Anopheles bengalensis*
Trinidad	*C. seriostriatus*	*Culex portesi*
United States		
Alaska	*C. borealis* var. *borealis*	*Aedes excrucians*
Connecticut	*C. lativittatus*	*Anopheles punctipennis*
Florida	*C. arcellaneus*	*Aedes taeniorhynchus*
Georgia	*C. bisymmetricus*	*Anopheles crucians*
	C. cribrosus	*Anopheles crucians*
		Anopheles punctipennis
		Anopheles sp.
	C. dodgei	*Anopheles crucians*
	C. keilini	*Anopheles crucians*
	C. lativittatus	*Anopheles crucians*
	C. pentangulatus	*Culex erratus psorophora ciliata*
	C. psorophorae	*Anopheles crucians*
	C. punctatus	*Anopheles crucians*
	C. punctatus	*Anopheles quadrimaculatus*
	C. quadrangulatus var. *irregularis*	*Anopheles punctipennis*
	C. quadrangulatus var. *quadrangulatus*	*Anopheles georgianus*
	C. sculptosporus	*Anopheles penctipennis*
	C. triangulatus	*Anopheles punctipennis*
	C. uranotaeniae	*Uranotaenia sappharina*
Louisiana	*C. arcellaneus*	*Aedes sollicitans*
	C. macleayae	*Aedes triseriatus*
	C. pentangulatus	*Culex erraticus*
		Culex peccator

TABLE VI. (Continued)

Collection site	Species	Host(s)
	C. psorophorae var. *halophilus*	*Aedes sollicitans*
	C. punctatus	*Anopheles bradleyi* *Anopheles crucians*
	C. quadrangulatus var. *lamborni*	*Aedes (Stegomyia) scutellaris*
Maine	*C. borealis* var. *borealis*	*Aedes communis*
Michigan	*C. lativittatus*	*Anopheles punctipennis*
Minnesota	*C. dodgei*	*Anopheles earlei*
	C. lativittatus	*Anopheles earlei*
	C. quadrangulatus var. *quadrangulatus*	*Anopheles walkeri*
	C. sculptosporus	*Anopheles walkeri*
North Carolina	*C. carolinianus*	*Anopheles punctipennis*
	C. cribrosus	*Anopheles crucians*
	C. lativittatus	*Anopheles crucians* *Anopheles punctipennis* *Anopheles* sp.
	C. psorophorae var. *halophilus*	*Aedes sollicitans*
	C. psorophorae var. *halophilus*	*Aedes taeniorhynchus*
	C. punctatus	*Anopheles punctipennis*
Ohio	*C. lativittatus*	*Anopheles punctipennis*
Pennsylvania	*C. lativittatus*	*Anopheles* sp.
Utah	*C. utahensis*	*Aedes atropalpus nielseni* *Aedes atropalpus epactius* *Culex tarsalis* *Culiseta incidens*
Virginia	*C. triangulatus*	*Anopheles punctipennis*
U.S.S.R.	*C. dubitskii*	*Culex modestus*
	C. iliensis var. *iliensis*	*Culex pipiens*
	C. iliensis var. *iliensis*	*Culex orientalis*
	C. iliensis var. *iliensis*	*Culex modestus*
	C. iliensis var. *irregularis*	*Culex modestus*
	C. quadrangulatus var. *quadrangulatus*	*Aedes* sp. *(Aerossicus)*
	C. quadrangulatus var. *quadrangulatus*	*Aedes vexans*

Blastocladiales. Although recognized to have features typical for other members of the Blastocladiales (i.e., posteriorly uniflagellate zoospores and thick-walled, resting sporangia with a prominent dehiscence slit), members of the Coelomomycetaceae were described as distinct in their parasitic mode of nutrition and unique vegetative hyphae lacking a cell wall. While these same characteristics are still recognized as distinctive for the Coelomomycetaceae, further studies now make it possible to supplement the family/generic description.

A. The Family

COELOMOMYCETACEAE

Couch ex Couch, *J. Elisha Mitchell Sci. Soc.* **78,** 126–138 (1962)

Couch, *J. Elisha Mitchell Sci. Soc.* **61,** 128 (1945).

Sporophytic thallus obligately parasitic in the body cavity of aquatic arthropods (primarily larvae of mosquitoes and chironomids); hyphae filamentous to cellular, branched, coenocytic, lacking rhizoids, holocarpic; differentiating into resting zoosporangia having a variously ornamented, brownish outer wall layer and a homogeneous inner layer; and, in some species, thin-walled, hyaline zoosporangia (mitosporangia). Sporangial dehiscence by cracking of the outer wall layer along a preformed dehiscence slit and extrusion of the zoospores into an evanescent vesicle. Zoospores posteriorly uniflagellate and structurally similar to those of other members of the order. Meiozoospores infecting copepods or ostracods.

Gametophytic thallus (in the forms for which it is known) obligately parasitic in the body cavity of copepods or ostracods; hyphae, coenocytic, holocarpic, differentiating into male and female gametangia (dioecious). Both gametangia hyaline in some forms, the male orange and the female hyaline in others. Sexual reproduction via fusion of posteriorly uniflagellate isogametes that are structurally similar to zoospores. Zygotes biflagellate, infecting larvae of mosquitoes or chironomids.

B. The Genus

Coelomomyces

Keilin, *emend.* Couch, *J. Elisha Mitchell Sci. Soc.* **61,** 128 (1945)

Coelomomyces Keilin, *Parasitology* **13,** 226 (1921)

Zoografia Bogoyavlensky, *Arch. Soc. Russe Protistenkd.* **1,** 113 (1922)

Characteristics: Those of the family.

Type Species: *Coelomomyces stegomyiae* Keilin, *Parasitology* **13,** 225 (1921)

C. Key to Species of the Genus *Coelomomyces*

1. Resting sporangia in Chironomidae 2
1′. Resting sporangia in Culicidae 4
2. Sporangial surface smooth .... ***C. tuzetae*** (page 271)
2′. Sporangial surface ornamented 3
3. Sporangial surface ornamented with regular polygonal areas containing irregular grooves or striae. Sporangial dimensions 18–45 × 27–42 μm .... ***C. chironomi*** (page 141)
3′. Sporangial surface ornamented with circular to irregular depressions containing deep radiating furrows .... ***C. canadensis*** (page 134)
4. Sporangial wall lacking ornamentation 5
4′. Sporangial wall ornamented with punctae, striae, bands, ridges, grooves or combinations of these features 6
5. Sporangia appearing five-angled in end view, measuring 12–18 × 18–40 μm .... ***C. pentangulatus*** (page 215)
5′. Sporangia oval, rarely flattened on one side, measuring 30–46.5 × 45–65 μm .... ***C. walkeri*** (page 278)
6. Sporangial wall furrowed to form uniform, contiguous, bands or with bands on one sporangial face and irregular striae/grooves on the other 7
6′. Sporangial surface ornamented with punctae, striae, ridges separated by depressed areas, or combinations of these features 11
7. Sporangia ellipsoid or flattened dorso–ventrally to appear spherical in some views. Irregular punctae/striae often present on the ventral sporangial face. Ridges as bands of consistent height 8
7′. Sporangia ellipsoid. Ridges/bands concentric to irregular, or beginning from a central dome on the ventral sporangial face and continuing, symmetrically and concentrically to converge on the dorsal face. Ridging alternating as high/low bands 10
8. Sporangia dorso–ventrally flattened to appear spherical in some views, measuring 28–39 μm wide × 33.5–54 μm long × 22–30 μm thick .... ***C. fasciatus*** (page 159)
8′. Sporangia ellipsoid 9
9. Sporangia with concentric ridges/bands on dorsal surface and on sides, breaking up to appear as irregular punctae/striae on the ventral surface .... ***C. dodgei*** (page 152)
9′. Sporangia with concentric ridges/bands only .... ***C. lativittatus*** (page 195)
10. Ridges/bands regular and concentric sporangia bisymmetrical .... ***C. bisymmetricus*** (page 126)
10′. Ridges/bands circular to irregular 11
11. Ridges modified to enclose low ridges as 3–20 circular depressed areas, often bearing 1–2 striae .... ***C. cribrosus*** (page 147)
11′. Raised and low ridges forming an irregular pattern .... ***C. sculptosporus*** (page 255)
12. Sporangial surface rugose or ornamented with regular to irregular grooves, punctae, and/or striae 13
12′. Sporangial surface ornamented with regular to irregular ridges or bands enclosing depressed areas of varying patterns and bearing grooves, punctae, striae, and/or papillae 28
13. Sporangial surface with punctae or with irregular depressed areas appearing as narrow to broad striae 14

13′. Sporangial surface rugose . . . .*C. rugosus* (page 250)

14. Sporangial surface with regular to irregular depressed areas appearing as narrow to broad striae 15

14′. Sporangial surface with circular to irregular punctae that may or may not have internal differentiation 18

15. Striae as irregular, broad lacune and/or punctae (1–2 μm wide × 1–3 μm long and 1–4 μm apart). 16

15′. Striae narrow, irregularly scattered, linear in concentric rows running parallel with long axis of the sporangium, or in regular rows running perpendicular to the long axis of the sporangium 17

16. Dehiscence slit bordered by lacunae, not central along prominent dorsal ridge . . . .*C. lacunosus* var. *lacunosus* (page 188)

16′. Dehiscence slit central along prominent dorsal ridge . . . .*C. lacunosus* var. *madagascaricus* (page 191)

17. Striae irregularly scattered or linear in concentric rows that are parallel to the long axis of the sporangium . . . .*C. punctatus* (page 234)

17′. Striae in regular rows running perpendicular to the long axis of the sporangium . . . .*C. seriostriatus* (page 256)

18. Resting sporangia discoid, flattened dorso–ventrally, appearing circular in face view. Dorsal and ventral sides with punctae, vertical striae around edges . . . .*C. arcellaneus* (page 123)

18′. Resting sporangia ellipsoid 19

19. Sporangial surface ornamented with simple punctae lacking internal differentiation 20

19′. Sporangial surface ornamented with simple to irregular punctae bearing internal differentiation as striae, bars, pits, or papillae 23

20. Dehiscence slit bordered by punctae, not central along a longitudinal low . . . . .ridge 21

20′. Dehiscence slit central along a low, longitudinal ridge . . .*C. utahensis* (page 275)

21. Resting sporangia from hosts occurring in fresh water . . . .*C. psorophorae* var. *psorophorae* "complex" (page 218)

21′. Resting sporangia from hosts occurring in brackish water to seawater 22

22. Resting sporangia occurring in *Aedes sollicitans* and/or *Aedes taeniorhynchus* . . . .*C. psorophorae* var. *halophilus* (page 229)

22′. Resting sporangia occurring in *Aedes australis* or *Opifex fuscus* . . . .*C. psorophorae* var. *tasmaniensis* (page 231)

23. Punctae with internal striae radiating out from the center to give a stellate appearance . . . .*C. keilini* (page 187)

23′. Punctae with central bars (bridges), pits, or papillae 24

24. Punctae bisected by a central bar or bridge . . . .*C. ponticulus* (page 217)

24′. Punctae with a central pit or papilla 25

25. Punctae circular and with a central pit 26

25′. Punctae irregular and with a central papilla. 27

26. Sporangia ellipsoid, bipolar pointing of sporangial ends rare . . . .*C. stegomyiae* var. *stegomyiae* "complex" (page 258)

26′. Sporangia ellipsoid to fusiform, 50% or more exhibiting bipolar pointing . . . .*C. stegomyiae* var. *chapmani* (page 261)

27. Sporangia measuring 33.5–71 × 63–126 μm . . . . . .*C. borealis* var. *borealis* (page 126)

27′. Sporangia measuring 76–134 × 109–182 μm . . . .*C. borealis* var. *giganteus* (page 131)

28. Resting sporangia asymmetrical; in face view, semicircular with one side flattened,

in side view, biconvex; surface ornamented with concentric to irregular ridges separated by depressed areas bearing transverse striae .... *C. anophelesicus* (page 119)

28′. Resting sporangia symmetrical, usually broadly to narrowly ellipsoid 29

29. Sporangia with three or four broad ridges running parallel to the long axis of the sporangium so as to give the sporangium a three- or four-angled appearance in end view. Ridges separated by depressed areas with transverse or irregular striae ... 30

29′. Wall otherwise 34

30. Wall with four ridges, giving sporangia a quadrangulate appearance in end view. Transverse striae between ridges.. 31

30′. Wall with three ridges, giving sporangia a triangulate appearance in end view. Striae between ridges irregular .... *C. triangulatus* (page 269)

31. Sporangia measuring 26–31 × 36–58 μm. *C. quadrangulatus* var. *lamborni* (page 240)

31′. Sporangia 8.8–21 × 12.9–40 μm.. 32

32. Sporangia 8.8–15.6 × 12.9–21.4 μm ... *C. quadrangulatus* var. *parvus* (page 241)

32′. Sporangia generally larger than 12 × 19 μm........................... 33

33. Outer sporangial wall regular ..... *C. quadrangulatus* var. *quadrangulatus* (page 236)

33′. Outer wall crenate and very irregular... *C. quadrangulatus* var. *irregularis* (page 239)

34. Wall with steep, narrow, anastomosing ridges (spinelike in optical section), separated by depressed areas with punctae or striae......................... 35

34′. Wall with rounded ridges or mounds (domelike in optical section), separated by depressed areas with punctae, striae, or papillae........................ 38

35. Ridges interconnected by transverse striae .... *C. uranotaeniae* (page 273)

35′. Ridges interconnected by depressed areas containing pits 36

36. Sporangial faces with same ornamentation.............. *C. finlayae* (page 163)

36′. Sporangial faces differently ornamented 37

37. Ridges of one face enclosing hexagonal to irregular areas, inner ridges of other face strongly convoluted................................... *C. cairnsensis* (page 131)

37′. Ridges of one face enclosing irregular polygonal areas, some ridges of the other face longitudinal .... *C. macleayae* (page 198)

38. Ridging irregular, discontinuous, or elongate on the dorsal face; on the ventral face, enclosing circular areas with flattened domes the height of the ridges 39

38′. Ridging regular and generally continuous, elongate, running parallel to the long axis of the sporangium or anastomosing to enclose circular to irregular depressed areas with punctae, striae, or papillae 40

39. Ridging discontinuous, irregular; separated by depressed areas with punctae .... *C. madagascaricus* (page 201)

39′. Ridging elongate on the dorsal face, enclosing circular areas with domes the height of the ridging on the ventral face .... *C. tuberculatus* (page 270)

40. Ridging anastomosing to enclose elongate, circular, irregular or polygonal depressed areas with punctae, striae, or papillae.......................... 41

40′. Ridging elongate, running parallel to the long axis of the sporangium, in concentric rings around the sporangium or as elongate wings. Ridges separated by punctae and/or striae 53

41. Ridges enclosing irregular to elongate depressed areas with punctae or striae.... 42

41′. Ridges enclosing circular, depressed areas with punctae, striae, or papillae 47

42. Depressed areas with very fine central punctae which are difficult to discern via light microscopy .... *C. thailandensis* (page 267)

42′. Depressed areas with prominent punctae/striae 43

43. Internal, vertical striae prominent in midoptical sections of the wall. Dehiscence slit bordered by a depressed area with punctae/striae .... ***C. dubitskii*** (page 155)

43′. Internal, vertical striae not evident. Dehiscence slit along a middorsal ridge..... 44

44. Depressed areas mostly elongate to irregular 45

44′. Depressed areas circular on the ventral sporangial face and sides; sporangia measuring 9–24 μm × 18.5–35 μm .... ***C. africanus*** (page 114)

45. Depressed areas with central striae; sporangia appearing quadrangulate in end view .... ***C. celatus*** (page 139)

45′. Depressed areas filled and bordered with punctae/striae...................... 46

46. Sporangia measuring 18–26 × 33–49 μm ... ***C. raffaelei*** var. ***raffaelei*** (page 242)

46′. Sporangia measuring 12–21 × 21–34 μm ..... ***C. raffaelei*** var. ***parvus*** (page 245)

47. Depressed areas with papillae .... ***C. elegans*** (page 157)

47′. Depressed areas with punctae and/or striae................................ 48

48. Depressed areas with striae .... ***C. orbiculostriatus*** (page 211)

48′. Depressed areas with punctae 49

49. Dehiscence slit central along an elongate depressed area with punctae .... ***C. orbicularis*** (page 208)

49′. Dehiscence slit along a middorsal ridge; not bordered by punctae 50

50. Ridging in midoptical section, edge view as truncate domes interconnected by wings 51

50′. Ridging in midoptical section, edge view, as low, rounded domes.......... 52

51. Sporangia 27–55 × 43–80 μm.............. ***C. reticulatus*** var. ***reticulatus*** (page 247)

51′. Sporangia 22–35 × 33.5–55 μm................ ***C. reticulatus*** var. ***parvus*** (page 250)

52. Punctae in depressed areas indistinct, occurring only in the center. Dehiscence slit bordered by circular to polygonal depressed areas with punctae .... ***C. lairdi*** (page 191)

52′. Punctae in depressed areas prominent, filling the entire area. Dehiscence slit bordered on either side by an elongate depressed area........ ***C. couchii*** (page 144)

53. Ridges encircling the dehiscence slit and continuing in concentric rings around sporangial periphery; in some forms, ridges anastomosing at poles or over the entire sporangial surface to enclose circular to irregular depressed areas with striae 54

53′. Ridges as rounded domes or wings that do not encircle the dehiscence slit, running parallel to the long axis of the sporangium, anastomosing at the poles; in some forms, anastomosing on the ventral surface to enclose irregular to polygonal depressed areas with punctae/striae.. 61

54. Ridging anastomosing at the poles or occasionally over entire sporangia to enclose circular to irregular depressed areas with striae .... ***C. indicus*** (page 180)

54′. Ridging rarely anastomosing to enclose circular or irregular depressed areas 55

55. In midoptical section, end view, ridges appearing as 12–20 domes encircling the sporangium.. ***C. musprattii*** (page 205)

55′. In midoptical section, end view, ridges appearing as 7–9 domes encircling the sporangium 56

56. Sporangia elongate–ellipsoid in face view; ridges separated by depressed areas greater than 5 μm wide and bearing prominent transverse striae; internal, vertical striae not prominent in ridges; in midoptical section, face view, transverse striae as prominent teeth around sporangial periphery .... ***C. iyengarii*** (page 185)

56′. Sporangia broadly ellipsoid in face view, biconvex in side view, flattened dorso–ventrally; internal vertical striae prominent in ridges 57

57. Sporangia regular; ridging in concentric rows encircling the sporangia; ridging not as isolated domes at the poles 58

57′. At least some sporangia irregular, with frequently anastomosing ridges enclosing irregular depressed areas with striae; or sporangia with ridges as isolated domes at the poles 60

58. Ridges 5–10 μm apart 59

58′. Ridges less than 5 μm apart . . . . ***C. iliensis*** var. ***iliensis*** (page 166)

59. Internal vertical striae faint; transverse striae in depressed areas fine (1–1.5 μm wide); depressed areas lacking punctae . . . . ***C. iliensis*** var. ***indus*** (page 174)

59′. Internal vertical striae prominent; transverse striae in depressed areas prominent (1.5–2 μm wide); by light microscopy depressed areas often bearing a central row of punctae . . . . ***C. iliensis*** var. ***striatus*** (page 178)

60. At least some sporangia irregular, ridges anastomosing frequently and enclosing irregular depressed areas with transverse striae . . . . ***C. iliensis*** var. ***irregularis*** (page 176)

60′. Sporangia with ridges as isolated domes at the poles . . . . ***C. iliensis*** var. ***australicus*** (page 173)

61. Ridges separated by depressed areas with transverse striae 62

61′. Ridges separated by depressed areas with punctae 63

62. Dehiscence slit in a groove along a middorsal ridge; in midoptical section, end view, ridges as 8–10 domes encircling the sporangium . . . . . ***C. grassei*** (page 164)

62′. Dehiscence slit central along a middorsal ridge, not in a groove; in midoptical section, end view, ridges as 4–5 domes encircling the sporangium . . . . ***C. carolinianus*** (page 136)

63. Dehiscence slit central along a middorsal ridge; punctae in depressed areas often paired; ridges in side view distinctly serrate . . . . ***C. dentialatus*** (page 149)

63′. Dehiscence slit central along a depressed area with punctae which are not paired; ridges in side view not distinctly serrate . . . . ***C. sulcatus*** (page 263)

D. Descriptions of Species (Latin Translations by Donald P. Rogers)*

The following species descriptions are arranged alphabetically according to specific epithet, and include previously described and new species. Certain forms placed previously in the genus *Coelomomyces* are here excluded from this genus and considered under Section III,E. These forms include *C. ascariformis, C. beirnei, C. ciferrii, C. dubitskii, C. milkoi, C. notonectae,* and *C. solomonis.*

In each description, collection data for all specimens examined are given along with similar data for other reported collections that were not examined by us (indicated by brackets). Collections are listed alphabetically according to continent or country of origin, with collection data being given in the following order: host, collection site, collector, date of collection, depository, reference. A question mark indicates that informa-

* Professor of Botany, Emeritus, Department of Botany, University of Illinois at Urbana, Champaign, Urbana, Illinois 61801. Present address: 1809 20th Street NE, Auburn, Washington 98002.

tion was unavailable. The following list gives the complete names and addresses for the depositories abbreviated in the species descriptions.

AFRIMS	Armed Forces Research Institute for Medical Science (formerly SEATO Medical Laboratory) Bangkok, Thailand
BISH	Herbarium Bernice P. Bishop Museum P. O. Box 19000-A Honolulu, Hawaii 96817
BPI	National Fungus Collection Agricultural Research Center—West Beltsville, Maryland 20705
CAES	Connecticut Agricultural Experiment Station 123 Huntington Street New Haven, Connecticut 06504
CIH	Commonwealth Institute of Health University of Sydney New South Wales, Australia
CM	Dominion Museum Wellington, New Zealand
IP	Institut Pasteur 25 Rue du Dr. Roux Paris, France
K	The Herbarium Royal Batanic Gardens Kew, Richmond, Surrey TW9 3AB England
MUN	Memorial University of Newfoundland St. John's Newfoundland, Canada ALC-557
NCU	Herbarium Department of Botany University of North Carolina Chapel Hill, North Carolina 27514
QIMR	Queensland Institute of Medical Research Brisbane, Australia
SMCG	State Mosquito Control of Georgia Savannah, Georgia 31406
TNS	Department of Botany National Science Museum Hyakunin—cho 3-23-1, Shinjuku—ku Tokyo, Japan
TPI	Tawian Provincial Institute of Infectious Diseases Taipei, Taiwan
UQA	Department of Entomology University of Queensland Queensland, Australia
US	School of Public Health and Tropical Medicine University of Sydney Sydney, Australia

Coelomomyces africanus Walker ex Couch & Bland

Coelomomyces africanus Walker, *Ann. Trop. Med. Parasitol.* **32,** 231 (1938)

Figures 175–181

DESCRIPTION: Hyphae 4.0–11.3 μm thick, irregularly branched. Resting sporangia ellipsoid to allantoid, frequently flattened on side with dehiscence slit, 9–24 × 18.5–35 μm; surface ornamented with mostly circular to irregular depressions (2–4 × 2–12 μm) with scattered punctae/striae, and with a band of punctae/striae usually present on one or both sides of the dehiscence slit. In midoptical section; face view, wall with low ridges alternating with depressed areas; similar in end view but with dehiscence slit in prominent ridge. Thin-walled sporangia not observed.

TYPE HOST AND COLLECTION SITE: *Anopheles gambiae* Theo.; Sierra Leone, Africa (collected by A. J. Walker; see Walker, 1938).

DEPOSITORY FOR TYPE: BPI.

LATIN DIAGNOSIS: Hyphae 4.0–11.3 μm crassae, inaequaliter ramosae. Sporangia perennantia ellipsoidea vel in formam allantoideam vergentes, crebre ad latus unum applanata et incisura dehiscentiae instructa, 9–24 × 18.5–35 μm; superficie depressionibus plerumque rotundis vel irregularibus 2–4 × 2–12 μm decorata atque punctis striisque sparsis, etiam vitta punctorum striarumque plerumque ad latus unum vel utrumque incisurae; quasi in sectione media inspecta a latere, tunica ab liris humilibus cum areoleis depressis alternantibus insignia, ab cacumine, similia sed incisuram inter liram prominentem inclusam notabilia. Sporangia tenuiter tunicata non observata.

SPECIMENS EXAMINED AND OTHER COLLECTIONS (indicated by brackets):

Type

Africa

[*Anopheles argenteolobatus* (Gough); Calai, Angola; Ribeiro; 1981; ?; (Ribeiro *et al.*, 1981).]

[*Anopheles cydippis* DeMeillon; Calai, Angola; Ribeiro; 1981; ?; (Ribeiro *et al.*, 1981).]

[*Anopheles distinctus* (Newstead & Carter); Calai, Angola; Ribeiro; 1981; ?; (Ribeiro *et al.*, 1981).]

Anopheles funestus Giles; Balatva, Liberia; Giglioli; 1956; NCU.

Anopheles funestus Giles; Bailma, Liberia; Giglioli; 1956; NCU.

[*Anopheles funestus* Giles; Kisumu, Kenya; Haddow; 1941; ?; (Haddow, 1942).]

[*Anopheles gambiae* Giles; Kisumu; Kenya; Haddow; 1941; ?; (Haddow, 1942).]

Anopheles gambiae Giles; St. Paul River, Liberia; Darwish; 1960; NCU.

Anopheles gambiae Giles; Mbu, Nigeria; Darwish; 1962; NCU.

Anopheles gambiae Giles; Kaduna, Nigeria; Brown; 1974; NCU.

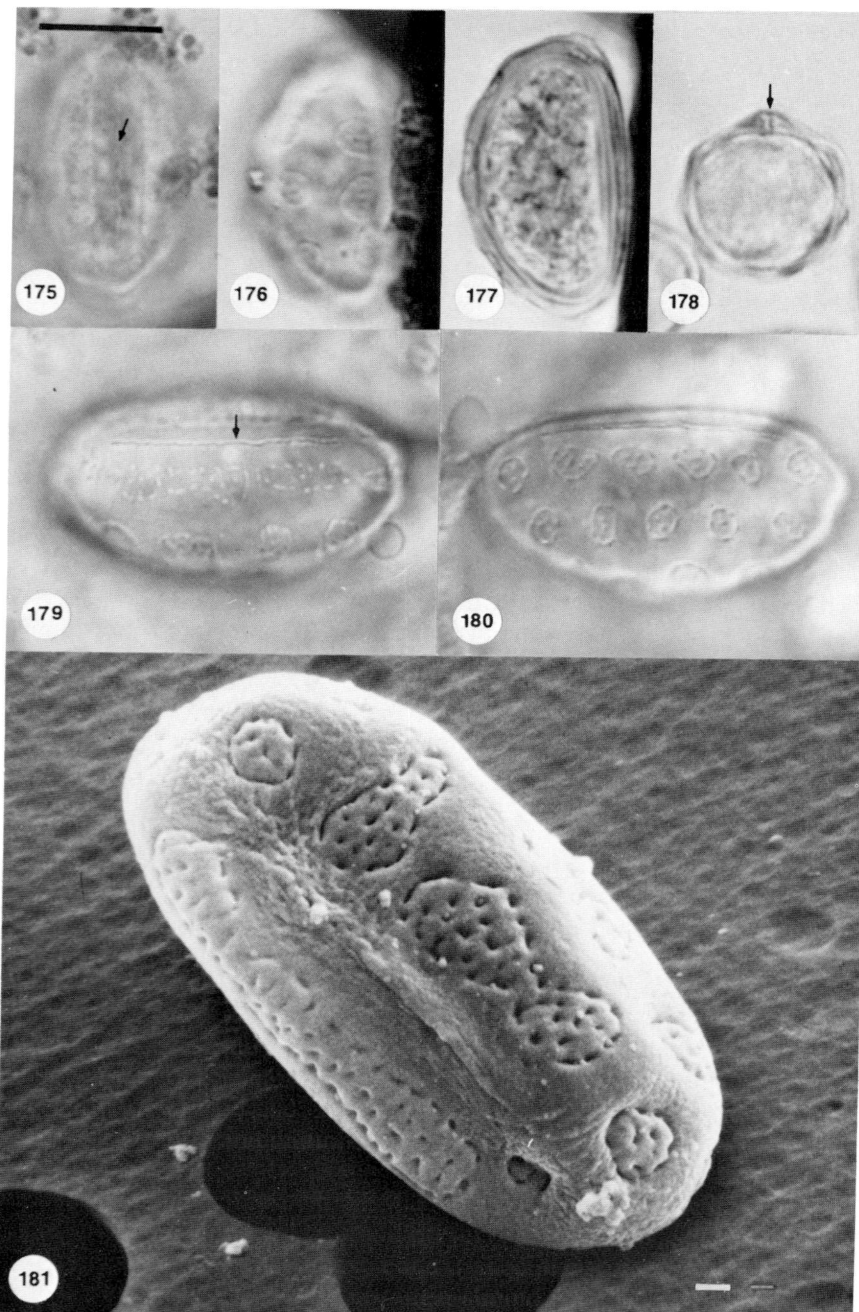
175
176
177
178
179
180
181

Anopheles gambiae Giles; Kaduna, Nigeria; Brown; 1975; NCU.

Anopheles gambiae Giles; Kaduna, Nigeria; Prasertphon; 1976, 1978, 1979; NCU.

Anopheles gambiae Giles; Upper Volta; Rodhain; 1967; NCU; (Rodhain and Brenques, 1974).

[*Anopheles rufipes* (Gough); Upper Volta; Rodhain; 1971; ?; (Rodhain and Gayral, 1971).]

Anopheles squamosus Theo.; Harare, Zimbabwe; Reid; 1962; NCU.

[*Anopheles squamosus* Theo.; Calai, Angola; Ribeiro; 1981; ?; (Ribeiro *et al.*, 1981).]

Madagasgar

Anopheles squamosus Theo.; Tananarive, Alasora; Grjebine; 1953; NCU.

Thailand

Anopheles barbirostris Wulp; Prochin Buri, Khao Dan Ngiu; Kol; 1966; NCU.

Anopheles campestris Reid; Patoomtani, Wat Bong Nang Boom; Kol; 1963; NCU.

COMMENTS: In a study of fungal infections of mosquitoes from Sierra Leone, Africa, Walker (1938) observed four types of sporangia that he ascribed to the genus *Coelomomyces*. For the most common of these (Type 3) occurring on larvae of *Anopheles gambiae* (*A. costalis*) he proposed the name *C. africanus*. However, limited diagnostic features were provided and there was no Latin diagnosis; thus the species lacked valid publication. Based on original specimens of Walker's *C. africanus* sent to J. N. C. and on other specimens collected in Africa and elsewhere, this species is here redescribed.

Coelomomyces africanus may be separated from other species of *Coelomomyces* by its unusually small sporangia and by the small, circular to irregular, punctate depressions occurring on its sporangial wall. However, similar sporangial size and ornamentation occur in specimens of *C. raffaelei* and *C. raffaelei* var. *parvus*. Distinction of *C. africanus* from

---

Figs. 175–178. *Coelomomyces africanus*. Resting sporangia of type from *Anopheles gambiae* collected by Walker in Sierra Leone, Africa. Bar = 10 μm. Fig. 175. Face view showing dehiscence slit (arrow) bordered by elongate depressed areas with punctae. Fig. 176. Surface view of sporangial side. Fig. 177. Midoptical section, side view. Fig. 178. Midoptical section, end view; dehiscence slit (arrow). Scale as for Fig. 175.

Figs. 179–180. *Coelomomyces africanus*. Resting sporangia from *Anopheles gambiae* collected by *Prasertphon* in Kaduna, Nigeria, Africa. Scale as for Fig. 175. Fig. 179. Face view showing dehiscence slit (arrow). Fig. 180. Side view of sporangial surface.

Fig. 181. *Coelomomyces africanus*. Scanning electron micrograph of resting sporangium of type. Bar = 1 μm.

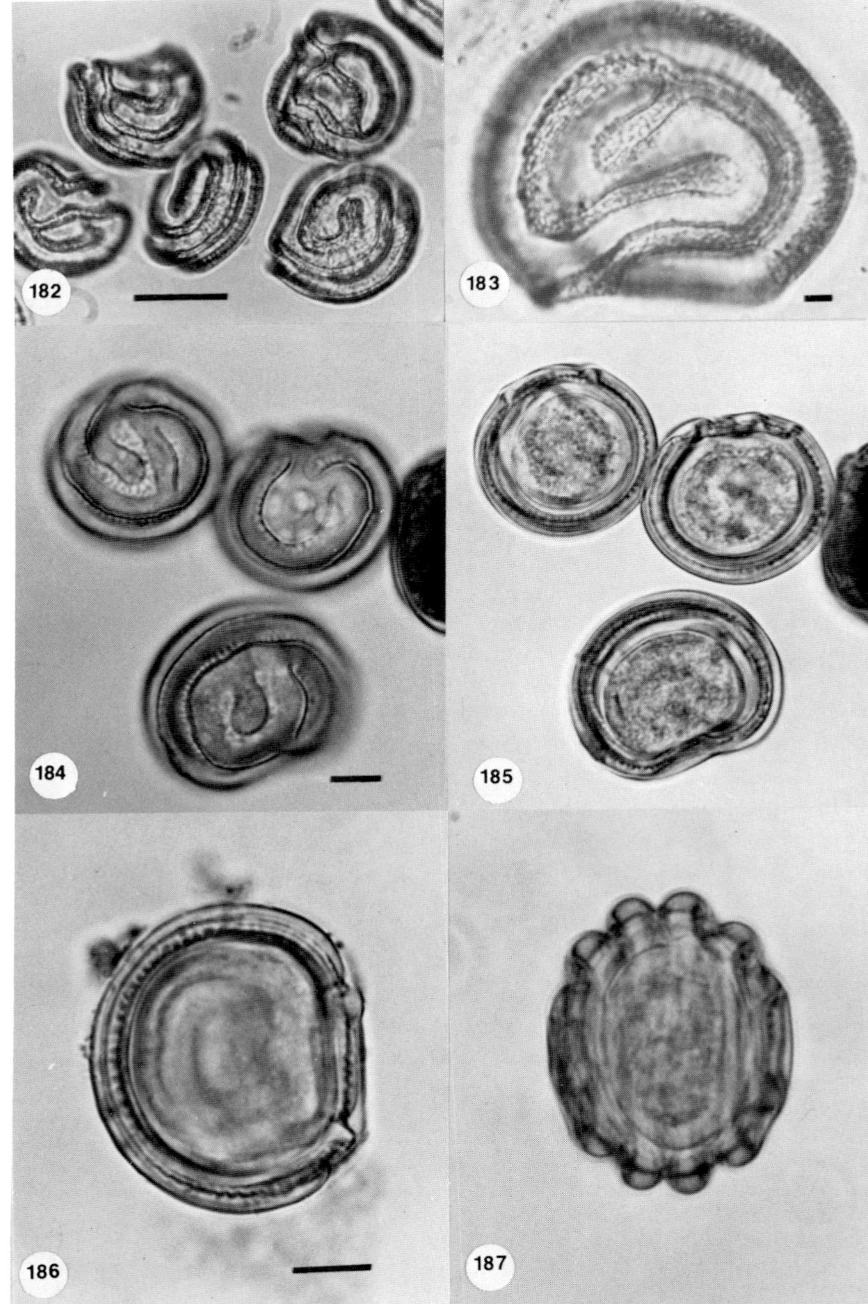
182
183
184
185
186
187

each of these may be made in that sporangia of *C. raffaelei* and *C. raffaelei* var. *parvus* are ornamented with large, irregular, depressed punctate areas and by prominent striae/punctae aligned perpendicular to the dehiscence slit so as to form a band on either side of this raised portion of the sporangial wall. Additionally, in optical section, sporangia of *C. raffaelei* and *C. raffaelei* var. *parvus* are more prominently ridged and in side view have an obvious wing along one side due to the raised area of the wall bearing the dehiscence slit.

Coelomomyces anophelesicus Iyengar ex Iyengar, *J. Elisha Mitchell Sci. Soc.* **78,** 134 (1962)

Coelomomyces anophelesica Iyengar, *Parasitology* **27,** 440 (1935)

Figures 182–194

DESCRIPTION: Hyphae 5–12 μm thick, branched. Resting sporangia asymmetrical; in face view semicircular with one side flattened, 34–46.5 μm deep (measuring across from flattened side) × 33.5–54 μm wide; in side view biconvex, 20.5–39 μm thick; surface ornamented with concentric to irregular ridges (3–5 μm high) separated by depressed areas (3–5 μm wide) with transverse striae, ridges often beginning irregularly on one sporangial face, continuing concentrically around the sides of the sporangium, and ending irregularly or concentrically on the opposite face. Dehiscence slit in depressed area along convex side of sporangium opposite flattened side. In optical section: face view, ridges appearing at sporangial periphery as wing with internal, vertical striae; side view, ridges as a series of 8-12 knobs encircling the sporangium. Wall 1.5–3.5 μm thick. Thin-walled sporangia ellipsoid to allantoid, 26–33.5 × 33.5–48 μm.

Figs. 182–183. *Coelomomyces anophelesicus.* Resting sporangia from neotype in *Anopheles subpictus* from Sonarpur, India. Sporangia mounted in Canada balsam. Fig. 182. Bar = 10 μm. Fig. 183. Bar = 1 μm.

Figs. 184–187. *Coelomomyces anophelesicus.* Resting sporangia from *Anopheles vagus* collected in Trivandrum, India. Figs. 184 and 185. Surface (Fig. 184) and midoptical section, face view (Fig. 185) of same three sporangia. Bar = 10 μm. Fig. 186. Midoptical section face view. Note prominent teeth at periphery. Bar = 10 μm. Fig. 187. Midoptical section side view.

Figs. 188–190. *Coelomomyces anophelesicus.* Resting sporangia from *Anopheles vagus* collected in Trivandrum, India. Figs. 188 and 189. Side views showing location of dehiscence slit (arrows). Bar = 10 μm. Fig. 190. Side view of go-stage sporangium. Bar = 10 μm.

Figs. 191–192. *Coelomomyces anophelesicus.* Scanning electron micrographs of resting sporangia of neotype. Note location of dehiscence slit evident in Fig. 192 (arrow). Bar = 1 μm.

188
189
190
191
192

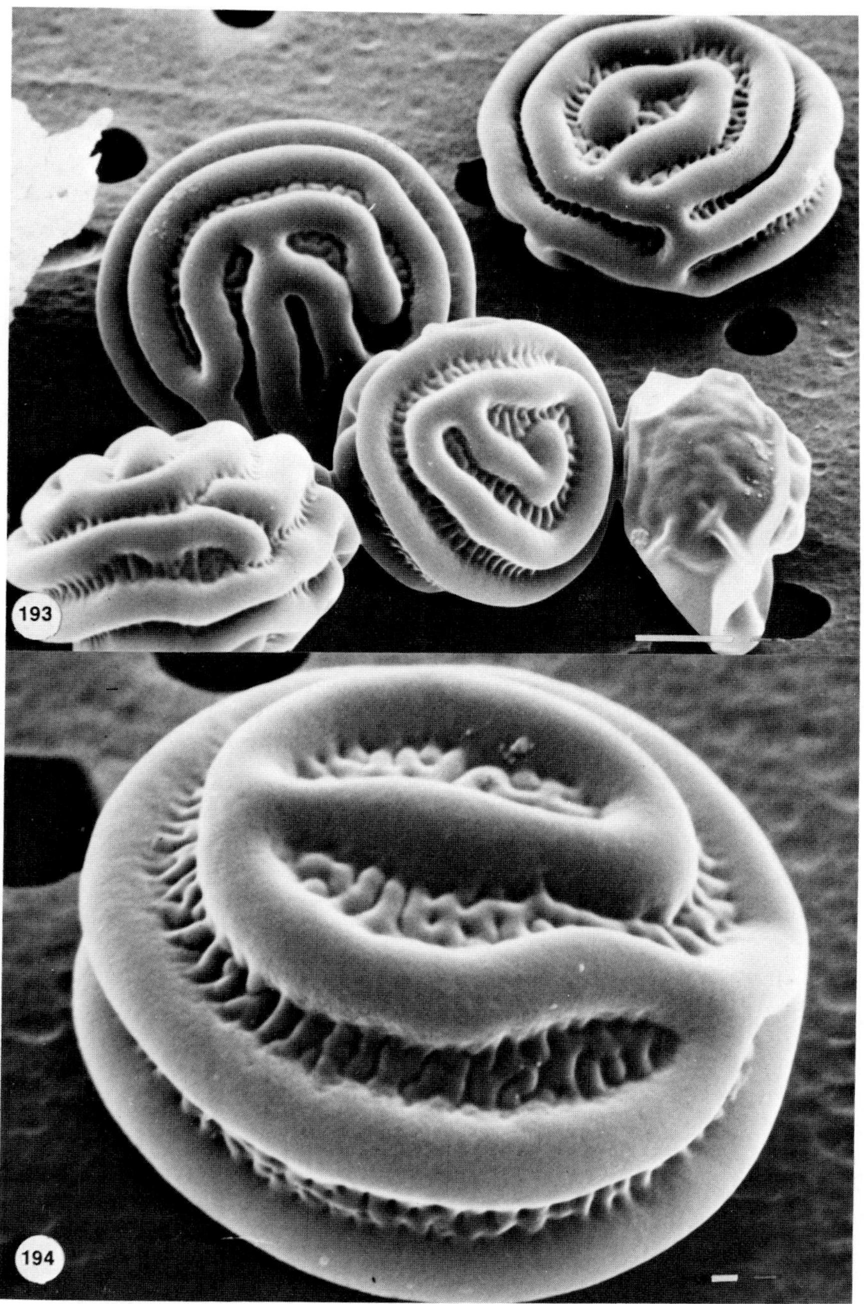

Figs. 193 and 194. *Coelomomyces anophelesicus.* Scanning electron micrographs of resting sporangia from *Anopheles vagus* collected in Trivandrum, India. Fig. 193. Bar = 10 μm. Fig. 194. Bar = 1 μm.

Type Host and Collection Site: The original description (Iyengar, 1935) of *C. anophelesicus* was based on specimens from larvae of *Anopheles subpictus* Grassi, *Anopheles vagus* Donitz, *Anopheles annularis* Wulp, and *Anopheles varuna* Iyengar collected from three locations in Bengal, India (Sonarpur, Krishnagar, and Meenglas). Although the collection site for the type was indicated to be Sonarpur, the type host was not designated (Iyengar, 1962). Since the type, retained in the private collection of Iyengar (decreased) is unavailable, a specimen collected by Iyengar on larvae of *Anopheles subpictus* Grassi from Sonarpur and identified by him as *C. anophelesicus* is here designated a neotype.

Depository for Type: A neotype (see above) is deposited at BPI.

Specimens Examined and Other Collections (indicated by brackets):

Type specimens (examined by Couch in 1962 and returned to Iyengar), **neotype**

India

Anopheles jamesi Theo.; Trivandrum; Iyengar; 1969; NCU.
Anopheles pallidus Theo.; Trivandrum; Iyengar; 1968; NCU.
Anopheles subpictus Grassi; ?; Iyengar; ?; NCU.*
Anopheles subpictus Grassi; Pondicherry; Rajagopalon; 1976–1977; NCU; (Chondrahas and Rajagopalon, 1979).
Anopheles vagus Donitz; Madras; Iyengar; 1970; NCU.
Anopheles vagus Donitz; Bangalore; Iyengar; 1966; NCU.
Anopheles vagus Donitz; Sonarpur; Iyengar; 1970; NCU.
Anopheles vagus Donitz; Trivandrum; Iyengar; 1966, 1968, 1970; NCU.
Anopheles vagus Donitz; Pondicherry; Rajagopalon; 1976–1977; ?; (Chondrahas and Rajagopalon, 1979).

Comments: Although sporangia of *C. anophelesicus* have ornamentation (ridges alternating with depressed areas with striae) similar to that of *C. indicus* and others, this species may be distinguished from other species of *Coelomomyces* by the unique shape of its sporangia. One should note, however, that the original description of this species was based on sporangia that were mounted in balsam, resulting in collapse of the sides and some shrinkage. Therefore, the description is here modified to reflect observations made on living specimens as well as specimens preserved in lactophenol.

* This specimen may be a portion of the type material designated by Iyengar (1935).

Coelomomyces arcellaneus Couch & Lum sp. nov.

Figures 195–200

DESCRIPTION: Hyphae not observed. Resting sporangia discoid; circular in face view, 21–42 μm in diameter, ellipsoid in side view but with dorsal surface slightly flattened, 18–25 × 21–42 μm; surface ornamented with circular punctae on both dorsal and ventral sides (punctae 0.5 μm in diameter and 2–5 μm apart); edges with parallel, vertical, striae running from dorsal to ventral face (striae approximately 3 μm apart); dehiscence slit in prominent, low ridge that is devoid of punctae. During preservation and/or mounting, dorsal face generally collapsing inward to give the sporangium a "bowl shape." Wall 2–3 μm thick. Thin-walled sporangia not observed.

TYPE HOST AND COLLECTION SITE: *Aedes taeniorhynchus* (Wiede.) ♀; Vero Beach, Florida (collected by Lum).

DEPOSITORY FOR TYPE: BPI.

LATIN DIAGNOSIS: Hyphae non visae. Sporangia perennantia discoidea, adversus frontem rotunda, 21–42 μm diametro, adversus latus ellipsoidea, superficie dorsali paulo applanata, 18–25 × 21–42 μm; superficie in regionibus et dorsalibus et ventralibus punctis rotundis 0.5 μm diametro, distantibus 2–5 μm inter se, marginibus striis parallelis, verticalibus ex facie dorsali ad ventralem currentibus, distantibus circa 3 μm inter se, decorata; incisura dehiscentiae in lira prominente, humili, punctis carente in clusa. Per conservationem vel per patulum ad observandum facies dorsalis plerumque introrsus collapsa, quare sporangium crateriforme. Tunica 2–3 μm crassa. Sporangia tenuiter tunicata non visa.

SPECIMENS EXAMINED OTHER THAN TYPE:

United States

- *Aedes taeniorhynchus* (Wiede.)*: Vero Beach, Florida; Lum; 1959; 1960; NCU.
- *Aedes sollicitans* (Walker); Cameron, Louisiana; Chapman; 1966; NCU.

COMMENTS: The unique shape and ornamentation of sporangia of *C. arcellaneus* distinguish it from all known species of *Coelomomyces*. This fungus, along with *C. psorophorae* var. *halophilus*, is also of interest in having been collected only on mosquito hosts from salt marshes. In one of

* Adults and larvae.

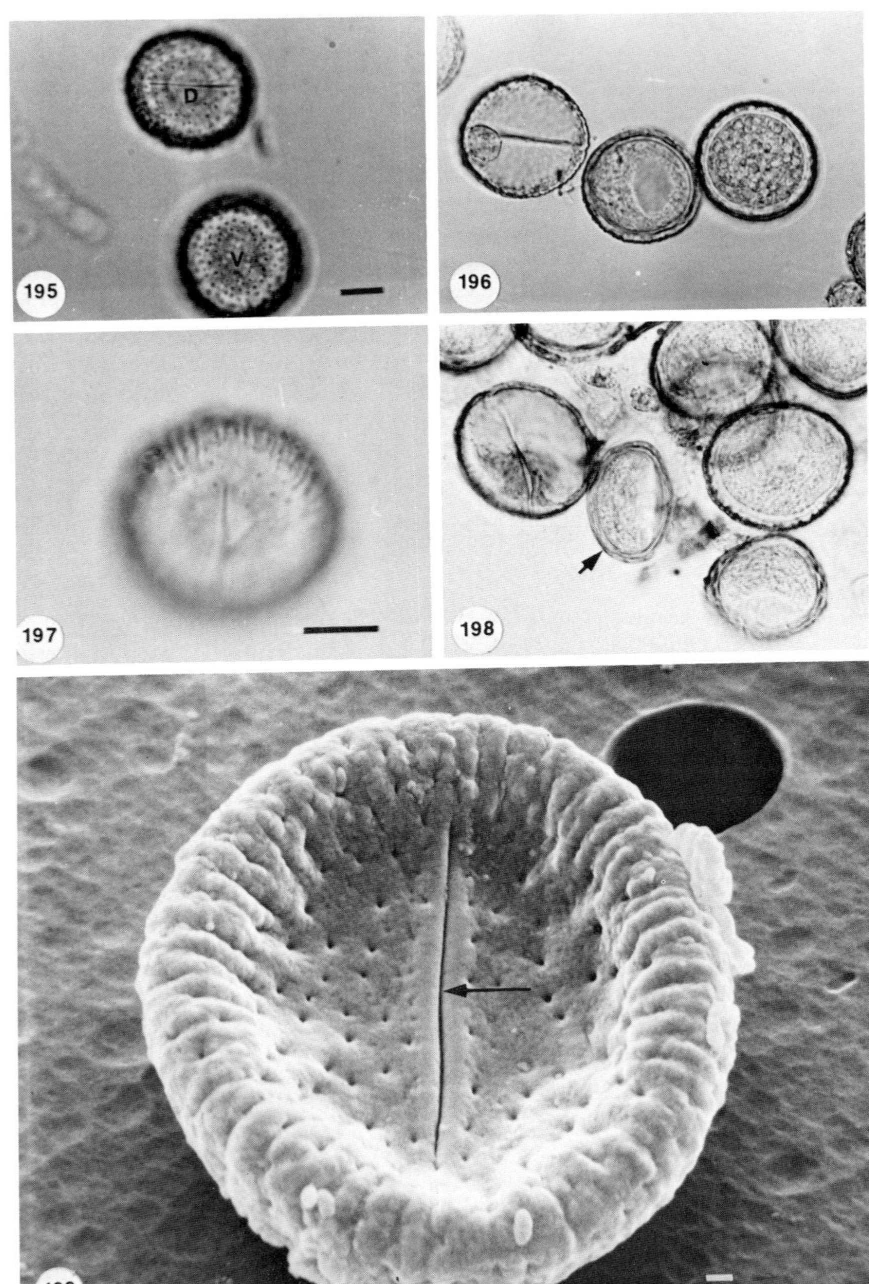

195
D
V
196
197
198
199

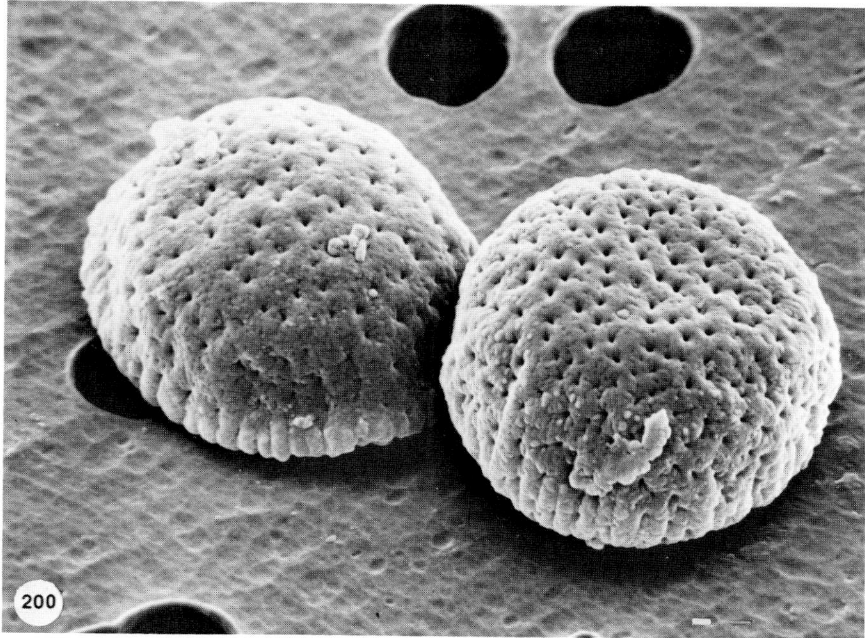

Fig. 200. *Coelomomyces arcellaneus.* Scanning electron micrograph of resting sporangia of type, ventral view. Bar = 1 μm.

these hosts, *Aedes taeniorhynchus,* P. T. M. Lum (personal communication) has observed that in addition to infecting larvae, both male and female adults may be parasitized. For the latter, infection is not restricted to the ovaries, as in most other instances of adult infection, but includes also the entire coelomic cavity. The specific epithet of *C. arcellaneus* is in reference to its resting sporangia, which in some views resemble the protozoan *Arcella.*

Figs. 195–199. *Coelomomyces arcellaneus.* Resting sporangia of type from *Aedes taeniorhynchus* collected in Florida. Fig. 195. Surface views of dorsal (D) and ventral (V) sporangial faces. Bar = 10 μm. Fig. 196. Midoptical section, face view. Scale as for Fig. 195. Fig. 197. Surface view of sporangial edge showing striae. Bar = 10 μm. Fig. 198. Midoptical section, edge view (arrow). Scale as for Fig. 197. Fig. 199. Scanning electron micrograph of collapsed resting sporangium, face view. Note dehiscence slit (arrow). Bar = 1 μm.

Coelomomyces bisymmetricus Couch & Dodge ex Couch, *J. Elisha Mitchell Sci. Soc.* **78,** 135 (1962)

Coelomomyces bisymmetricus Couch & Dodge, *J. Elisha Mitchell Sci. Soc.* **63,** 69 (1947)

Figures 201–210

DESCRIPTION: Hyphae 3.0–10.5 μm thick, sparingly branched and of uneven diameter. Resting sporangia ellipsoid, 23–28 × 34–48 μm; surface ornamented with circular and longitudinal, alternating high and low ridges, originating from a circular plug on ventral side of the sporangium and continuing to the opposite side to end in a single median longitudinal ridge or dorsal side; resulting in a bisymmetrical sporangium with the dehiscence slit occurring along the median longitudinal ridge. Thin-walled sporangia not observed.

TYPE HOST AND COLLECTION SITE: *Anopheles crucians* Wiede.; Moultri, Georgia (collected by S.M.C.G.).

DEPOSITORY FOR TYPE: BPI.

SPECIMENS EXAMINED AND OTHER COLLECTIONS (indicated by brackets):

Type

United States
Anopheles crucians Wiede.; Georgia; SMCG; 1945; NCU.
Anopheles crucians Wiede.; Georgia; ?; 1964; NCU.

COMMENTS: *Coelomomyces bisymmetricus* may be easily recognized by the unique bisymmetry of its sporangia. The infrequent occurrence of irregularly ornamented sporangia in this species may create confusion between it and *C. sculptosporus*. However, in *C. sculptosporus* sporangia are rarely symmetrically ornamented.

Coelomomyces borealis sp. nov. var. ***borealis*** Couch & Service

Figures 211–215

DESCRIPTION: Hyphae not observed. Resting sporangia broad–fusiform, with dorsal side distinctly flattened, 33.5–71 × 63–126 μm; surface ornamented with circular punctae (1–2 μm in diameter and 1–2 μm apart) containing 1 to rarely 2–3 central papillae, which are recessed within the

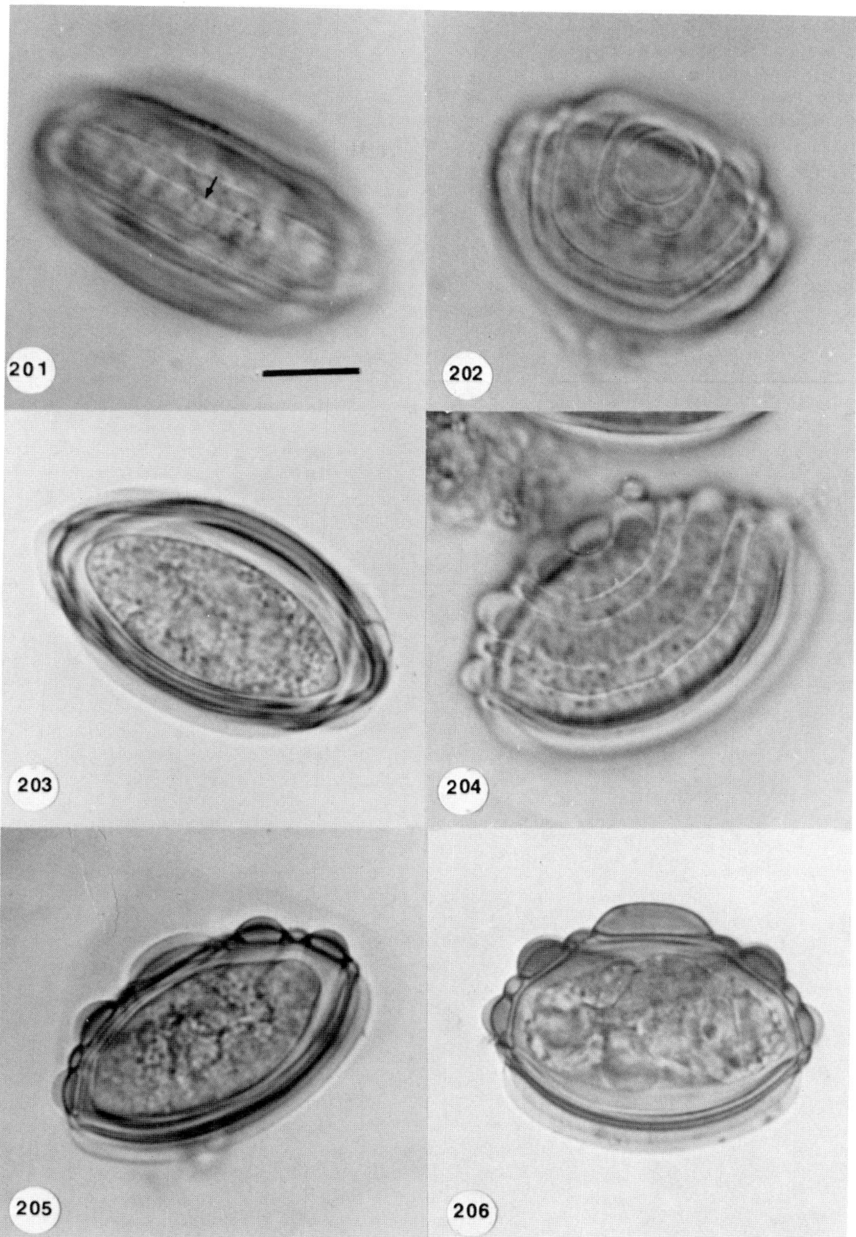

Figs. 201–206. *Coelomomyces bisymmetricus.* Resting sporangia of type from *Anopheles crucians* collected in Georgia. Bar = 10 μm. Fig. 201. Surface view, dorsal side. Note the dehiscence slit (arrow). Fig. 202. Surface view, ventral side. Fig. 203. Midoptical section, face view. Fig. 204. Surface view of side. Figs. 205 and 206. Midoptical section, side view.

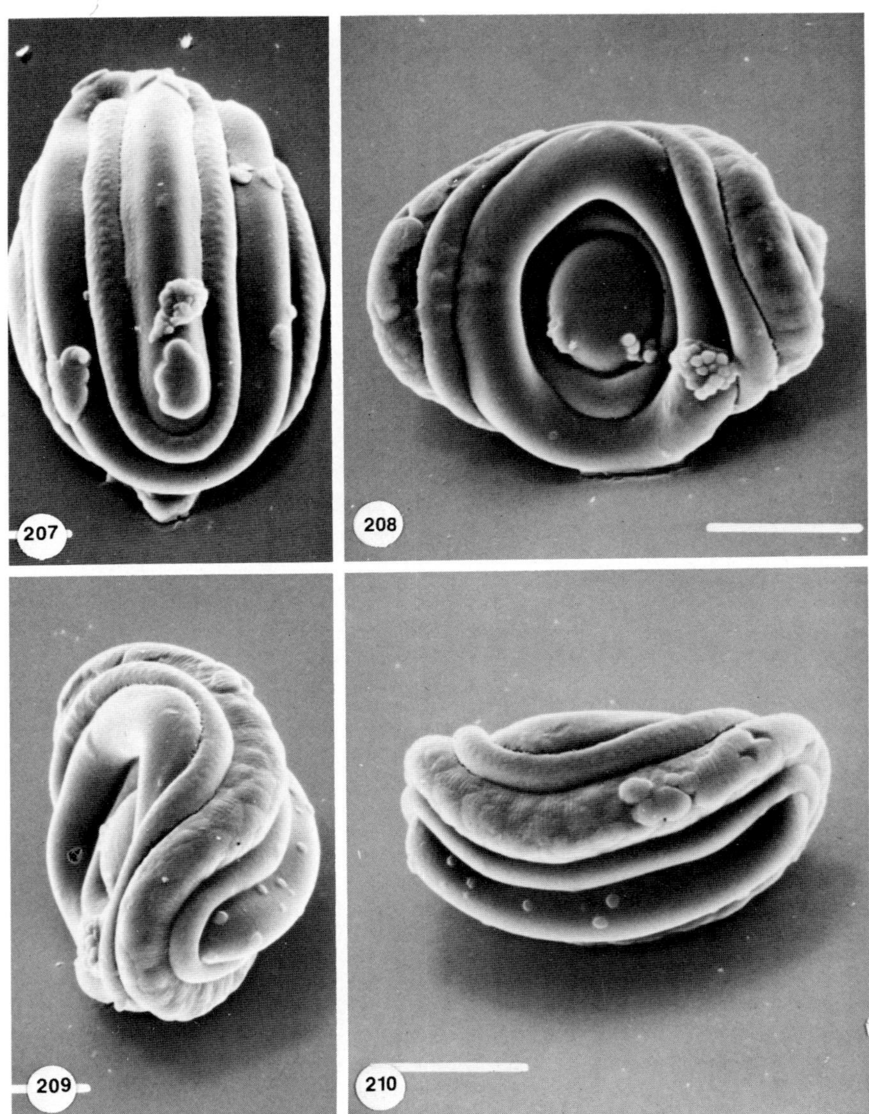

Figs. 207–210. *Coelomomyces bisymmetricus.* Scanning electron micrographs of resting sporangia of type. Fig. 207. Dorsal view. Fig. 208. Ventral view. Fig. 209. End view. Fig. 210. Side view. Fig. 208, bar = 10 μm, applying also to Figs. 207, 209, and 210. Partial scales visible in Figs. 207, 209, and 210 should be ignored.

punctae; punctae, often in rows perpendicular to the dehiscence slit, and in optical section immediately below the wall surface, appearing to have irregular margins. Wall 2.5–6.5 μm thick, with internal, vertical striae when viewed in midoptical section. Thin-walled sporangia not observed.

TYPE HOST AND COLLECTION SITE: *Aedes cantans* (Meigen) ♀; Monks Wood, Huntington, England (collected by Service).

DEPOSITORY FOR TYPE: BPI.

LATIN DIAGNOSIS: Hyphae non visae. Sporangia perennantia late fusiforma, latere dorsali distincte applanato, 33.5–71 × 63–126 μm; superficie punctis rotundis 1–2 μm diametro, 1–2 μm inter se distantibus, papillam centralem unam, raro 2–3, inter puncta conditam includentis; puncta saepe in seriebus recte ad incisuram dehiscentiae ordinata et sectione opticali prope sub tunicae superficie sita, specie marginibus enormibus definita. Tunica 2.5–6.5 μm crassa, in sectione media inspecta striis internis, verticalibus notata. Sporangia tenuiter tunicata non observata.

SPECIMENS EXAMINED OTHER THAN TYPE:

England

Aedes cantans (Meigen) ♀; Monks Wood, Huntington; Service; 1968; NCU; (Service, 1974).

United States

Aedes communis (Deg.) ♀; Orono, Maine; McDaniel; 1972, 1973; NCU.

Aedes excrucians (Walk.) ♀; College, Alaska; Sommerman; 1969; NCU.

COMMENTS: Although sporangial ornamentation of *C. borealis* varieties is similar to that of *C. psorophorae* and *C. stegomyiae,* sporangia of all varieties of *C. borealis* are distinct in being ornamented by punctae with one to three central papillae. In contrast, punctae of *C. psorophorae* are simple and those of *C. stegomyiae* are bordered.

The specific epithet of *C. borealis* is after *boreas* L. (north wind) in recognition of this species having been collected only from northern latitudes.

Coelomomyces borealis var. ***giganteus*** Couch & Bellamy var. nov.

Figures 216–218

DESCRIPTION: Characteristics as for the typical variety except resting

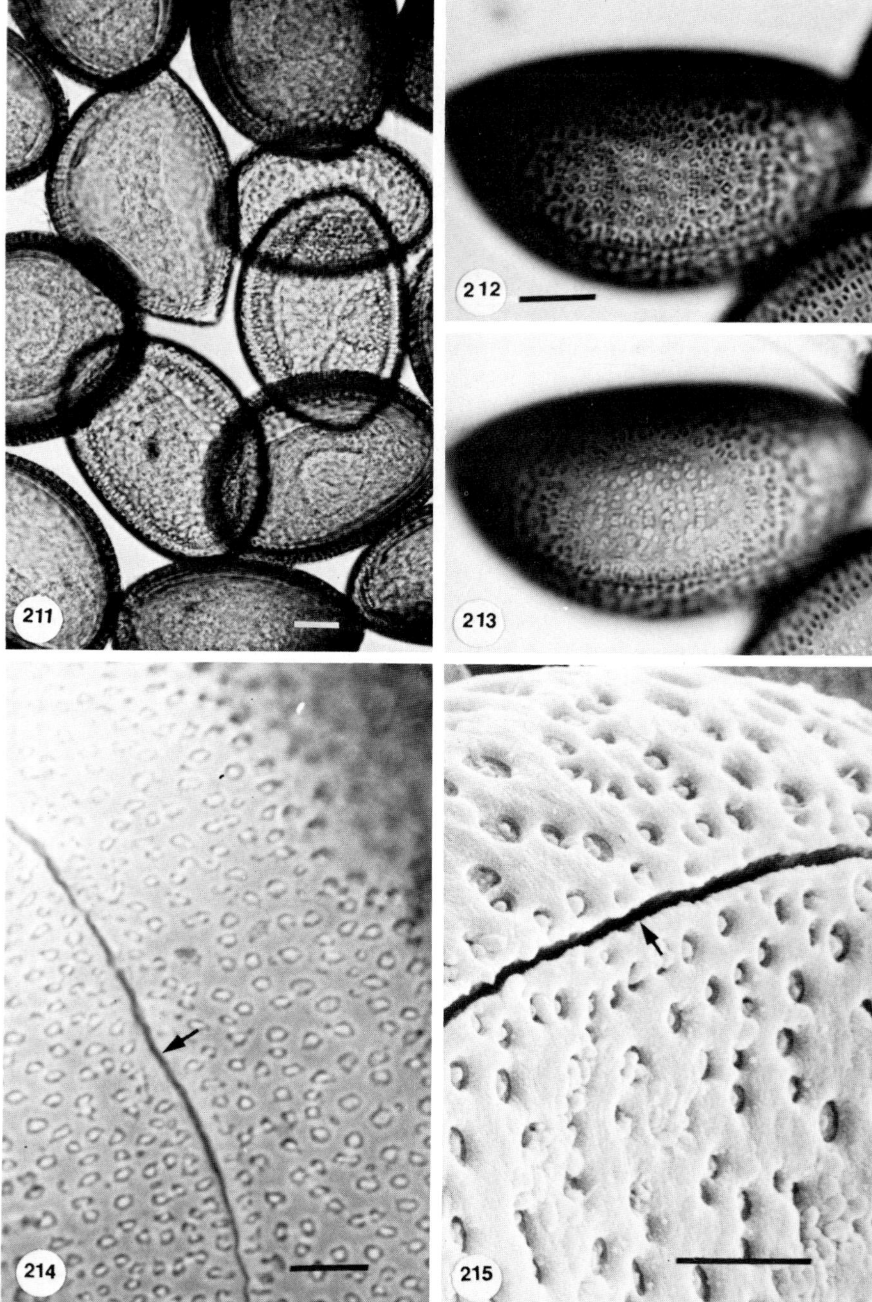
211
212
213
214
215

sporangia 76–134 × 109–182 μm, punctae 1–2 μm wide and 2–5 μm apart, and wall 5.5–7.5 μm thick.

Type Host and Collection Site: *Aedes fitchii* (Felt & Young) ♀; Belleville, Canada (collected by Bellamy).

Depository for Type: BPI. An isotype is on deposit at NCU.

Latin Diagnosis: Varietati typicae par, sporangiis perennantibus 76–134 × 109–182 μm, punctis 1–2 μm latis, 2–5 μm inter se distantibus decoratis, et tunica 5.5–7.5 μm crassa exceptis.

Specimens Examined Other than Type and Isotype:

Canada
Aedes stimulans (Walk.) ♀; Belleville; Ballamy; 1971; NCU.

Comments: This variety is distinguished from the typical variety on the basis of the former's larger sporangia with thicker walls and more widely spaced punctae. Both, however, occur only on mosquitoes from northern latitudes and have practically identical sporangial ornamentation.

The varietal name of *C. borealis* var. *giganteus* is from *giganteus* L. (large) in accord with its having resting sporangia significantly larger than the typical variety.

Coelomomyces cairnsensis Laird ex Laird, *J. Elisha Mitchell Sci. Soc.* **78,** 132 (1962)

Coelomomyces cairnsensis Laird, *Bull. R. Soc. N. Z.* **6,** Wellington, 1 (1956a)

Figures 219–222

Description: Hyphae not observed. Resting sporangia elliptical to elongate ellipsoid, 26.5–35.5 μm broad × 42–61 μm long × 14–26 μm thick; surface ornamented with narrow ridges (1.5–2.5 μm wide × 3.0–5.0 μm high) enclosing hexagonal to polygonal, depressed areas measuring approximately 5–12 μm across (7–10 complete areas and portions of others visible in surface view) and filled with randomly arranged punctae (punctae approximately 0.7 μm in diameter and spaced 1.2–2.5 μm apart). In midoptical section, ridges at sporangial periphery appearing in end

---

Figs. 211–215. *Coelomomyces borealis* var. *borealis.* Resting sporangia from *Aedes commumis* collected in Orono, Maine. Fig. 211. Low-magnification view showing general sporangial morphology. Bar = 10 μm. Figs. 212 and 213. Bar = 10 μm. Surface view of side of the same sporangium showing differing appearance obtained by varying the plane of focus. Figs. 214 and 215. Surface view (Fig. 214) and scanning electron micrograph of resting sporangia of type (Fig. 215) from *Aedes cantans* collected in Monks Wood, England. Dehiscence slit (arrow). Bar = 5 μm.

Figs. 216–218. *Coelomomyces borealis* var. *giganteus.* Resting sporangia of type from *Aedes fitchii* collected in Belleville, Canada. Bar = 10 μm. Figs. 216 and 217. Surface views of sporangial face. Dehiscence slit (arrow). Fig. 218. Scanning electron micrograph.

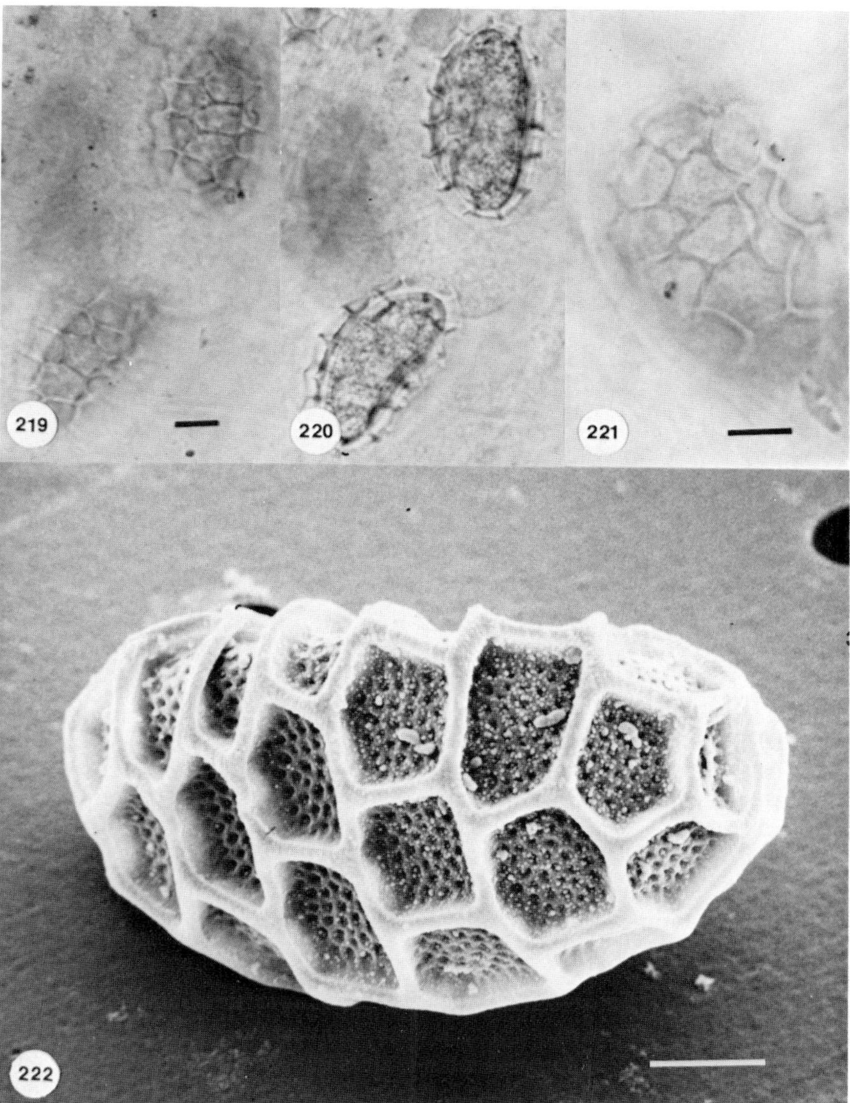

Figs. 219–222. *Coelomomyces cairnsensis.* Resting sporangia of paratype from *Anopheles farauti,* collected in Queensland, Australia. Fig. 219. Surface view of side. Bar = 10 μm. Fig. 220. Same specimen as in Fig. 219, but viewed in midoptical section. Scale as for Fig. 219. Fig. 221. Surface view showing punctae in depressed areas. Bar = 10 μm. Fig. 222. Scanning electron micrograph, ventral view. Bar = 10 μm.

view as triangular teeth and in side view as a wing interconnecting the teeth. Thin-walled sporangia not observed.

TYPE HOST AND COLLECTION SITE: *Anopheles farauti* Loveran; Cairns, Queensland, Australia.

DEPOSITORY FOR TYPE: The type slide, deposited in the collection of the National Museum of New Zealand, Wellington, New Zealand, has been lost (R. K. Dell, personal communication). Paratypes are on deposit at the Department of Entomology, University of Queensland, St. Lucia, Queensland, Australia, and at BPI.

SPECIMENS EXAMINED AND OTHER COLLECTIONS:

Paratypes (UQA and BPI).

COMMENTS: Sporangial ornamentation in *C. cairnsensis* most closely resembles that of *C. finlayae* Laird (1959). Distinction of the two is possible in that sporangia of *C. finlayae* are smaller (17.8–27.5 × 31.5–47.5 μm) and have lower ridges (1.4–2.9 μm high). Similarities exist also in sporangial ornamentation of *C. cairnsensis* and *C. macleayae*. The two may be separated in that ridges of *C. maclaeyae* are parallel on one side of the sporangium, and depressed areas between ridges of sporangia of *C. macleayae* have pits arranged in rows to give the impression of striae.

While *C. cairnsensis* is undoubtedly a valid and easily recognizable species, the description (Laird, 1956a, 1962) of one sporangial face as concave and with convoluted ridges is a questionable diagnostic feature since this is exactly the appearance such a surface would assume if it collapsed, a common phenomenon in preserved sporangia mounted in Canada balsam or Euparal (as was the case for those of *C. cairnsensis* on which the species description was based). Considering this, and after thorough study of the paratype, the description is here modified to exclude this feature. Additionally, Laird's suggestion (1959b) that the plasma membrane in this species is persistent and conspicuous in optical section is thought to be a misinterpretation of a side view of the ridges, which in optical section appear as wings at the periphery of the sporangium.

Coelomomyces canadensis (Weiser & McCauley) Nolan, *Can. J. Bot.* **56,** 2303 (1978)

Coelomomyces chironomi var. *canadense* Weiser & McCauley, *Can. J. Zool.* **48,** 65 (1971); **50,** 365 (1972)

Figures 223–226

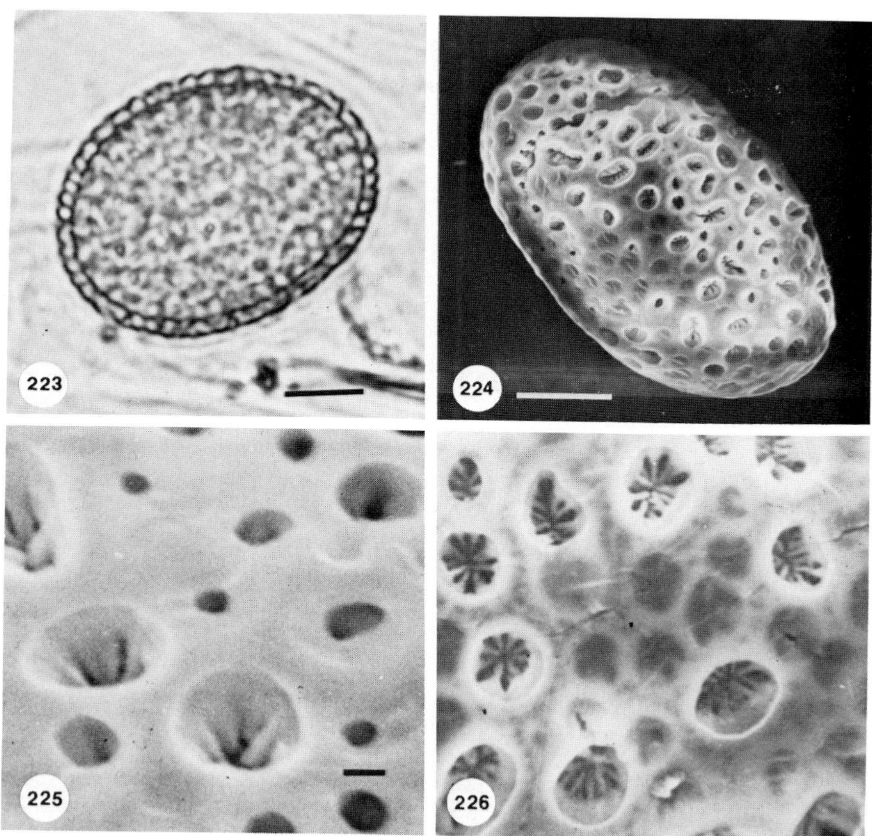

Figs. 223–226. *Coelomomyces canadensis.* Resting sporangia from the chironomid *Psectrocladius* sp. G. collected in British Columbia, Canada. Fig. 223. Midoptical section. Figs. 224–226. Scanning electron micrographs. Bar: Figs. 223 and 224, 10 μm; Figs. 225 and 226, 1 μm. [Micrographs from Nolan (1978).]

DESCRIPTION: Hyphae not observed. Resting sporangia ellipsoid, broadly ellipsoid, to rarely subglobose; 33.8–68 × 26.7–45 μm; wall 2.3–2.9 μm thick; surface ornamented with a network of ridges which delineate irregular, polygonal areas of 5–6 sides, measuring 1.4–2.9 μm across, and possessing deep furrows radiating out from a common center or an elongated central furrow (SEM only). Thin-walled sporangia not observed.

TYPE HOST AND COLLECTION SITE: *Psectrocladius* sp. G.; collected by J. E. McCauley, Marion Lake, British Columbia, Canada.

DEPOSITORY FOR TYPE: Insect pathogen collection of J. Weiser, Institute of Entomology, Czechoslovak Academy of Sciences, Prague, Czechoslovakia. An isotype is deposited at BPI.

SPECIMENS EXAMINED:

Isotype (BPI).

COMMENTS: Resting sporangia of *C. canadensis* are most similar to those of *C. chironomi* Rašín in that both are onamented with ridges enclosing polygonal areas. Distinction of the two via light microscopy may be made in that sporangia of *C. chironomi* are smaller (25 × 40 μm), have a thinner wall (1.2 μm), and have larger, more regular polygonal areas (3 × 3–5 μm across). Scanning electron microscopy reveals the depressed polygonal areas of *C. canadensis* to be more rounded than those of *C. chironomi* and to contain deep furrows radiating from a central pit or furrow, rather than a reticulate pattern of grooves as typical of *C. chironomi*.

Coelomomyces canadensis is one of a small group of *Coelomomyces* spp. that have been collected only from larval chironomids. Other species collected solely from chironomids include: *C. beirnei* (see Incompletely Known and Excluded Species), *C. chironomi,* and *C. tuzetae*.

Nolan (1978) reports that one potential complication in studies of this fungus is the fact that the membrane within which the sporangium is formed is often retained around the sporangium, resulting in an impression of simple pits in the sporangial wall.

The specific epithet of *C. canadensis* is here gramatically corrected from the original designation of *C. canadense* (Nolan, 1978).

Coelomomyces carolinianus Couch, Umphlett, & Bond sp. nov.

Figures 227–233

Figs. 227–233. *Coelomomyces carolinianus.* Resting sporangia of type from *Anopheles punctipennis* collected in Chapel Hill, North Carolina. Bar = 10 μm. Fig. 227. Face view showing dehiscence slit (arrow). Figs. 228 and 229. Surface view of side. Bar = 10 μm. Fig. 230. Midoptical section, face view. Scale as for Fig. 228. Fig. 231. Midoptical section, side view. Scale as for Fig. 228. Fig. 232. Midoptical section, end view. Note dehiscence slit in prominent dorsal ridge (arrow). Scale as for Fig. 228. Fig. 233. Scanning electron micrograph, dorsal views. Bar = 10 μm.

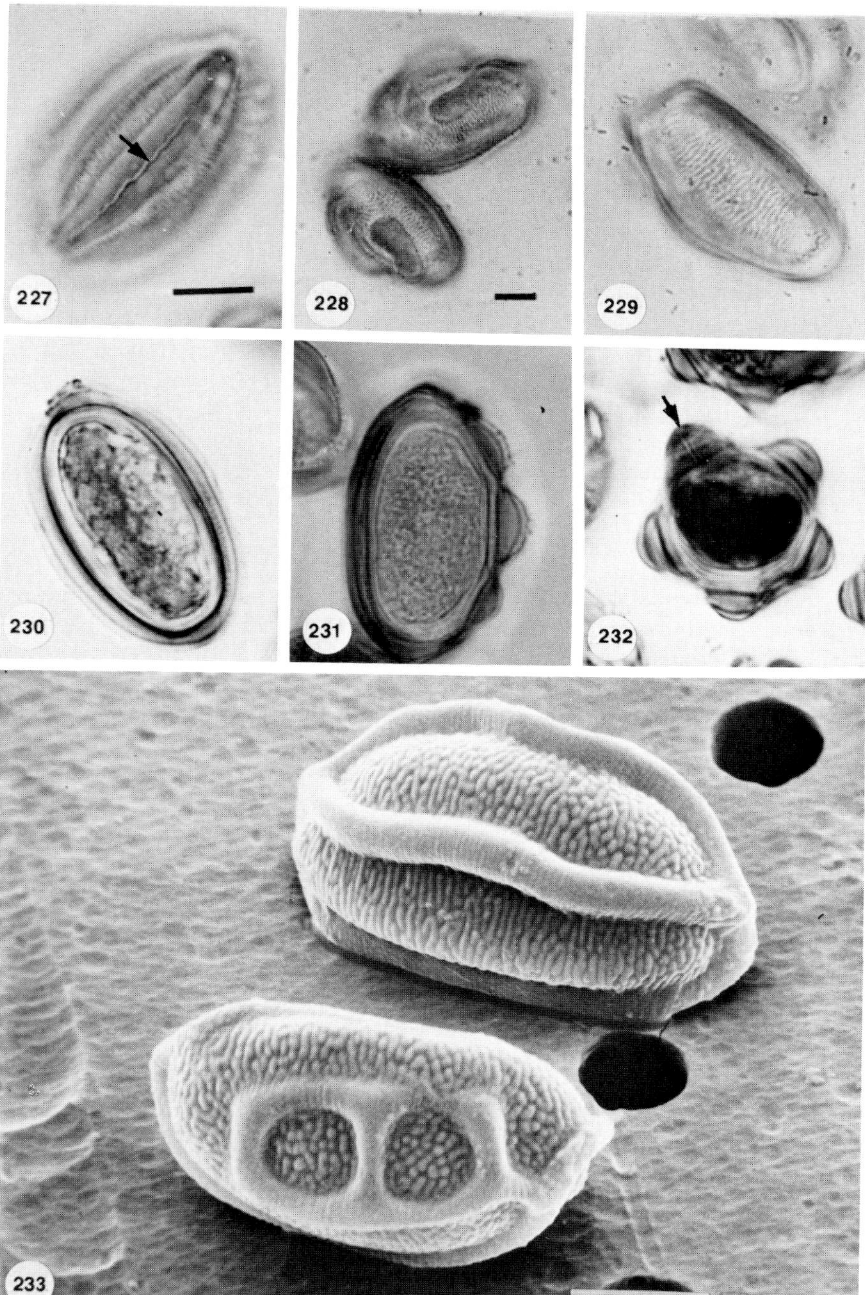
227
228
229
230
231
232
233

DESCRIPTION: Hyphae too distorted for measurement. Resting sporangia ellipsoid, 17–26 × 32–50 μm; surface ornamented with elongate to irregular, anastomosing to isolated ridges (3.5–4.5 μm high and 3.5–5 μm wide) enclosing elongate to irregular depressed areas (5–10 μm wide) with prominent, irregular to anastomosing transverse striae; dehiscence slit centrally located along an elongate ridge; ridges on dorsal surface mostly elongate and anastomosing at poles, becoming irregular and/or isolated on ventral surface. In midoptical section: side view, ridges appearing as a wing along dorsal edge and as short-to-elongate domes along ventral edge; end view, ridges as four to five domes at sporangial periphery giving the sporangium a rarely quadrangulate to usually pentangulate appearance. Thin-walled sporangia not observed.

TYPE HOST AND COLLECTION SITE: *Anopheles punctipennis* (Say); Chapel Hill, North Carolina (collected by Umphlett and Bond).

DEPOSITORY FOR TYPE: BPI.

LATIN DIAGNOSIS: Hyphae non mensae quia nimis distortae. Sporangia parennantia ellipsoidea, 17–26 × 32–50 μm, superficie liris elongatis vel enormibus, anastomosantibus vel sejunctis, 3.5–4.5 μm altis, 3.5–5 μm latis, includentis areolas elongatas vel enormes depressas 5–10 μm latas, striis transversis enormibus vel anastomosantibus decoratas; incisura dehiscentiae per lirae elongatae medium locata; in superficie dorsali lirae plerumque elongatae et ad polos anastomosantes, in ventrali enormes vel sejunctae evadentes; quasi in sectione media inspecta a latere liris instar alae per marginem dorsalem, et tholorum brevium vel elongatorum per ventralem; ab cacumine instar tholorum quattuor vel quinque ad marginem sporangii speciem efficientium raro quadrangularem vel plerumque quinquangularem. Sporangia tenuiter tunicata non observata.

SPECIMENS EXAMINED OTHER THAN TYPE:

United States

Anopheles punctipennis (Say); Chapel Hill, North Carolina; Umphlett & Bond; 1963, 1964, 1965; NCU.

COMMENTS: Although appearing somewhat similar to those of *C. quadrangulatus*, sporangia of *C. carolinianus* may be distinguished on the basis of their irregular to isolated pattern of ridges occurring on the ventral surface. Ridging in *C. quadrangulatus* is generally uniform resulting in symmetrical, quadrangulate sporangia. However, the irregular ridging of *C. carolinianus* results most often in sporangia that appear pentangulate when observed in end view.

The specific epithet of *C. carolinianus* is in recognition of this species having been collected only in North Carolina.

Coelomomyces celatus Couch & Hembree sp. nov.

Figures 234–240

DESCRIPTION: Hyphae not observed. Resting sporangia ellipsoid, slightly flattened on dorsal side, 20.5–26 × 33.5–39 μm; surface ornamented with elongate to reticulate ridges (1–2.5 μm high and 2–3 μm wide) enclosing elongate to smoothly polygonal depressed areas (8–12 μm at widest point and 14–39 μm long) with central, irregular punctae/striae (punctae/striae absent from a 3–4 μm wide border at periphery of depressed areas); dehiscence slit central along a longitudinal ridge. In mid-optical section; side view, ridges as a continuous wing at sporangial periphery or as a wing along dorsal edge and scattered domes at poles and along ventral edge. Thin-walled sporangia not observed.

TYPE HOST AND COLLECTION SITE: *Uranotaenia campestris* Leic.; Ban Pa Daeng, Lampang, Thailand [collected by AFRIMS; see Hembree (1979), host/pathogen association #15].

DEPOSITORY FOR TYPE: BPI. An isotype is deposited at NCU.

LATIN DIAGNOSIS: Hyphae non visae. Sporangia perennantia ellipsoidea, in latera dorsali paululum applanata, 20.5–26 × 33.5–39 μm; superficie liris elongatis vel reticulatis, 1–2.5 μm altis, 2–3 μm latis, decorata, depressiones elongatas vel aeque polygonas ad 8–12 μm latas, 14–39 μm longas, punctis striisque irregularibus, centralibus, ab depressionum marginibus 3–4 μm latis absentibus, notatas includentibus; incisura dehiscentiae per lirae longitudinalis medium locata, quasi in sectione media inspecta a latere, lirae instar alae continuae ad sporangii marginem vel per marginem dorsalem instar alae et ad polos et per marginem ventralem instar tholorum sparsorum. Sporangia tenuiter tunicata non observata.

SPECIMENS EXAMINED OTHER THAN TYPE AND ISOTYPE: None.

COMMENTS: The description of *C. celatus* is based on a single collection of this species on a larva of *U. campestris* from a small pool in a dry stream bed (Hembree, 1979). As originally mounted in balsam, resting sporangia of this species were badly distorted and unsuitable for study. Subsequent remounting of some sporangia in lacophenol and preparation of others for scanning electron microscopy yielded specimens exhibiting

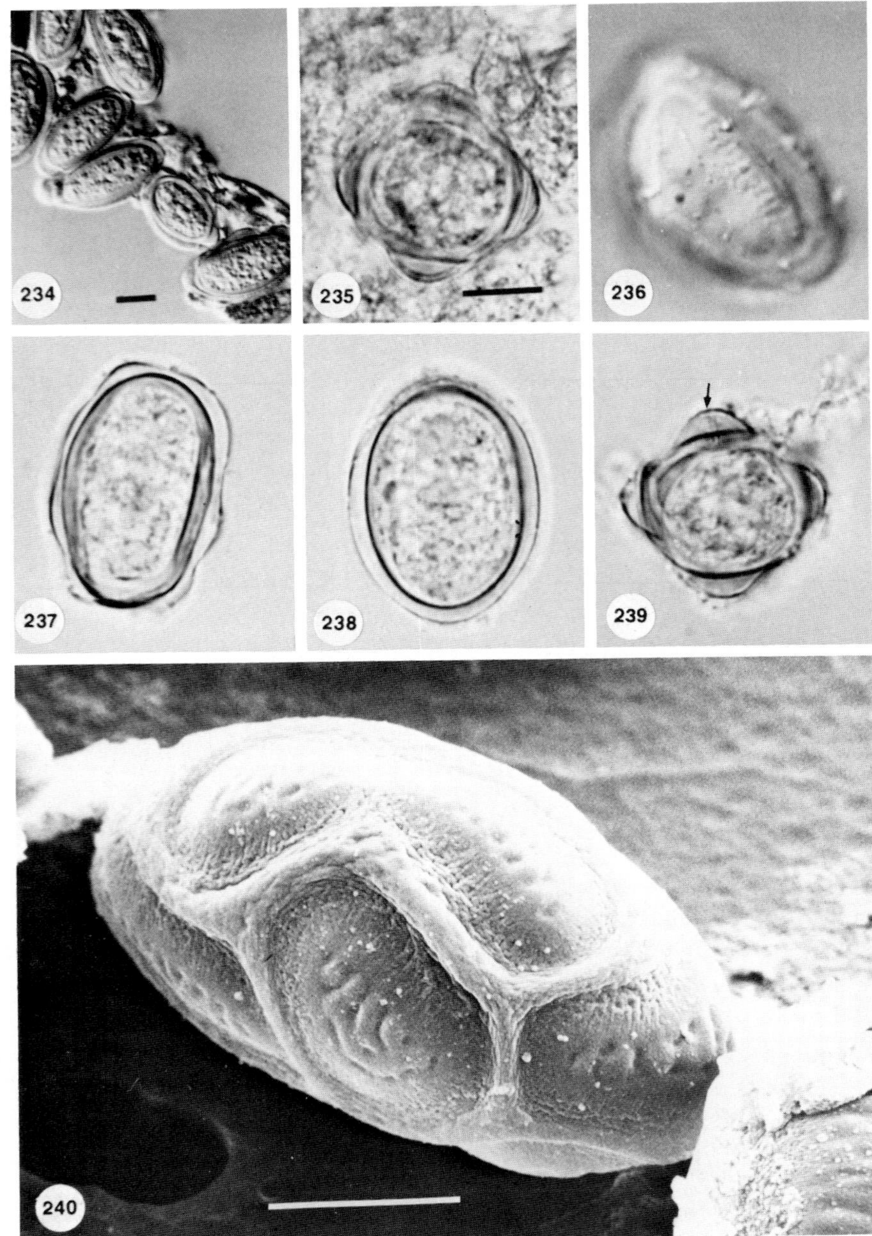
234
235
236
237
238
239
240

sporangial characteristics distinguishing them as belonging to a new species. However, sporangial morphology in the type material of this species is extremely difficult to discern via light microscopy, perhaps due to the small size of its sporangia or possibly to handling procedures. The specific epithet for *C. celatus* is from *celatus* L. (hidden or concealed), in recognition of this difficulty in distinguishing sporangial ornamentation.

Although sporangial ornamentation in *C. celatus* is somewhat similar to that of *C. thailandensis*, distinction from the latter may be made in that sporangia of *C. celatus* are smaller and have a different pattern of ridges enclosing depressed areas.

Coelomomyces chironomi Răsín, *Věstn. Sjezdu Česk. Přírodozp.* **3,** 146 (1928)

Figures 241–247

DESCRIPTION: Hyphae compact, 20–100 μm thick, during early development; becoming branched and filamentous, 10–40 μm thick. Resting sporangia ellipsoid, 18–25 × 27–42 μm; surface ornamented with a reticulum of anastomosing ridges enclosing regular polygonal depressions, measuring 2–3.5 μm across, and containing irregular, shallow grooves or striae (only faintly visible by LM). In midoptical section, ridges appearing as triangular teeth around periphery of the sporangium. Dehiscence slit obvious, bordered by polygonal depressions. Wall approximately 1–1.5 μm thick. Thin-walled sporangia not observed.

NEOTYPE HOST AND COLLECTION SITE: *Chironomus paraplumosus* (L.); Radov, CSSR (collected by J. Weiser and designated neotype; examined by Răsín).

DEPOSITORY FOR NEOTYPE: Insect pathogen collection of J. Weiser, Institute of Entomology, Czechoslovak Academy of Sciences, Prague, Czechoslovakia.

A specimen from the neotype "series"* collected on C. *paraplumosus*

* So designated by Weiser.

Figs. 234–240. *Coelomomyces celatus.* Resting sporangia of type from *Uranotaenia campestris* collected in Lampong, Thailand. Fig. 234. Low-magnification view showing general sporangial morphology in midoptical section. Bar = 10 μm. Fig. 235. Midoptical section, end view. Dehiscence slit (arrow). Bar = 10 μm. Fig. 236. Face view of side showing striae between ridging. Scale as for Fig. 235. Figs. 237 and 238. Midoptical sections, face view. Scale as for Fig. 235. Fig. 239. Midoptical section, end view. Fig. 240. Scanning electron micrograph, dorsal view. Bar = 10 μm.

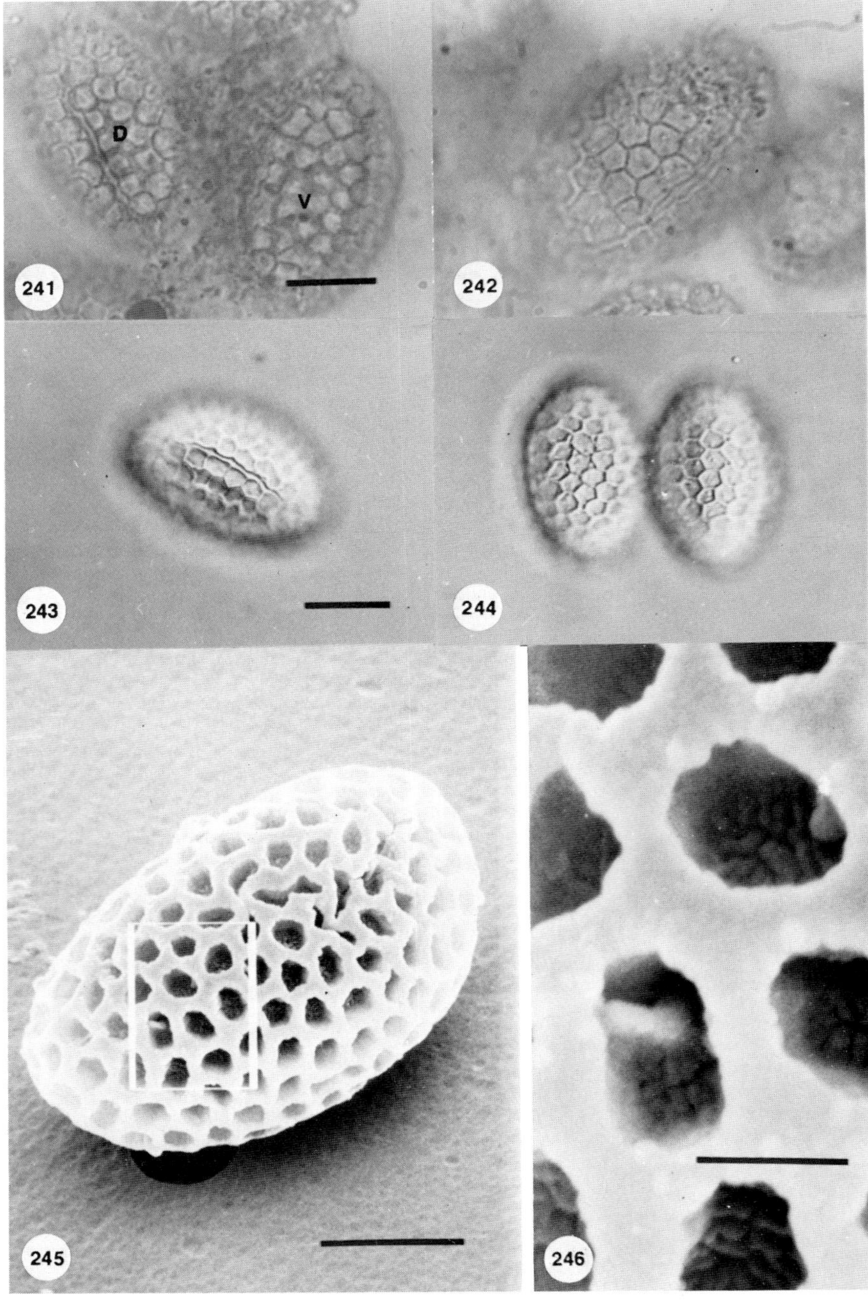
D
V
241
242
243
244
245
246

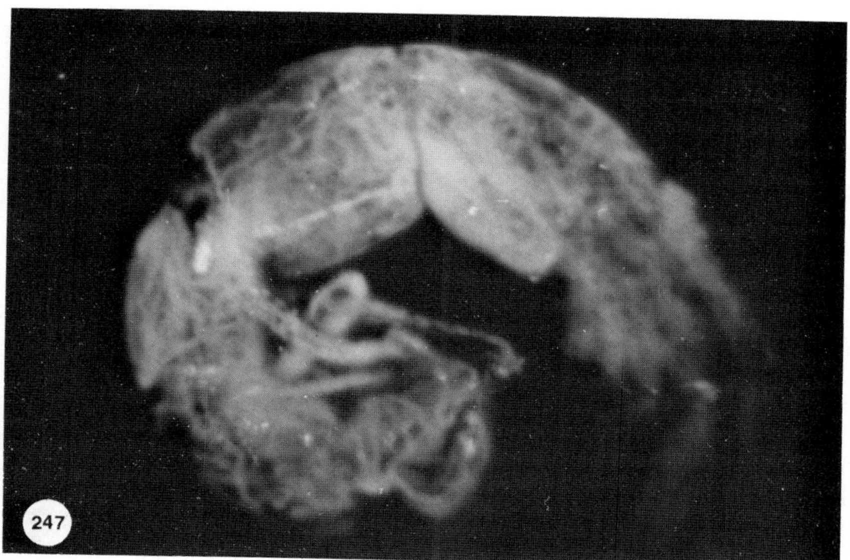

Fig. 247. *Coelomomyces chironomi.* Infected larva of *Chironomus plumosus.* Note hyphae visible inside the larva. [Photo courtesy of J. Weiser.]

from Radov, CSSR, by J. Weiser and examined by Răsín, has been deposited in the BPI.

ALTERNATE HOST: *Heterocypris incongruens* Sars (Weiser, 1976).

SPECIMENS EXAMINED AND OTHER COLLECTIONS (indicated by brackets):

Neotype "series"* (BPI)

Czechoslovakia

[*Chironomus paraplumosus* (L.); Brno; Răsín; 1928; ?; (Răsín, 1928)].

Chironomus plumosus (L.); Blatna; Weiser; 1962; NCU; (Weiser and Vávra, 1964).

* So designated by Weiser.

Figs. 241–246. *Coelomomyces chironomi.* Resting sporangia of type from *Chironomus paraplumosus* collected in Radov, CSSR. Fig. 241. Surface views of dorsal (D) and ventral (V) surface of two sporangia. Bar = 5 μm. Fig. 242. Side view of sporangium. Bar = 5 μm. Fig. 243 and 244. Face view dorsal (Fig. 243) and ventral (Fig. 244) surface. Note dehiscence slit in Fig. 243. Bar = 10 μm. Figs. 245 and 246. Scanning electron micrographs. The area indicated on Fig. 245 (bar = 10 μm) is enlarged to Fig. 246 (bar = 2 μm).

Chironomus plumosus (L.); Bilek; Weiser; 1966; NCU; (Weiser, 1976, 1977a).

COMMENTS: Described originally by Răsín (1928), *C. chironomi* has been further described and illustrated in subsequent works (Răsín, 1929; Weiser and Vávra, 1964). Additionally, Weiser (1976, 1977) identified the ostracod *Heterocypris incongruens* as the intermediate host in the life cycle of *C. chironomi*.

Although sporangial ornamentation in *C. chironomi* is similar to that of *C. canadensis* (see Comments for *C. canadensis*), it is easy to distinguish *C. chironomi* from this and other species of *Coelomomyces* on the basis of its unique pattern of wall ornamentation: i.e., anastomosing ridges enclosing regular polygonal depressed areas containing grooves/striae. A possible complication in identifying this taxon has been reported by Weiser and Vávra (1964) to be the frequent occurrence of teratosporangia in some collections. Such teratosporangia are characterized by irregularities in the pattern of ridges covering the sporangial surface, including diagonally twisted to parallel ridges to totally irregular small areas on some spherical sporangia.

Coelomomyces couchii Nolan & Taylor, *J. Med. Entomol.* **16,** 297 (1979)

Figures 248–253

DESCRIPTION: Hyphae not observed. Resting sporangia ellipsoid, 32–27 × 48–59 μm, slightly flattened on side with dehiscence slit; surface ornamented with a reticulate pattern of low ridges (1.0–2.0 μm high × 2.2–4.9 μm wide) enclosing circular to elongate depressed areas (8.0 × 8.0–20 μm) with scattered punctae, but with an elongate depressed area with punctae surrounding the dehiscence slit (33–46 μm long), which is located in a ridge along the flattened side of the sporangium. In midoptical section: side view, ridging as an elongate, low dome along dorsal edge and as domes alternating with depressed areas at poles and along ventral edge; face view; ridging as low domes alternating with depressed areas around

---

Figs. 248–253. *Coelomomyces couchii.* Resting sporangium of type from *Anopheles farauti* collected on Gizo Island, Solomon Islands. Fig. 248. Specimen on original type slide mounted in Euparal. Note collapsed, boat-shape of the sporangium. Bar = 10 μm. Figs. 249–252. Specimens from type slide, but rehydrated and mounted in lactophenol. Figs. 249 and 250. Surface views of dorsal side showing dehiscence slit (arrow). Fig. 251. Surface view of ventral side. Fig. 252. Midoptical section, side view. Fig. 253. Scanning electron micrograph showing ventral surface and side.

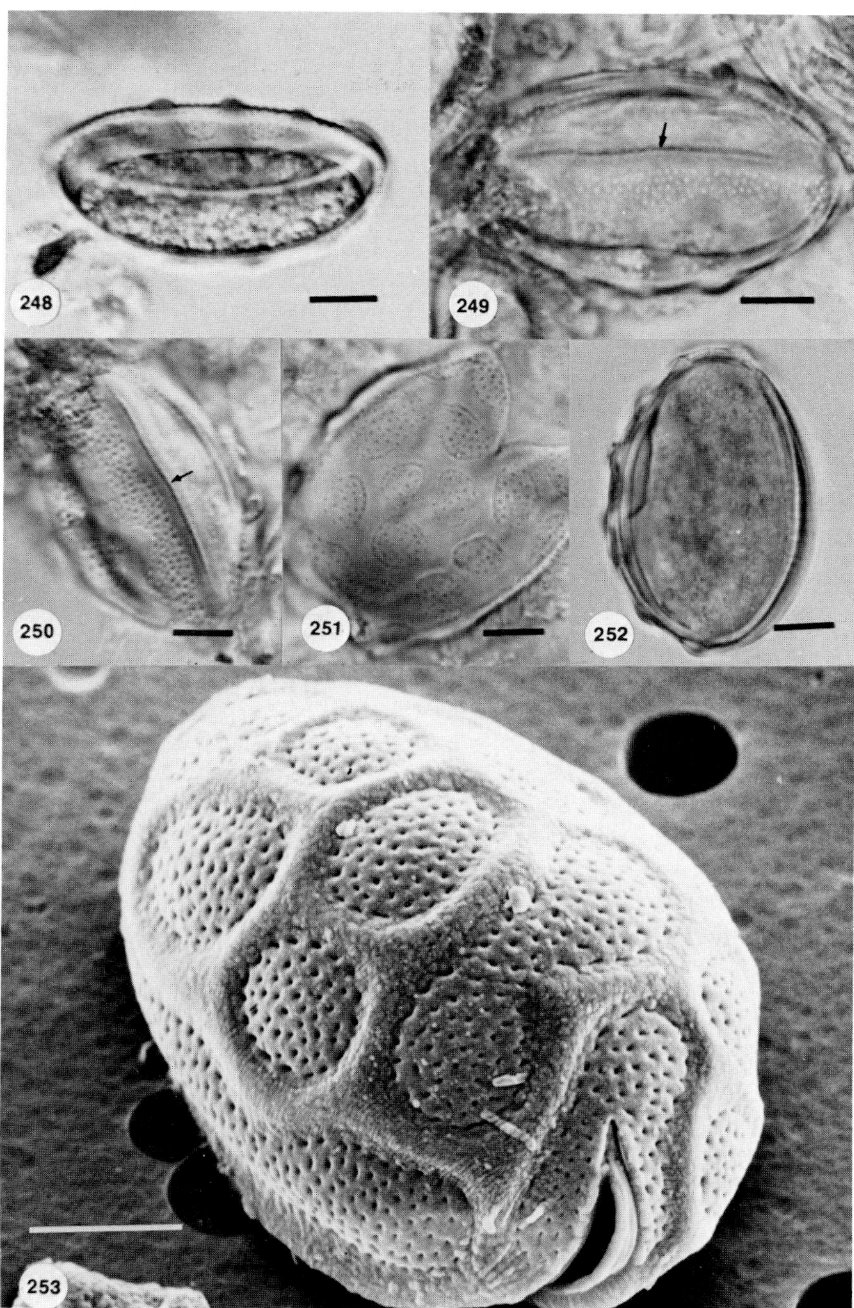
248
249
250
251
252
253

sporangial periphery. Wall approximately 1.0–2.0 μm thick. Thin-walled sporangia not observed.

TYPE HOST AND COLLECTION SITE: *Anopheles* (*Cellia*) *farauti* Loveran; New Manda, Gizo Island, Solomon Islands (Collected by B. Taylor).

DEPOSITORY FOR TYPE: BISH (Accession no. 11,503).

SPECIMENS EXAMINED AND OTHER COLLECTIONS:

Type and isotype (BISH)

Borneo

Culex fraudatrix Leo.; Jesselton, N. Borneo; Colless; 1956; ?; (Laird, 1959b).*

Culex t. summorosus Dyar; Jesselton, N. Borneo; Colless; 1956; ?; (Laird, 1959b).*

Singapore

Culex t. summorosus Dyar; Blakang Mati Island; Colless; 1955; NCU; (Laird, 1959a).*

Thailand

Anopheles navipes (Theo.); Ban Pan Nua, Lampoing; AFRIMS; ?; NCU; (Hembree, 1979).

COMMENTS: The boat-shaped sporangia of *C. couchii* described by Nolan and Taylor (1979) were no doubt a result of the partial collapse of the sporangial wall during mounting in Euparal. Remounting of sporangia from the type (see introduction to this chapter for technique) in water or lactophenol reveals them to be typically oval and to have dimensions somewhat greater than those reported previously (18.6–29.4 × 40.2–55.9 μm) by Nolan and Taylor (1979).

Resting sporangia of *C. couchii* appear similar to those of both *C. lairdi* and *C. reticulatus*. Distinction from each may be made in that in *C. couchii* the elongate ridge bearing the dehiscence slit is bordered on either side by elongate depressed areas, whereas in both *C. lairdi* and *C. reticulatus* this area bears circular to polygonal depressed areas. Also, ridging in sporangia of *C. couchii* is lower and broader than in those of either *C. lairdi* or *C. reticulatus*.

* Identified originally as *C. cribrosus* (Laird 1959b), but studied subsequently by Couch and determined to be *C. couchii*.

Coelomomyces cribrosus Couch & Dodge ex Couch, *J. Elisha Mitchell Sci. Soc.* **78,** 135 (1962)

Coelomomyces cribrosus Couch & Dodge, *J. Elisha Mitchell Sci. Soc.* **63,** 69 (1947)

Figures 254–259

DESCRIPTION: Hyphae 6–11 μm thick. Resting sporangia ellipsoid to slightly allantoid, 24–42 × 42–71 μm; surface ornamented with mostly circular to elongate/irregular, domelike, depressed areas, with one to several central punctae or striae (circular areas 2–6.4 μm in diameter; elongate areas 2–6.4 μm wide and of variable length, occasionally encircling the sporangium), separated by raised, bandlike areas of the wall usually 3–5 μm wide and without punctae/striae; elongate depressed areas usually bordering the dehiscence slit, which is located in the middle of a raised band along one side of the sporangium. In midoptical section, the periphery of the sporangium appears as a series of alternating high and low ridges. Thin-walled sporangia not observed.

TYPE HOST AND COLLECTION SITE: *Anopheles crucians* Wiede.; Rome, Georgia (collected by SMCG).

Paratypes on *Anopheles crucians* Wiede. and *Anopheles punctipennis* Say; Georgia (collected by SMCG).

DEPOSITORY FOR TYPE: BPI. Several paratypes are on deposit at NCU.

SPECIMENS EXAMINED AND OTHER COLLECTIONS:

Type, paratypes

United States

Anopheles crucians Wiede.; North Carolina; Roan and Parks; ?; NCU.
Anopheles sp.; Georgia; ?; 1964; NCU.

COMMENTS: While *C. cribrosus* may be easily distinguished from most other species of *Coelomomyces,* difficulty may arise in separating it from *C. sculptosporus*. In fact, collections of either of these species may contain individual sporangia that are indistinguishable from the other species. In general, however, sporangia of *C. sculptosporus* are more prominently sculptured and have few to usually no circular depressed areas, which, when present, occur only on one side.

Specimens of *Culex t. siamensis* collected from Singapore and of *Culex fraudatrix* and *Culex t. summorosus* from British North Borneo were reported by Laird (1959a,b, respectively) to be parasitized by *C. cribro-*

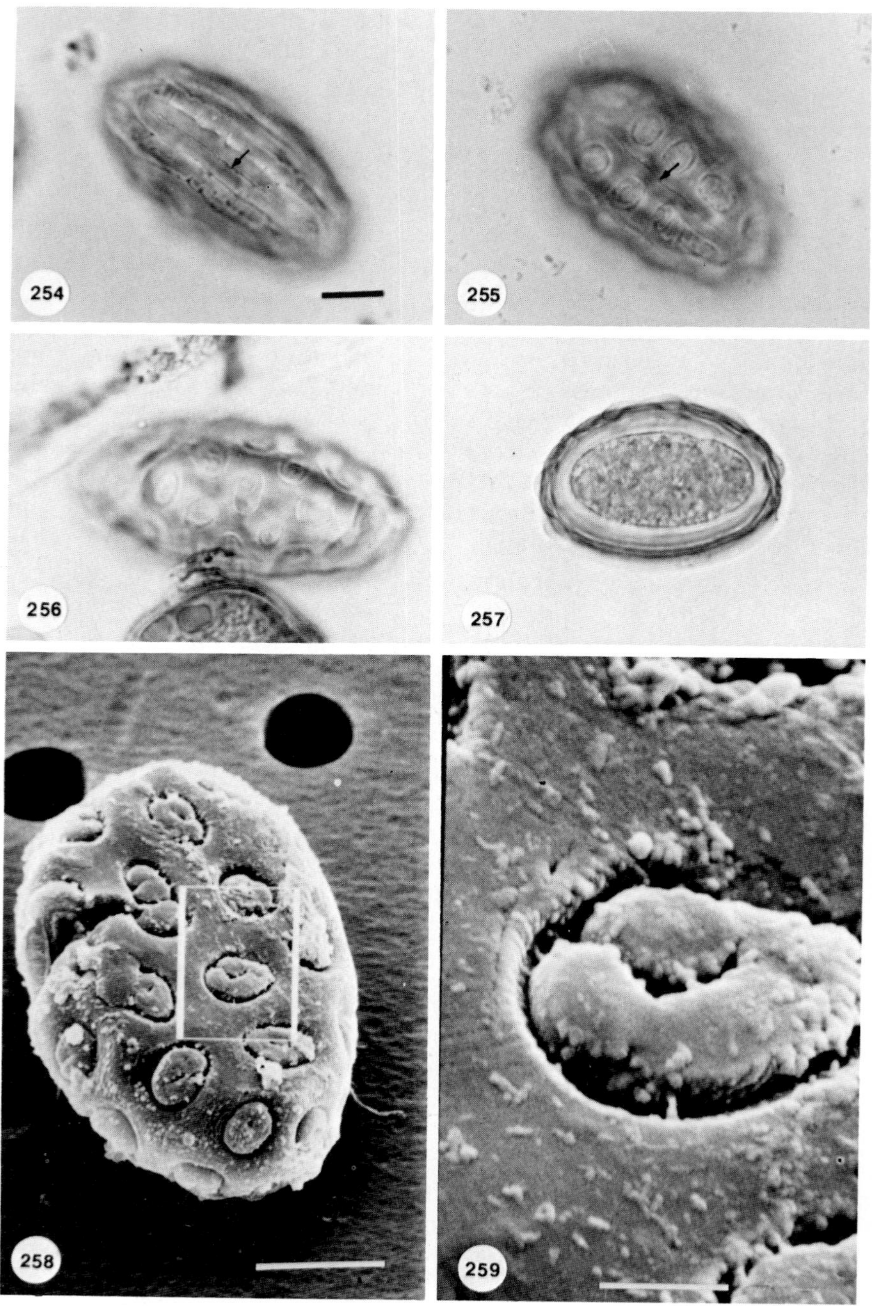
254
255
256
257
258
259

sus. However, examination of these specimens by J. N. C. revealed the species of *Coelomomyces* present to be not *C. cribrosus* but rather *C. couchii*.

Coelomomyces dentialatus Couch & Rajapaksa sp. nov.

Figures 260–266

DESCRIPTION: Hyphae 6–9 μm in diameter. Resting sporangia ellipsoid 19–37 × 33.5–68 μm; surface ornamented with elongate ridges (3–5 μm high and 3–5 μm wide, longitudinal on dorsal surface and anastomosing on ventral surface) with scattered papillae (0.5 μm in diameter); ridges separated by elongate, depressed areas (10–12 μm wide and up to the length of the sporangium) with scattered, slitlike punctae; punctae occurring usually in twos and oriented perpendicular to the ridges, causing the depressed areas to appear striate in some views; dehiscence slit central along a longitudinal ridge. In midoptical section: side view, ridges as finely dentate wing along sporangial periphery; ridges in edge view appearing at poles and on dorsal surface as truncate domes connected by a wing; end view, ridges giving the sporangium a four- to six-angled appearance. Thin-walled sporangia not observed.

TYPE HOST AND COLLECTION SITE: *Aedes* sp.; Kalutara, Sri Lanka (collected by Rajapaksa; see Rajapaksa, 1963, 1964.).

DEPOSITORY FOR TYPE: BPI. An isotype is on deposit also at NCU.

LATIN DIAGNOSIS: Hyphae 6–9 μm diametro. Sporangia perennantia ellipsoidea, 19–37 × 33.5–68 μm; superficie liris elongatis 3–5 μm altis et latis, per superficiem dorsalem longitudinalibus, per ventralem anastomosantibus atque papillis sparsis 0.5 μm diametro decorata; lirae a depressionibus elongatis 10–12 μm latis, usque ad sporangii longitudinem, punctis sparsis, ad instar incisurarum, notatis; puncta plerumque bina, ad liras recta, depressiones ex aspectis quibusdam speciem striatam imponentia; incisura dehiscentiae per lirae longitudinalis medium locata; quasi in sectione media inspecta a latere, lirae instar alae subtiliter dentatae per sporangii marginem, desuper ad polos et in superficie dorsali instar

---

Figs. 254–259. *Coelomomyces cribrosus*. Resting sporangia of type specimen from *Anopheles crucians* collected in Rome, Georgia. Figs. 254 and 255. Surface view, dorsal side. Note prominent dehiscence slit (arrow). Bar = 10 μm. Fig. 256. Surface view, ventral side. Scale as for Fig. 254. Fig. 257. Midoptical section, side view. Scale as for Fig. 254. Figs. 258 and 259. Scanning electron micrographs. Bar = 10 μm. Fig. 259. Enlargement of area indicated in Fig. 258. Bar = 2 μm.

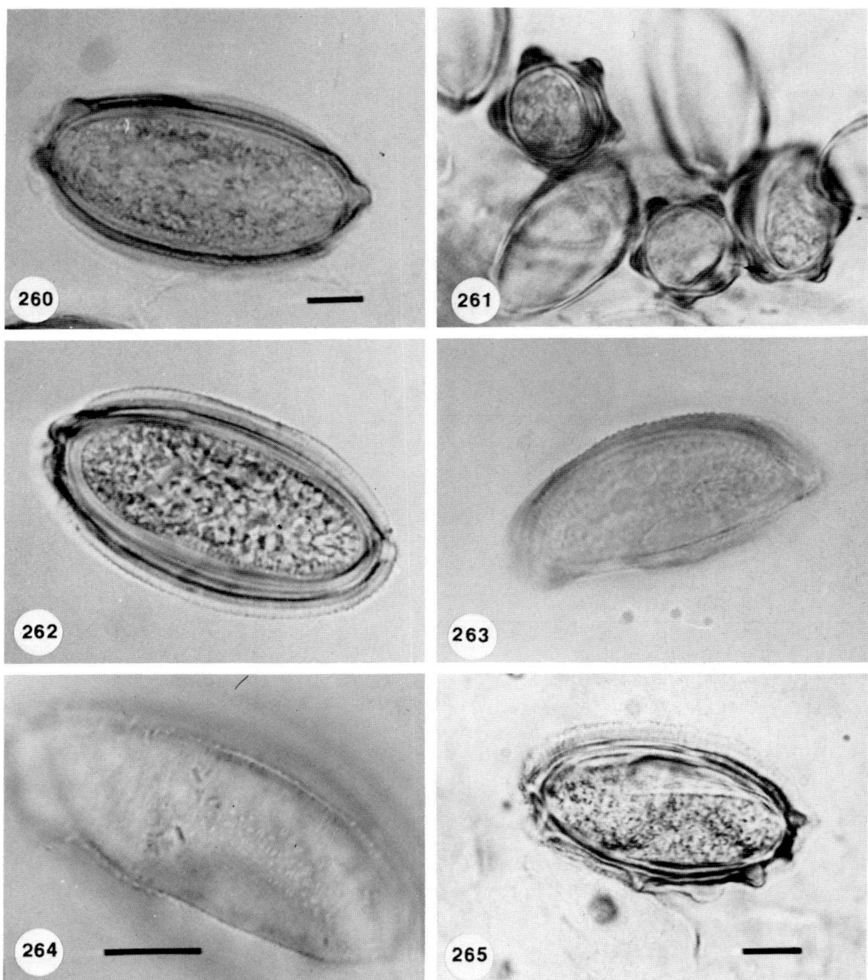

Figs. 260–265. *Coelomomyces dentialatus.* Resting sporangia of type specimen in *Aedes* sp. collected in Kalutara, Sri Lanka. Fig. 260. Midoptical section, side view. Bar = 10 μm. Fig. 261. Midoptical section, end view. Note dehiscence slit (arrow) on one of the specimens. Scale as for Fig. 260. Fig. 262. Midoptical section, face view. Note prominent teeth around lateral ridge. Scale as for Fig. 260. Figs. 263 and 264. Surface view of sporangial side to show bumps on longitudinal ridges and punctae in depressed areas between ridges. Bar = 10 μm. Fig. 265. Midoptical section, side view. Bar = 10 μm.

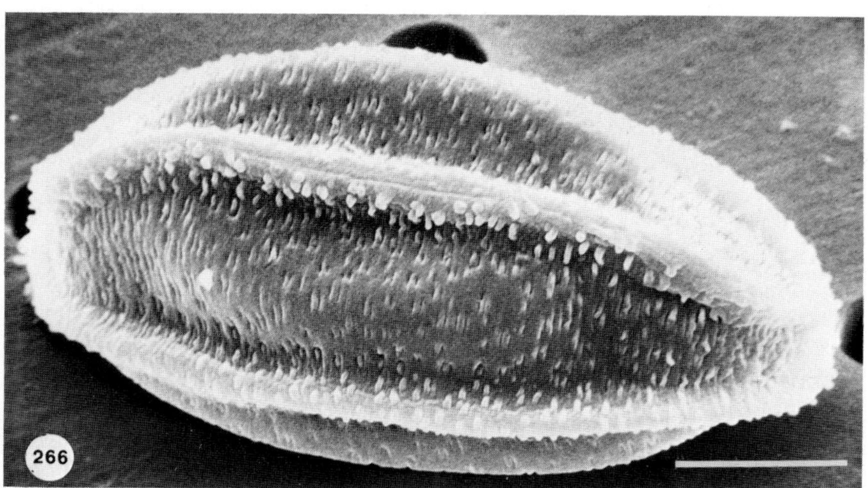

Fig. 266. *Coelomomyces dentialatus.* Scanning electron micrograph of resting sporangia. Note prominent bumps along longitudinal ridges and paired punctae in the depressed areas between ridges. The dehiscence slit is visible along the central longitudinal ridge. Bar = 10 μm.

tholorum truncatorum ab ala junctae; ab cacumine lirae specimen 4–6-angulatam imponentes. Sporangia tenuiter tunicata non observata.

SPECIMENS EXAMINED OTHER THAN TYPE AND ISOTYPE:

Madagascar
Aedes aegypti (L.); Anamakia; Rodhain; 1980; NCU.
Sri Lanka (Ceylon)
Aedes albopictus (Skuse); Ambalongada, Rajapaksa; 1962; NCU.
Aedes albopictus (Skuse); Katunayake; Rajapaksa; 1963; NCU.
Aedes aegypti (L.); Maharagama; Rajapaksa; 1963; NCU.
Armigeres obturbans (Walker); Telawala; Rajapaksa; 1962; NCU.

COMMENTS: Although *C. dentialatus* is somewhat similar to *C. macleayae,* the former is distinguished by its resting sporangia with papillate ridges appearing as dentate wings in side view, and by its depressed areas with double, slitlike punctae. Also, ridges on sporangia of *C. macleayae* anastomose more frequently than those of *C. dentialatus.*

To date, *C. dentialatus* has been collected only from mosquito larvae occurring in discarded receptacles and spent nuts in Sri Lanka, and from a tree hole in Madagascar. Its specific epithet is derived from *dens* L. (tooth) and *alatus* L. (winged) in recognition of its papillate ridges which appear in side view as finely toothed wings.

Coelomomyces dodgei Couch & Dodge ex Couch, *J. Elisha Mitchell Sci. Soc.* **78,** 135 (1962)

Coelomomyces dodgei Couch emend Couch & Dodge, *J. Elisha Mitchell Sci. Soc.* **63,** 69 (1947)
Coelomomyces dodgei Couch, *J. Elisha Mitchell Sci. Soc.* **61,** 124 (1945)

Figures 267–273

DESCRIPTION: Hyphae 4–17 μm thick, irregularly branched. Resting sporangia ellipsoid, 27–42 × 37–65 μm; dorsal surface ornamented with linear depressions appearing as generally continuous concentric striae, surrounding the dehiscence slit and continuing in rows to encircle the sporangium, shortening to become irregular punctae on the ventral surface; concentric striae separated by 3–4 μm giving sporangia a banded appearance in some views. In midoptical section, bands and punctae as low ridges around the periphery of the sporangium. Wall 1.5–4.2 μm thick. Thin-walled sporangia not observed.

TYPE HOST AND COLLECTION SITE: *Anopheles crucians* Wiede.; Thomasville, Georgia (collected by SMCG).

DEPOSITORY FOR TYPE: BPI.

ALTERNATE HOST: *Cyclops vernalis* Fisher (Federici and Chapman, 1977).

SPECIMENS EXAMINED AND OTHER COLLECTIONS (indicated by brackets):

Type

United States
Anopheles crucians Wiede.; Georgia; SMCG; 1945; NCU.
Anopheles crucians Wiede.; Georgia; Bond; 1962; NCU.
[*Anopheles crucians* Wiede.; Georgia; Chapman; 1972; ?; (Chapman and Glenn, 1972).]
Anopheles earlei Vargas; Minnesota; Barr; 1950; NCU; (Barr, 1958; Laird, 1961); ?; North Carolina; Roane; 1966; NCU.

Figs. 267–272. *Coelomomyces dodgei.* Resting sporangia of type on *Anopheles crucians* collected in Thomasville, Georgia. Fig. 267. Surface view, dorsal side (dehiscence slit). Fig. 268. Surface view, ventral side of same sporangium as seen in Fig. 267. Fig. 269. Midoptical section, side view. Fig. 270. Surface view of sporangial side showing longitudinal striae. Fig. 271. Go-stage sporangium. Fig. 272. Go-stage sporangium showing lipid plates in forming zoospores. Fig. 267, bar = 10 μm, applies to Figs. 267–269. Fig. 270, bar = 10 μm, applies to Figs. 270 and 271. Fig. 272, bar = 5 μm.

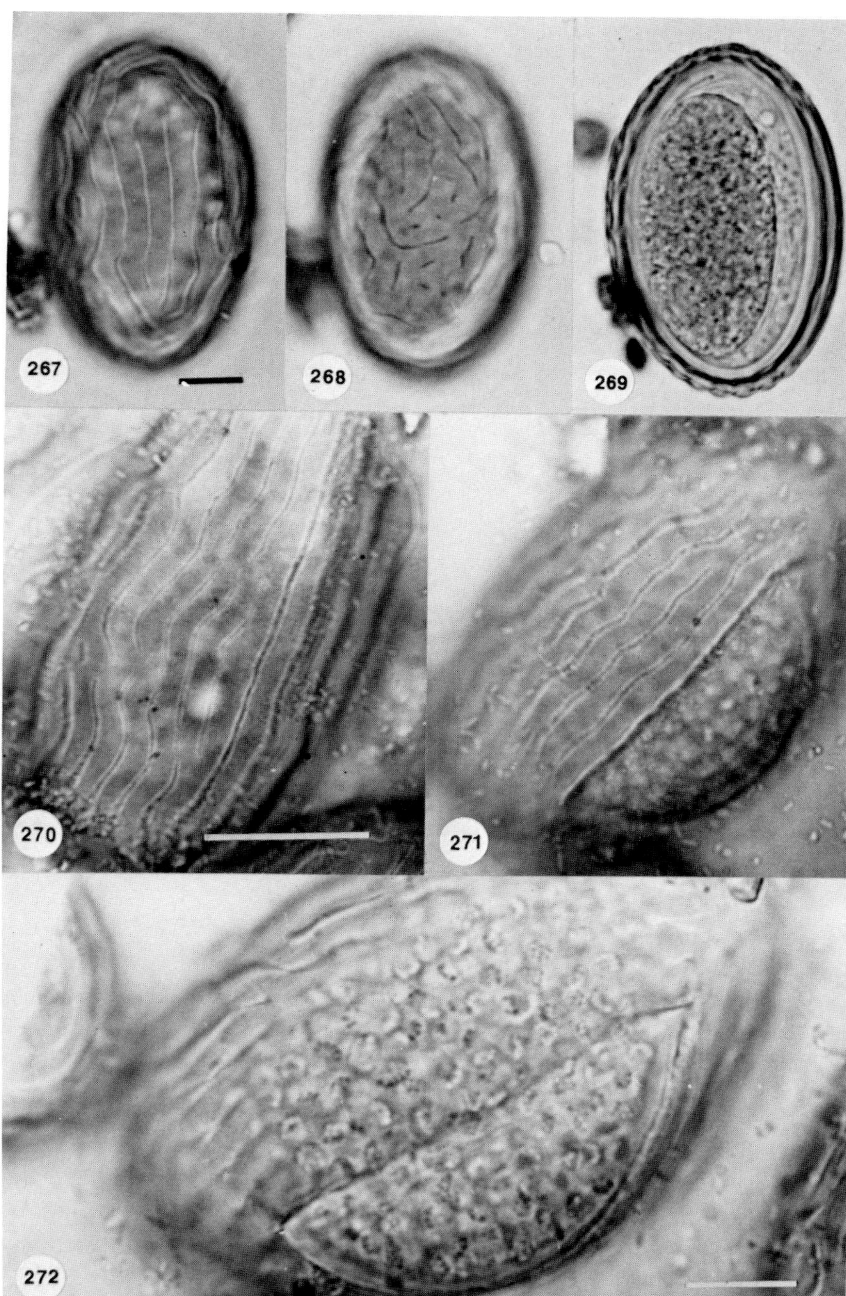
267
268
269
270
271
272

Fig. 273. *Coelomomyces dodgei.* Scanning electron micrograph of resting sporangium. Bar = 10 μm.

COMMENTS: Sporangia of *Coelomomyces dodgei* are most similar to those of *C. punctatus* and *C. lativittatus,* all of which were described originally (Couch, 1945) as characteristic of the single species, *C. dodgei.* However, Couch and Dodge (1947), recognizing three constant sporangial types ("resting sporangia with pits and narrow bands, with only pits, and with wide bands") occurring selectively on different hosts, emended the description of *C. dodgei* and recognized two new species. As redefined, *C. dodgei* includes forms with pits (punctae) and narrow bands and occurring on *An. crucians, C. punctatus* includes forms with punctae and occurring primarily on *An. quadrimaculatus,* and *C. lativittatus* includes forms with wide bands and occurring on *An. crucians*. While these characteristics are fairly constant, some confusion may arise in that specimens of any one of the three species may contain some sporangia identical to either of the other two. Therefore, species determinations must be based on the prominent type of sporangium present. Separation of *C. dodgei* from *C. punctatus* via a physiological response has been observed by B. A. Federici (personal communication) in that release of gametes of *C. punctatus* occurs 8 hr after onset of darkness whereas gamete release *C. dodgei* occurs approximately 17 hr after onset of darkness.

Coelomomyces dubitskii Couch & Bland sp. nov.

Figures 274–277

DESCRIPTION: Hyphae not seen. Resting sporangia broadly ellipsoid, 26–35 × 41–54 μm, surface ornamented with low, domelike ridges (4.5–7 μm wide × 2.5–3 μm high) anastomosing to enclose circular to elongate depressed areas (4–6 μm diameter when circular or 4–6 μm wide × up to 25 μm long) containing irregular punctae and striae. Dehiscence slit central along an elongate, depressed area with punctae/striae. Ridges containing internal, vertical striae appearing in optical section, faceview, as punctae, but as striae in midoptical section, side or end view. In midoptical section: side view, ridges as irregular domes along ventral edge and at poles, and as an elongate ridge along the dorsal surface. Thin-walled sporangia ellipsoid, 20–30 × 37–52 μm.

TYPE HOST AND COLLECTION SITE: *Culex modestus* Fica.; Kazachstan, U.S.S.R. (collected by Dubitskij).

DEPOSITORY FOR TYPE: BPI.

LATIN DIAGNOSIS: Hyphae non visae. Sporangia perennantia late ellipsoidea, 26–35 × 41–54 μm, superficie liris humilibus, tholis similibus 4.5–7 μm latis × 2.5–3 μm altis anastomosantibus, areas depressis rotundas 4–6 μm diametro vel elongatas, 4–6 μm latas × usque ad 25 μm longas, punctis atque striis enormibus notatas includentis, ornata; incisura dehiscentiae per aream elongatam, depressam punctis atque striis notatam media; lirae strias interiores, verticales, sectione opticali, ex aspectu frontali, instar punctorum, striarum autem ex aspectu laterali vel ab cacumine visas; sectione media opticali ex aspectu laterali lirae instar tholorum enormium per marginem ventralem atque ad polas visae, per superficiem dorsalem instar lairae elongatae. Sporangia tenuiter tunicata ellipsoidea, 20–30 × 37–52 μm.

SPECIMENS EXAMINED OTHER THAN TYPE: None.

COMMENTS: Resting sporangia of *C. dubitskii* are similar to those of *C. iliensis* var. *irregularis;* however, distinction may be made in that ridging

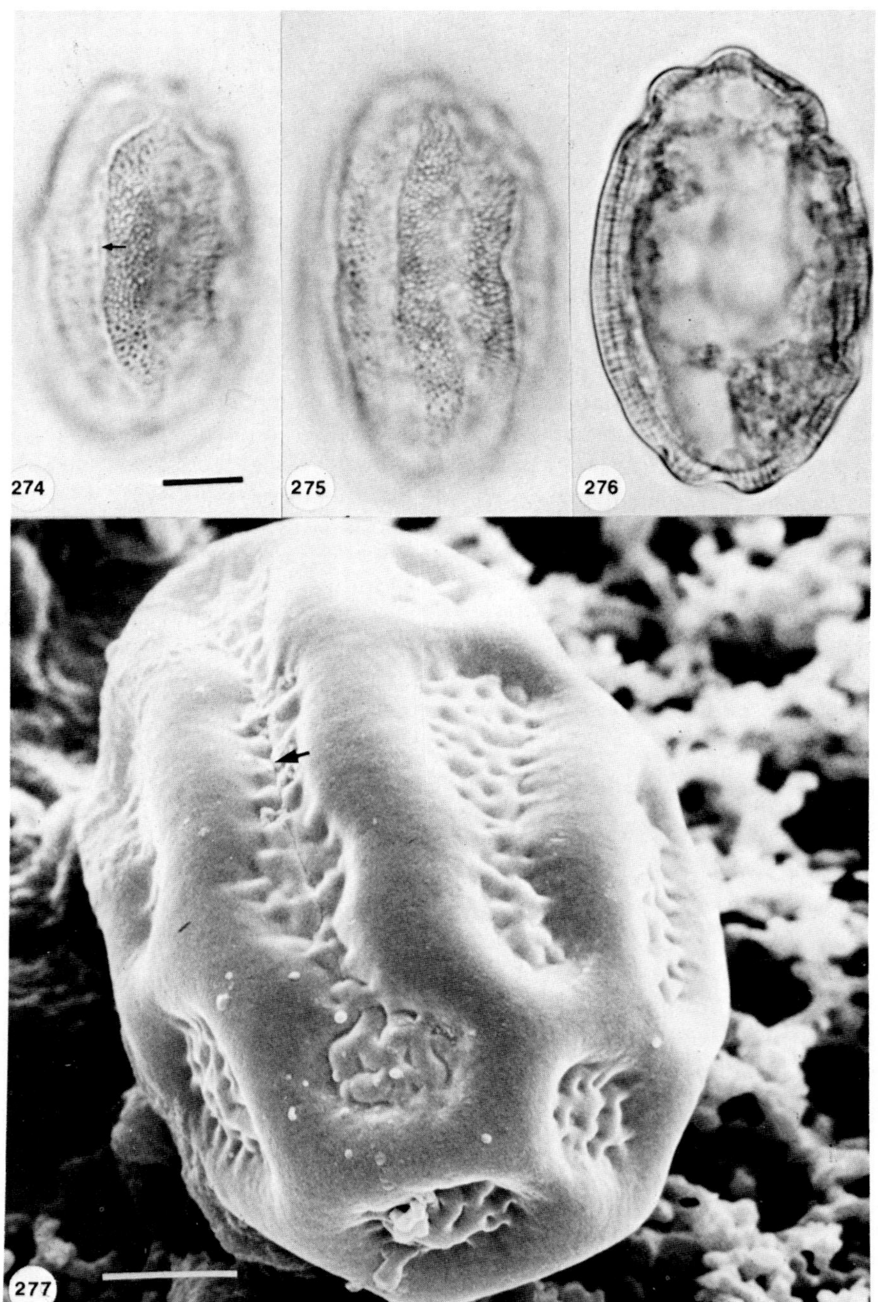
274
275
276
277

in *C. dubitskii* is broader and less abundant. In addition, depressed areas on sporangia of *C. dubitskii* have numerous punctae or short, irregular striae, whereas depressed areas on sporangia of *C. iliensis* var. *irregularis* have prominent, transverse striae.

The specific epithet of *C. dubitskii* is in recognition of the significant accomplishment of Dr. A. M. Dubitskij. Although this epithet was ascribed previously, but erroneously (see Comments for *C. iliensis* var. *iliensis* and under Excluded Taxa see Comments for *C. dubitskii*), to a specimen of *C. iliensis* (Shcherbak *et al.*, 1977), the organism considered by these authors to be *C. iliensis* is here recognized as a new species, *C. dubitskii*, thereby maintaining the wish to honor A. M. Dubitskij.

Coelomomyces elegans Couch & Rajapakse sp. nov.

Figures 278–282

DESCRIPTION: Hyphae not observed. Resting sporangia broadly ellipsoid, 39–61 × 48–71 μm; surface ornamented with anastomosing ridges (3–5 μm high and 3–5 μm wide) enclosing circular to smoothly polygonal, depressed areas (4.5–9 μm in diameter) with numerous (10–25), centrally clustered papillae (0.5–1 μm in diameter, often appearing as punctae via light microscopy); ridges appearing prominently punctate in surface views of the sporangial wall; dehiscence slit central along elongate, longitudinal depressed area with papillae. In midoptical section: ridges at sporangial periphery appearing in edge view as truncate domes with internal, vertical striae, interconnected by a similarly striate wing (ridges in side view). Thin-walled sporangia not observed.

TYPE HOST AND COLLECTION SITE: *Culex gelidus* Theo.; Matara, Sri Lanka (collected by Rajapakse; see Rajapakse, 1963, 1964).

DEPOSITORY FOR TYPE: BPI. Two isotypes are on deposit at NCU.

LATIN DIAGNOSIS: Hyphae non visae. Spongia perennantia late ellipsoidea, 36–91 × 48–71 μm; superficie liris anastomosantibus 3–5 μm altis et latis decorata, includentis areolas depressas rotundas vel aeque poly-

---

Figs. 274–277. *Coelomomyces dubitskii.* Resting sporangia from *Culex modestus* collected from Kazachstan, U.S.S.R. Fig. 274. Surface view of sporangial face. The dehiscence slit (arrow) is faintly visible between two of the ridges. Fig. 275. Surface view, side. Note internal vertical striae appearing as punctae in the ridging. Fig. 276. Midoptical section, side view. Fig. 277. Scanning electron micrograph of resting sporangium. The dehiscence slit (arrow) is visible between two longitudinal ridges. Fig. 274, bar = 10 μm, applies to Figs. 274–276. Fig. 277, bar = 5 μm.

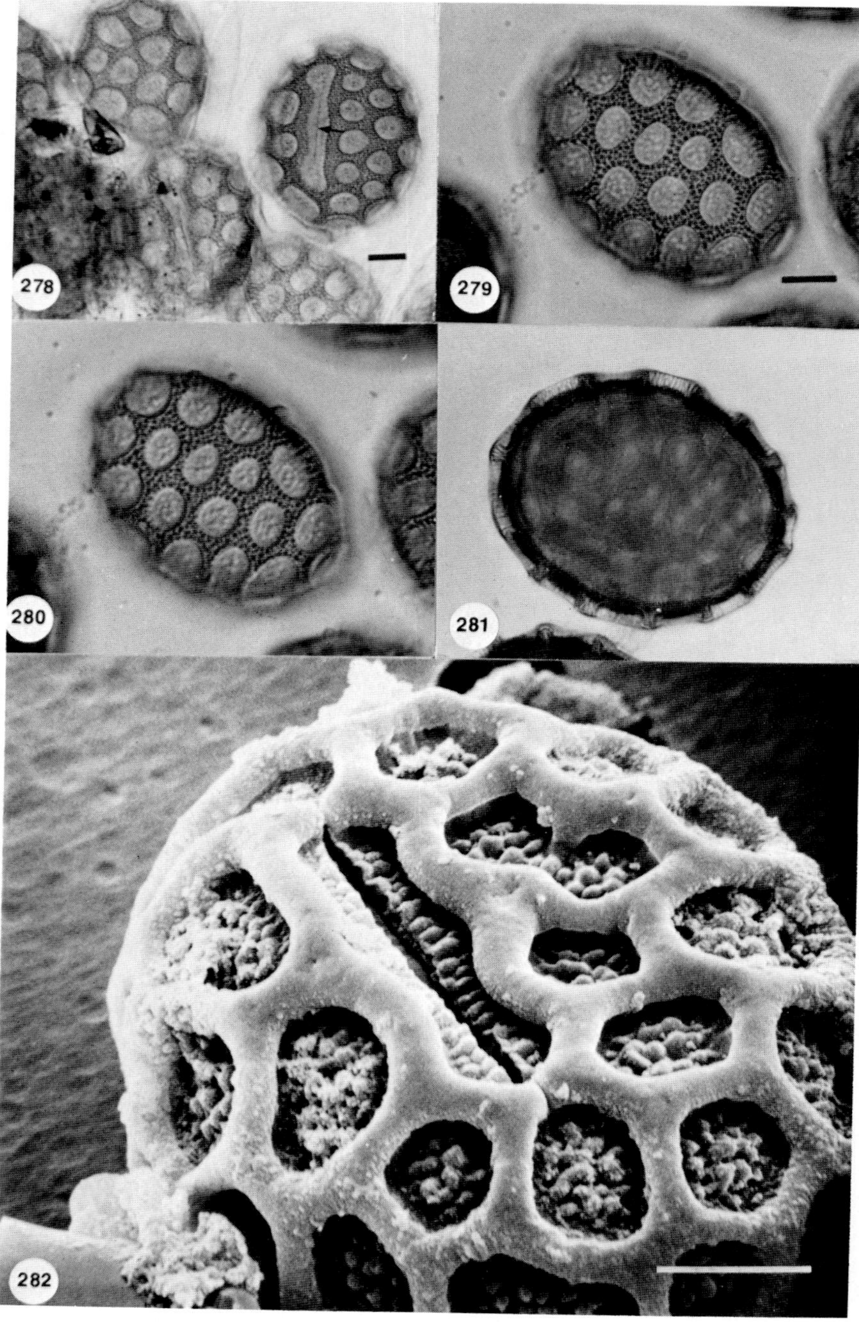
278
279
280
281
282

gonales 4.5–9 μm diametro, papillis numerosis prope medium congregatis, 0.5–1 μm diametro, sub microscopio lucido observatis speciem punctorum exhibentibus, notatas; liris specie conspicue punctatis in aspectu superficiali tunicae sporangialis; incisura dehiscentiae per depressionis longitudinalis medium locata, papillata; quasi in sectione media inspecta, liris ad sporangiorum marginem desuper instar tholorum truncatorum striis internis, verticalibus notatorum ab ala striata quae lirae a latere inspectae junctorum. Sporangia tenuiter tunicata non observata.

Specimens Examined Other than Type and Isotypes: None.

Comments: *Coelomomyces elegans* is distinguished from all known species of *Coelomomyces* by its resting sporangia ornamented by circular depressed areas with papillae. Other distinctive features include: dehiscence slit bordered by papillae in an elongate depressed area, and prominent vertical striae within ridges, appearing as punctae in surface view.

This species has been collected only from borrow pits near Matara, Sri Lanka. Its specific epithet is from *elegans* L. (elegant) in reference to its beautifully ornamented resting sporangia.

Coelomomyces fasciatus Couch & Iyengar sp. nov.

Figures 283–290

Description: Hyphae 4.5–9 μm in diameter. Resting sporangia broadly ellipsoid to spherical in some views, slightly flattened dorso–ventrally, 28–39 μm wide × 33.5–54 μm long × 22–30 μm thick; surface ornamented with contiguous, broad bands/ridges (5 μm wide), as concentric to irregular bands/domes on dorsal and ventral surfaces and as generally regular, concentric bands/ridges encircling sporangia at sides; dehiscence slit in central, elongate groove between ridges on dorsal surface. In midoptical section: face view, ridges as a wing encircling the sporangium; side view, ridges as contiguous domes at sporangial periphery. Thin-

Figs. 278–282. *Coelomomyces elegans.* Type specimen from *Culex gelidus* collected in Matara, Sri Lanka. Fig. 278. Surface view of sporangial face showing dehiscence slit (arrow). Figs. 279 and 280. Surface view, dorsal side of same sporangium viewed under differing optical conditions; bright field (Fig. 279) and interference–contrast (Fig. 280). Fig. 281. Midoptical section, face view. Fig. 282. Scanning electron micrograph of resting sporangium showing dehiscence slit in central depressed area. Fig. 278, bar = 10 μm. Fig. 279, bar = 10 μm, applies to Figs. 279–281. Fig. 282, bar = 10 μm.

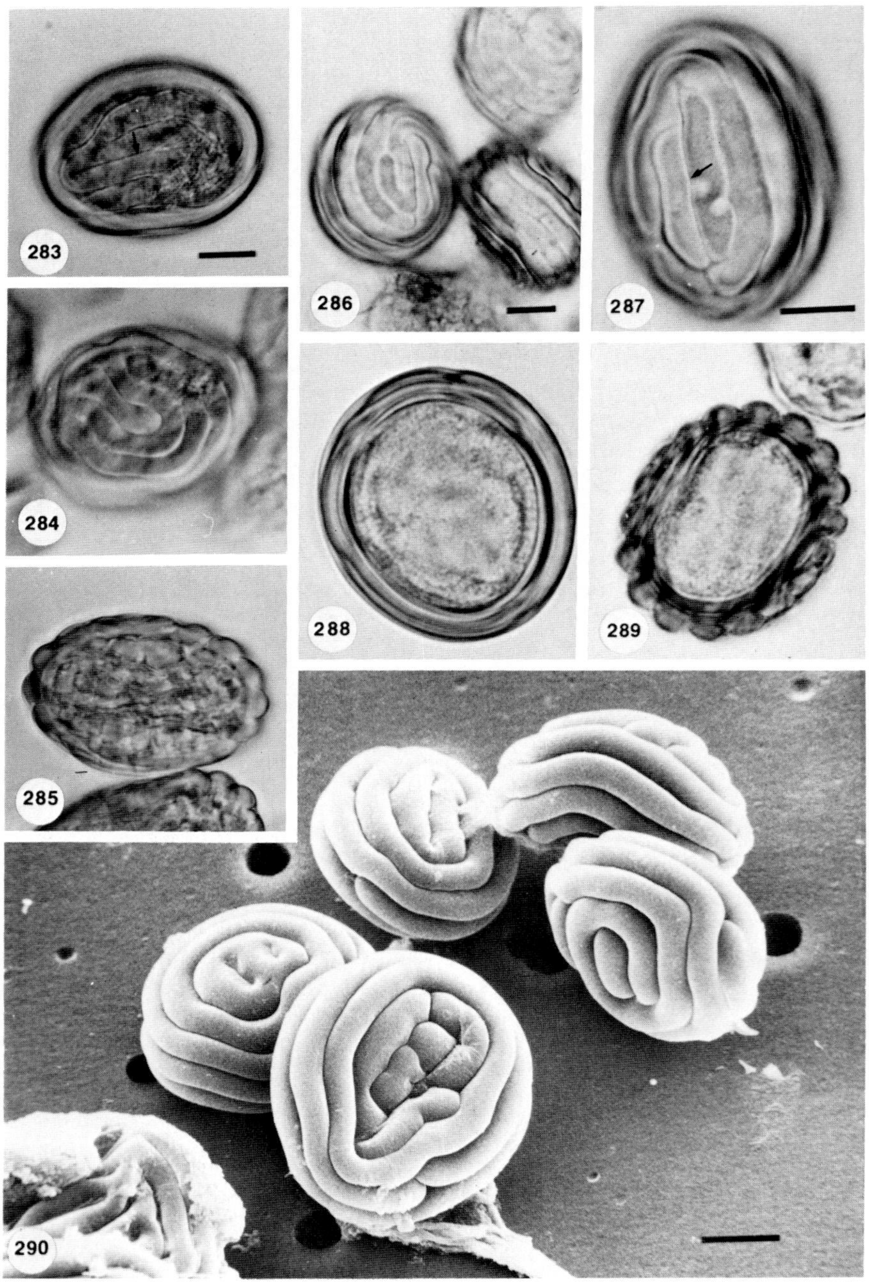
283
284
285
286
287
288
289
290

walled sporangia ellipsoid, slightly flattened on one side, 20.5–30 × 28–48 μm.

Type Host and Collection Site: *Anopheles jamesi* Theo. ♀; Wynaad, India (collected by Iyengar).

Depository for Type: BPI.

Latin Diagnosis: Hyphae 4.5–9 μm crassae. Sporangia perennantia late ellipsoidea vel speciebus nonnullis sphaerica, dorsi–ventraliter paululum applanata, 28–39 μm lata × 33.5–54 μm longa × 22–30 μm crassa; superficie vittis (vel liris) contiguis, 5 μm latis decorata, instar vittarum (vel tholorum) concentricarum vel enormium in superficie et dorsali et ventrali et plerumque regularium, concentricarum, sporangium cingentium ad latera; incisura dehiscentiae in canali centrali elongato inter liras superficiei dorsalis locata; quasi in sectione media inspecta ex aspectu frontali, liris instar alae sporangium amplectentis, e latere tholorum continuorum ad sporangii marginem. Sporangia tenuiter tunicata ellipsoidea, uno latere paululum applanata, 20.5–30 × 28–48 μm.

Specimens Examined Other than Type:

Anopheles gambiae Giles; N. Transvaal, South Africa; Service; 1975; BPI, NCU.

Comments: Resting sporangia of *C. fasciatus* are ornamented similarly to those of *C. sculptosporus*, *C. bisymmetricus*, and *C. lativittatus*. However, distinction from all three may be made in that resting sporangia of *C. fasciatus* are more broadly ellipsoid and have a distinct pattern of banding. A superficial resemblance exists also between resting sporangia of *C. fasciatus* and *C. anophelesicus*. However, the presence of transverse striae between ridges of sporangia of *C. anophelesicus* and their absence in those of *C. fasciatus* clearly separates the two.

The specific epithet of *C. fasciatus* is from *fasciatus* L., meaning banded.

Figs. 283–289. *Coelomomyces fasciatus.* Resting sporangia of type specimen from *Anopheles jamesi* collected in Wynaad, India. Bars = 10 μm. Fig. 283. Surface view, dorsal side. The dehiscence slit (arrow) is faintly visible in the center of the sporangium. Fig. 284. Surface view, ventral side. Fig. 285. Midoptical section, side view. Fig. 286. Surface view of the dorsal face and side of two different sporangia. Fig. 287. Surface of sporangial face showing central dehiscence slit (arrow). Fig. 288. Midoptical section, face view. Fig. 289. Midoptical section, side view.

Fig. 290. *Coelomomyces fasciatus.* Resting sporangia of specimens from *Anopheles gambiae* collected in Transvaal, South Africa. Bar = 10 μm.

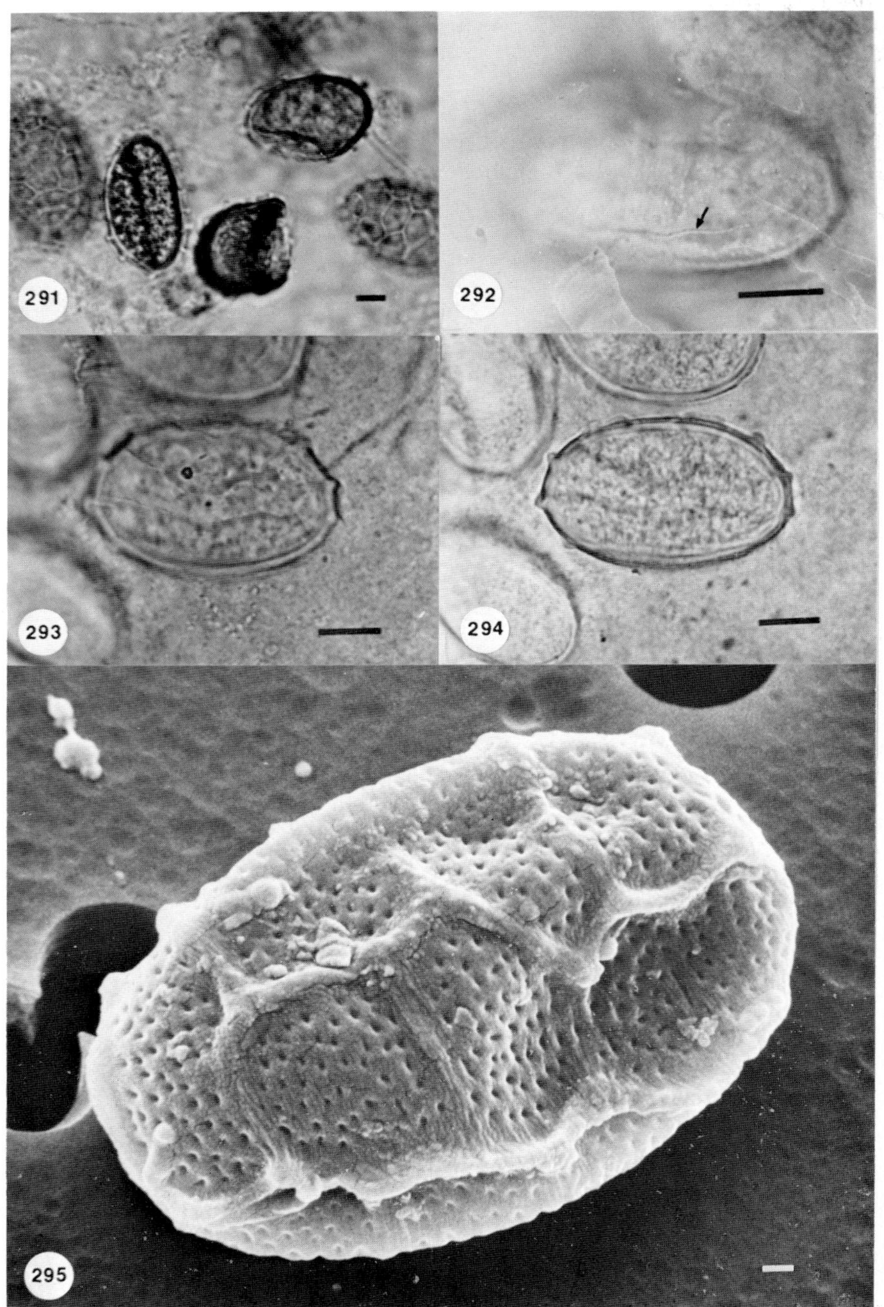

291
292
293
294
295

Coelomomyces finlayae Laird ex Laird, *J. Elisha Mitchell Sci. Soc.* **78,** 132 (1962)

Coelomomyces finlayae Laird, *Can. J. Zool.* **37,** 781 (1959)

Figures 291–295

DESCRIPTION: Hyphae not observed. Resting sporangia ellipsoid, 18–27.5 × 31.5–47.5 μm; surface ornamented with narrow ridges (1.4–2.9 μm high) anastomosing to enclose irregular polygonal depressed areas (6–12 μm wide) with numerous scattered punctae. In midoptical section, ridges at sporangial periphery appearing in end view as teeth and in side view as a wing interconnecting the teeth. Wall approximately 1 μm thick between ridges. Thin-walled sporangia not observed.

TYPE HOST AND COLLECTION SITE: *Aedes notoscriptus* (Skuse); S. Queensland, Australia (collected by E. N. Marks).

DEPOSITORY FOR TYPE: The type specimen deposited at the University of Queensland, Queensland, Australia has been lost. A cotype is deposited at the BPI.

SPECIMENS EXAMINED AND OTHER COLLECTIONS:

Type (sent J. N. C. by Laird, since lost from depository), **cotype**

New Hebrides

Aedes hebrideus Edwards; Hog Harbour; Rodhain; 1974; NCU; (Rodhain and Fauran, 1975).

COMMENTS: Resting sporangia of *C. finlayae* resemble closely those of *C. cairnsensis* and *C. macleayae*. Distinction from *C. cairnsensis* may be made in that sporangia of *C. cairnsensis* are larger and have more prominent ridges (3–5 μm high). Separation from *C. macleayae* is based on the fact that, in *C. macleayae,* resting sporangia have different ornamentation on two sides (ridges enclosing polygonal areas on the ventral side and parallel ridges on the other side), and have pits arranged in rows to give the appearance of striae within depressed areas of the wall.

Fig. 291. *Coelomomyces finlayae.* Resting sporangia of type specimen from *Aedes notoscriptus* collected in South Queensland, Australia. Bar = 10 μm.

Figs. 292–295. *Coelomomyces finlayae.* Resting sporangia from *Aedes hebrideus* collected in Hog Harbour, New Hebrides. Fig. 292. Surface view, dorsal face and side showing prominent dehiscence slit (arrow). Bar = 10 μm. Figs. 293 and 294. The same sporangium seen in surface view of side (Fig. 293), and midoptical section, side view (Fig. 294). Bars = 10 μm. Fig. 295. Scanning electron micrograph of resting sporangium. Bar = 1 μm.

Coelomomyces grassei Rioux & Pech ex Rioux & Pech, *J. Elisha Mitchell Sci. Soc.* **78,** 135 (1962)

Coelomomyces grassei Rioux & Pech, *Acta Trop.* **17,** 179 (1960)

Figures 296–303

DESCRIPTION: Hyphae 2–10 μm wide. Resting sporangia ellipsoid, 25–40 × 42–70 μm; surface ornamented with broad, longitudinal, parallel ridges (3–5 μm wide × 3.0–3.5 μm high) separated by depressed areas (3.0–3.5 μm wide) with transverse striae; ridges anastomosing at the poles. Dehiscence slit prominent, in central, depressed area of a longitudinal ridge. In midoptical section: side view, ridges at periphery of sporangia as lateral wings with internal, vertical striae; in end view, ridges as 8–10 semicircular domes surrounding the sporangium. Thin-walled sporangia ellipsoid to ovoid 23–33 × 27–50 μm.

TYPE HOST AND LOCATION: *Anopheles gambiae* Giles; North Chad, Africa.

DEPOSITORY FOR TYPE: Institut de Parasitologie, Montpellier, France. A paratype is on deposit at BPI.

SPECIMENS EXAMINED AND OTHER COLLECTIONS (indicated by brackets):

Paratype (sent J. N. C. by Rioux)

Africa

Anopheles gambiae Giles; Kisumu, Kenya; Service; 1971, 1974, 1975; NCU.

Anopheles gambiae Giles; Kruger National Park, S. Africa; Muspratt; 1978; NCU.

[*Anopheles gambiae* Giles ♀; Upper Volta, Tingrila; Rodhain and Brengues; 1974; ?; (Rodhain and Brengues, 1974).]

Anopheles squamosus Theo.; Nora Lisboa, Angola; Ribeiro; 1967; NCU.

COMMENTS: *Coelomomyces grassei* is distinguished from all known species of *Coelomomyces* by the unique appearance of its dehiscence slit in a depressed area of a longitudinal ridge.

While the paratype of this species includes sporangia that have shrunken and collapsed during mounting (Fig. 296), scanning electron microscopy of paratype specimens reveal their true form; the same morphology evident in other specimens of this species (Fig. 297).

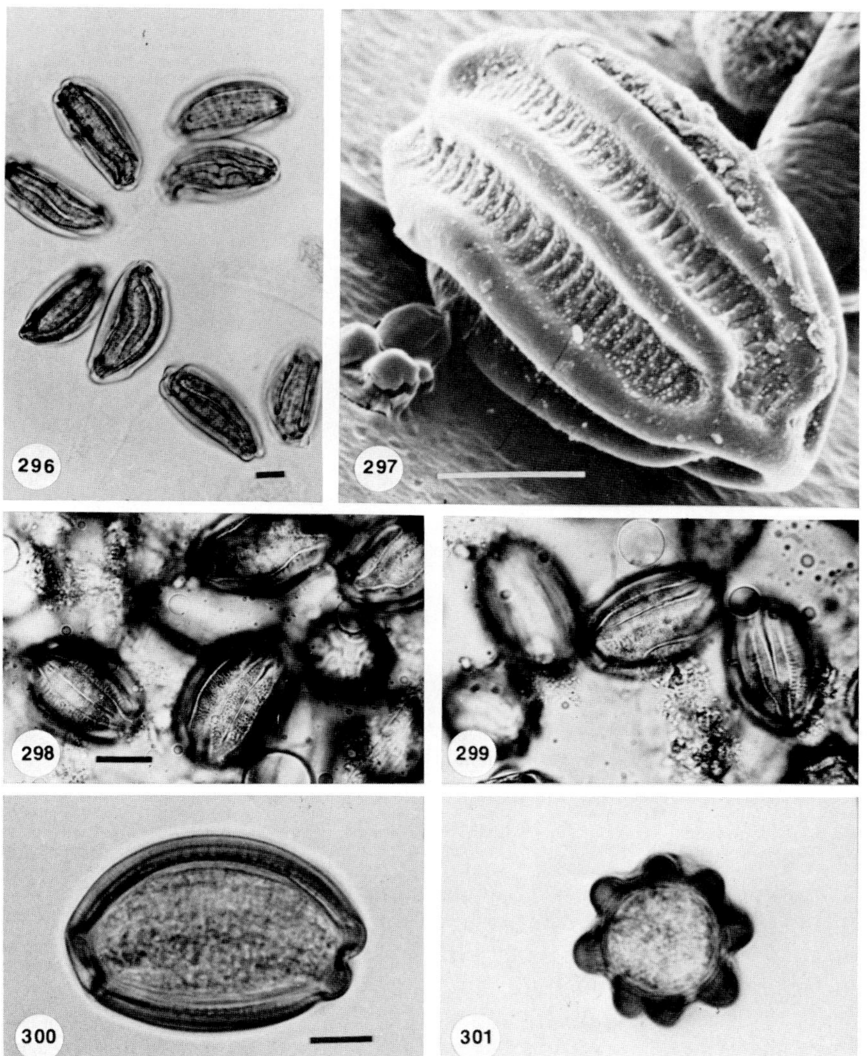

Figs. 296–297. *Coelomomyces grassei.* Paratype from *Anopheles gambiae* collected in North Chad, Africa. Fig. 296. Specimens on original slide of paratype showing severely distorted sporangia. Fig. 297. Scanning electron micrograph of specimen from original paratype slide that has been rehydrated and mounted for scanning electron microscopy. Note lack of sporangial distortion.

Figs. 298–301. *Coelomomyces grassei.* Resting sporangia of specimen from *Anopheles gambiae* collected in Kisumu, Kenya, Africa. Bars = 10 μm. Figs. 298 and 299. Surface view, dorsal face of several sporangia. The dehiscence slit lies between the two prominent longitudinal ridges. Bar = 10 μm. Fig. 300. Midoptical section, side view. Bar = 10 μm. Fig. 301. Midoptical section, end view. Scale as for Fig. 300.

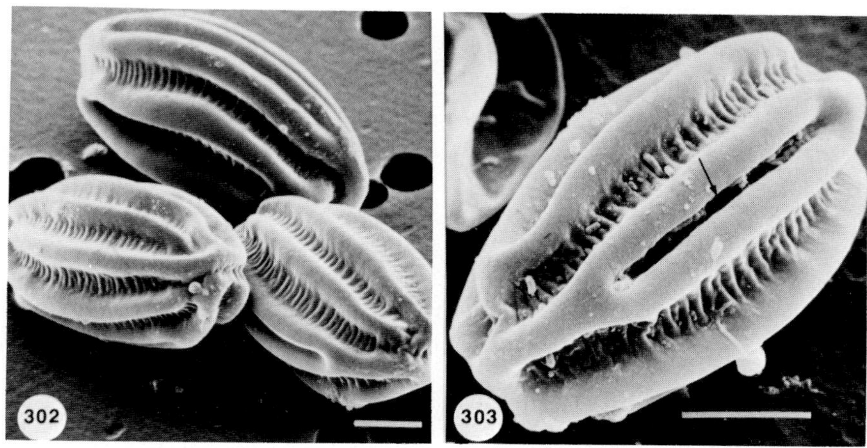

Figs. 302–303. *Coelomomyces grassei.* Scanning electron micrographs of resting sporangia collected in Kisumu, Kenya, Africa. The prominent dehiscence slit is visible in Fig. 303. Bars = 10 μm.

Coelomomyces iliensis var. ***iliensis*** Dubitskij *et al., Mikol. Fitopatol.* **7,** 136 (1973) emend Couch & Dubitskij

Coelomomyces omorii Laird *et al., J. Parasitol.* **61,** 539 (1975)
Coelomomyces dubitskii nomen nudum Shcherbak *et al., Dopov. Akad. Nauk Ukr. RSR, Ser. B: Geol., Khim. Biol. Nauki* **8,** 758 (1977)

Figures 304–318

DESCRIPTION: Hyphae 6–12 μm in diameter. Resting sporangia broadly ellipsoid in face view, 26–43 × 34–60 μm; biconvex in side view, 24–39 μm thick. Surface ornamented with prominent longitudinal ridges (3–5 μm wide and 5–7 μm high) of differing patterns on opposing sporangial faces; ridging on dorsal face often as an ellipse surrounding the dehiscence slit, with 2–3 additional, concentric ridges continuing around the sides of the sporangium, occasionally beginning adjacent to the dehiscence slit and following a spiral to rarely irregular course around the sporangium; ridging on ventral face spiral, irregular, to concentric–ellipti-

---

Figs. 304–310. *Coelomomyces iliensis* var. *iliensis.* Resting sporangia from *Culex orientalis* collected in far eastern USSR. Fig. 304. Surface view, dorsal face. Note the dehiscence slit (arrow). Bar = 10 μm, applies to Figs. 304–309. Fig. 305. midoptical section, face view. Fig. 306. Midoptical section, side view. Figs. 307 and 308. Surface views of sporangial side. Fig. 309. Midoptical section, end view. Fig. 310. Scanning electron micrograph. Bar = 1 μm.

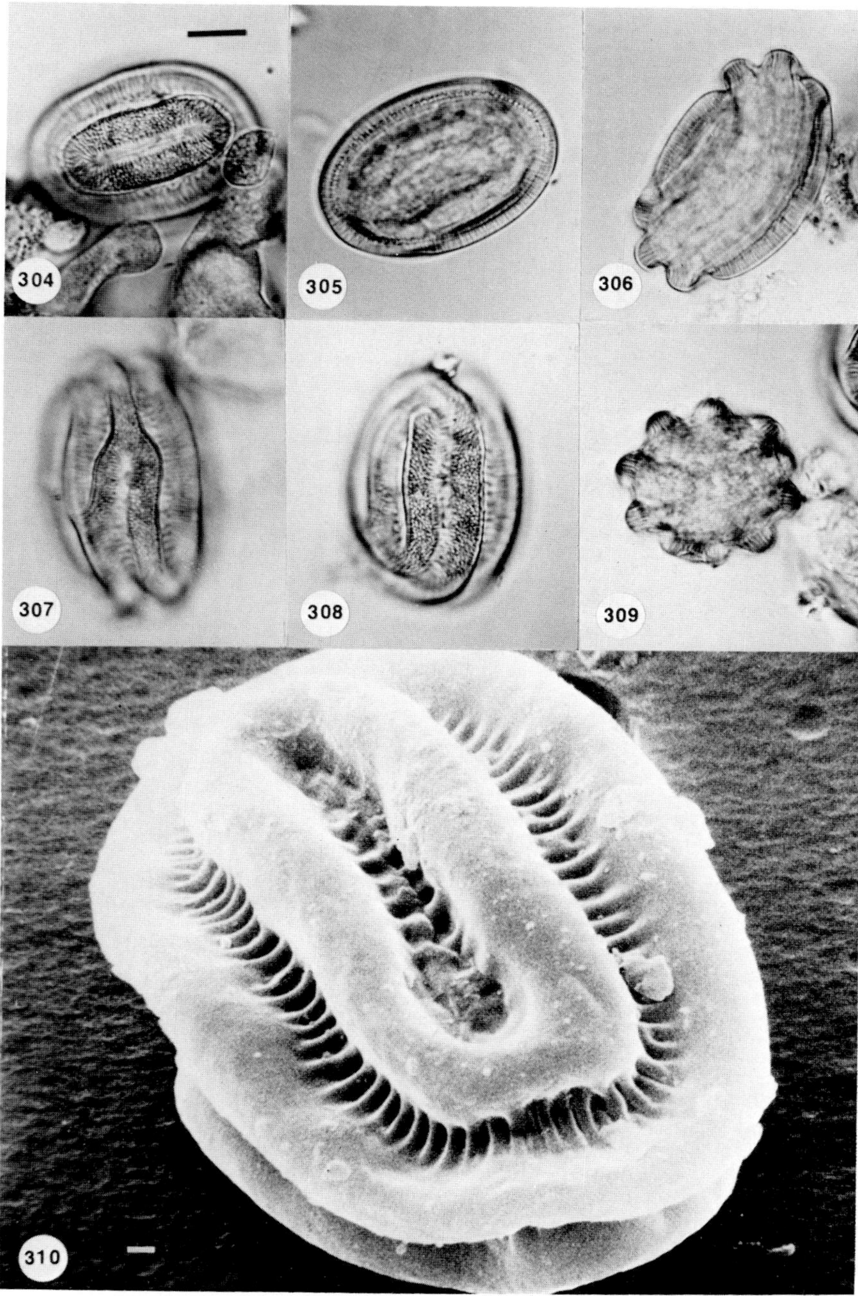
304
305
306
307
308
309
310

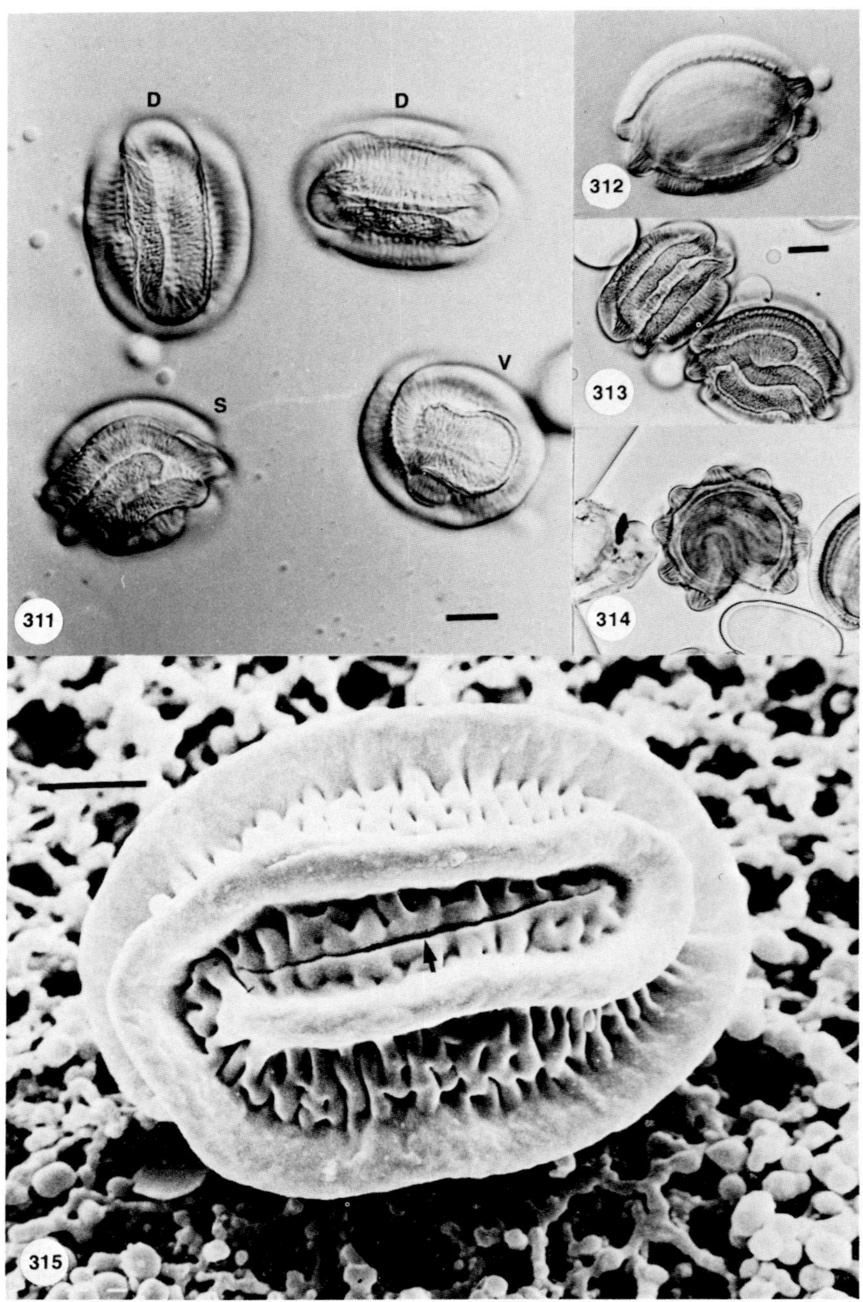
D
D
S
V
311
312
313
314
315

cal, a single linear to domelike ridge sometimes as a central island inside a final continuous, elliptical ridge. Ridges separated by depressed areas (4–5 μm) with prominent, transverse striae. Dehiscence slit prominent along a longitudinal depressed area with transverse striae and bordered or surrounded by longitudinal ridging. In midoptical section: face view, ridging as a vertically striate wing at sporangial periphery, transverse striae between ridges appearing as teeth along proximal edge of wing; side view, ridging as an elongate wing along dorsal edge, a shorter wing along the ventral edge, and as 2–4 simicircular domes at the poles; end view, ridging as 7–10 domes encircling the sporangium. Thin-walled sporangia ellipsoid, 16–26 × 24–46 μm.

TYPE HOST AND COLLECTION SITE: *Culex modestus* Fica. Alma-Ata, Kazakhstan, U.S.S.R. (collected by Dubitskij).

ALTERNATE HOSTS:

Cyclops (Acanthocyclops) viridis (Jurine)	(Nam, 1981)
Cyclops (Acanthocyclops) languidoides Lilljeborg	(Nam, 1981)
Microcyclops varicans Lilljeborg	(Nam, 1981)
Eucyclops agilis as *Cyclops serrulatus* (Fischer)	(Nam, 1981)
Cyclops strenuus Fischer	(Nam, 1981)*
Mesocyclops sp. Sars	(Nam, 1981)*

DEPOSITORY FOR TYPE: Institute of Zoology, Academy of Sciences, Kazakhstan, U.S.S.R.

LATIN DIAGNOSIS: Hyphae 6–12 μm diametro. Sporangia perennantia ex aspectu frontali late ellipsoidea, 26–43 × 34–60 μm, e latere biconvexa, 24–39 μm crassa. Superficies ab liris longitudinalibus prominentibus 3.5 μm latis, 5–7 μm altis, ornata, sporangiorum faciebus oppositis descriptione dissimilibus; in facie dorsali liris ellipsem incisuram dehiscentiae circumscribentem, atque liras 2–3 accessorias concentricas per latera sporangii productas, efformantibus, nonnumquam prope incisuram ortis et spiram vel raro cursum irregularem circum sporangium describen-

* Reported for *C. iliensis* var. *orientalis*, a nomen nudum.

Figs. 311–315. *Coelomomyces iliensis* var. *iliensis.* Resting sporangia from *Culex tritaeniorhynchus summorosus* collected in Nagasaki, Japan. These are sporangia of the type for *Coelomomyces omorii* as designated by Laird *et al.* (1975). Fig. 311. Surface views of four different resting sporangia; dorsal view (D), side view (S), ventral (V). Bar = 10 μm, applies also to Fig. 312. Fig. 312. Midoptical section, side view. Fig. 313. Surface view, side. Bar = 10 μm, applies also to Fig. 314. Fig. 314. Midoptical section, end view of ruptured sporangium. Fig. 315. Scanning electron micrograph (dehiscence slit, arrow). Bar = 5 μm.

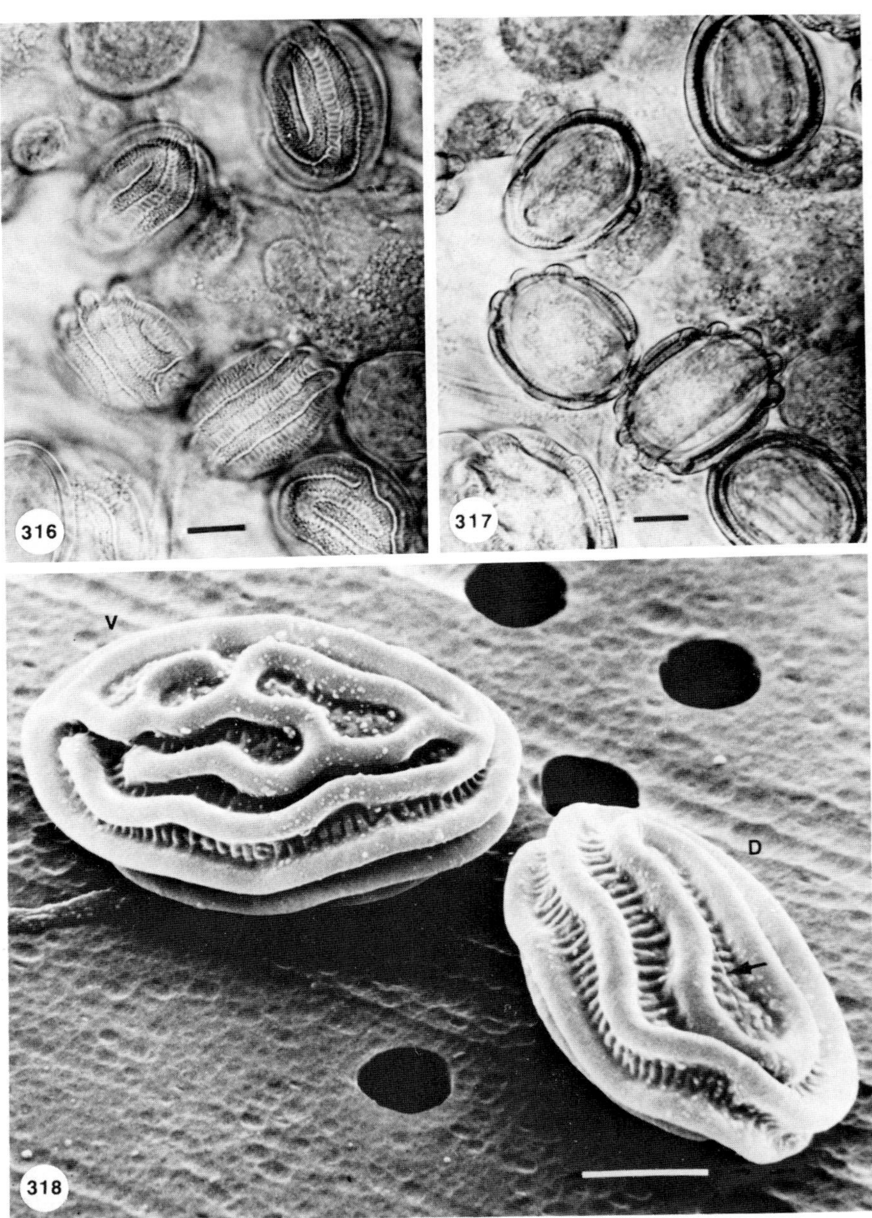

Figs. 316–318. *Coelomomyces iliensis* var. *iliensis.* Resting sporangia from *Culex antennatus* collected in Kisumu, Kenya, Africa. Bars = 10 μm. Figs. 316 and 317. Surface (Fig. 316) and midoptical (Fig. 317) views of the same sporangia. Fig. 318. Scanning electron micrograph showing dorsal (D) and ventral (V) faces of two different sporangia. The dehiscence slit (arrow) is visible on one of these sporangia.

tibus; in facie ventrali liris spiralibus, irregularibus, vel concentricis, ellipticis, nonnumquam lira una lineari vel thaliformi instar insulae centralis inter liram continuam, ellipticam constituente. Lirae a spatiis 4–5 μm depressis a striis transversis prominentibus notatis separatae. Incisura dehiscentiae e spatio longitudinali depresso transverse striato prominens, liris longitudinalibus marginata. In sectione media inspecta ex aspectu frontali, liris instar alae verticaliter striatae secundae sporangii peripheram, striis transversis inter liras instar dentium per alae marginem proximans; e latere, instar alae elongatae per marginem dorsalem, et brevioris per marginem ventralem, et tholorum 2–4 semi-circularium ad polos; e cacumine, liris instar tholorum 7–10 sporangium circumplectentibus. Sporangia tenuiter tunicata ellipsoidea, 16–26 × 24–46 μm.

SPECIMENS EXAMINED AND OTHER COLLECTIONS (indicated by brackets):

Africa

Culex antennatus Beck.; Kisumu, Kenya; Service; 1974; NCU.

Australia

Culex annulirostris Skuse; Cairns, Queensland; Kay; 1979; CIH, NCU.

Japan

Culex tritaeniorhynchus summorosus Dyar*; Nagasaki; Mogi; 1973; BPI, TNS. (Laird *et al.*, 1975; Mogi *et al.*, 1976).

Taiwan

[*Culex annulus;* various locations; Laird and collaborators; 1955, 1966, 1974; ?; (Laird *et al.*, 1980).]

U.S.S.R.

Culex modestus Fica; Alma-Ata, Kazakhstan; Dubitskij; 1980; NCU.

Culex pipiens L.; Alma-Ata; Kazakhstan; Dubitskij; 1969; ?; (Dubitskij *et al.*, 1970, 1973).

Culex orientalis Edwards; Vladivostok; Shipitsinc; 1962; UNC.

Culex orientalis Edwards; ?; Dubitskij; 1980; UNC.

COMMENTS: The unique shape and ornamentation of resting sporangia of *C. iliensis* distinguish this species from all known forms of *Coelomomyces*. However, as noted by Laird *et al.* (1975, 1980), a superficial resemblance does exist between sporangia of *C. iliensis* (*C. omorii*) and those of *C. anophelesicus*. Clear distinction may be made, however, in that resting sporangia of *C. anophelesicus* are characteristically flattened along one side of the narrow edge and have the dehiscence slit central along the narrow edge opposite the flattened side, rather than central between two prominent ridges on a broad sporangial face as in *C. iliensis*.

* Host for type of *C. omorii*.

Although resting sporangia of *C. iliensis* were described originally to be ornamented with longitudinal, nonanastomosing ridges, there was no indication that such ridges had internal vertical striae or were separated by depressed areas with transverse grooves. Laird *et al.* (1975), in describing a new species of *Coelomomyces* from Japan, *C. omorii,* recognized similarities in sporangial ornamentation between the new species and *C. iliensis*. Acknowledging the fact that "examination of fresh material by SEM could narrow the gap" between *C. omorii* and *C. iliensis,* distinction was made in that sporangia of *C. omorii* had ridges with internal vertical striae and were separated by depressed areas with transverse grooves. Recent examination of fresh specimens of *C. iliensis* (provided by Dubitskij) reveals sporangial ornamentation identical to that of *C. omorii;* the two therefore are conspecific and *C. omorii* becomes synonymous with *C. iliensis,* which has priority. The specimen described by Shcherbak *et al.* (1977) as *C. dubitskii* (a nomen nudum since no Latin diagnosis was provided) is felt to be *C. iliensis,* whereas the *C. iliensis* of the same authors (specimens of which were sent to J. N. C. and C. B. by Weiser) is described in this monograph as a new species, *C. dubitskii.* Through examination of fresh specimens of *C. iliensis* (provided by Dubitskij) and their comparison with published drawings and photographs of this species (Dubitskij *et al.*, 1973; Dubitskij, 1978), it is apparent that Shcherbak *et al.* (1977) were mistaken in their identification.

Nam (1981) recognized several varieties of *C. iliensis* from the U.S.S.R., including *C. iliensis, C. iliensis* var. *culicis, C. iliensis* var. *orientalis,* and *C. iliensis* var. *omorii*. However, no complete taxonomic descriptions were given for any of the forms, all therefore being invalid. Hopefully, all forms referred to by Nam (1981) have been included in the present treatment of this group, and can be identified.

In considering specimens of *C. iliensis* (*C. omorii*) from Taiwan, Laird *et al.* (1980) noted this form to have resting sporangia ranging from "broadly rounded ones inseparable from the Japanese variety to others very possibly inseparable from '*C. dubitskii*' " (*C. iliensis*). Examination of material from several collections of *C. iliensis* from a variety of locations substantiates the existance of some variation in sporangial dimensions and ornamentation between specimens from different collections. Although sporangia practically indistinguishable from one another are present among all collections, sufficient variation exists for designation of some forms as varieties. The primary factors used in delineating varieties include variations in sporangial size, pattern of ridging, nature of transverse grooves in depressed areas, and prominence of internal vertical striae in ridges. All forms are united, however, in having resting sporangia ornamented with ridging that begins around the dehiscence slit and continues spirally or concentrically around the sides of the sporangium to terminate in a concentric, insular, to irregular pattern on the ventral surface. In all varieties, face views of sporangia will normally reveal only four ridges.

An interesting feature regarding the occurrence of varieties of *C. iliensis* is that in most collections (Africa, India, Japan, and Taiwan), hosts were collected from rice fields. With the exception of *C. iliensis* var. *indus* and *C. iliensis* var *irregularis,* collected on anopheline as well as on a culicine host, all other collections of this species have been on hosts of the genus *Culex*.

Coelomomyces iliensis var. ***australicus*** var. nov. Couch & Russell

Figures 319–322

Description: Hyphae not observed. Resting sporangia as in typical variety except ridges often discontinuous, frequently appearing at poles as two to several isolated domes. Thin-walled sporangia as in typical variety.

Type Host and Collection Site: *Culex annulirostris* Skuse; Nhulunbuy Lagoon, Gove, Australia (collected by Russell).

Depository for Type: CIH. An isotype is on deposit at BPI.

Latin Diagnosis: Hyphae non visae. Sporangia perennantes ut in varietate typica; liris saepe discontinuis, saepe ad polos instar tholorum duorum vel plurium sejunctorum exceptis. Sporangia tenuiter tunicata ut in varietate typica.

Specimens Examined Other than Type and Isotypes: None.

Comments: *Coelomomyces iliensis* var. *australicus* may be distinguished from other varieties of this species by its resting sporangia ornamented with discontinuous ridges appearing often as domes at the poles. Some sporangia having ornamentation identical to that of the typical variety may be present, however.

As this variety is described from a single collection obtained in 1972, it is possible that additional sampling in Australia may yield specimens more closely aligning it with *C. iliensis* var. *iliensis* or possibly necessitating its description as a new species. For now, however, the basically similar pattern of sporangial ornamentation in *C. iliensis* var. *australicus* and the typical variety dictates its inclusion in this species.

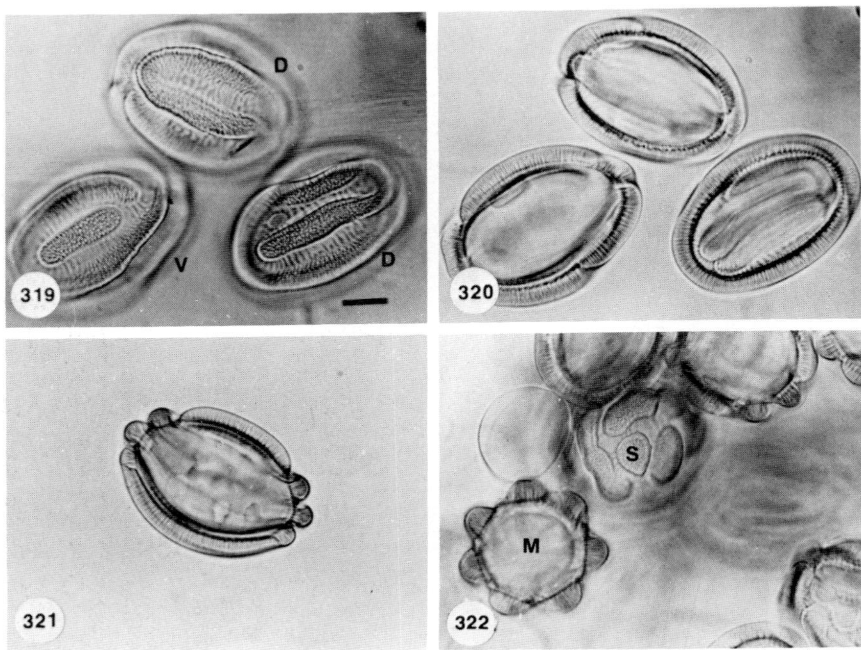

Figs. 319–322. *Coelomomyces iliensis* var. *australicus.* Resting sporangia of type specimen from *Culex annulirostris* collected in Gove, Australia. Figs. 319 and 320. Surface (Fig. 319) and midoptical section (Fig. 320) views of the same three sporangia. Dorsal view (D); ventral view (V). Bar = 10 μm. Fig. 321. Midoptical section, side view. Scale as for Fig. 319. Fig. 322. Midoptical section, end view (M) and surface view (S) at the pole of a sporangium. Scale as for Fig. 319.

Coelomomyces iliensis var. *indus* var. nov. Couch & Iyengar

Figures 323–328

DESCRIPTION: Hyphae not observed. Resting sporangia as in typical variety except measuring 32–41 × 43–67 × 30–39 μm thick and with regular ridges separated by depressed areas 5–10 μm wide and with numerous, fine (1–1.5 μm wide), parallel transverse striae or rarely punctae; ridging on dorsal surface rarely anastomosing, usually concentric around sides and ending on the dorsal surface as a continuous elliptical ridge, which may or may not enclose a central, insular ridge. Thin-walled sporangia ellipsoid to ovid, 20.5–33.5 × 43–48 μm.

TYPE HOST AND COLLECTION SITE: *Anopheles nigerrimus* Giles; Trivandrum, India (collected by Iyengar).

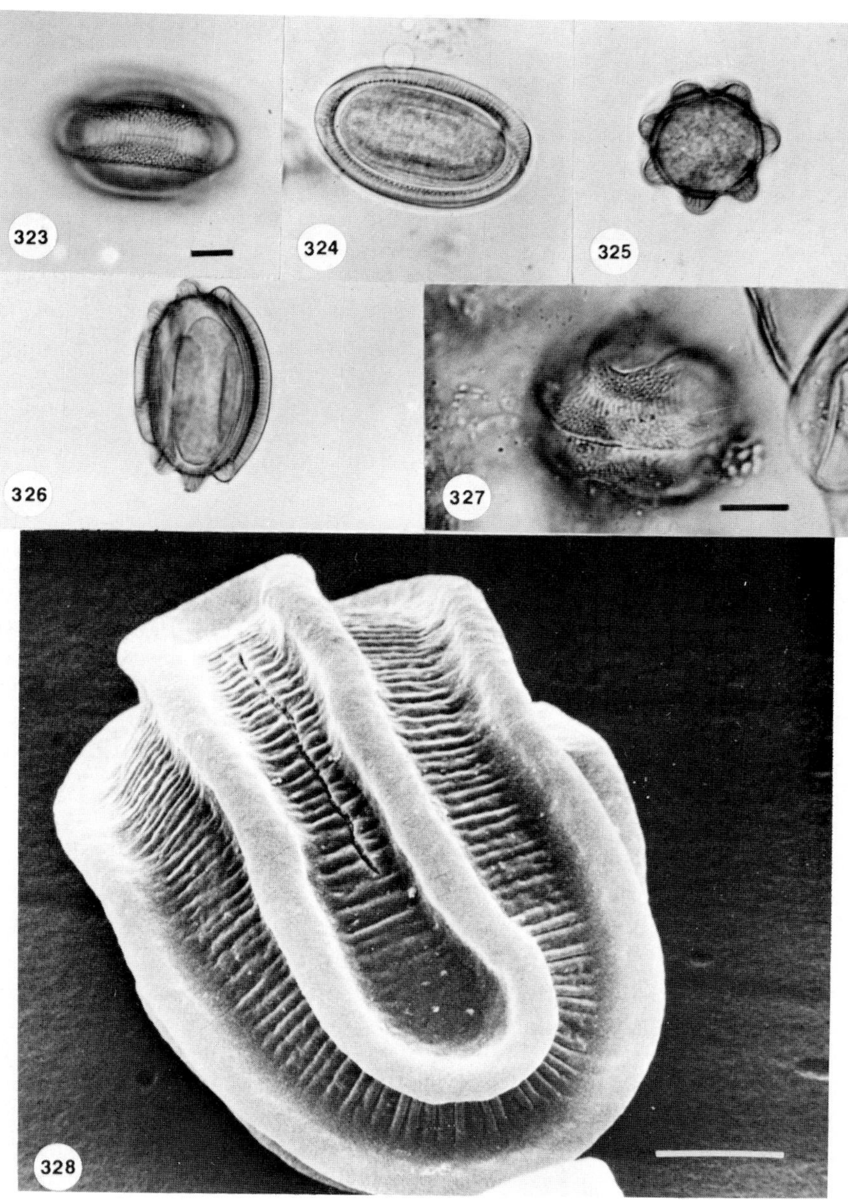

Figs. 323–328. *Coelomomyces iliensis* var. *indus*. Resting sporangia of type specimen on *Anopheles nigerrimus* collected in Trivandrum, India. Fig. 323. Surface view, dorsal side (dehiscence slit, arrow). Bar = 10 μm, applies to Figs. 323–326. Fig. 324. Midoptical section, face view. Fig. 325. Midoptical section, end view. Fig. 326. Midoptical section, side view. Fig. 327. Surface view of sporangial end. Bar = 10 μm. Fig. 328. Scanning electron micrograph. Note prominent dehiscence slit along depressed area on dorsal side. Bar = 5 μm.

DEPOSITORY FOR TYPE: BPI.

LATIN DIAGNOSIS: Hyphae non visae. Sporangia perennantia ut in varietate typica, sed 32–41 × 43–67 × 30–39 μm et liris regularibus decorata, ab depressionibus 5–10 μm latis striis numerosis subtilibus parallelis transversis, 1–1.5 μm latis notatis vel raro punctatis; liris in superficie dorsali raro anastomosantibus, plerumque secundum latera concentricis et in superficie dorsali terminatis ut lira continua elliptica liram centralem insularem includens vel non. Sporangia tenuiter tunicata ellipsoidea vel ovoidea, 20.5–33.5 × 43–48 μm.

SPECIMENS EXAMINED OTHER THAN TYPE:

India

Anopheles nigerrimus Giles; Trivandrum; Iyengar; 1970; NCU.

COMMENTS: *Coelomomyces iliensis* var. *indus* may be distinguished from other varieties of this species on the basis of its generally larger resting sporangia (with the exception of *C. iliensis* var. *striatus*) with ridges spaced 5–10 μm apart. Additionally, the transverse striae in depressed areas between ridges appear narrower and straighter than in other varieties, a characteristic obvious via SEM, but, although visible, somewhat difficult to discern via light microscopy (LM). Clear distinction from *C. iliensis* var. *striatus* may be made in that resting sporangia of *C. iliensis* var. *striatus* are larger, have more irregular ridging on the ventral surface, and have more distinct internal vertical striae.

Coelomomyces iliensis var. *irregularis* Couch & Dubitskij var. nov.

Figures 329–334

DESCRIPTION: Hyphae not observed. Resting sporangia ellipsoid, 24–36 × 37–54 μm, ridging frequently appearing irregular, often anastomosing to enclose elongate to polygonal depressed areas with transverse striae/punctae, or ornamentation as in typical variety. In midoptical section: face view, ridging as a scalloped, vertically striate wing encircling the sporangium; side view, ridging as a scalloped wing along dorsal and ventral surfaces and as 3–5 domes at the poles; end view, ridging as 8–12 domes encircling the sporangium. Thin-walled sporangia ovoid, 19–28 × 28–46.5 μm.

TYPE HOST AND COLLECTION SITE: *Culex modestus* Fica.; Alma-Ata, Kazakhstan, U.S.S.R. (collected by Dubitskij).

DEPOSITORY FOR TYPE: BPI.

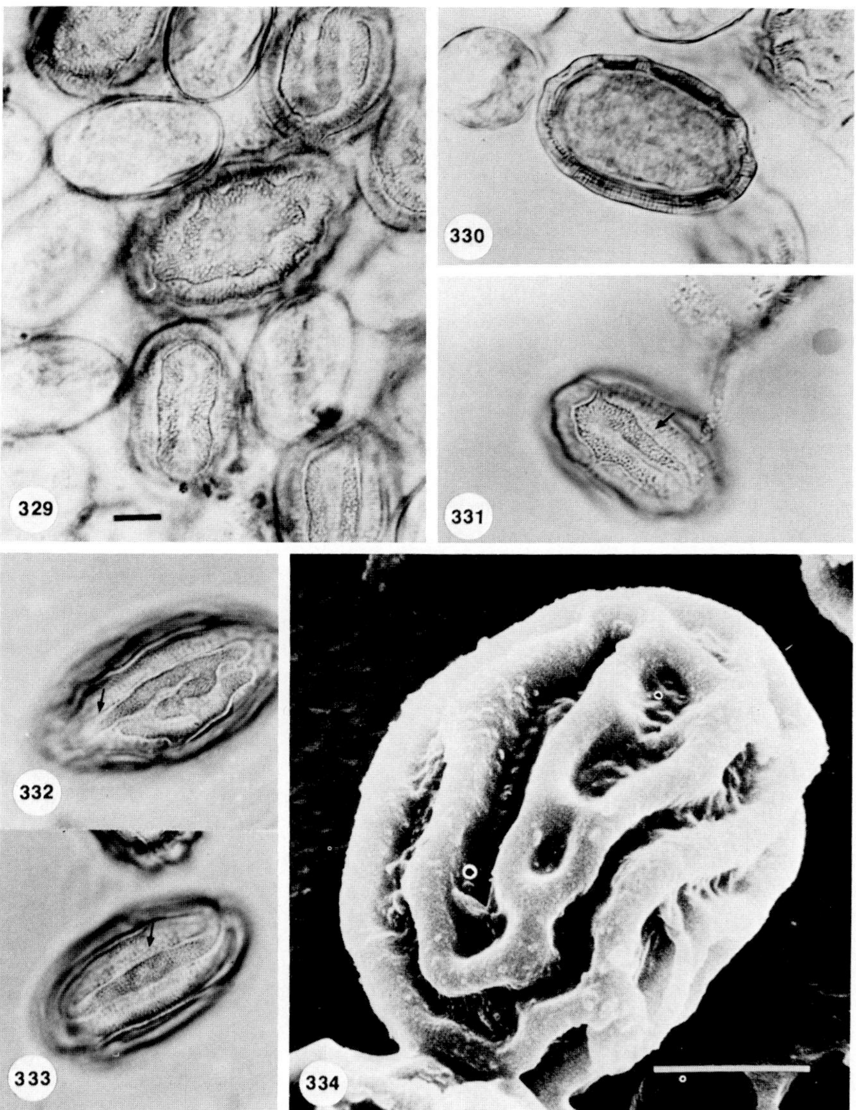

Figs. 329–331. *Coelomomyces ilinesis* var. *irregularis*. Resting sporangia of type specimen from *Culex modestus* collected in Kazakhstan, U.S.S.R. Fig. 329. Surface view of sporangia showing highly irregular forms and one regular sporangium. Bar = 10 μm, applies to Figs. 329–331. Fig. 330. Midoptical section, face view. Fig. 331. Optical section, face view showing dehiscence slit (arrow).

Figs. 332–334. *Coelomomyces iliensis* var. *irregularis*. Resting sporangia from *Anopheles aconitus* collected in Bengal, India. Fig. 332. Surface view, dorsal face (dehiscence slit, arrow) of a highly irregular sporangium. Scale as for Fig. 329. Fig. 333. Surface view, dorsal side (dehiscence slit, arrow) of a regular sporangium. Scale as for Fig. 329. Fig. 334. Scanning electron micrograph showing dorsal face of an irregular sporangium. Bar = 10 μm.

LATIN DIAGNOSIS: Hyphae non visae. Sporangia perennantia ellipsoidea, 24–36 × 37–54 μ, liris saepe enormibus aspectu, saepe anastomosantibus et depressiones elongatas vel polygonales, striis punctisque notatas includentibus vel ut in varietate typica decoratas; quasi in sectione media inspecta in aspectu frontali, liris instar alae introrsus sinuatae verticaliter striatae sporangium cingentis, e latere alae introrsus sinuatae secundum superficiem et dorsalem et ventralem extensae et ad polos instar tholorum 3–5, e cacumine instar tholorum 8–12 sporangium cingentium. Sporangia tenuiter tunicata ovoidea, 19–28 × 28–46.5 μm.

SPECIMENS EXAMINED OTHER THAN TYPE:

India
Anopheles aconitus Dön; Bengal; Iyengar; 1968; NCU.
Anopheles ramsayi Cov.; Bengal; Iyengar; 1968, 1969; NCU.
Anopheles varuna Iyen; Bengal; Iyengar; 1968; NCU.
U.S.S.R.
Culex modestus Fica.; Alma-Ata, Kazakhstan; Dubitskij; 1980; NCU.

COMMENTS: *Coelomomyces iliensis* var. *irregularis* may be distinguished from other varieties within this species on the basis of its resting sporangia with highly irregular ridging. In fact, ridging on some sporangia may be so irregular as to make identification virtually impossible. However, in all collections examined thus far there are present not only highly irregular sporangia but also sporangia practically indistinguishable from those of the typical *C. iliensis,* as well as intermediate forms. Although there are slight variations between the specimens collected in the U.S.S.R. on a culicine host and those collected in India on anopheline hosts (the primary difference being that resting sporangia of specimens from the U.S.S.R. have slightly more prominent internal wall striae), wall markings on resting sporangia of collections from both sites are identical; thus they are considered here as the same taxon.

Coelomomyces iliensis var. ***striatus*** Couch & Iyengar var. nov.

Figures 335–341

Figs. 335–341. *Coelomomyces iliensis* var. *striatus.* Resting sporangia of type specimen from *Culex tritaeniorhynchus* collected in Trivandrum, India. Fig. 335. Face view, dorsal surface (dehiscence slit, arrow). Bar = 10 μm, applies to Figs. 335–340. Fig. 336. Midoptical section, face view. Fig. 337. Surface view of sporangial side. Fig. 338. Midoptical section, side view. Fig. 339. Surface view of sporangial side. Fig. 340. Midoptical section, end view. Fig. 341. Scanning electron micrograph of sporangium showing dorsal face. Bar = 10 μm.

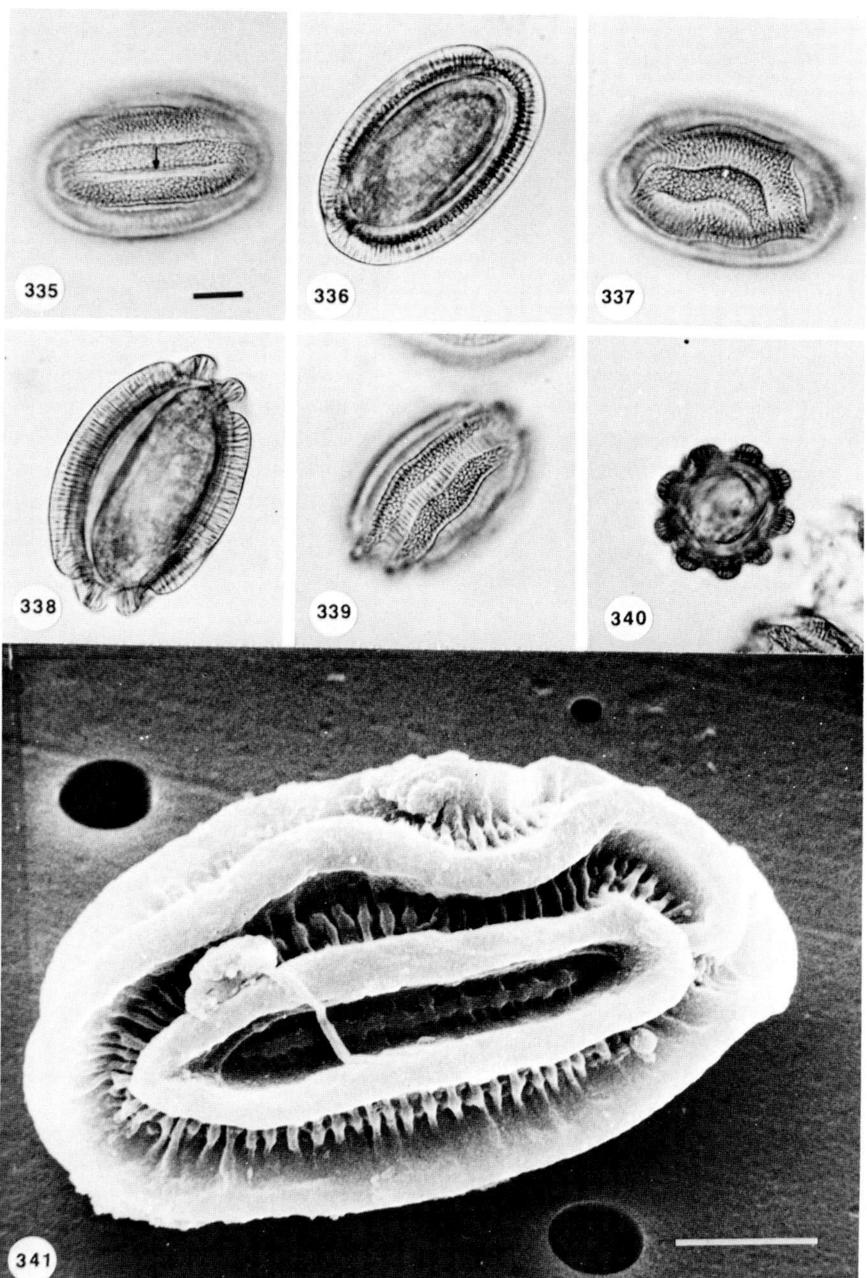
335
336
337
338
339
340
341

DESCRIPTION: Hyphae 10.5–19.5 μm in diameter. Resting sporangia as in typical variety, except measuring 37–52 × 46.5–82 μm, and with ridges measuring 5–10 μm wide, 6–10 μm high, and having very prominent internal vertical striae appearing as circular punctae in edge view of ridges. Depressed areas between ridges appearing by LM to have a central row of circular punctae. Thin-walled sporangia not observed.

TYPE HOST AND COLLECTION SITE: *Culex tritaeniorhynchus* Giles; Trivandrum, India (collected by Iyengar).

DEPOSITORY FOR TYPE: BPI.

LATIN DIAGNOSIS: Hyphae 10.5–19.5 μm diametro. Sporangia perennantia ut in varietate typica, sed 37–52 × 46.5–82 μm, liris 5–10 μm latis, 6–10 μm altis, striis internis verticalibus valde prominentibus, desuper inspectis specie punctis rotundis notatis; depressiones inter liras sub microscopio lucido inspectae specie serie centrali punctorum rotundorum notatae. Sporangia tenuiter tunicata non observata.

SPECIMENS EXAMINED OTHER THAN TYPE:

India

Culex tritaeniorhynchus Giles; Trivandrum; Iyengar; 1969; NCU.
Culex tritaeniorhynchus Giles; Pondicherry; Rajagopalan; 1980; NCU.
Culex vishnui (Theo.); Trivandrum; Iyengar; 1970; NCU.

COMMENTS: Although having ornamentation similar to that of the typical variety, *C. iliensis* var. *striatus* is distinguished by its larger resting sporangia and significantly more prominent internal vertical striae. See Comments for *C. iliensis* var. *indus*.

Coelomomyces indicus Iyengar ex Iyengar, *J. Elisha Mitchell Sci. Soc.* **78,** 134 (1962)

Coelomomyces indiana Iyengar, *Parasitology* **27,** 440 (1935)

Figures 342–348

Figs. 342–348. *Coelomomyces indicus.* Neotype specimen from *Anopheles subpictus* collected in Bangalore, India. Figs. 342 and 343. Surface view, dorsal side. Note the prominent dehiscence slit visible in the center of both figures. Fig. 342, bar = 10 μm, applies to Figs. 342–346. Figs. 344 and 345. Midoptical section, face view. Fig. 346. Thin-walled sporangia. Fig. 347. Thin-walled sporangia and thick-walled sporangia. Bar = 10 μm. Fig. 348. Scanning electron micrograph showing resting sporangia. Bar = 5 μm.

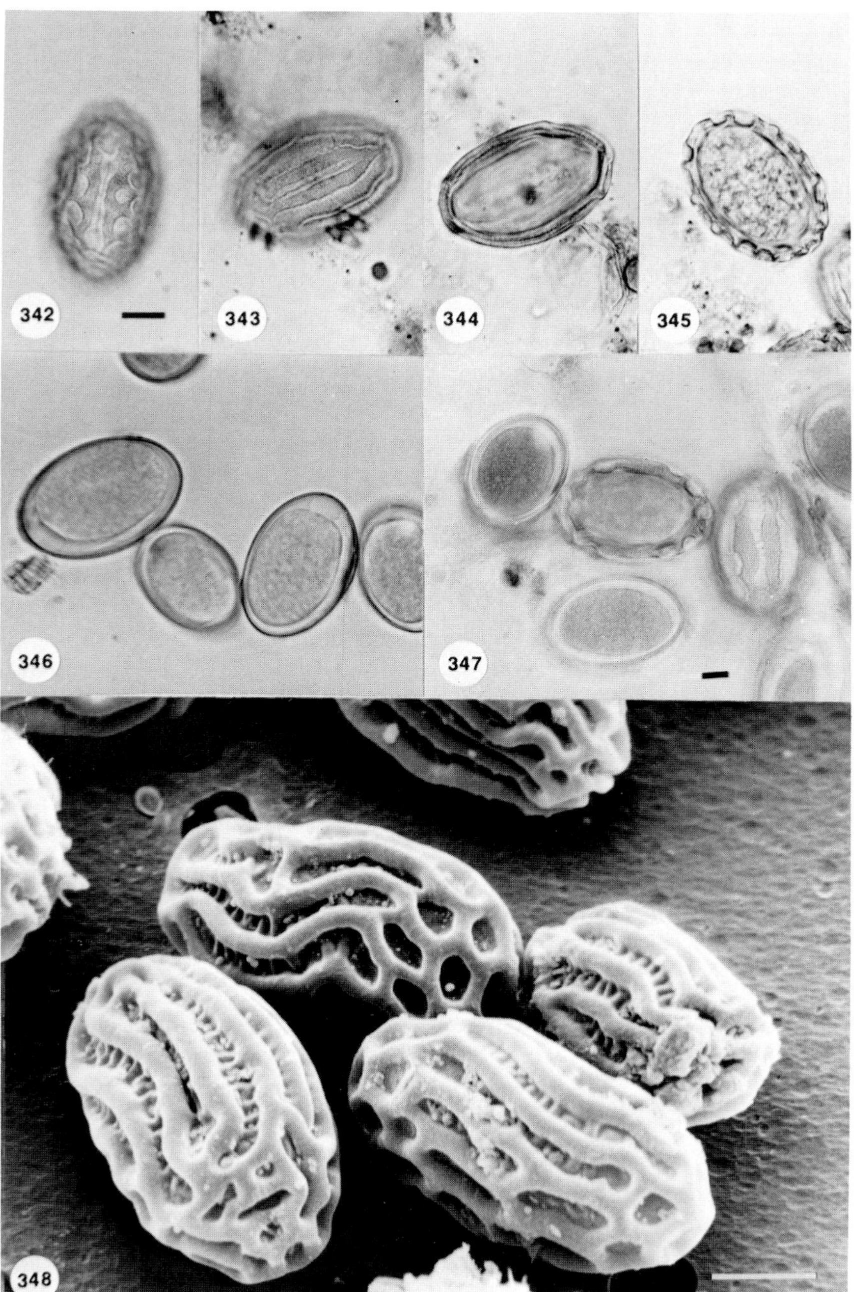
342
343
344
345
346
347
348

Description: Hyphae 7–14 μm in diameter. Resting sporangia ellipsoid, 25–41 × 29–65.5 μm; surface ornamented with elongate to frequently anastomosing ridges (3–5 μm wide and 3–5 μm high) enclosing elongate, polygonal, to circular depressed areas (3–4 μm wide and up to the length of the sporangium when elongate, 3–5 μm in diameter when circular) with transverse striae; dehiscence slit central along an elongate, longitudinal depressed area with transverse striae. Ridging usually anastomosing at poles to enclose circular depressed areas, or reticulate and delineating circular depressed areas over the entire sporangial surface. In midoptical section: face view, ridging as a vertically striate wing encircling the sporangium, wing appearing entire or frequently scalloped depending on frequency with which the ridges anastomose; side view, ridging as a striate wing along dorsal edge and as an entire to scalloped wing at poles and on ventral surface; end view, ridging as 10–16 striate domes encircling the sporangium. Thin-walled sporangia ovoid to ellipsoid, 15–30 × 25–43 μm.

Type Host and Collection Site: The original description (Iyengar, 1935) of *C. indicus* was based on specimens from larvae of *Anopheles barbirostris* Wulp, *Anopheles hyrcanus* var. *nigerrimus* Giles, *Anopheles subpictus* Grassi, *Anopheles aconitus* Dön, *Anopheles varuna* Iyengar, *Anopheles ramsayi* Covell, *Anopheles annularis* Wulp, and *Anopheles jamesi* Theo. collected from three locations in Bengal, India (Sonarpur, Krishnagar, and Meenglas). Although the collection site for the type was indicated to be Sonarpur, the type host was not designated (Iyengar, 1962). Since the type, retained in the private collection of Iyengar (deceased), is unavailable, a specimen collected by Iyengar on larvae of *Anopheles subpictus* Grassi from Bangalore and identified by him as *C. indicus,* is here designated a neotype.

Depository for Type: A neotype (see above) is deposited at BPI.

Specimens Examined and Other Collections (indicated by brackets):

Type (examined by Couch in 1962 and returned to Iyengar), **neotype**

Africa

[*Anopheles funestus* Giles; Livingstone, Zambia; Muspratt; 1941–1945; ?; (Muspratt, 1946a).]

[*Anopheles gambiae* Giles; Angola; Ribeiro; 1981; ?; (Ribeiro *et al.*, 1981).]

Anopheles gambiae Giles; Kisumu, Kenya; Service; 1973, 1974; NCU.

Anopheles gambiae Giles; Enugu, Nigeria; Service; 1975; NCU.

Anopheles gambiae Giles; Livingstone, Zambia; Muspratt; 1941, 1962; NCU; (Muspratt, 1946a,b, 1962, 1963a,b).

[*Anopheles gambiae* Giles*; Livingstone, Zambia; Madelin; ?; (Madelin, 1968).]

Anopheles pharoensis Theo.*; Cairo, Egypt; Gad; 1966, 1972; NCU; (Gad *et al.*, 1967; Gad and Sadek, 1968).

Anopheles pharoensis Theo.; Giza, Egypt; Gad; 1966; NCU; (Gad *et al.*, 1967).

Anopheles pretoriensis Theo.*; Livingstone, Zambia; Muspratt; 1943; NCU; (Muspratt, 1946a, 1962).

[*Anopheles rivulorum* Leeson; Livingstone; Zambia; Muspratt; 1941–1945; ?; (Muspratt, 1946a, 1962).]

[*Anopheles rufipes* (Gough); Livingstone, Zambia; Muspratt; 1941–1945; ?; (Muspratt, 1946a, 1962).]

[*Anopheles squamosus* Theo.; Angola; Ribeiro; 1981; ?; (Ribeiro *et al.*, 1981).]

[*Anopheles squamosus* Theo.; Livingstone, Zambia; Muspratt; 1941–1945; ?; (Muspratt, 1946a).]

[*Culex simpsoni* Theo.; Livingstone, Zambia; Muspratt; 1941–1945; ?; (Muspratt, 1946a).]

Asia

Anopheles subpictus Grassi; Phum Swai, Cambodia; Colless, NCU; (Laird, 1959b).

Anopheles vagus Dön; Chiangmai, Thailand; AFRIMS; 1972; NCU; (Hembree, 1979).

Australia

[*Aedomyia catastica* Knab; Innisfail, Queensland; Laird; 1956; ?; (Laird, 1956a).]

India†

Anopheles culicifacies Giles; Bangalore; Iyengar; 1963; NCU.

Anopheles jeyporensis James; Wynaad; Iyengar; 1941; NCU.

Anopheles pallidus Theo.; Bangalore; Iyengar; 1963; NCU.

Anopheles splendidus Koid.; Bangalore; Iyengar; 1963; NCU.

Anopheles stephensi Liston; Bangalore; Iyengar; 1962, 1963; NCU.

Anopheles subpictus Grassi; Bangalore; Iyengar; 1961, 1962, 1963, 1970; NCU.

Anopheles subpictus Grassi; Ghaziabad; Gugnani; 1964; NCU; (Gugnani *et al.*, 1965).

* Infection induced experimentally.

† Excluding specimens cited by Iyengar (1935).

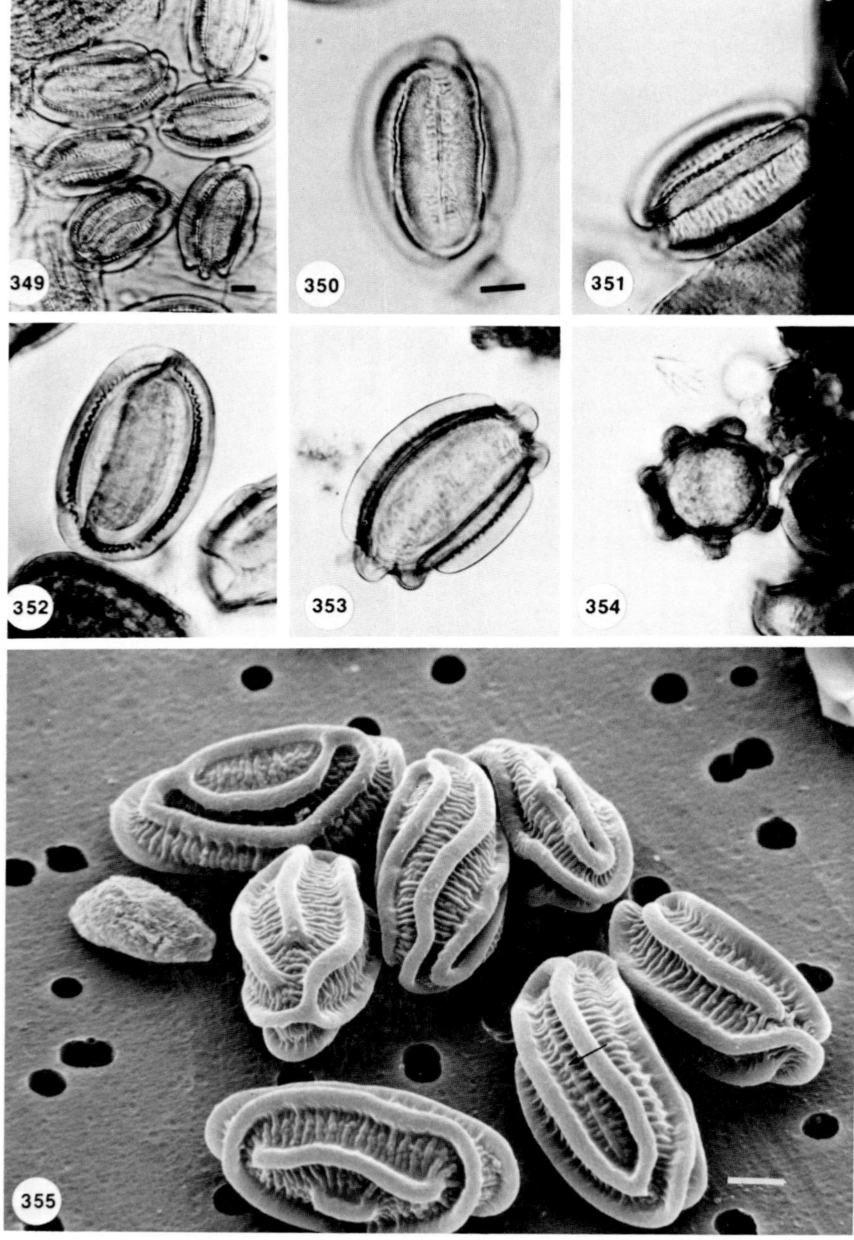

Figs. 349–355. *Coelomomyces iyengarii.* Resting sporangia of type specimen from *Anopheles vagus* collected in Trivandrum, India. Fig. 349. surface view of several sporangia. Bar = 10 μm. Fig. 350. Surface view, dorsal side (dehiscence slit, arrow). Bar = 10 μm, applies to Figs. 350–354. Fig. 351. Surface view of side. Fig. 352. Midopti-

Anopheles subpictus Grassi; Pondicherry; Rajagopalan; 1977; NCU.
Anopheles vagus Dön.; Bangalore; Iyengar; 1961, 1963; NCU.
[*Anopheles vagus* Dön.; Pondicherry; Rajagopalan; 1976–1977; ?; (Chandrahas and Rajagopalan, 1979).]

Philippines

Anopheles subpictus indefinitus (Lud.); San Pablo City; Villacarlos; 1972; NCU; (Villacarlos and Gabriel, 1974).
Anopheles subpictus indefinitus (Lud.); Los Banos; Chapman; 1979; NCU.

COMMENTS: Although several species of *Coelomomyces* are characterized by resting sporangia ornamented with ridges separated by depressed areas with striae, distinction of *C. indicus* may be made in that resting sporangia of this species are ornamented with not only elongate ridges, but also ridges anastomosing to form a reticulum at the poles or over the entire sporangial surface. While the degree of anastomosis of sporangial ridges varies from collection to collection, some sporangia of all types (elongate ridges only, elongate ridges with polar anastomoses, and ridges anastomosing as a reticulum over all or most of the sporangial surface) are usually present in all collections. Additionally, although the SEM shows transverse striae to always be present in depressed areas between sporangial ridges, such striae are at times difficult to discern via LM. Careful scrutiny at magnifications of ×400 or above should, however, reveal them.

The report by Gugnani *et al.* (1965) of *C. indicus* occurring in specimens of *Daphnia* sp. is considered questionable, since examination of one of the specimens on which this determination was based showed sporangia to be confined to the gut. As the specimens of *Daphnia* sp. were collected from a site in which infected larvae of *Anopheles subpictus* had been collected also, it is likely that the resting sporangia had been released from moribund larvae and subsequently ingested.

Coelomomyces iyengarii Couch sp. nov.

Figures 349–355

DESCRIPTION: Hyphae 8–15 μm in diameter. Resting sporangia elongate–ellipsoid, 37–46.5 × 48–76 μm; surface ornamented with prominent,

---

cal section, face view. Note prominent teeth at periphery. Fig. 353. Midoptical section, side view. Fig. 354. Midoptical section, end view. Fig. 355. Scanning electron micrograph showing several sporangia. Note the prominent dehiscence slit visible in one of the sporangia (arrow). Bar = 10 μm.

elongate to rarely anastomosing ridges (5–7 μm wide and 5–7 μm high) separated by depressed areas (8–12 μm wide) with prominent transverse striae (2.5 μm wide); ridging surrounding the elongate dehiscence slit or beginning along one side of the slit and continuing concentrically or spirally around the sides to end on the ventral surface as isolated elongate ridges, a continuous elliptical ridge, or rarely in an irregular pattern; four ridges generally visible in face views of sporangia; dehiscence slit central in depressed area with transverse striae, bordered by ridging. In midoptical section: face view, ridging as an internally striate wing surrounding the sporangium, transverse striae in depressed areas appearing in edge view as teeth along proximal edge of the wing; side view, ridging as an elongate, striate wing along dorsal and ventral surfaces and as two striate domes at the poles; end view, ridging as 6–8 striate domes alternating with depressed areas at periphery. Thin-walled sporangia not observed.

Type Host and Collection Site: *Anopheles vagus* Dön.; Trivandrum, India (collected by Iyengar).

Depository for Type: BPI.

Latin Diagnosis: Hyphae 8–15 μm diametro. Sporangia perennantia elongato–ellipsoidea, 37–46.5 × 48–76 μm; superficie liris prominentibus, elongatis, raro anastomosantibus 5–7 μm latis et altis, decoratis ab depressionibus 8–12 μm latis, striis prominentibus transversis 2.5 μm latis notatis; liris incisuram dehiscentiae elongatam cingentibus vel secundum latus incisurae unum ortis et concentrice vel spiraliter productis circum latera et in superficie ventrali ut lirae sejunctae elongatae, vel lira continua elliptica, vel raro in descriptione enormi; adversus frontem sporangii lirae quattuor plerumque visibiles; incisura dehiscentiae in depressione centralis, striis transversis notata, ab liris marginata; quasi in sectione media inspecta adversus frontem liris instar alae intus striatae sporangium cingentis, in depressionibus striae ad instar dentium secundum marginem alae proximum observatae; a latere, liris instar alae elongatae, striatae secundum superficiem et dorsalem et ventralem, et ad polos tholorum striatorum duorum; ab cacumme, liris instar 6–8 tholorum striatorum ad marginem cum depressionibus alternantium. Sporangia tenuiter tunicata non observata.

Specimens Examined Other than Type:

Asia

Anopheles aconitus Dön.; Chiangmai, Thailand; AFRIMS; 1971; NCU; (Hembree, 1979).

Anopheles vagus Dön.; Chiangmai, Thailand; AFRIMS; 1972; NCU; (Hembree, 1979).

Anopheles vagus Dön.; Gia-tink, South Viet Nam; Nguyen-Dang-Que; 1964; NCU.

India

Anopheles jamesi Theo.; Trivandrum; Iyengar; 1969, 1971; NCU.
Anopheles pallidus Theo.; Trivandrum; Iyengar; 1971; NCU.
Anopheles subpictus Grassi; Trivandrum; Iyengar; 1969, 1971; NCU.
Anopheles vagus Dön.; Trivandrum; Iyengar; 1968, 1969, 1970, 1971; NCU.

COMMENTS: Although resting sporangia of *C. iyengarii* are ornamented with ridging in patterns resembling those of sporangia of *C. iliensis* varieties, distinction of *C. iyengarii* may be made in that resting sporangia of this species are more elongate–ellipsoid and generally larger than those of *C. iliensis,* have broader, more prominent ridging, and wider, more prominent transverse striae between ridges. Another distinctive feature of sporangia of *C. iyengarii* is the characteristic appearance in optical section of the transverse striae as large teeth at the sporangial periphery. Although similar teeth are evident on sporangia of *C. iliensis,* in *C. iyengarii* such teeth are much more obvious.

The specific epithet of *C. iyengarii* is recognition of the significant contributions to this group made by Dr. M. O. T. Iyengar.

Coelomomyces keilini Couch & Dodge ex Couch, *J. Elisha Mitchell Sci. Soc.* **78,** 135 (1962)

Coelomomyces keilini Couch & Dodge, *J. Elisha Mitchell Sci. Soc.* **63,** 69 (1947)

Figures 356–358

DESCRIPTION: Hyphae 5–25 μm thick × 60–170 μm long, subdichotomously branched. Resting sporangia ellipsoid, usually slightly flattened on one side, 34–46 × 58–71 μm; surface ornamented with "slit-like" punctae appearing "star-like" in optical section as one focuses immediately below the surface; punctae 0.5–1.5 μm wide and 0.5–1.5 μm apart. In midoptical section, punctae appearing "spindle-shaped" and wall with fine, vertical striae between the punctae. Wall 3–7 μm thick. Thin-walled sporangia not observed.

TYPE HOST AND COLLECTION SITE: *Anopheles crucians* Wiede.; Valdosta, Georgia (collected by SMCG).

DEPOSITORY FOR TYPE: The type slide, deposited at NCU, has been lost.

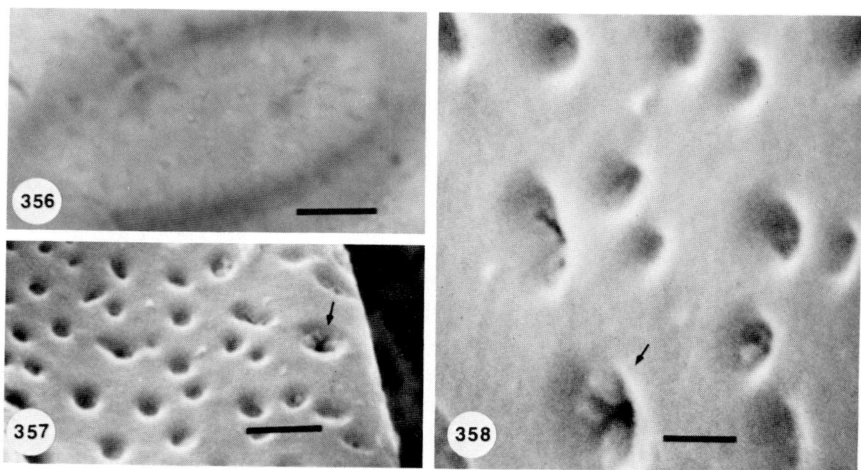

Figs. 356–358. *Coelomomyces keilini.* Resting sporangia of type from *Anopheles crucians* collected in Valdosta, Georgia. Fig. 356. Surface view of sporangial side showing "star-like" punctae. Bar = 10 μm. Figs. 357 and 358. Scanning electron micrographs showing sporangial surface. Note the occurrence of both regular punctae and "star-like" punctae (arrows). Fig. 357, bar = 5 μm. Fig. 358, bar = 1 μm.

SPECIMENS EXAMINED AND OTHER COLLECTIONS:

Type

COMMENTS: The "slit-like" to "star-like" punctae of *C. keilini* distinguish it from all other species of *Coelomomyces*. However, since it has only been collected once, and since the type specimen has been lost, further study of this taxon must await additional collections.

Coelomomyces lacunosus sp. nov. var. *lacunosus* Couch & Sousa

Figures 359–363

Figs. 359–363. *Coelomomyces lacunosus* var. *lacunosus*. Resting sporangia of the type from *Trichoprosopon longipes* collected in Darien, Panama. Figs. 359 and 360. Surface view of resting sporangia. Bars = 10 μm, applies to Figs. 359–362. The dehiscence slit (arrow) is visible in two of the specimens. Fig. 261. Midoptical section, side view. Fig. 362. Midoptical section, end view. Note that the dehiscence slit is visible along the dorsal side of the sporangium. Fig. 363. Scanning electron micrograph showing two partially collapsed resting sporangia. Bar = 10 μm.

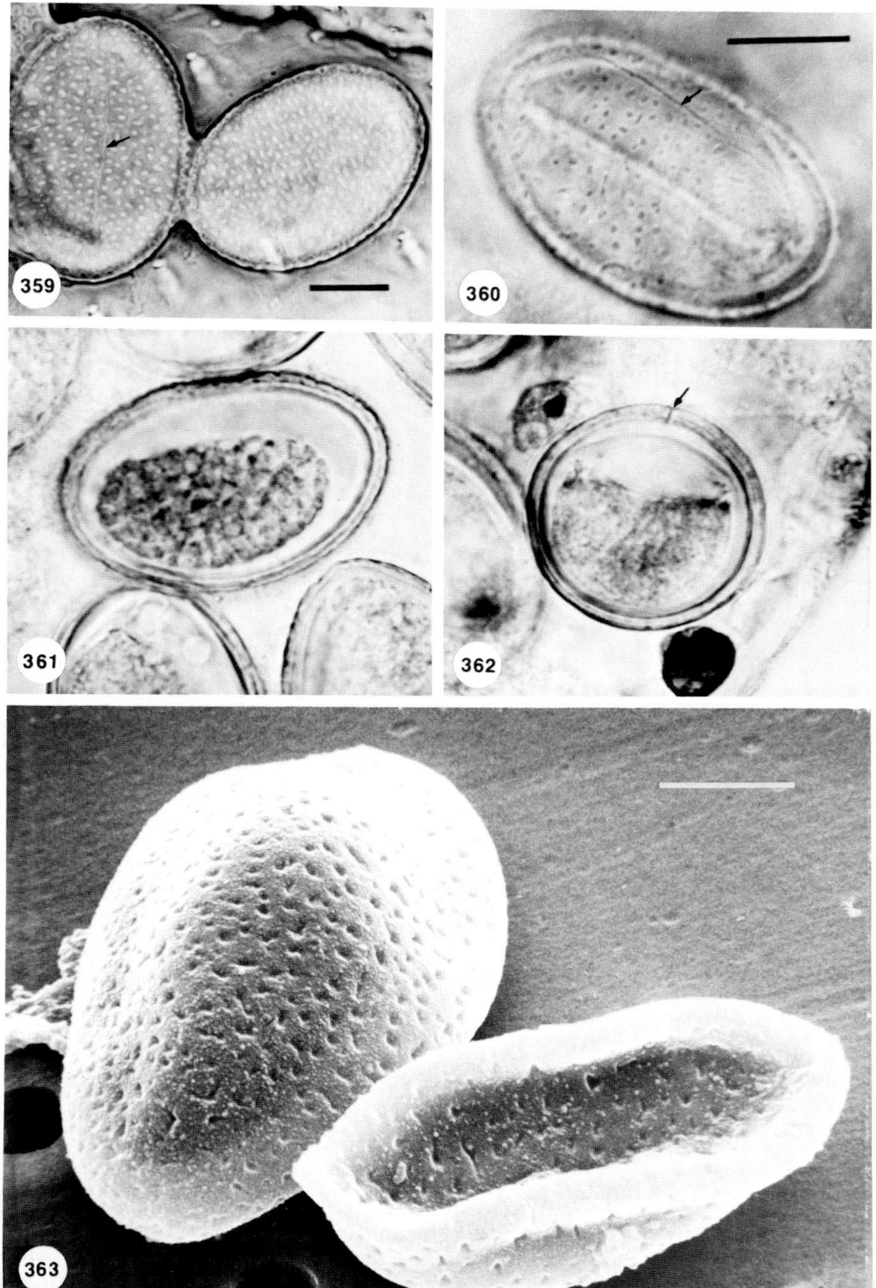

359
360
361
362
363

DESCRIPTION: Hyphae not observed. Resting sporangia broadly ellipsoid, 17–33 × 25–57 μm; surface ornamented with numerous, irregularly shaped, shallow lacunae and punctae (1–2 μm wide × 1–3 μm long and 1–4 μm apart); dehiscence slit bordered by lacunae. In midoptical section: sporangial periphery appearing irregular, scalloped to dentate as a result of wall lacunae. Thin-walled sporangia not observed.

TYPE HOST AND COLLECTION SITE: *Trichoprosopon longipes* (Fab.) ♀, La Laguna, Pucro, Darien, Panama (sent by Gorgas Memorial Laboratory, Darien, Panama to Department of Biology, University of Calfornia, Los Angeles, California, thence to NCU).

DEPOSITORY FOR TYPE: BPI. Three isotypes are on deposit at NCU.

LATIN DIAGNOSIS: Hyphae non visae. Sporangia perennantia late ellipsoidea, 17–33 × 25–57 μm; superficie lacunis et punctis numerosis enormibus haud altis, 1–2 μm latis × 1–3 μm longis, inter se 1–4 μm distantibus decorata; incisura dehiscentiae a lacunis marginata; quasi in sectione media inspecta margine sporangii aspectu enormi, ob lacunas tunicae introrsus sinuato vel dentato. Sporangia tenuiter tunicata non observata.

SPECIMENS EXAMINED OTHER THAN TYPE AND ISOTYPES:

Panama

Sabethes cyaneus (Fab.) ♀; Carti, San Blas Territory; Sousa; 1974; NCU.

COMMENTS: Although sporangial ornamentation in *C. lacunosus* vars. is similar to that of *C. canadensis,* depressed areas (lacunae) on the sporangial surface of *C. lacunosus* vars. are smaller than in *C. canadensis* and lack the internal, radiating striae found in depressed areas of this species. Similarity in ornamentation exists also between resting sporangia of *C. lacunosus* var. and *C. punctatus* and *C. utahensis*. However, resting sporangia of both *C. punctatus* and *C. utahensis* are generally larger than in *C. lacunosus* vars. Also, punctae/striae of *C. punctatus* are usually linearly arranged, as opposed to random and irregular in *C. lacunosus* vars., and those of *C. utahensis* are finer and more regular than in *C. lacunosus* vars.

Coelomomyces lacunosus var. *lacunosus* has been collected only twice in Panama, then only from mosquitoes characteristically inhabiting plant cavities—a *Dieffenbachia* plant axil in the case of the type specimen. The specific epithet for this new species is from *lacunosus* L. (covered with depressions), in recognition of the unique wall markings on resting sporangia of this species.

Coelomomyces lacunosus var. *madagascaricus* Couch & Rodhain var. nov.

Figures 364–368

Description: Hyphae not observed. Resting sporangia as in typical variety except dehiscence slit in prominent ridge that is devoid of lacunae/punctae. In midoptical section: side view, ridge with dehiscence slit as a wing along dorsal surface; end view, ridge as a single dome at dorsal edge, dehiscence slit generally evident as a single vertical stria in center of dome. Thin-walled sporangia not observed.

Type Host and Collection Site: *Uranotaenia* sp. Périnet, Madagascar (collected by Rodhain).

Depository for Type: BPI.

Latin Diagnosis: Hyphae non visae. Sporangia perennantia ut in varietate typica, incisura dehiscentiae in lira prominenti lacunis et punctis carenti excepta; quasi in sectione media inspecta a latere, lira incisuram includenti instar alae per superficiem dorsalem, ab cacumine, lira instar tholi singularis ad marginem dorsalem, incisura plerumque manifesta instar striae singularis verticalis in tholi medio inclusae. Sporangia tenuiter tunicata non observata.

Specimens Examined Other than Type:

Madagascar

Uranotaenia douceti Grje.; Sokobé, Nossi-Bé, Majunga; Grjebine; 1955; NCU.

Uranotaenia sp.; Maroantsetra; Rodhain; 1979; NCU.

Comments: *Coelomomyces lacunosus* var. *madagascaricus* may be distinguished from the typical variety in that resting sporangia of the former are characterized by the dehiscence slit being central along a prominent longitudinal ridge, whereas a similar ridge is lacking in the typical variety.

As for the typical variety, *C. lacunosus* var. *madagascaricus* has been collected only on mosquito hosts typically occupying tree cavities.

The varietal designation for *C. lacunosus* var. *madagascaricus* is in recognition of this variety having been collected only in Madagascar.

Coelomomyces lairdi Maffi & Nolan, *J. Med. Entomol.* **14,** 29–32 (1977)

Figures 369–372

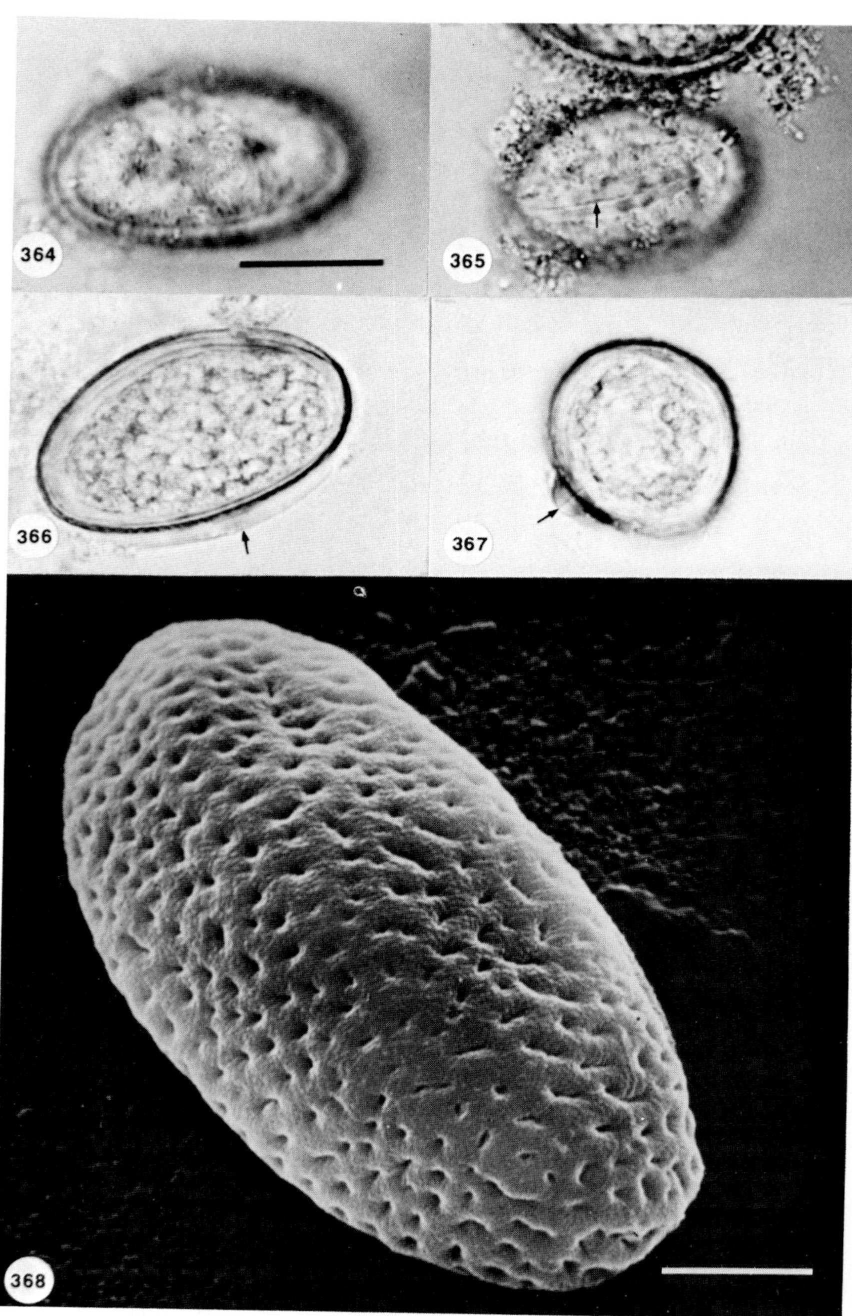
364
365
366
367
368

DESCRIPTION: Hyphae not observed. Resting sporangia ellipsoid, 22–33 × 35–57 μm surface ornamented with anastomosing ridges (2–2.5 μm wide × 1.2–2.5 μm high), enclosing circular to rarely polygonal depressed areas (5–9 μm in diameter) with scattered central punctae; dehiscence slit central along an elongate ridge bordered by circular to polygonal depressed areas. In midoptical section, ridges as hyaline domes alternating with depressed areas around sporangial periphery. Wall 2.0–2.5 μm thick. Thin-walled sporangia not observed.

TYPE HOST AND COLLECTION SITE: *Anopheles* (*Cellia*) *punctulatus* "complex"; Irian Java, Indonesian, New Guinea.

DEPOSITORY FOR TYPE: BISH (slide #740111/3).

SPECIMENS EXAMINED AND OTHER COLLECTIONS:

Type

Solomon Islands

Anopheles farauti Laveran; Varagona, Guadalcanal; Maffi; 1968; NCU; (Maffi and Genga, 1970).

COMMENTS: Although the resting sporangia of *C. lairdi* were described originally as being "boat-shaped, with a strongly convex surface and a moderately concave one" (Maffi and Nolan, 1977), the sporangia on which this description was based were obviously collapsed due to their having been mounted in Euparal. Examination of sporangia that have been rehydrated and mounted in lactophenol reveals their more likely ellipsoid shape and slightly greater dimensions than reported originally. Also, careful observation by both LM and SEM reveals the depressed sporangial wall areas of *C. lairdi* to have scattered, central punctae rather than lacking either pits or striae as described originally (Maffi and Nolan, 1977). It is not known whether the papillae revealed via SEM to occur at the periphery of depressed areas in the sporangial wall of *C. lairdi* are real or artifactual. They are therefore not included here as a species characteristic.

Figs. 364–368. *Coelomomyces lacunosus* var. *madagascaricus.* Resting sporangia of type specimen from *Uranotaenia* sp. collected from Perinet, Madagascar. Fig. 364. Surface view of sporangial side. Bar = 10 μm, applies to Figs. 364–367. Fig. 365. Surface view, dorsal side (dehiscence slit, arrow). Fig. 366. Midoptical section, side view. Note that the dorsal ridge appears as a wing along one side of the sporangium (arrow). Fig. 367. Midoptical section, end view. The dehiscence slit is visible in the center of the middorsal ridge (arrow). Fig. 368. Scanning electron micrograph showing side view of a resting sporangium. Bar = 5 μm. [Photograph courtesy of Station Centrale de Microscopie Electronique, Institut Pasteur de Paris.]

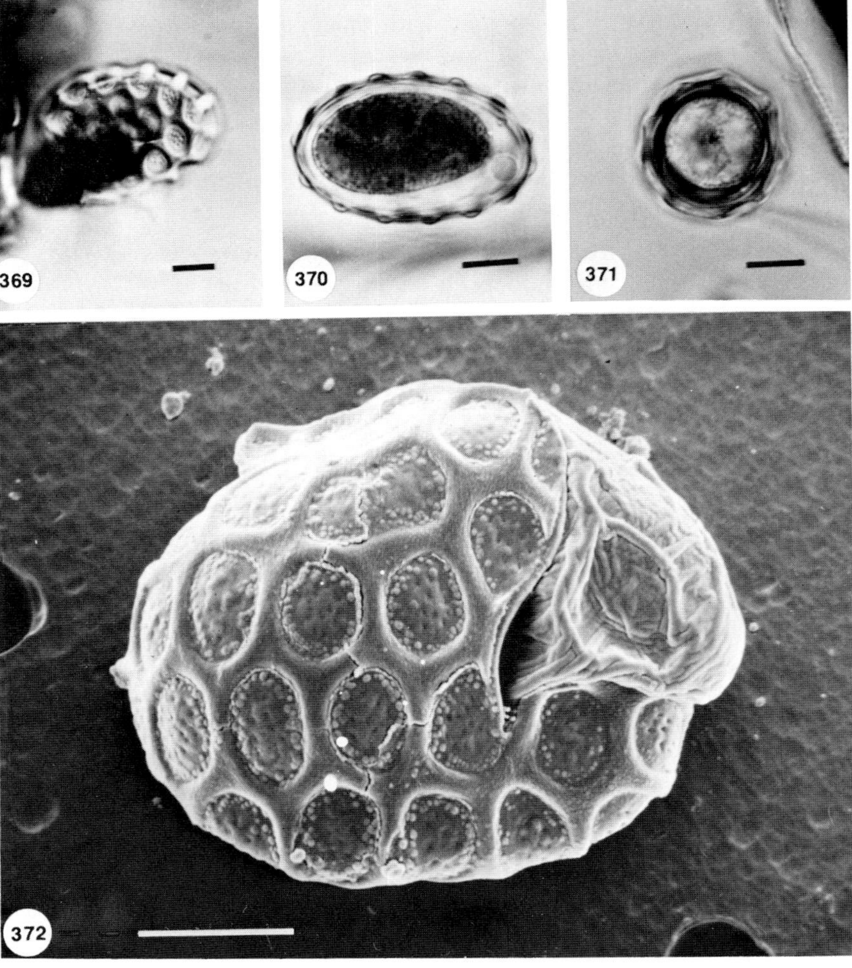

Figs. 369–372. *Coelomomyces lairdi.* Resting sporangia of type specimen collected from *Anopheles punctulatus* "after the punctulatus complex" from Indonesian, Guinea. Bars = 10 μm. Fig. 369. Surface view of resting sporangium showing punctae in depressed areas visible under interference contrast optics. Fig. 370. Midoptical section, side view. Fig. 371. Midoptical section, end view. Fig. 372. Scanning electron micrograph of ruptured sporangium.

Sporangial ornamentation in *C. lairdi* is very similar to that of both *C. couchii* and *C. reticulatus*. Distinction from both may be made in that depressed areas in the sporangial wall of *C. lairdi* bear only a few, scattered, central punctae, whereas depressed areas in *C. couchii* and *C. reticulatus* are uniform and prominently punctate. Also, in sporangia of *C. reticulatus* and *C. lairdi*, the ridge bearing the dehiscence slit is bordered by circular depressed areas, whereas that of *C. couchii* is bordered by elongate, depressed areas.

Coelomomyces lativittatus Couch & Dodge ex Couch, *J. Elisha Mitchell Sci. Soc.* **78,** 135 (1962)

Coelomomyces lativittatus Couch & Dodge, *J. Elisha Mitchell Sci. Soc.* **63,** 69 (1947)

Coelomomyces dodgei Couch, *J. Elisha Mitchell Sci. Soc.* **61,** 124 (1945)

Figures 373–380

DESCRIPTION: Hyphae 4–10 μm thick, irregularly branched. Resting sporangia ellipsoid, occasionally flattened on one side, 29–36 × 40–77; dorsal surface ornamented with linear depressions, beginning as generally continuous concentric striae, surrounding the dehiscence slit and continuing to encircle the sporangium, becoming irregular to rarely punctate on ventral surface; striae separated by 3.0–6.3 μm, giving the sporangia a distinctly banded appearance. In midoptical section: side view, bands as 2–5 ridges at poles; end view, bands as ridges encircling the sporangium. Wall 1.5–4.0 μm thick. Thin-walled sporangia not observed.

TYPE HOST AND COLLECTION SITE: *Anopheles crucians* Wiede.; Moultrie, Georgia (collected by SMCG).

DEPOSITORY FOR TYPE: BPI. An isotype is on deposit at NCU.

SPECIMENS EXAMINED AND OTHER COLLECTIONS:

Type, Isotype

United States

Anopheles crucians Wiede.; Georgia; Walker; 1944; NCU.
Anopheles crucians Wiede.; Georgia; SMCG; 1945; NCU.
Anopheles crucians Wiede.; North Carolina; Roane; 1967; NCU.
Anopheles earlei Vargas; Minnesota; Barr; 1954; NCU; (Barr, 1958; Laird, 1961).
Anopheles punctipennis (Say); North Carolina; Roane; 1967; NCU.

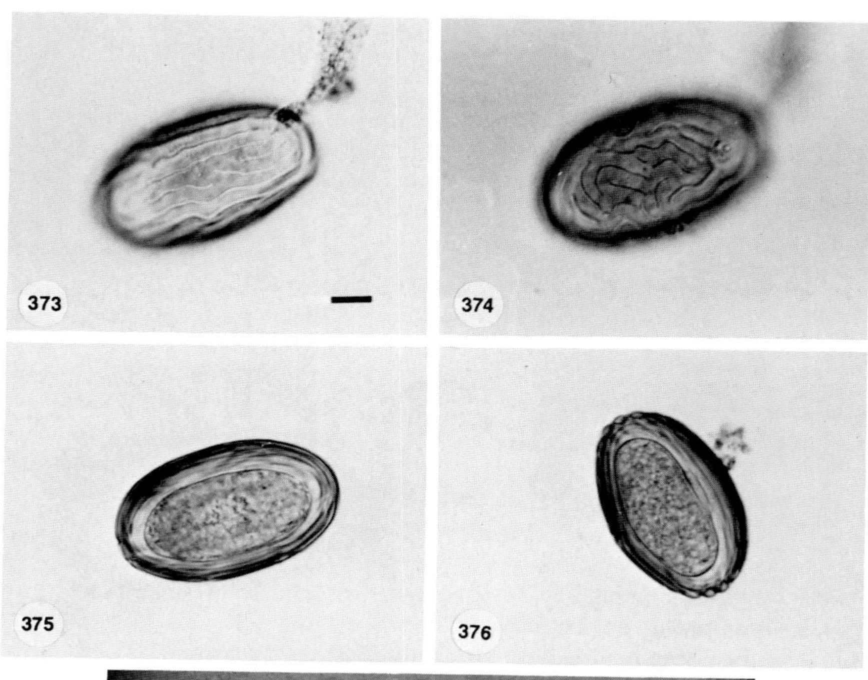
373
374
375
376

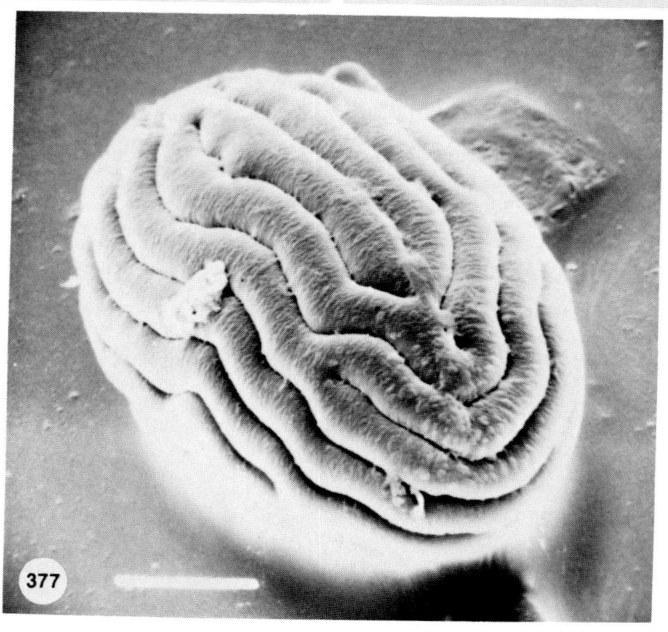
377

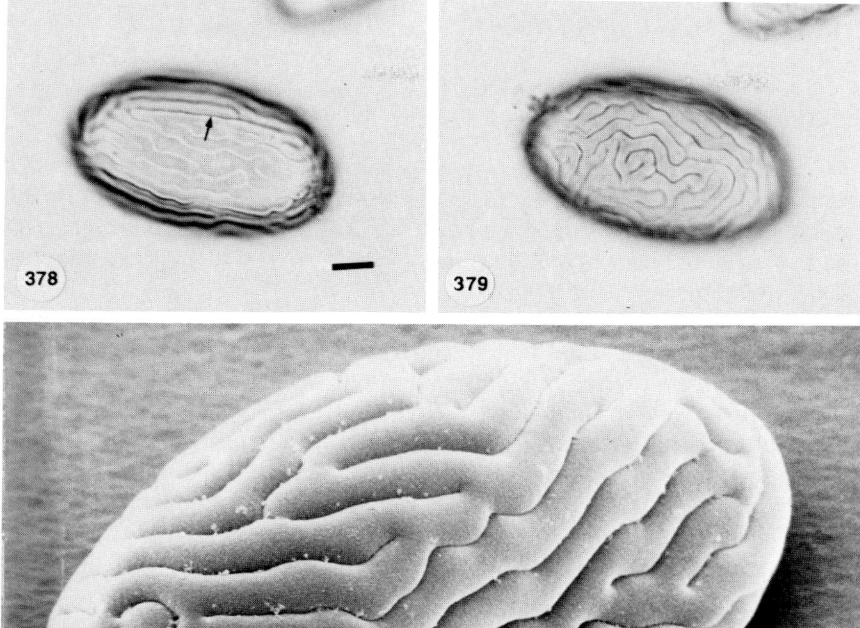

Figs. 378–380. *Coelomomyces lativittatus.* Resting sporangia from *Anopheles punctipennis* collected in Connecticut. Bars = 10 μm. Figs. 378 and 379. Surface views of the dorsal face (Fig. 378) and ventral face (Fig. 379) of the same sporangium. Note the dehiscence slit (arrow) visible in the dorsal face of the sporangium shown in Fig. 378. Fig. 380. Scanning electron micrograph showing side view.

Figs. 373–377. *Coelomomyces lativittatus.* Resting sporangia of type specimen from *Anopheles crucians* collected in Moultrie, Georgia. Figs. 373 & 374. Dorsal (Fig. 373) and ventral (Fig. 374) views of the same sporangium. Bar = 10 μm, applies to Figs. 373–376. Fig. 375. Midoptical section, face view. Fig. 376. Midoptical section, side view. Fig. 377. Scanning electron micrograph. Bar = 10 μm.

Anopheles punctipennis (Say); Ohio; Mead; 1947; NCU.
Anopheles punctipennis (Say); Connecticut; Anderson; 1965; NCU.
Anopheles punctipennis (Say); Michigan; Broome; 1978; NCU.
Anopheles sp.; North Carolina; Bond; 1962; 1964; NCU.
Anopheles sp.; Pennsylvania; Martin; 1979; NCU.

COMMENTS: Although sporangial ornamentation in *C. lativittatus* is similar to that of *C. dodgei, C. lativittatus* may be separated from this species in that the sporangial wall of *C. dodgei* is ornamented with narrow bands *and* punctae, whereas that of *C. lativittatus* is very rarely punctate (see Comments for *C. dodgei*).

While early collections of *C. lativittatus* included only forms with relatively small sporangia and widely separated striae giving the appearance of broad bands, more recent collections from Georgia, Connecticut, Ohio, and Minnesota have included specimens with sporangia slightly larger than the type and with less widely separated striae, giving the appearance of smaller, more numerous bands. Since recognition of the two forms is usually quite easy, it is possible that the larger form should be described as a variety of *C. lativittatus*. However, characteristics for the two do overlap. For this reason, the species description is here broadened to include both forms. Additional collections will, no doubt, reveal whether the two should be separated or retained as a single species.

In studies concerning experimental hybridization of *C. dodgei* and *C. punctatus,* Federici (1979) reported that such crosses often produced sporangia resembling those of *C. lativittatus,* indicating this species may be a naturally occurring hybrid. However, later studied by B. A. Federici (personal communication) revealed meiospores from hybrid sporangia to be incapable of infecting copepods, thus nonviable. The taxonomic significance of this is that, if meiospores of naturally occurring *C. lativittatus* can infect copepods, it is a valid species and not a hybrid (B. A. Federici, personal communication).

Coelomomyces macleayae Laird ex Laird, *J. Elisha Mitchell Sci. Soc.* **78,** 135 (1962)

Coelomomyces macleayae Laird, *Can. J. Zool.* **37,** 781 (1959)

Figures 381–391

DESCRIPTION: Hyphae 6–10 μm wide. Resting sporangia ellipsoid, 17–37 × 35–62 μm; surface ornamented with narrow, irregular ridges (1–2 μm high), anastomosing to longitudinal, often extending from pole to pole, to occasionally reticulate on ventral surface, enclosing elongate to polygonal depressed areas; depressed areas 6–9 μm wide, with linearly arranged, transverse punctae appearing as transverse striae. In midoptical section: side view, ridges (edge view) as teeth interconnected by a wing (ridges in side view) encircling the sporangium; end view, ridges as 5–10 triangular teeth around sporangial periphery. Wall 1–14 μm thick. Thin-walled sporangia not observed.

TYPE HOST AND COLLECTION SITE: *Aedes* (*Macleaya*) sp.; Palm Island, North Queensland, Australia (collected by Dr. M. J. Mackerras).

DEPOSITORY FOR TYPE: The type (deposited in the Department of Entomology, University of Queensland, Queensland, Australia) has been lost. A cotype is on deposit at BPI and NCU. Also, a paratype, prepared by Pillai from a specimen provided by Laird, is on deposit at the Queensland Institute of Medical Research, Brisbane, Australia.

SPECIMENS EXAMINED AND OTHER COLLECTIONS (indicated by brackets):

Cotypes (BPI and NCU), **paratype** (QIMR)

Burma

Aedes aegypti (L.); Rangoon; Muspratt; 1963, NCU; (Muspratt, 1964).

Fiji

Aedes polynesiensis Marks; Koro Is.; Pillai; 1968; ?; (Pillai and Rakai, 1970).

Aedes polynesiensis Marks; Waieyo, Taveuni; Pillai; 1970; NCU.

Aedes polynesiensis Marks; Korods.; Pillai; 1970; NCU.

Taiwan

Aedes (*Stegomyia*) *alcasida* Huang; Hangchun, Pingtung; ?; 1955; ?; (Laird *et al.*, 1980).]

[*Aedes* (*Finlaya*) *japonicus shintienensis* Tsai and Lein; Tajun, Tartung; ?; 1956; (Laird *et al.*, 1980).]

[*Tripteroides bambusa* (Yamada); Wulai, Taipei; ?; 1976; ?; (Laird *et al.*, 1980).]

Thailand

Aedes albopictus (Skuse); Kanchanaburi; AFRIMS; ?; NCU; (Hembree, 1979).

Ae. mediopunctatus Theo.; Kanchanaburi; AFRIMS; ?; NCU; (Hembree, 1979).

Ae. pseudalbopictus Borel; Kanchanaburi; AFRIMS; ?; NCU; (Hembree, 1979).

Armigeres longipalpis (Leic.); Nan; AFRIMS; ?; NCU; (Hembree, 1979).

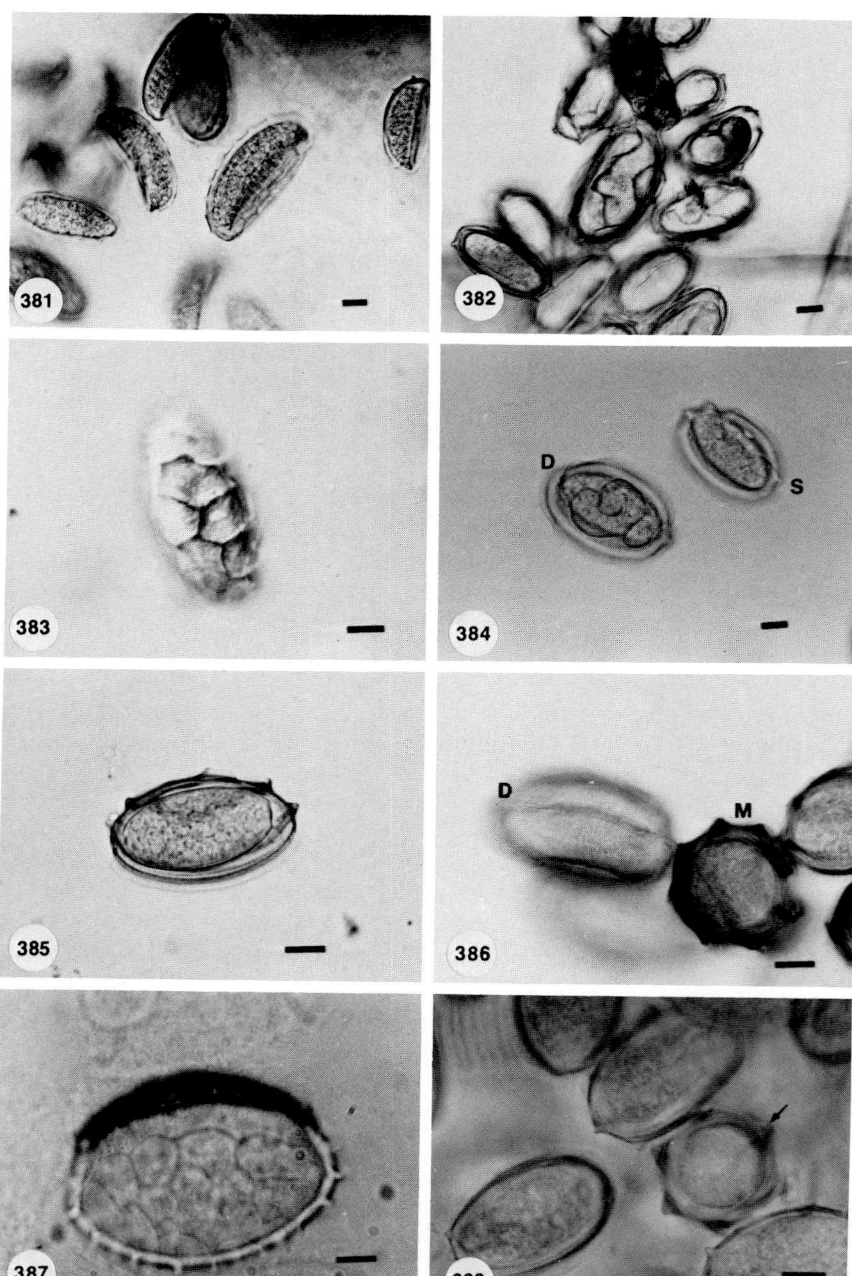
381
382
383
D
S
384
385
D
M
386
387
388

Toxyrhynchites gravelyi (Ed.); Kanchanaburi; AFRIMS; ?; NCU; (Hembree, 1979).

Toxyrhynchites sp.; Kanchanaburi; AFRIMS; ?; NCU; (Hembree, 1979).

United States

Aedes triseriatus (Say); Louisiana; Chapman; 1968; NCU; (Chapman *et al.*, 1969).

[*Toxyrhynchites rutilus septentrionalis* (Dyar and Knab); Louisiana; Chapman; 1971; MUN; (Nolan *et al.*, 1973).]

COMMENTS: Sporangia of *C. macleayae* are similar to those of both *C. cairnsensis* and *C. finlayae* in that all are characterized by ornamentation as ridges enclosing irregular polygonal areas with punctae. *Coelomomyces macleayae* may be distinguished from each, however, in that its sporangia have elongate ridges (always present on dorsal face, frequently on both faces) and punctae arranged in rows giving the appearance of striae between the ridges. One should note also the highly variable pattern formed by the ridges on the sporangial surface in *C. macleayae,* ranging from mostly reticulate to mostly irregularly elongate and extending from pole to pole. Sporangia exhibiting both types of ornamentation, including intermediate variations, are usually present in any sample, although one form may be more abundant.

Coelomomyces madagascaricus Couch & Grjebine sp. nov.

Figures 392–395

Fig. 381. *Coelomomyces macleayae.* Resting sporangia of type specimen from *Aedes* sp. from North Queensland, Australia. Note that these specimens appear boat-shaped due to their having been mounted in Canada balsam. Bar = 10 μm.

Fig. 382. *Coelomomyces macleayae.* Resting sporangia from *Aedes triseriatus* collected in Louisiana. Bar = 10 μm.

Figs. 383–386. *Coelomomyces macleayae.* Resting sporangia from *Aedes aegypti* collected in Rangoon, Burma. Fig. 383. Surface view of side. Fig. 384. Dorsal view (D) and side view (S). Fig. 385. Midoptical section, side view. Fig. 386. Surface view, dorsal side (D) and midoptical section, end view (M). Bars = 10 μm.

Fig. 387. *Coelomomyces macleayae.* Resting sporangium from *Armigeres longipalpis* collected in Nan, Thailand. This is a surface of the side of the sporangium with the dorsal face directed upward in the micrograph. Bar = 10 μm.

Fig. 388. *Coelomomyces macleayae.* Resting sporangia from *Aedes aegypti* collected in Rangoon, Burma. The sporangium seen in the center of the micrograph is a midoptical section, end view. Note that the dehiscence slit is visible along one of the lateral ridges (arrow). Bar = 10 μm.

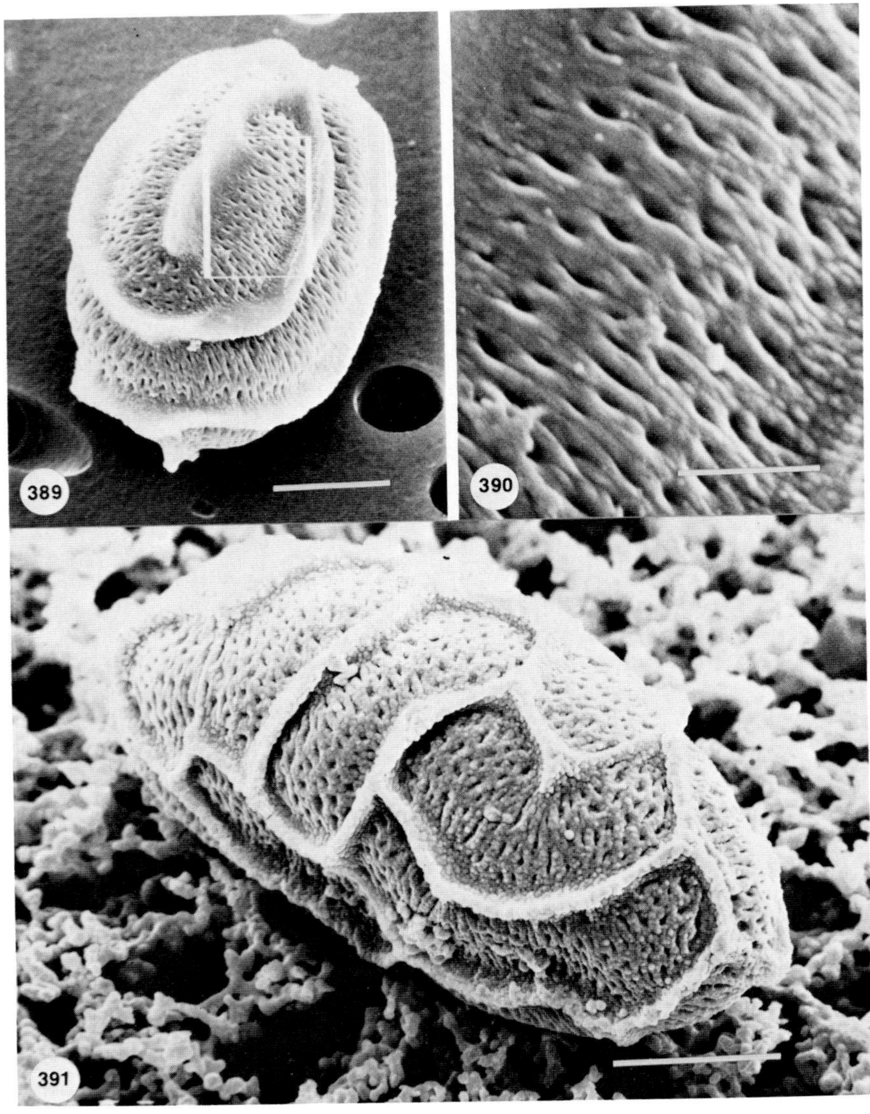

Figs. 389 and 390. *Coelomomyces macleayae.* Resting sporangia from *Aedes aegypti* collected in Rangoon, Burma. Note that Fig. 390 (bar = 2.5 μm) is an enlargement of the area indicated in Fig. 389 (bar = 10 μm).

Fig. 391. *Coelomomyces macleayae.* Resting sporangium from *Aedes pseudoalbopictus* collected in Kanchanaburi, Thailand. Bar = 10 μm.

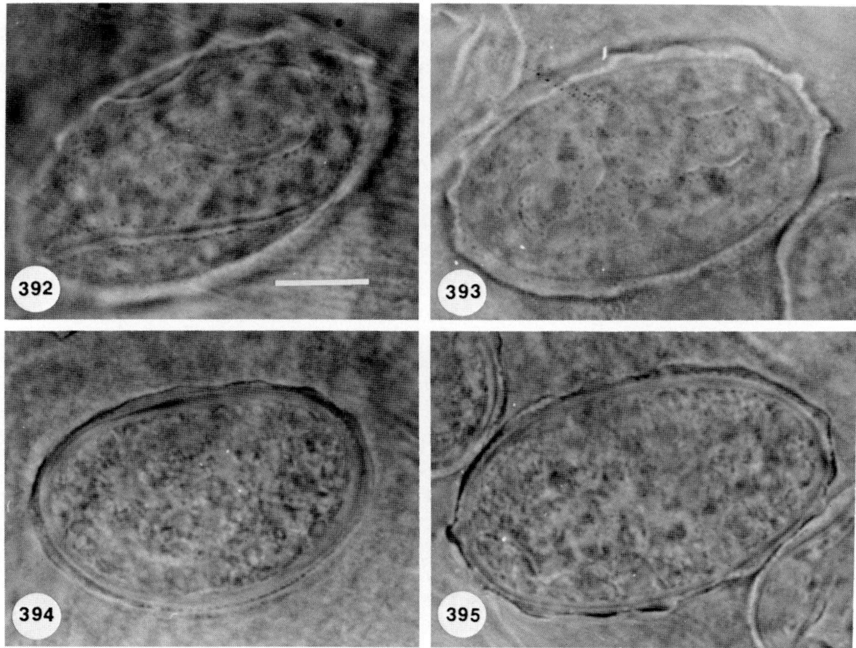

Figs. 392–395. *Coelomomyces madagascaricus*. Resting sporangia of type specimen collected from *Ingramia roubaudi* collected in Majunga, Madagascar. Fig. 392. Surface view, dorsal side. Bar = 10 μm, applies to Figs. 392–395. Fig. 393. Surface view, ventral side. Figs. 394 and 395. Midoptical section, side view.

DESCRIPTION: Hyphae not observed. Resting sporangia broadly ellipsoid, 15–30 × 28–38 μm; surface ornamented with irregular, isolated ridges and papillae (1.5–2.5 μm wide, 0.5–2 μm high, and 2.5–25 μm long; individual ridges of variable height, i.e., with "peaks" and "valleys") separated by depressed areas (2–8 μm wide) with scattered punctae (0.5 μm in diameter and 0.5–1 μm apart); dehiscence slit central along longitudinal ridge. In midoptical section: ridges as domes alternating with pitted depressed areas along sporangial periphery. Thin-walled sporangia not observed.

TYPE HOST AND COLLECTION SITE: *Ingramia* (*Ravenalites*) *roubaudi* Doucet; Sokobé, Nossi-bé, Majunga, Madagascar (collected by Grjebine).

DEPOSITORY FOR TYPE: BPI.

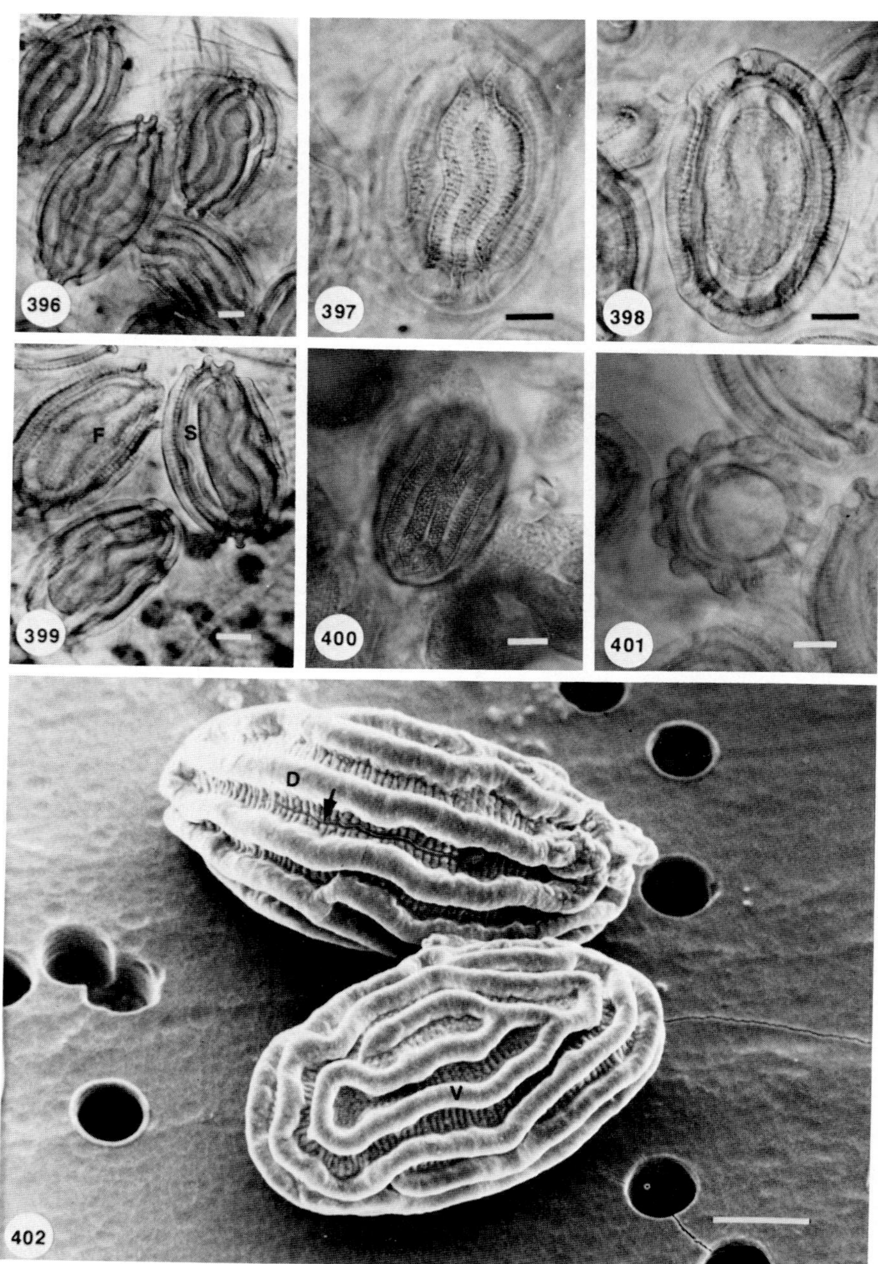
396
397
398
399
F
S
400
401
402
D
V

LATIN DIAGNOSIS: Hyphae non visae. Sporangia perennantia late ellipsoidea, 15–30 × 28–38 μm; superficie liris enormibus et sejunctis, papillisque 1.5–2.5 μm latis, 0.5–2 μm altis, 2.5–25 μm longis, ornata; liris singulis altitudine variabilibus, videlicet specie cacuminibus vallibusque alternantibus definitis, ab areolis depressis 2–8 μm latis, punctis sparsis 0.5 μm diametro, inter se 0.5–1 μm distantibus notatis; incisura dehiscentiae per liram longitudinalem media; quasi in sectione media inspecta, liris specie tholorum cum areolis depressis punctatis alternantium per ambitum apparentibus. Sporangia tenuiter tunicata non observata.

SPECIMENS EXAMINED OTHER THAN TYPE:

Madagascar

Rovenolats nonberuli; Sokobé, Nossi-bé, Majunga; Grjebine; 1955; NCU.

COMMENTS: The species description of *C. madagascaricus* is based on two specimens that were mounted in an unknown substance insoluble in water, ethanol, or xylene. Although it is apparent that this mounting medium caused some shrinkage of resting sporangia, surface ornamentation was sufficiently distinct for description of this form as a new and unique species of *Coelomomyces*. The specific epithet reflects the sole collection site for this species.

Coelomomyces musprattii Couch sp. nov.

Figures 396–413

DESCRIPTION: Hyphae 5–9 μm in diameter. Resting sporangia ellipsoid, 26–45 × 37–70 μm, often flattened on ventral surface; surface ornamented with elongate, anastomosing ridges (3.5–5 μm wide and 3–5 μm high) extending to and continuous around poles, 5–7 ridges usually visible in surface view; ridges often concentric on dorsal surface, becoming irregular and anastomosing at sides and on ventral surface, separated by de-

---

Figs. 396–402. *Coelomomyces musprattii.* Resting sporangia of type specimen from *Culex theileri* collected in Livingstone, Zambia. Bars = 10 μm. Fig. 396. Low-magnification micrograph showing general sporangial morphology. Fig. 397. Surface view of sporangial side. Fig. 398. Midoptical section, face view. Fig. 399. Midoptical section of three sporangia. Note that one sporangium is seen in face view (F) and that the other is seen in side view (S). Fig. 400. Surface view, dorsal side. Fig. 401. Midoptical section, end view. Fig. 402. Scanning electron micrograph showing one sporangium from dorsal side (D) and the other sporangium from the ventral side (V). Note that the dehiscence slit (arrow) is visible on one of the sporangia.

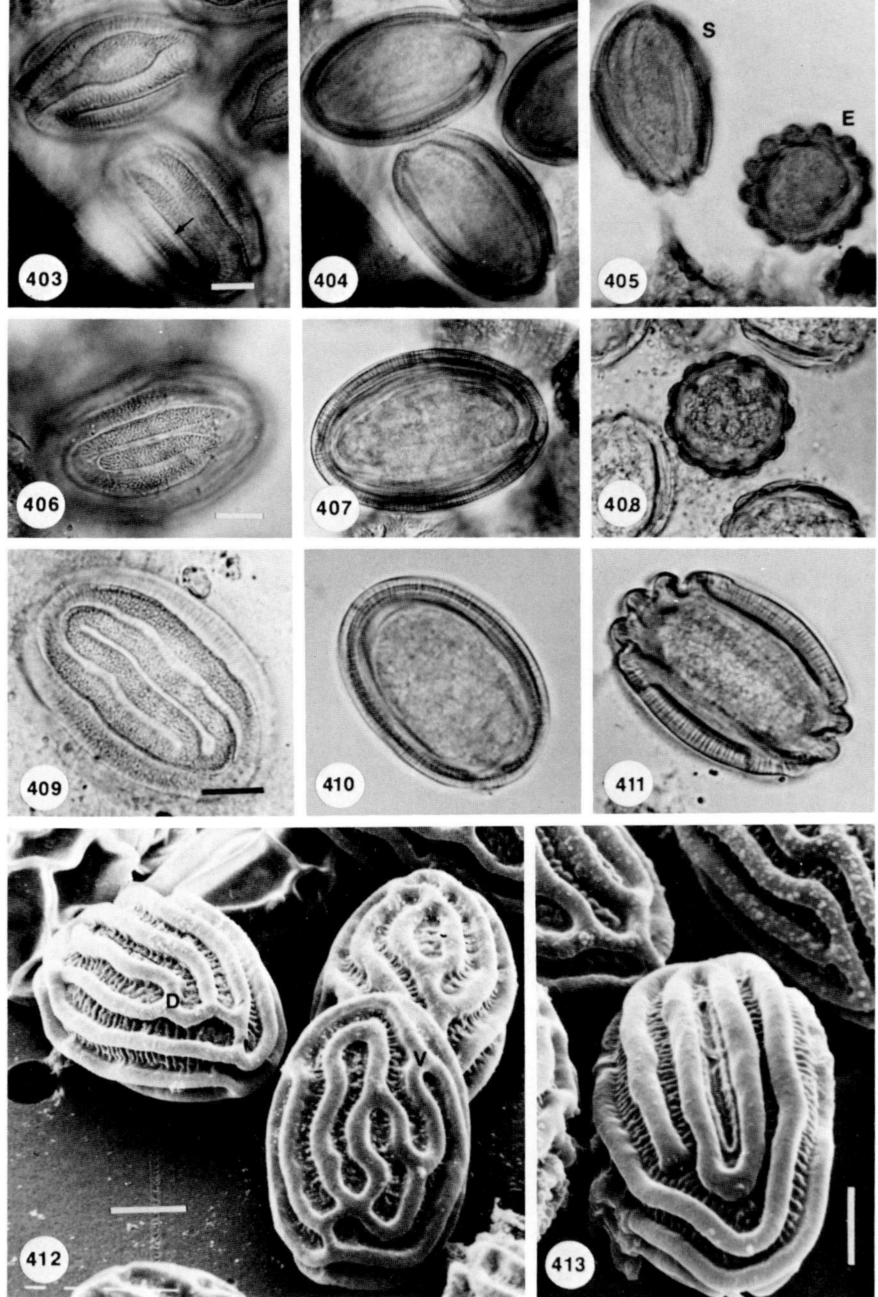

403
404
S
E
405
406
407
408
409
410
411
D
V
412
413

pressed areas (3–5 μm wide) with prominent transverse striae and central punctae; dehiscence slit prominent in depressed area bordered by an elongate, continuous, elliptical ridge or by two parallel elongate ridges, one of which may continue spirally around sporangial periphery. In mid-optical section: face view, ridging as a vertically striate wing encircling the sporangium; side view, ridging as a striate wing along dorsal and ventral surfaces and as 3–5 domes at the poles; end view, ridging, as 12–20 domes encircling the sporangium. Thin-walled sporangia ellipsoid, 20–30 × 30–48 μm.

TYPE HOST AND COLLECTION SITE: *Culex theileri* Theo.; Livingstone, Zambia (collected by Muspratt).

DEPOSITORY FOR TYPE: BPI.

LATIN DIAGNOSIS: Hyphae 5–9 μm crassae. Sporangia perennantia ellipsoidea, 26–45 × 37–70 μm, saepe in superficie ventrali applanta; superficie liri elongatis, anastomosantibus 3.5–5 μm latis, 3–5 μm altis polos contingentes et circum eos continuis, decorata, in aspectu superficiali plerumque 5–7 liris visibilibus, crebre concentricis in superficie dorsali, per latera et in superficie ventrali enormibus et anastomosantibus, ab areolis depressis 3–5 μm latis, striis transversis et punctis centralibus notatis, separatis; incisura dehiscentiae conspicua areola depressa a lira elongata, continua, elliptica vel duobus parellis marginata, quarum una interdum circum circulum sporangii in spira protracta; quasi in sectione media adversus inspecta, liris instar alae verticaliter striatae sporangium cingentis, a latere instar alae striatae per superficies et dorsalem et ventralem, ad polos 3–5 tholorum; ab cacumine instar tholorum 12–20 per

---

Figs. 403–405. *Coelomomyces musprattii.* Resting sporangia from *Culex fraudatrix* collected in Mandya, India. Fig. 403. Surface view of two sporangia. The dehiscence slit (arrow) is visible in one of the sporangia. Bar = 10 μm, for Figs. 403–405. Fig. 404. Midoptical section of the two sporangia shown in Fig. 403. Fig. 405. Midoptical section of two sporangia, one in side view (S), the other in end view (E).

Figs. 406–408. *Coelomomyces muspratti.* Resting sporangia from *Culex antennatus* collected in Giza, Egypt. Fig. 406. Surface view, dorsal side. Bar = 10 μm, for Figs. 406–408. Fig. 407. Midoptical section, face view. Fig. 408. Midoptical section, end view.

Figs. 409–411. *Coelomomyces muspratti.* Resting sporangia from *Culex fatigans* collected in New South Wales, Australia. Fig. 409. Surface view, dorsal side. Bar = 10 μm for Figs. 409–411. Fig. 410. Midoptical section, face view. Fig. 411. Midoptical section, side view.

Figs. 412 and 413. *Coelomomyces muspratti.* Scanning electron micrographs of resting sporangia. Fig. 412. Specimen collected from *Culex fatigans* in New South Wales, Australia. Both dorsal (D) and ventral (V) views of the sporangia are shown. Bar = 10 μm. Fig. 413. Specimen from *Culex fraudatrix* collected in Mandya, India. Note the prominent dehiscence slit visible in this dorsal view. Bar = 5 μm.

sporangii circumferentiam distributorum. Sporangia tenuiter tunicata ellipsoidea, 20–30 × 30–48 μm.

SPECIMENS EXAMINED OTHER THAN TYPE:

Africa
Culex antennatus Beck.; Giza, Egypt; Gad; 1966; NCU; (Gad *et al.*, 1967; Gad and Sadek, 1968).
Culex simpsoni Theo.; Livingstone, Zambia; Muspratt; 1945; NCU.
Culex theileri Theo.; Livingstone, Zambia; Muspratt; 1962; NCU.
Culex univittatus Theo.; Mata, Bechunaland, South Africa; Muspratt; 1946, 1975; NCU.

Australia
Culex basicinctus Ed.; Cunnamulla, Queensland; Lee; 1951; CIH.
Culex quinquefasciatus; Moree, New South Wales; Russell; 1980; CIH, NCU.

India
Anopheles vagus Dön.; Trivandrum; Iyengar; 1970; NCU.
Culex bitaeniorhynchus Giles; Faridpur; Iyengar; 1970; NCU.
Culex bitaeniorhynchus Giles; Mandya; Iyengar; 1970; NCU.
Culex fraudatrix Theo.; Mandya; Iyengar; 1970; NCU.
Culex fuscocephalus Theo.; Faridpur; Iyengar; 1966, 1971; NCU.
Culex fuscocephalus Theo.; Mandya; Iyengar; 1969; NCU.
Culex tritaeniorhynchus Giles; Mandya; 1970; NCU.
Culex sp.; Faridpur; Iyengar; 1942; NCU.

COMMENTS: Although sporangial ornamentation in *C. musprattii* resembles somewhat that of *C. indicus,* distinction may be made in that resting sporangia of *C. musprattii* are generally larger and are ornamented with ridges which never anastomose to form a reticulum enclosing circular depressed areas. Instead, sporangial ridges in *C. musprattii* divide infrequently and connect with adjacent ridges. Also, LM observation of sporangia of *C. musprattii* reveals a central row of punctae in depressed areas between sporangial ridges. Such punctae are not seen in similar observation of sporangia of *C. indicus.*

The specific epithet of *C. musprattii* is in recognition of the many contributions of J. Muspratt to the study of this group.

Coelomomyces orbicularis Couch & Muspratt sp. nov.

Figures 414–419

DESCRIPTION: Hyphae 6–24 μm in diameter. Resting sporangia ellipsoid, 30–50 × 39–75 μm; surface ornamented with a reticulum of broad ridges (2 μm high and 2–3 μm wide) enclosing circular to rarely polygonal or elongate depressed areas (2.5–4 μm in diameter when circular or polygonal; 2.5–4 μm wide and up to 35 μm long when elongate) with fine punctae (<0.5 μm in diameter and 0.5–1 μm apart); dehiscence slit central in an elongate, longitudinal, depressed area. In midoptical section: ridges in end view as truncate domes alternating with depressed areas around sporangial periphery, ridges in side view as a wing interconnecting the domes. Thin-walled sporangia oval, 22–33.5 × 28–58 μm.

TYPE HOST AND COLLECTION SITE: *Anopheles squamosus* Theo.; Livingstone, Zambia, Africa (collected by Muspratt; see Muspratt, 1946a).

DEPOSITORY FOR TYPE: BPI.

LATIN DIAGNOSIS: Hyphae 6–24 μm crassae. Sporangia perennantia ellipsoidea, 30–50 × 39–75 μm; superficie reticulo lirarum laturam 2 μm altarum, 2–3 μm latarum consistente, areolas depressas rotundas vel raro polygonales, 2.5–4 μm diametro, vel elongatas, ad 35 μm longas, punctis subtilibus minus quam 0.5 μm diametro, 0.5–1 μm inter se distantibus notatas, ornata; incisura dehiscentiae in areola depressa elongata, longitudinali media; quasi in sectione media inspecta liris in aspectu cacuminis instar tholorum truncatorum cum areolis depressis alternantium per sporangii circulum, a latere instar alae tholos connectentis. Sporangia tenuiter tunicata ovalia, 22–33.5 × 28–58 μm.

SPECIMENS EXAMINED OTHER THAN TYPE:

Africa

Anopheles gambiae Giles; Kisumu, Kenya; Service; 1974, 1975, 1977; NCU.

Culex univittatus Theo.; Pretoria, South Africa; Muspratt; 1975; NCU.

COMMENTS: Resting sporangia of *C. orbicularis* are ornamented similarly to those of *C. reticulatus* varieties and *C. orbiculostriatus*. Distinction from both may be made in that the dehiscence slit as sporangia of *C. orbicularis* is always central along a depressed area, whereas in both *C. reticulatus* and *C. orbiculostriatus* the dehiscence slit is along a longitudinal ridge. Additionally, depressed areas on sporangia of *C. orbicularis* are small and finely punctate, whereas those of *C. reticulatus* varieties are larger and prominently punctate, and those of *C. orbiculostriatus* are larger and prominently striate.

Occurring on larvae collected from ground pools and rice fields, *C. orbicularis* often occurs in joint infections with *C. grassei*. Examples of this have been observed for both of the hosts listed above.

Because of the prominent, circular depressed areas characteristically

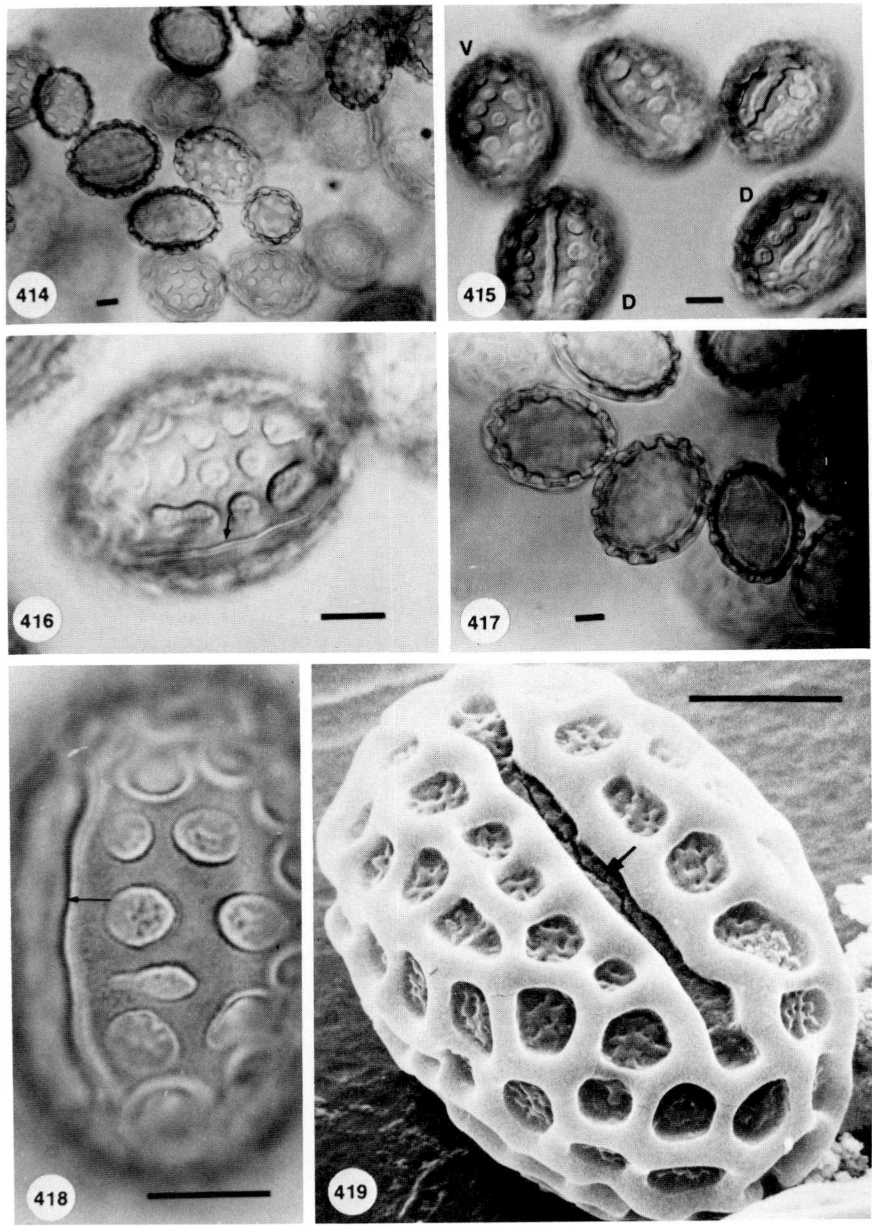
414
415
V
D
D
416
417
418
419

observed on sporangia of this species, it is here named *C. orbicularis* from the diminutive *orbiculus* (L.), meaning small circle.

Coelomomyces orbiculostriatus Couch & Prasertphon sp. nov.

Figures 420–427

DESCRIPTION: Hyphae 6–15 μm in diameter. Resting sporangia ellipsoid, 32–61 × 46.5–108 μm; surface ornamented with broad, low ridges (2.5–3 μm high and 3.5 to usually 5–10 μm wide) enclosing circular to occasionally elongate, bordered, depressed areas (3–10 μm in diameter and with 1.5–2.5 μm wide borders) with irregular anastomosing striae and punctae (punctae and striae 1–2 μm apart); dehiscence slit in central, longitudinal, low ridge. In midoptical section: ridges as low, bilayered domes alternating with depressed areas around sporangial periphery, striae/punctae in depressed areas appearing in edge view as fine teeth. Thin-walled sporangia not observed.

TYPE HOST AND COLLECTION SITE: *Culex vishnui* (Theo.); Trivandrum, India (collected by Iyengar).

DEPOSITORY FOR TYPE: BPI.

LATIN DIAGNOSIS: Hyphae 6–15 μm crassae. Sporangia perennantia ellipsoidea, 32–61 × 46.5–108 μm; superficie liris latis, humilibus 2.5–3 μm altis, 3.5 vel plerumque 5–10 μm latis, areolas depressas rotundas vel interdum elongatas, marginatas, 3–10 μm diametro, marginibus 1.5–2.5 μm latis, striis enormibus, anastomosantibus punctisque inter se 1–2 μm distantibus notatas includentis, ornata; incisura dehiscentiae in lira humili media inclusa; quasi in sectione media inspecta lirae instar tholorum bistratorum humilium, cum areolis depressis circum sporangii circulum al-

---

Figs. 414 and 417. *Coelomomyces orbicularis.* Resting sporangia from *Culex univittatus* collected in Pretoria, South Africa. Bars = 10 μm. Fig. 414. Low-magnification micrograph showing general sporangial morphology. Fig. 417. Midoptical section, face view.

Figs. 415, 416, 418, and 419. *Coelomomyces orbicularis.* Resting sporangia from *Anopheles gambiae* collected in Kisumu, Kenya. Bars = 10 μm. Fig. 415. Surface views of several sporangia, some seen in dorsal view (D) and some in ventral view (V). Fig. 416. Surface view of sporangial side. Note that the dehiscence slit (arrow) is visible along one side of the sporangium. Fig. 418. Surface view of sporangial side, again showing the dehiscence slit along one edge. Fig. 419. Scanning electron micrograph showing dorsal side of a sporangium. Note the dehiscence slit (arrow).

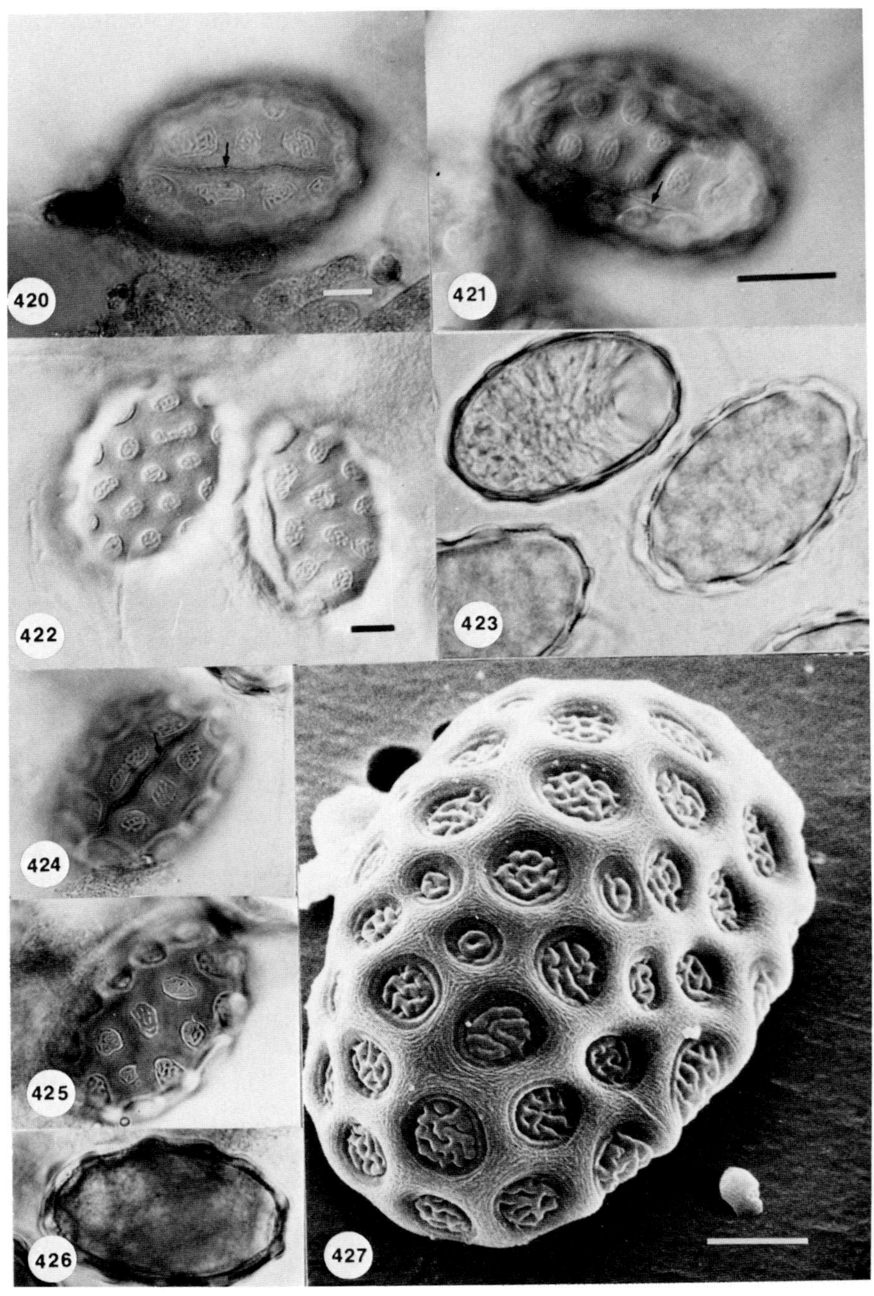
420
421
422
423
424
425
426
427

ternantium, striis punctisque in areolis depressis desuper observatis instar denticulorum subtilium. Sporangia tenuiter tunicata non observata.

SPECIMENS EXAMINED OTHER THAN TYPE:

Africa

Anopheles gambiae Giles; Dan Kande, Nigeria; Prasertphon; 1976; NCU.

Anopheles gambiae Giles*; Pala, Upper Volta; Rodhain; 1967, 1968; IP; (specimens HV1507 and HV1508 of Rodhain and Gayral, 1971; Rodhain and Brengues, 1974.)

Anopheles gambiae Giles; Dan Kande, Nigeria; Federici; 1980; ?.

Thailand

Culex vishnui (Theo.); Ban Phra, Chonburi; AFRIMS; 1966; NCU; (Hembree, 1979).

COMMENTS: *Coelomomyces orbiculostriatus* is distinguished from all known species of *Coelomomyces* by its resting sporangia ornamented with circular to elongate depressed areas with scattered striae/punctae. In other species having sporangia with similar, regular, circular depressed areas, such areas bear only punctae and are not bordered.

Although sporangial dimensions for the type specimen of *C. orbiculostriatus* are somewhat greater than for the other specimens studied, all are here treated as the same species because of their identical ornamentation and since the type was from a single collection. It is, however, possible that data from additional collections may necessitate the subdivision of this species into different varieties.

* Parasite identified originally as *C. ascariformis*.

Fig. 420. *Coelomomyces orbiculostriatus.* Resting sporangium from *Anopheles gambiae* collected in Nigeria, Africa. Face view, dorsal side (dehiscence slit, arrow). Bar = 10 μm.

Fig. 421. *Coelomomyces orbiculostriatus.* Resting sporangia of type specimen from *Culex vishnui* collected in Trivandrum, India. Surface view of sporangial side. Note that the dehiscence slit (arrow) is visible along one edge of the sporangium. Bar = 10 μm.

Figs. 422 and 423. *Coelomomyces orbiculostriatus.* Resting sporangia from *Culex vishnui* collected in Chonburi, Thailand. Fig. 422. Surface view of sporangial side. Bar = 10 μm, applies to Figs. 422–426. Midoptical section, face view.

Figs. 424–426. *Coelomomyces orbiculostriatus.* Resting sporangia from *Anopheles gambiae* collected in Upper Volta, Africa. Scale as in Fig. 422. Fig. 424. Surface view, dorsal side (dehiscence slit, arrow). Fig. 425. Surface view, ventral side. Fig. 426. Midoptical section, face view.

Fig. 427. *Coelomomyces orbiculostratus.* Scanning electron micrograph showing ventral side of type. Bar = 10 μm.

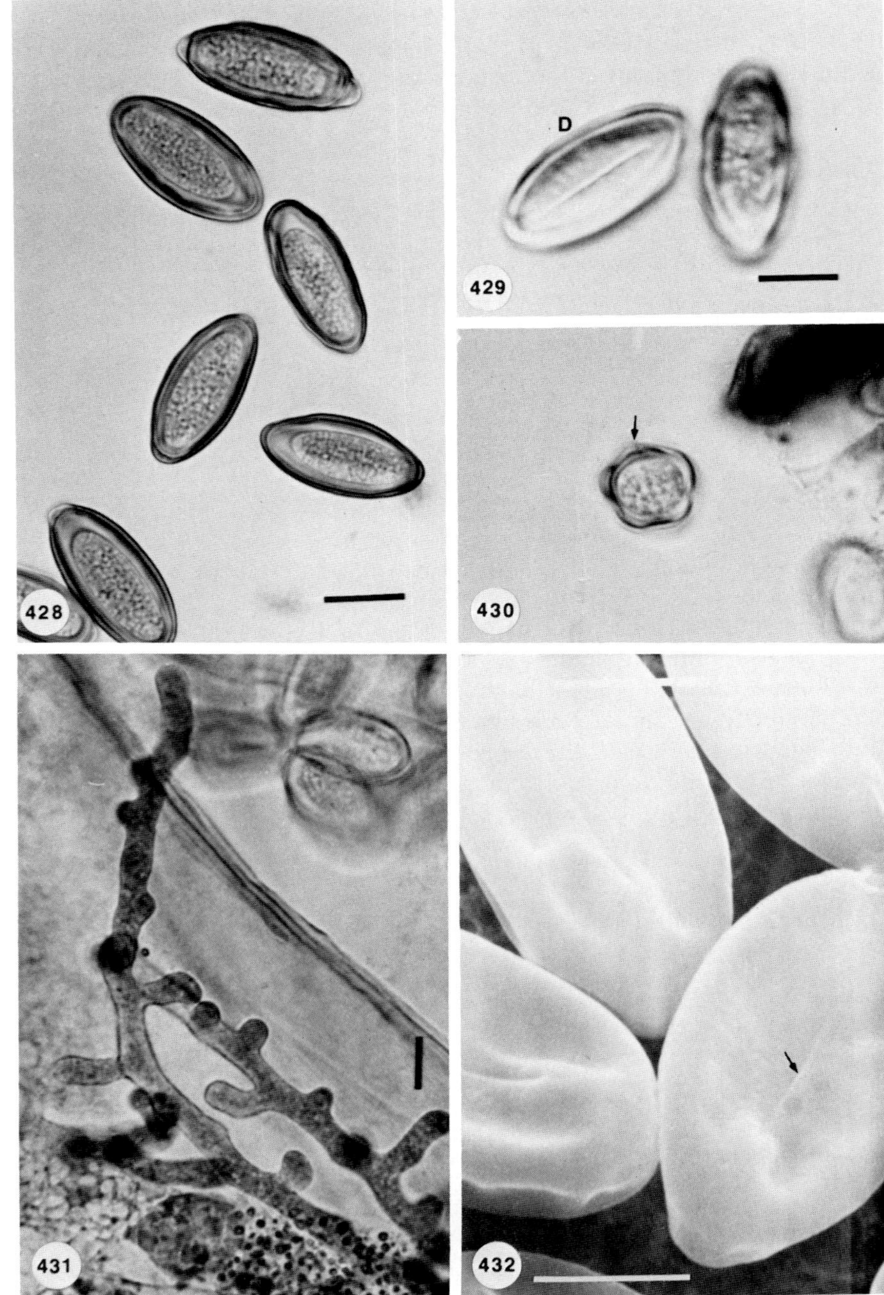
D
428
429
430
431
432

As thought by Radhain and Gayral (1971) and Rodhain and Brengues (1974), resting sporangial ornamentation in *C. orbiculostriatus* resembles somewhat that of the "Type 4" illustrated by Walker (1938) and described by van Thiel (1954, 1962) as *C. ascariformis*. However, due to the confusion surrounding the true nature of *C. ascariformis* (see Section III,E), it is impossible to determine conclusively whether they are the same taxon. Because of this, *C. orbiculostriatus* is here described as a new species.

The specific epithet of *C. orbiculostriatus* is derived from the diminutive *orbiculus* L. (small circle) and *stria* L. (line or furrow), in recognition of the numerous, circular, depressed areas with striae occurring on resting sporangia of this species.

Coelomomyces pentagulatus Couch ex Couch, *J. Elisha Mitchell Sci. Soc.* **78,** 135 (1962)

Coelomomyces pentagulata Couch, *J. Elisha Mitchell Sci. Soc.* **61,** 124 (1945)

Figures 428–432

DESCRIPTION: Hyphae 4.2–10.0 μm thick. Resting sporangia ellipsoid, elongate–oval, to elliptical, 12–18 × 18–40 μm, with 5 to rarely 6 longitudinal ridges causing sporangia to appear five- to six-angled in end view; surface otherwise devoid of ornamentation. Dehiscence slit centrally located along one of the longitudinal ridges. Wall 2–3 μm thick. Thin-walled sporangia not observed.

TYPE HOST AND COLLECTION SITE: *Culex erraticus* (Dyar and Knab); Valdosta, Georgia (collected by SMCG).

DEPOSITORY FOR TYPE: BPI. Several paratypes collected in Georgia and on the same host as the type are on deposit at NCU.

SPECIMENS EXAMINED AND OTHER COLLECTIONS:

Type, Paratypes

Figs. 428–432. *Coelomomyces pentangulatus*. Type specimen from *Culex erraticus* collected in Valdosta, Georgia. Fig. 428. Low-magnification micrograph showing general sporangial shape. Bar = 10 μm. Fig. 429. Surface view, dorsal side (D). Bar = 10 μm. Fig. 430. Midoptical section, end view (dehiscence slit, arrow). Scale as for Fig. 429. Fig. 431. Vegetative hyphae. Bar = 5 μm. Fig. 432. Scanning electron micrograph (dehiscence slit, arrow). Bar = 5 μm.

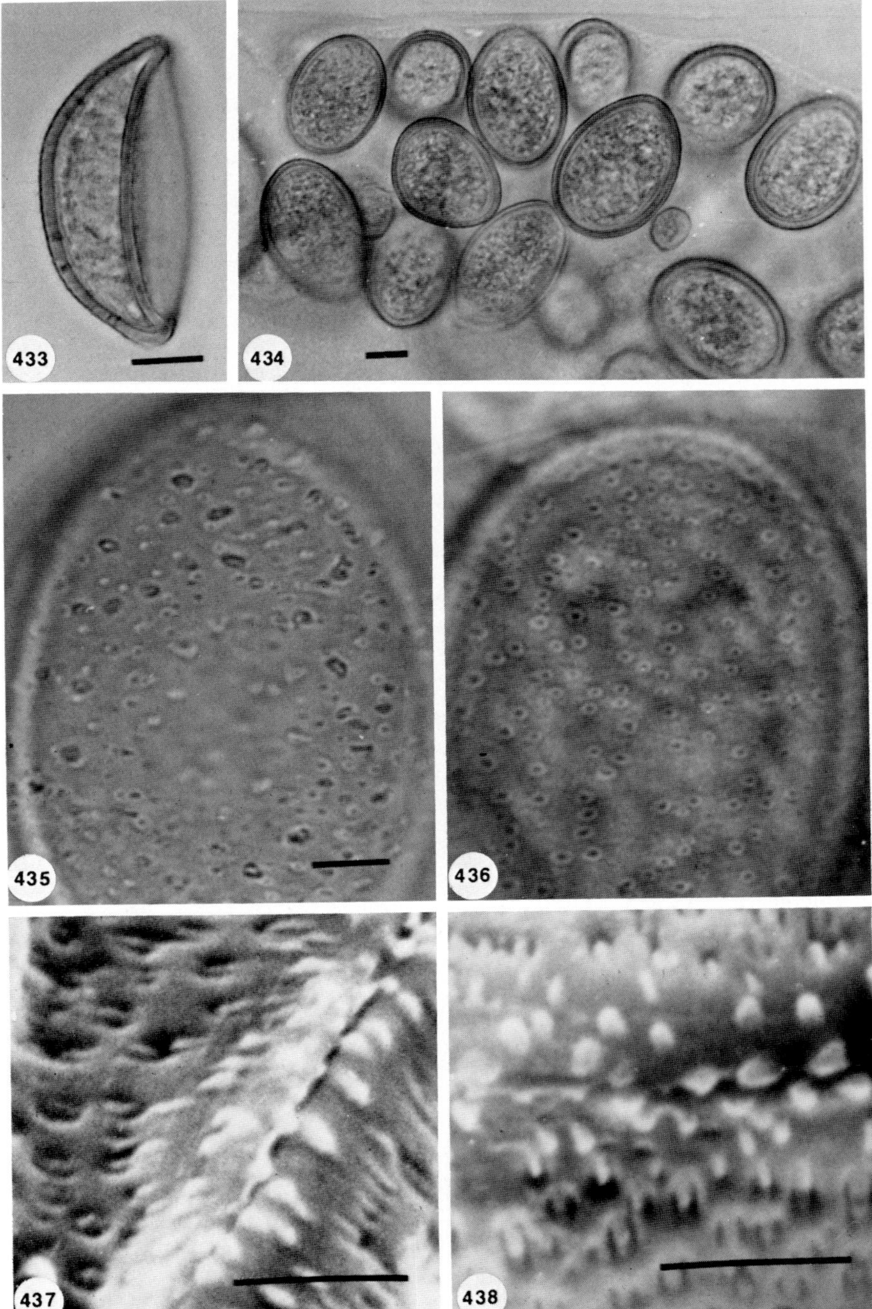
433
434
435
436
437
438

United States

Culex erraticus (Dyar & Knab); Georgia; SMCG; 1945; NCU.
Culex erraticus (Dyar & Knab); Louisiana; Chapman; 1970; NCU.
Culex peccator (Dyar & Knab); Louisiana; Chapman; 1970; NCU.

COMMENTS: *Coelomomyces pentangulatus* is easily recognized on the basis of its small sporangia with smooth, relatively thin walls. The five-angled appearance of the sporangia in end view is a constant feature, but often difficult to discern because of the infrequency with which sporangia are oriented so as to provide an end view. Because of its thin sporangial walls lacking ornamentation, the sporogenic cytoplasm within sporangia is often clearly visible in optical section as a highly granular, vacuolate mass. Also, separation of sporangial wall layers at the poles often results in hyaline knobs at this location. Although this species has been collected from several localities in Georgia and Louisiana, it has not been reported elsewhere.

Coelomomyces ponticulus Nolan & Mogi, *J. Med. Entomol.* **17,** 333 (1980)

Figures 433–438

DESCRIPTION: Hyphae not observed. Resting sporangia ellipsoid, 19–34 × 35–56 μm; surface ornamented with elongate pits, 0.5–1.0 μm in diameter and 0.5–1.0 μm apart, with a central elongate bridge dividing the pit lengthwise into approximately equal halves; via SEM, small papillae are evident adjacent to the dehiscence slit and in clusters or running lengthwise elsewhere on the wall; in midoptical section, pits giving the surfaces of the sporangium an irregular appearance. Wall 1.0–1.5 μm thick. Thin-walled sporangia not observed.

TYPE HOST AND COLLECTION SITE: *Aedes* (*Stegomyia*) *flavopictus** Yamada; Mikura-Jima Island, Japan (collected by Mogi).

* Combined infection with *C. stegomyiae* var. *stegomyiae*.

Fig. 433. *Coelomomyces ponticulus.* Resting sporangium of type from *Aedes flavopictus* collected on Mikura-Jima Island, Japan. Sporangia of the type specimen appear "boat-shaped" due to their having collapsed in the mounting medium. Bar = 10 μm.

Figs. 434 and 435. *Coelomomyces ponticulus.* Resting sporangia of a specimen from the type collection no. 15L29. Fig. 434. Low-magnification micrograph showing general sporangial shape. Bar = 10 μm. Fig. 435. Surface view of sporangial side. Bar = 5 μm.

Fig. 436. Surface view of resting sporangium of *Coelomomyces stegomyiae* found in the same host as sporangia of *Coelomomyces ponticulus.* Scale as for Fig. 436.

Figs. 437 and 438. *Coelomomyces ponticulus.* Scanning electron micrographs of resting sporangia of type. Bars = 5 μm. [Micrographs from Nolan and Mogi (1980).]

DEPOSITORY FOR TYPE: BISH. An isotype is on deposit at the Botanical Section, National Science Museum, Tokyo, Japan.

SPECIMENS EXAMINED AND OTHER COLLECTIONS:

Type

Japan

Aedes (*Stegomyia*) *flavopictus* Yamada; Mikura-Jima Island, Japan; Mogi; 1976; NCU; (Nolan and Mogi, 1980).

COMMENTS: Although sporangia of *C. ponticulus* were described originally as appearing "boat- or bowl-shaped" in side view (Nolan and Mogi, 1980), such morphology can be attributed to collapse of the sporangial wall as a consequence of the technique(s) used in preservation. Indeed, examination of specimens from the type collection (#15-L29) that were originally preserved in ethanol, but subsequently rehydrated, showed sporangia of this species to be typically oval.

Resting sporangia of *C. punticulus* most closely resemble those of *C. stegomyiae*. However, distinction may be made in that punctae of *C. stegomyiae* are bordered whereas those of *C. ponticulus* are irregular and have a central bridge. Distinction of the two may be complicated, however, since all collections of *C. ponticulus* have thus far been in combination with *C. stegomyiae*.

Coelomomyces psorophorae var. ***psorophorae*** Couch ex Couch, *J. Elisha Mitchell Sci. Soc.* **78,** 135 (1962)

Coelomomyces psorophorae Couch, *J. Elisha Mitchell Sci. Soc.* **61,** 124 (1945)

Figures 439–471

DESCRIPTION: Hyphae 7.5–10.0 μm thick. Resting sporangia ellipsoid, 25–80 × 56–164 μm; frequently flattened on one side, rarely spherical, 40–78 μm in diameter; surface ornamented with numerous, randomly arranged punctae, 0.5–1.0 μm in diameter and 0.5–5.0 μm apart. In mid-optical section, punctae as anastomosing to parallel, vertical striae within the wall. Wall 2–10 μm thick. Thin-walled sporangia not observed.

TYPE HOST AND COLLECTION SITE: *Psorophora ciliata* (Fab.); Moultrie, Georgia (collected by SMCG).

DEPOSITORY FOR TYPE: BPI.

ALTERNATE HOST
Cyclops vernalis Fischer (Whisler *et al.*, 1975).
Mesocyclops sp. Sars (Nam, 1981).

SPECIMENS EXAMINED AND OTHER COLLECTIONS:

Type

Note: Because of high variability in sporangial size, wall thickness and frequency of punctae within the *C. psorophorae* var. *psorophorae* "complex," complete descriptive data for all collections is given in Table VII, with illustrations of many forms included in Fig. 439–471.

COMMENTS: Described originally (Couch, 1945) from a single larva of *Psorophora ciliata* collected near Moultrie, Georgia, representatives of *C. psorophorae* var. *psorophorae* have since been collected from many hosts and habitats from around the world. Although all resting sporangia of this variety are large and have a finely punctate wall, much variation has been observed in sporangial wall thickness, size and frequency of punctae, and pattern of channels within the wall. As a result of variations in pattern of wall channels among different forms, punctae often appear different at various levels of focus: i.e., circular, stellate, lunate, irregular. At the wall surface, however, punctae in most forms appear roughly circular. Figures 439–471 and Table VII illustrate and document some of the variation observed in the *C. psorophorae* var. *psorophorae* "complex." Note that the type specimen (Figs. 439–441) is characterized by sporangia with rather thick walls (3–10 μm) and distinct (0.5–1.0 μm in diameter), widely separated (1–3 μm apart) punctae. Since sporangial morphology of specimens from many collections differs considerably

---

Figs. 439–441. *Coelomomyces psorophorae* var. *psorophorae.* Resting sporangia of type specimen from *Psorophora ciliata* collected in Moultrie, Georgia. Fig. 439. Low-magnification micrograph showing general sporangial shape. Bar = 100 μm. Fig. 440. Surface view, sporangial side. Bar = 10 μm. Fig. 441. Scanning electron micrograph showing sporangial side. Bar = 10 μm.

Figs. 442 and 443. *Coelomomyces psorophorae* var. *psorophorae.* Resting sporangia from *Phorophora ciliata* collected from Valdosta, Georgia. Fig. 442. Surface view of sporangial side. Fig. 443. Midoptical section, side view of same sporangia as seen in Fig. 442. Scale as in Fig. 441.

Fig. 444. *Coelomomyces psorophorae* var. *psorophorae.* Scanning electron micrograph of resting sporangia from *Aedes vexans* collected in Moultrie, Georgia. Scale as in Fig. 441.

Figs. 445 and 446. *Coelomomyces psorophorae* var. *psorophorae.* Resting sporangia from *Psorophora ciliata* collected in Thomasville, Georgia. Scale as in Fig. 441. Fig. 445. Surface view of sporangial side. Fig. 446. Midoptical section, face view.

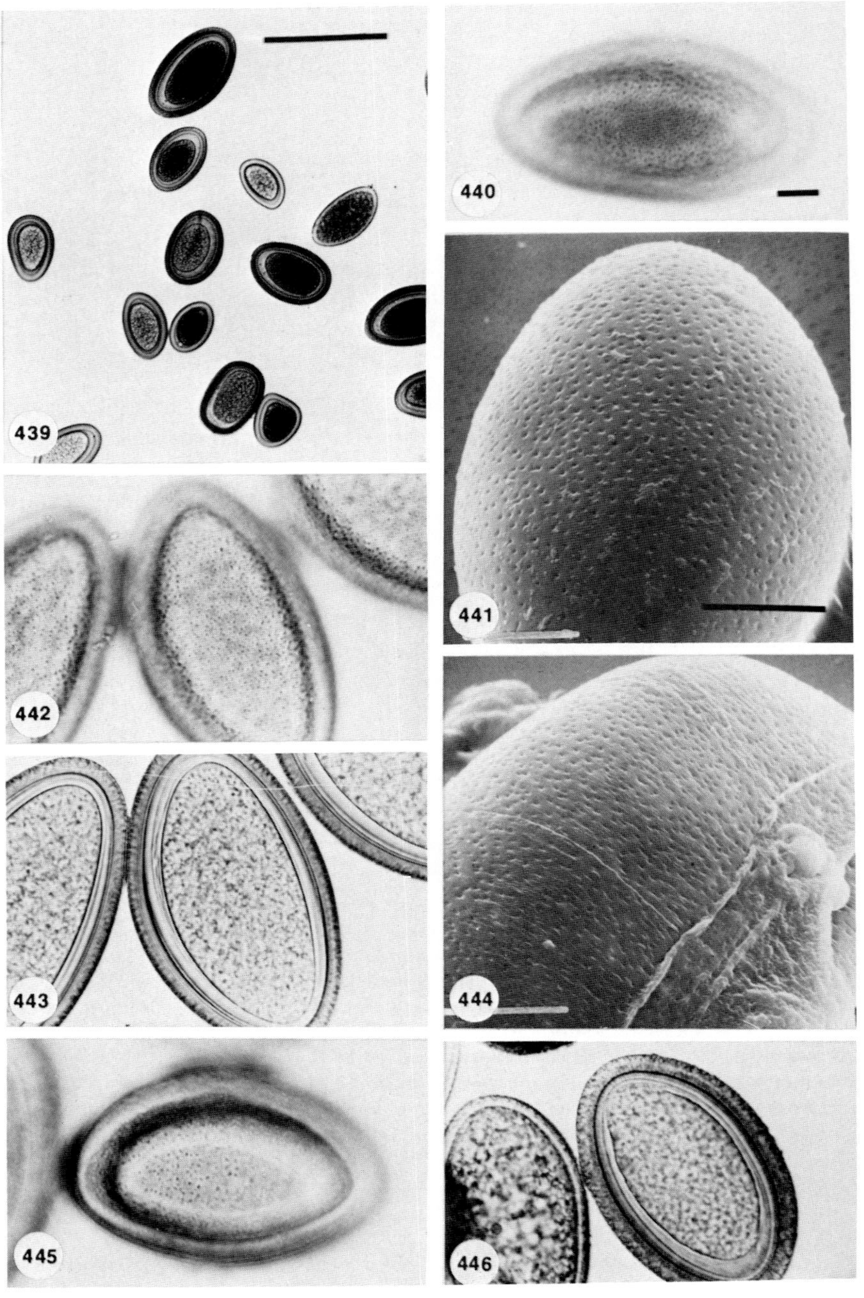
439
440
441
442
443
444
445
446

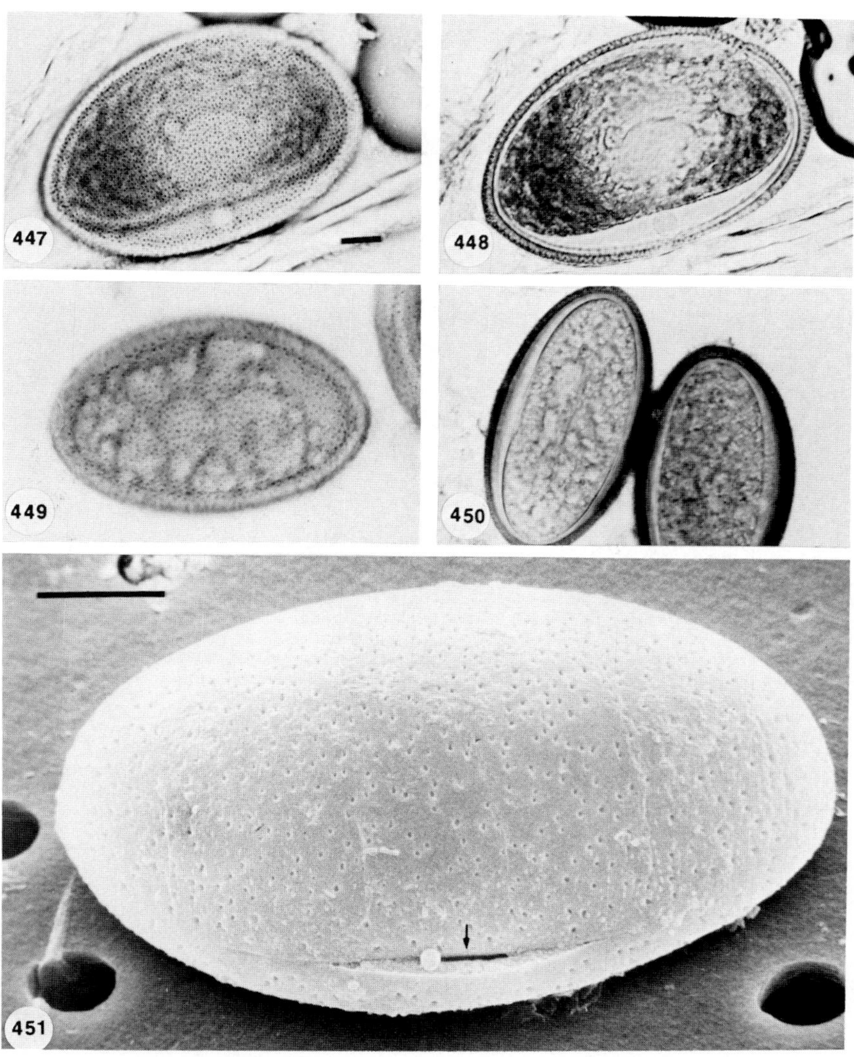

Figs. 447 and 448. *Coelomomyces psorophorae* var. *psorophorae.* Resting sporangia from *Culiseta incidens* collected in California. Fig. 447. Surface view of sporangial side. Bar = 10 μm. Fig. 448. Midoptical section, side view of same sporangium as seen in Fig. 447. Scale as in Fig. 447.

Figs. 449 and 450. *Coelomomyces psorophorae* var. *psorophorae.* Resting sporangia from *Psorophora howardii* collected in Florida. Scale as in Fig. 447. Fig. 449. Surface view of sporangial side. Fig. 450. Midoptical section, face view.

Fig. 451. *Coelomomyces psorophorae* var. *psorophorae.* Scanning electron micrograph of resting sporangium from *Psorophora howardii* collected in Lake Charles, Louisiana. Note the dehiscence slit (arrow) is visible along one side of the sporangium. Bar = 5 μm.

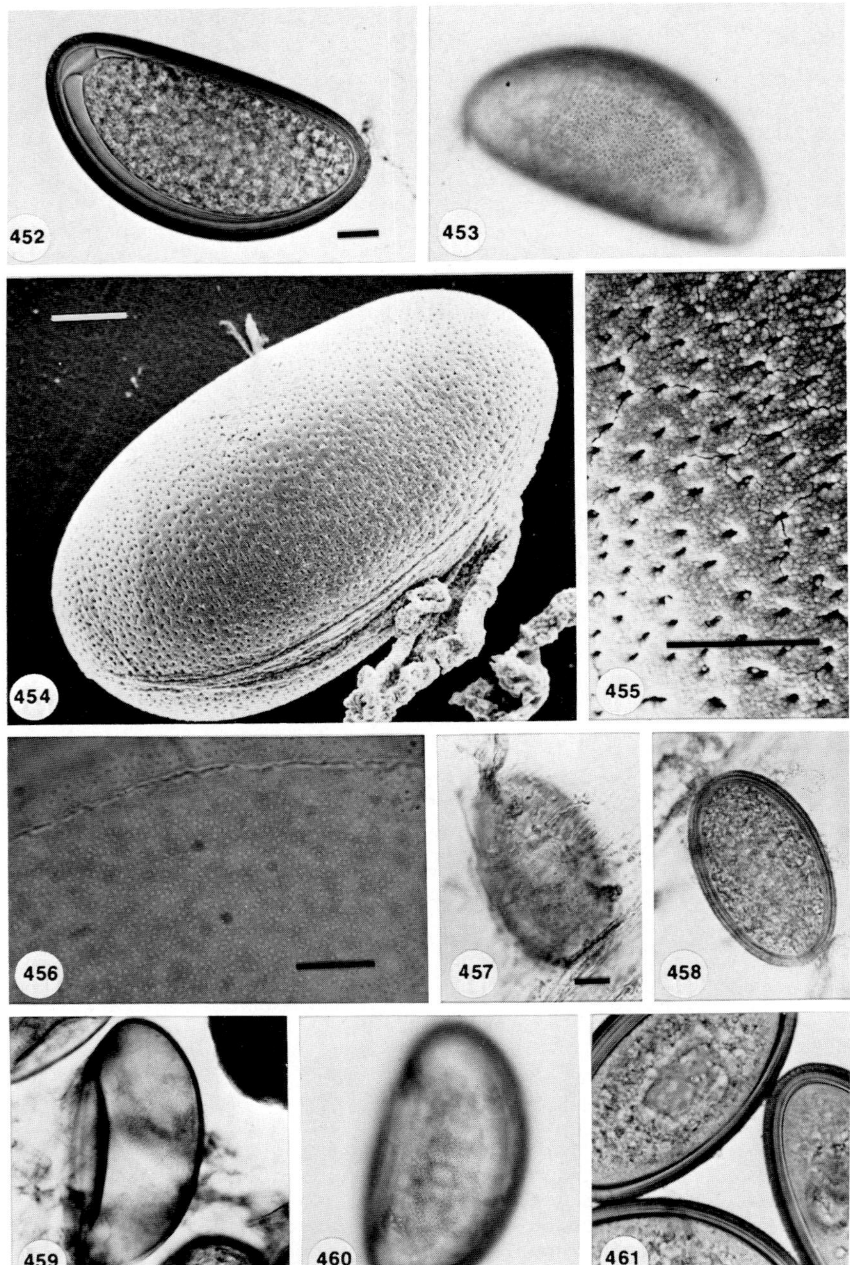
452
453
454
455
456
457
458
459
460
461

from that of the type, it is possible that several varieties or possibly species are included in this group. However, because the extreme overlap of characters from collection to collection has made futile all attempts to group these forms on the basis of host, wall thickness, and diameter and spacing of punctae, it is prudent, for the present, to refer most forms of this morphology to the *C. psorophorae* var. *psorophorae* "complex." In so doing, an attempt has been made to compile complete collection and morphological data for all forms considered in the preparation of this monograph as well as those described by others. Hopefully, such information will be beneficial in any future comprehensive treatment of this species. A few distinctive forms having some characteristics of the *C. psorophorae* var. *psorophorae* "complex" are described as varieties of this species (*C. psorophorae* var. *tasmaniensis* var. nov., *C. psorophorae* var. *psorophorae* as given under Comments for each.

Sporangia of all varieties of *C. psorophorae* are similar to those of *C. stegomyiae* in that sporangia of both are punctate. In *C. stegomyiae*, however, the punctae are more widely spaced (averaging 3 μm apart) and are bordered by a 0.5-μm-wide circular depression (Fig. 549). In a further distinction via SEM, scattered "bumps" (0.5–1.0 μm in diameter) are evident on the sporangial wall of *C. stegomyiae*, but lacking on that of *C. psorophorae* varieties.

Coelomomyces psorophorae is one of the few species of *Coelomomyces* for which full details of its life cycle have been elucidated. For a complete description of this, see Chapter 2.

The variety of *C. psorophorae* described by Muspratt (1946a) on larvae

---

Figs. 452 and 453. *Coelomomyces psorophorae* var. *psorophorae.* Resting sporangia from *Culiseta inornata* collected in Alberta, Canada. Fig. 452. Midoptical section, side view. Bar = 10 μm. Fig. 453. Surface view of sporangial side. Scale as in Fig. 452.

Figs. 454 and 455. *Coelomomyces psorophorae* var. *psorophorae.* Scanning electron micrographs showing sporangial surface. Fig. 454. Bar = 10 μm. Fig. 455. Bar = 5 μm.

Fig. 456. *Coelomomyces psorophorae* var. *psorophorae.* Surface view of sporangium from *Aedes vexans* collected in Bladtna, Czechoslovakia. Bar = 10 μm.

Figs. 457 and 458. *Coelomomyces psorophorae* var. *psorophorae.* Resting sporangia from *Aedes funestus* collected in New Forest, Hants, England. Fig. 457. Surface of sporangial side. Bar = 10 μm. Fig. 458. Midoptical section, face view. Scale as in Fig. 457.

Fig. 459. *Coelomomyces psorophorae* var. *psorophorae.* Surface view of resting sporangia from *Aedes togoi* collected in Japan. Scale as in Fig. 457.

Figs. 460 and 461. *Coelomomyce psorophorae* var. *psorophorae.* Resting sporangia from *Aedes cinereus* collected in the Rhone Valley, Switzerland. Scale as in Fig. 457. Fig. 460. Surface view of sporangial side. Fig. 461. Midoptical section, side view showing sporangial wall.

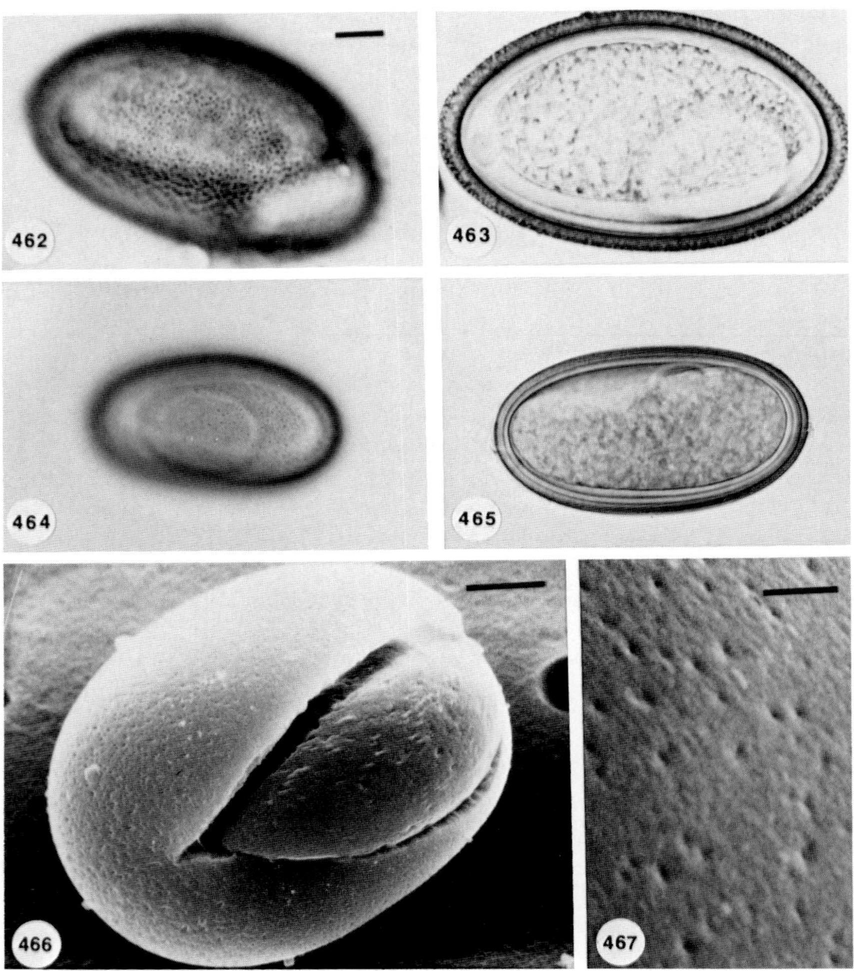

Figs. 462 and 463. *Coelomomyces psorophorae* var. *psorophorae.* Resting sporangia from *Aedes vexans* from Nebraska. Fig. 462. Surface view of sporangial side. Bar = 10 μm. Fig. 463. Midoptical section, side view. Scale as in Fig. 462.

Figs. 464–467. *Coelomomyces psorophorae* var. *psorophorae.* Resting sporangia from *Culiseta melaneura* collected in New York. Fig. 464. Surface view of sporangial side. Scale as in Fig. 462. Fig. 465. Midoptical section, side view. Scale as in Fig. 462. Figs. 466 and 467. Scanning electron micrographs. Note in Fig. 466 (bar = 5 μm) that the sporangial wall has ruptured along the dehiscence slit and that the sporogenic cytoplasm is projecting through the slit. Fig. 467 is a close up of the sporangial wall. Bar = 1 μm.

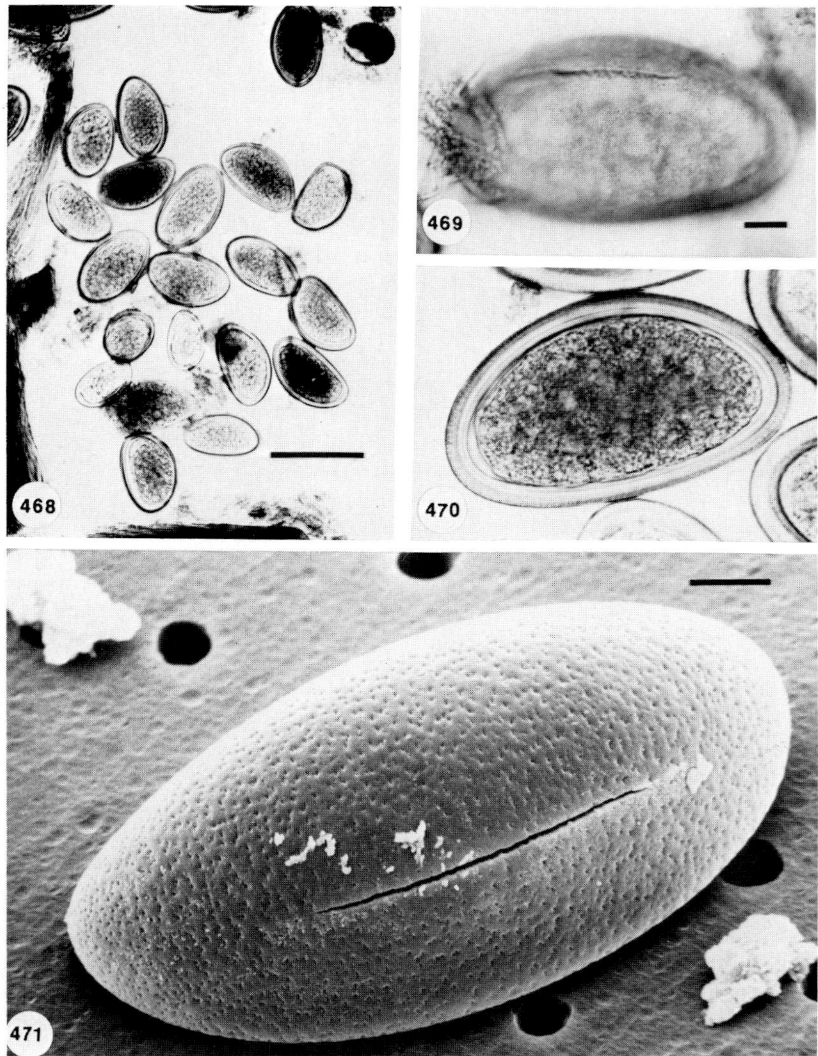

Figs. 468–470. *Coelomomyces psorophorae* var. *psorophorae.* Resting sporangia from *Aedes cinereus* from Moscow, U.S.S.R. Fig. 468. Low-magnification micrograph showing general sporangial shape. Bar = 100 μm. Fig. 469. Surface view of sporangial side. Bar = 10 μm. Fig. 470. Midoptical section, side view. Scale as in Fig. 469.

Fig. 471. *Coelomomyces psorophorae* var. *psorophorae.* Scanning electron micrograph of resting sporangium from *Aedes vexans* from Moscow, U.S.S.R. The dehiscence slit is visible along one side of the sporangium. Bar = 10 μm.

TABLE VII. *Coelomomyces psorophorae* var. *psorophorae*—Collections and Data[a]

Collection site	Host	Collector	Date	Sporangial size (μm)	Wall thickness (μm)	Punctae spacing (μm)	Punctae diameter (μm)	Location and habitat	Comments, references and depository (in parentheses)	Figure numbers
Canada										
Alberta	*Culiseta inornata* (Williston)	Shemanchuk	1956–1959	37–58 × 61–100	3–5	1–2.5	~0.5	Irrigated districts of southern Alberta	Shemanchuk, 1959; Anthony *et al.*, 1971; Whisler *et al.*, 1972, 1974, 1975; Travland, 1979; Zebold *et al.*, 1979 (NCU)	452–455
Alberta	[*Aedes vexans* Meig.	Shemanchuk	1979	—	—	—	—	—	Zebold *et al.*, 1979	]
Manitoba	[*Aedes trivittatus* (Coq.)♀	Taylor	1978–1979	105.6 × 60.7[b]	—	—	—	Winnipeg	Infection of larvae in artificial pools successful. Taylor, 1980 (NCU)	]
Czechoslovakia	*Aedes vexans* Meig.	Weiser	1963	55–71 × 80–104	3–6.5	~1	~0.5	Blatna	Weiser and Vávra, 1964 (NCU)	456
England	*Aedes funestus* Giles	Who 1024	1965	33–54 × 63–93	3–4	~1	~0.5	New Forest, Hants	(NCU)	457–458
France	[*Aedes cinereus* Meig.	Eckstein	1922	—	—	—	—	Strasbourg	Ekstein, 1922	]
	[*Aedes vexans* Meig.	Eckstein	1922	—	—	—	—	—	Ekstein, 1922	]
Japan	*Aedes togoi* (Theo.)♀	Omoril	1965	40–60 × 70–98	~3	~1	~0.5	—	Found in ovaries only	459
Switzerland	*Aedes cinereus* Meig.	Rabaud	1977	41–63 × 74–117	3–4	0.5–1	~0.5	Rhone Valley	(NCU)	460–461
United States										
California	*Aedes melanimon* Dyar♀	Kellen	1962	60–80 × 85–130	2–4	~1	~0.5	Dos Palos, Merced County	Sporangia found in ovaries only. Kellen, 1963; Kellen *et al.*, 1963 (NCU)	
California	[*Culiseta incidens* (Thompson)♀	Kellen	1963	—	—	—	—	—	Found in ovaries only. Kellen, 1963	447 and 448
Connecticut	*Aedes cantator*	Andreadis	1982	36–48 × 50–82	—	0.6–2.2	0.1–0.5	Guilford, salt marshes	Andreadis and Magnarelli, 1984 (CAES and NCU)	
Connecticut	*Aedes sollicitans* (Walker)	Andreadis	1982	31–36 × 48–60	—	0.3–2.2	0.4–0.6	Guilford, salt marshes	Collected on single larva (CAES)	

Florida	*Psorophora howardii* Coq.	Lum	1960	42–69 × 63–94	3–4	1–3	~1	Salt marsh near Indian River	Punctae lunate in optical section. Lum, 1963; Anthony *et al.*, 1971 (NCU)	449 and 450
Georgia	*Aedes vexans* Meig.	SMCG	1945	42–66 × 63–94	4–7	0.5–1	~0.5	Moultrie	(NCU)	444
Georgia	*Psorophora ciliata* (Fab.)	SMCG	1945	37–67 × 46–100	3–10	1–1.5	~0.5	Moultrie	Type–Couch, 1945 (BPI)	439–441
Georgia	*Psorophora ciliata* (Fab.)	SMCG	1945	40–72 × 68–164	2–10.5	1–1.5	~0.5	Thomasville	Couch and Dodge, 1947 (NCU)	445 and 446
Georgia	*Psorophora ciliata* (Fab.)	SMCG	1945	45–68 × 72–100	5–10	1–1.5	~0.5	Valdosta	Punctae lunate in optical secion. Couch and Dodge, 1947 (NCU)	442 and 446
Louisiana	*Aedes vexans* Meig.	Chapman	1965	45–70 × 72–102	3–10	0.5–1	~0.5	Numerous wooded habitats	Chapman and Woodard, 1966 (NCU)	
Louisiana	*Aedes vexans* Meig.	Chapman	1965	47–61 × 82–127	2–5	0.5–1	~0.5	From a temporary pool in a wooded area near Sulphur	Chapman and Woodard, 1966 (NCU)	
Louisiana	*Culex restuans* Theo.	Chapman	1965 and 1975	38–60 × 70–110	2.5–4	0.5–1	~0.5	From a wooded area near Lake Charles	Chapman and Woodard, 1966 (NCU)	
Louisiana	*Culex salinarius* Coq.	Chapman	1965	42–70 × 84–125	3–10	—	—	From a salt marsh near Cameron	Punctae difficult to see. Chapman and Woodard, 1966 (NCU)	
Louisiana	*Culiseta inornata* (Williston)	Chapman	1964	35–63 × 60–121	3–5	0.5–1	~0.5	Haymarket	Chapman and Woodard, 1966 (NCU)	
Louisiana	*Culiseta inornata* (Williston)	Chapman	1965	35–63 × 60–121	3–5	0.5–1	~0.5	From a salt marsh near Cameron	Chapman and Woodard, 1966 (NCU)	
Louisiana	*Culiseta inornata* (Williston)	Chapman	1965	35–60 × 55–100 (some 100 × 159)	3–5	0.5–1	~0.5	From a wooded area near Big Lake	Chapman and Woodard, 1966 (NCU)	
Louisiana	*Psorophora ciliata* (Fab.)	Esslinger	1963	50–55 × 85–93	3–5	0.5–1	~0.5	Near New Orleans	(NCU)	
Louisiana	[*Psorophora howardii* Coq.	Chapman	1965	—	—	—	—	From a temporary pool near Lake Charles	Chapman and Woodard 1966	451]
Louisiana	Theobaldia (Culiseta) *inornata* (Williston)	Walker	1944	~58 × 119	—	—	—	Port Sulphur	Specimen in poor condition. (NCU)	

(*continued*)

TABLE VII. (Continued)

Collection site	Host	Collector	Date	Sporangial size (μm)	Wall thickness (μm)	Punctae spacing (μm)	Punctae diameter (μm)	Location and habitat	Comments, references and depository (in parentheses)	Figure numbers
Minnesota	[*Aedes vexans* Meig.	Barr	1955	45–61 × 77–100	5–7	—	—	St. Paul	Laird, 1961	]
Minnesota	[*Culiseta morsitans* (Theo.)	Barr	1958	—	—	—	—	Itasca State Park	Barr, 1958 Specimen thought referable to *C. psorophorae* by Laird, 1961	]
Mississippi	[*Psorophora ciliata* (Fab.)	Peterson	1943	41–67 × 65–98	~3.6	~1.5	—	Greenwood	Barr, 1958; Laird, 1961	]
Nebraska	*Aedes vexans* Meig.	Mitchell	1976	41–71 × 61–112	4–6	1–1.5	~0.5	Bazile Creek	(NCU)	462 and 463
New York	*Culiseta melaneura* (Coq.)	Mehter	1976	25–40 × 50–73	2.5–3	0.15	~0.5	Oswego Co.	Punctae very difficult to see (NCU)	464–467
South Carolina	*Psorophora howardii* Coq.	Walker	1942	—	—	—	—	Charleston	Specimen in poor condition. Couch and Dodge, 1947 (UNC)	
U.S.S.R.[c]	[*Aedes caspius* (Pallas)♀	Shcherban	1971	—	—	—	—	Uzebekistan	—	]
U.S.S.R.	*Aedes cinereus* Meig.	Beklemishev	1960	46–60 × 98–114	3–5.2	~0.5	~0.5	Moscow	(NCU)	468–470
U.S.S.R.	*Aedes togoi* (Theo.)♀	Fedder	1871	42–67 × 70–130	2–3	—	—	Vladivostok	Punctae very difficult to see due to processing. Fedder *et al.*, 1971 (NCU)	
U.S.S.R.	*Aedes vexans* Meig.	Beklemishev	1957	40–70 × 72–121	2–4	0.5–3	~0.5	Moscow	Punctae often clustered. (NCU)	471
U.S.S.R.	[*Aedes vexans* Meig.	Morozov	1967	—	—	—	—	Krasnodar	Morozov, 1967	]
U.S.S.R.	[*Aedes vexans* Meig.	Kuznetzov	1970	—	—	—	—	Soviet Far East	Kuznetzov and Mikheeva, 1970	]
U.S.S.R.	[*Aedes vexans* Meig. ♀	Zharov	1973	—	—	—	—	Astrakhan	Found in ovaries. Zharov, 1973	]

[a] Bar indicates information unavailable or specimen unsuitable for measurements. Brackets indicate specimens not examined in the present study.
[b] Data from Taylor (1980).
[c] For other references to collections in the U.S.S.R., see Nam (1981).

of *Aedes scatophagoides* has been studied and determined to be *C. stegomyiae* var. *stegomyiae*.

Coelomomyces psorophorae var. ***halophilus*** var. nov. Couch & Lum

Figures 472–478

DESCRIPTION: Hyphae 8–13 μm wide, highly branched. Resting sporangia ellipsoid, elongate–ellipsoid, to allantoid (in side view); 26–50 × 37–90 μm, frequently flattened to slightly concave on dorsal side; surface ornamented with fine punctae, 0.5 μm in diameter and 1.0–2.5 μm apart. Wall 1.5–2.0 μm thick, with internal vertical striae appearing parallel in midoptical section. Thin-walled sporangia not observed.

LATIN DIAGNOSIS: Hyphae 8–13 μm latae, valde ramosae. Sporangia perennantia ellipsoidea, elongato–ellipsoidea, vel ab aspectu laterali allantoidea, 26–50 × 37–90 μm, crebre applanata vel in dorso paulum concava; superficie punctis subtilibus 0.5 μm diametro, inter se 1.0–2.5 μm crassa, distantibus, ornata; tunica 1.5–2.0 μm crassa, striis internis verticalibus in sectione media opticali specie parallelibus notata. Sporangia tenuiter tunicata non observata.

TYPE HOST AND COLLECTION SITE: *Aedes taeniorhynchus* (Wiede.); Vero Beach, Florida (collected by Lum in 1960; see Lum, 1963).

DEPOSITORY FOR TYPE: BPI. Several isotypes are on deposit at NCU.

SPECIMENS EXAMINED OTHER THAN TYPE AND ISOTYPES:

Cayman Islands

Aedes taeniorhynchus (Wiede.); Grand Cayman; Giglioli; 1966; NCU.

United States

Aedes sollicitans (Walker); Louisiana; Chapman; 1965; NCU; (Chapman and Woodard, 1966).

Aedes sollicitans (Walker); North Carolina; Kiser; 1974; NCU; (Kiser, 1976).

Aedes taeniorhynchus (Wiede.); North Carolina; Romney; 1974; NCU.

Aedes taeniorhynchus (Wiede.); North Carolina; McNitt; 1975, 1976; NCU.

COMMENTS: *Coelomomyces psorophorae* var. *halophilus* is separated from other varieties of *C. psorophorae* on the basis of its thinner-walled resting sporangia, which often appear allantoid in side view. Further distinction may be made in that representatives of this variety have been

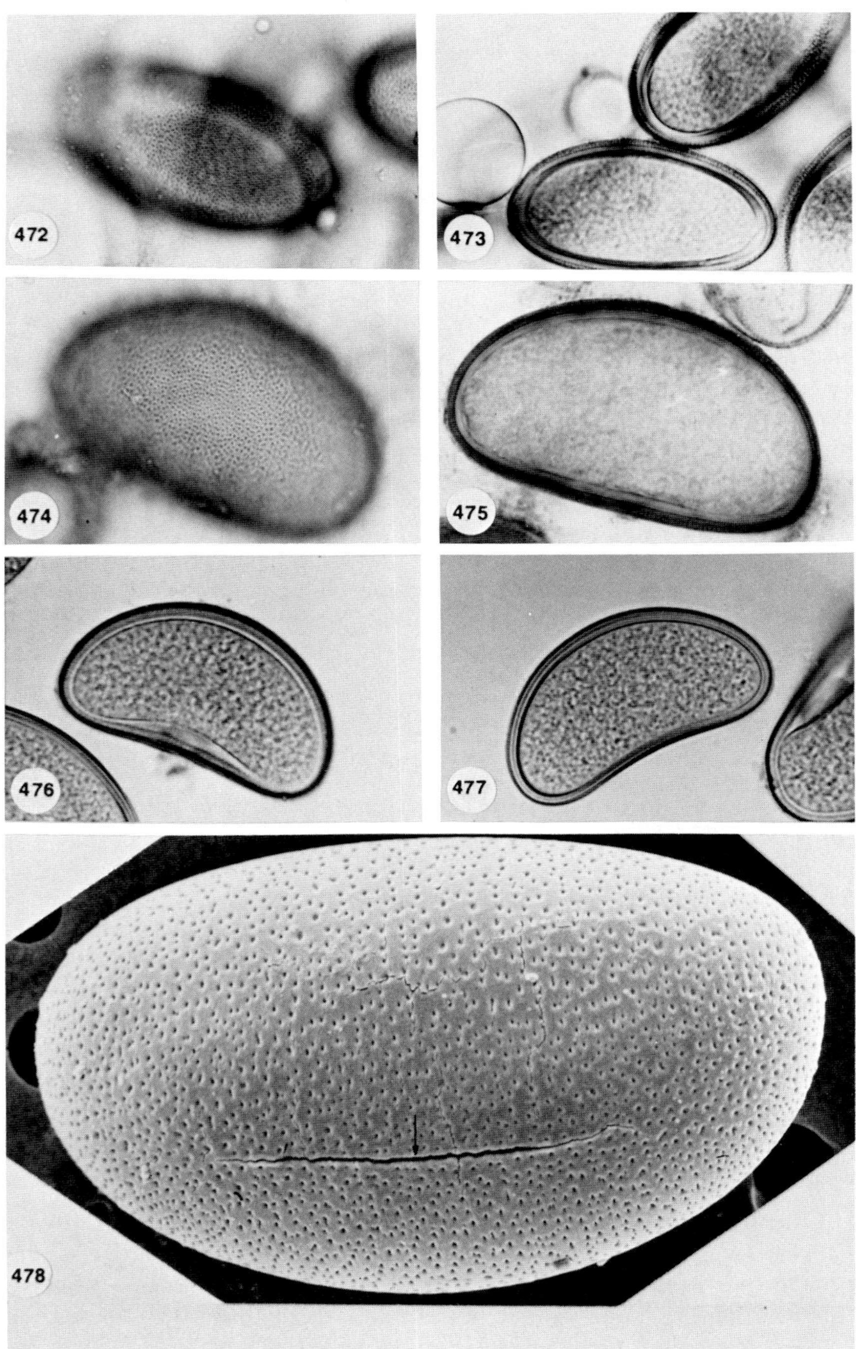
472
473
474
475
476
477
478

collected only on hosts occurring in brackish water. Most collections have included both parasitized larvae and adult females in which infection was confined to the ovaries.

Although it cannot be stated with certainty, it is likely that the variety of *C. psorophorae* (on *Aedes taeniorhynchus* and *Aedes sollicitans* collected by Chapman in Louisiana) used in the infection experiments of Federici and Roberts (1975) and the culture experiments of Shapiro and Roberts (1976) is the same as that described here.

Coelomomyces psorophorae var. ***tasmaniensis*** var. nov. Couch & Laird

Coelomomyces tasmaniensis Laird, *J. Parasitol.* **42,** 53 (1956b)
Coelomomyces opifexi Pillai & Smith, *J. Invertebr. Pathol.* **11,** 316 (1968)

Figures 479–482

Description: Hyphae 3–15 μm in diameter. Resting sporangia oval, 35–83 × 50–179 μm, frequently pointed at poles and flattened on dorsal side, occasionally spherical, 25–62 μm in diameter; surface ornamented with fine punctae, 0.25–0.5 μm in diameter and 0.5–1.0 μm apart. Wall usually 2–4 μm (occasionally 7 μm) thick, appearing striate in midoptical section due to continuation of punctae into the wall. Thin-walled sporangia not observed.

Latin Diagnosis: Hyphae 3–15 μm crassae. Sporangia perennantia ovalia, 35–83 × 50–179 μm, crebre ad polos acuta et dorsaliter applanata, interdum sphaerica, 25–62 μm diametro, inter se 0.5–1.0 μm distantibus, ornata; tunica plerumque 2–4 μm, nonnumquam 7 μm, crassa, quasi in sectione media inspecta ob continuationem punctorum in tunicam specie striata. Sporangia tenuiter tunicata non observata.

Type Host and Collection Site: *Aedes* (*Pseudokusea*) *australis* (Erickson); Trumpeter Island, Port Dorey, Tasmania (collected by E. N. Marks).

Figs. 472 and 473. *Coelomomyces psorophorae* var. *halophilus*. Resting sporangia of type from *Aedes taeniorhynchus* collected in Florida. Fig. 472. Surface view of sporangial side. Fig. 473. Midoptical section, side view.

Figs. 474–478. *Coelomomyces psorophorae* var. *halophilus*. Resting sporangia from *Aedes sollicitans* collected in Florida. Fig. 474. Surface view of sporangial side. Figs. 475–477. Midoptical section, side view showing general sporangial shape. Fig. 478. Scanning electron micrograph. Note the dehiscence slit (arrow) along one side of the sporangium.

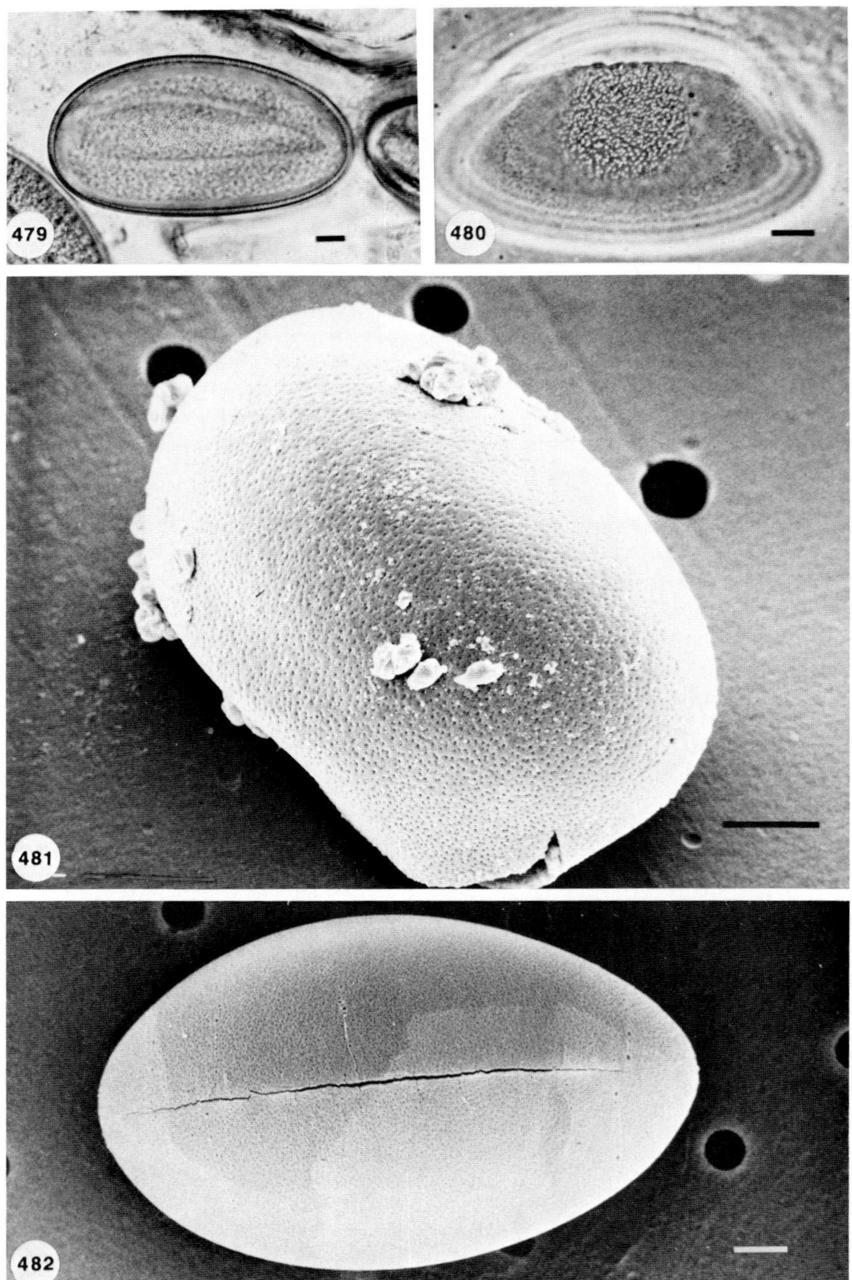
479
480
481
482

Depository for Type: Type specimens of both *C. tasmaniensis* and *C. opifexi* are deposited at the Dominion Museum, Wellington, New Zealand. Paratypes of *C. tasmaniensis* are deposited at the Department of Entomology, University of Queensland, Brisbane, Australia, and BPI. The type specimen for *C. tasmaniensis* (deposited at the DM) therefore becomes the type for *C. psorophorae* var. *tasmaniensis,* of which there are two paratypes deposited at UQA and BPI.

Alternate Host: *Tigriopus* sp. near *T. angulatus* (Pillai *et al.*, 1976).

Specimens Examined and Other Collections: Both paratypes of *C. tasmaniensis* were studied, as well as the following collections of *C. opifexi.*

New Zealand

Opifex fuscus Hutton; Otago Peninsula; Pillai; 1968; Pillai, BPI, and NCU; (Pillai, 1969).

[(*Aedes australis* (Erickson); Otago Peninsula; Pillai; 1967–1970; Pillai; (Pillai, 1971).]

Aedes australis (Erickson); Otago Peninsula; Pillai; 1979; BPI and NCU.

Comments: Two previously described species of *Coelomomyces, C. tasmaniensis* and *C. opifexi,* are here combined as a variety of *C. psorophorae.* Justification for this combination follows. As described originally (Laird, 1956b), *C. tasmaniensis* was separated from *C. psorophorae* in that the former "has more slender hyphae," "infects larvae occurring in a brackish habitat," and has sporangia in which one side is always "either flat or concave." With the exception of the characteristic of sporangia with a concave side, which is an artifact of mounting, each of the other criteria for separation from *C. psorophorae* has since been observed in members of the *C. psorophorae* "complex." *Coelomomyces opifexi,* described by Pillai and Smith (1968), was separated from both *C. psorophorae* and *C. tasmaniensis* on the basis of its "constant wide range and size of the sporangia, abundance of spherical sporangia,

---

Figs. 479–481. *Coelomomyces psorophorae* var. *tasmaniensis.* Resting sporangia of paratype of *Coelomomyces tasmaniensis* from *Aedes australis* collected in Port Dorie, Tasmania. Bars = 10 μm. Figs. 479 and 480. Surface view of sporangial side. Fig. 481. Scanning electron micrograph.

Fig. 482. *Coelomomyces psorophorae* var. *tasmaniensis.* Scanning electron micrograph of resting sporangia of *Coelomomyces opifexi* from *Opifex fuscus* collected in New Zealand. Note the prominent dehiscence slit along the center of the sporangium. Bar = 10 μm.

and the occurrence of sporangia with both smooth and crenated walls." Of these criteria for separation, the first two have been negated in that similar features have since been observed frequently in collections within the *C. psorophorae* "complex." The last is considered to be based on an artifact of preservation, as noted by Laird (1956b), for sporangia of *C. tasmaniensis,* or on teratogenic sporangia, a frequent occurrence within species of *Coelomomyces*. Based on similarities in sporangial ornamentation, hosts, habitat, and distribution, *C. tasmaniensis* and *C. opifexi* are felt to be the same taxon. Having characteristics tying them closely to the *C. psorophorae* "complex," they are here united as *C. psorophorae* var. *tasmaniensis*. Separation of this variety from others in this species is made primarily on the basis of its occurrence on *Opifex fuscus* and *Aedes australis* from brackish water habitats, but also on the basis of its generally large, relatively thin-walled resting sporangia with fine, closely spaced punctae.

Coelomomyces punctatus Couch & Dodge ex Couch, *J. Elisha Mitchell Sci. Soc.* **78,** 135 (1962)

Coelomomyces punctatus Couch & Dodge, *J. Elisha Mitchell Sci. Soc.* **63,** 69 (1947)
Coelomomyces dodgei Couch, *J. Elisha Mitchell Sci. Soc.* **61,** 124 (1945)

Figures 483–487

DESCRIPTION: Hyphae 7–14 μm thick, irregularly branched. Resting sporangia ellipsoid 32–41 × 42–75 μm, occasionally flattened on dorsal side; surface ornamented with rounded to elongate punctae (striae), 0.5 μm in diameter when rounded and 0.5–4.0 μm or more when elongate, approximately 2–4 μm apart, often linearly arranged. Thin-walled sporangia not observed.

TYPE HOST AND COLLECTION SITE: *Anopheles quadrimaculatus* Say; Dublin, Georgia (collected by SMCG).

DEPOSITORY FOR TYPE: BPI. Because of the poor condition of the type, a paratype has been deposited also at the BPI.

ALTERNATE HOST: *Cyclops vernalis* Fisher (Federici and Roberts, 1976).

SPECIMENS EXAMINED AND OTHER COLLECTIONS (indicated by brackets):

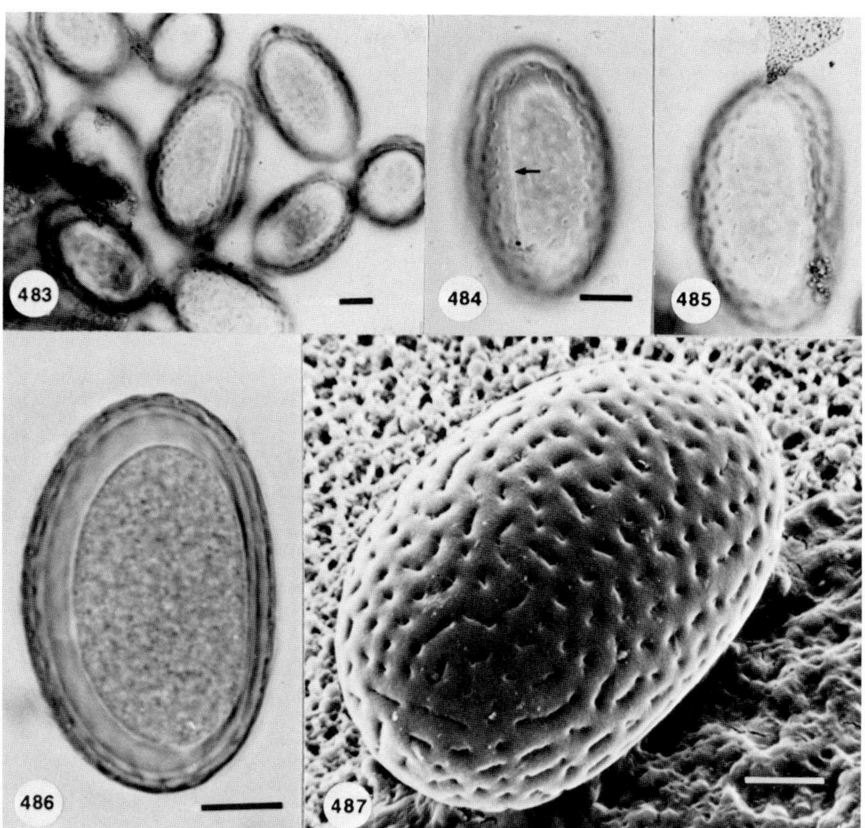

Figs. 483–487. *Coelomomyces punctatus.* Resting sporangia of type specimen from *Anopheles quadrimaculatus* collected in Georgia. Fig. 483. Low magnification micrograph showing general sporangial shape. Bar = 10 μm. Fig. 484. Surface view of dorsal side showing dehiscence slit (arrow). Bar = 10 μm. Fig. 485. Surface view, dorsal side. Scale as in Fig. 484. Fig. 486. Midoptical section, side view. Bar = 10 μm. Fig. 487. Scanning electron micrograph. Bar = 5 μm.

Type, paratype

United States

Anopheles bradleyi King; Louisiana; Chapman; 1963; NCU.

Anopheles crucians Wiede.; Georgia; SMCG; 1944; NCU.

Anopheles crucians Wiede.; Louisiana; Chapman; 1966, 1967; NCU; (Chapman and Glenn, 1972).

[*Anopheles freeborni* Aitken*; California; Federici; 1977; ?; (Federici, 1977).]
Anopheles punctipennis (Say); North Carolina; Roane; 1965–1969; NCU.
Anopheles quadrimaculatus Say; Georgia; SMCG; 1944; NCU.
Anopheles quadrimaculatus Say; Louisiana; Chapman; 1966; NCU.
Anopheles quadrimaculatus Say; North Carolina; Roane; 1965–1969; NCU.
Anopheles quadrimaculatus Say; Virginia; Martin; 1972; NCU.
[*Anopheles stephensi* Liston*; Georgia; Federici; 1980; ?; (Federici *et al.,* 1975).]

COMMENTS: Collected only from the southeastern United States on species of *Anopheles, C. punctatus* is distinguished by its resting sporangia ornamented with regular to elongate punctae. Although sporantia of *C. punctatus* resemble those of *C. dodgei,* those of the latter are characterized by having punctae on one sporangial face and elongate striae on the other. While larvae infected with *C. punctatus* may contain sporangia resembling not only *C. punctatus* but also *C. dodgei* and *C. lativittatus,* identification should be made on the basis of the predominant type of sporangium present. The occasional occurrence of all three types of sporangia in the same larva no doubt reflects the close relationship between these three species, considered originally (Couch, 1945) to be the same taxon, *C. dodgei.* B. A. Federici (personal communication) reports that a physiological distinction of *C. punctatus* from *C. dodgei* is possible in that meiospore release in *C. punctatus* begins 8 hr after onset of darkness, whereas release in *C. dodgei* begins approximately 17 hr after onset.

Federici (1979) obtained experimental hybridization of *C. dodgei* and *C. punctatus.* Sporangia resulting from crosses resembled primarily *C. dodgei* or in some instances *C. lativittatus.*

Coelomomyces quadrangulatus var. ***quadrangulatus*** Couch ex Couch, *J. Elisha Mitchell Sci. Soc.* **78,** 135 (1962)

Coelomomyces quadrangulata Couch, *J. Elisha Mitchell Sci. Soc.* **61,** 124 (1945)

Figures 488–494

DESCRIPTION: Hyphae 9.6–25.0 μm thick. Resting sporangia ellipsoid, 12–21 $\times$ 19–40 μm; surface ornamented with four prominent longitudinal

* Laboratory infection.

ridges that are interconnected by reticulae, transverse striae; dorsal ridge, with centrally located dehiscence slit, connecting at poles with lateral ridges that encircle the sporangium; ventral ridge usually insular or rarely as 2–3 lobes. In midoptical section: face view, lateral ridges as vertically striate wing that encircles the sporangium; end view, ridges as oppositely opposed lobes giving the sporangium a four-angled appearance; side view, ridges appearing as a knob at either pole and a wing along the dorsal and ventral surface, or ventral ridge appearing rarely as 2–3 elongate domes. Wall varying in thickness from approximately 1 μm in depressed areas to 3–5 μm at ridges. Thin-walled sporangia are observed.

TYPE HOST AND COLLECTION SITE: *Anopheles crucians* Wiede.; Georgia (collected by SMCG).

DEPOSITORY FOR TYPE: BPI. Isotypes have been deposited also at the BPI and NCU.

SPECIMENS EXAMINED AND OTHER COLLECTIONS (indicated by brackets):

Type, isotypes

United States

Anopheles crucians Wiede.; Georgia; SMCG; 1945; NCU.

Anopheles georgianus King; Georgia; SMCG; 1945; NCU; (Couch and Dodge, 1947).

Anopheles punctipennis (Say); Georgia; SMCG; 1944, 1945; NCU.

Anopheles punctipennis (Say); North Carolina; Roane and Martin; 1965, 1968; NCU.

Anopheles quadrimaculatus Say; Georgia; SMCG; 1945; NCU.

Anopheles walkeri Theo.; Minnesota; Barr; 1954; NCU; (Barr, 1958; Laird, 1961).

Taiwan

[*Uranotaenia* sp. (near *recondita* Edwards); Tartung, Taiwan; Chapman; 1971; MUN; (Laird *et al.*, 1980).]

U.S.S.R.

[*Culex pipiens modestus* Forskal; Ukraine; (Lavitskaya, 1967; as reported by Dubitskij *et al.*, 1973).]

[*Aedes* sp. (*Ae. rossicus* or *Ae. geniculatus*); Ukraine; (Lavitskaya, 1967, as reported by Dubitskij *et al.*, 1973).]

[*Aedes vexans* Meig.; Ukraine; (Lavitskaya, 1967, as reported by Dubitskij *et al.*, 1973).]

COMMENTS: *Coelomomyces quadrangulatus* var. *quadrangulatus* is distinguished by its uniquely lobed resting sporangia, appearing four-an-

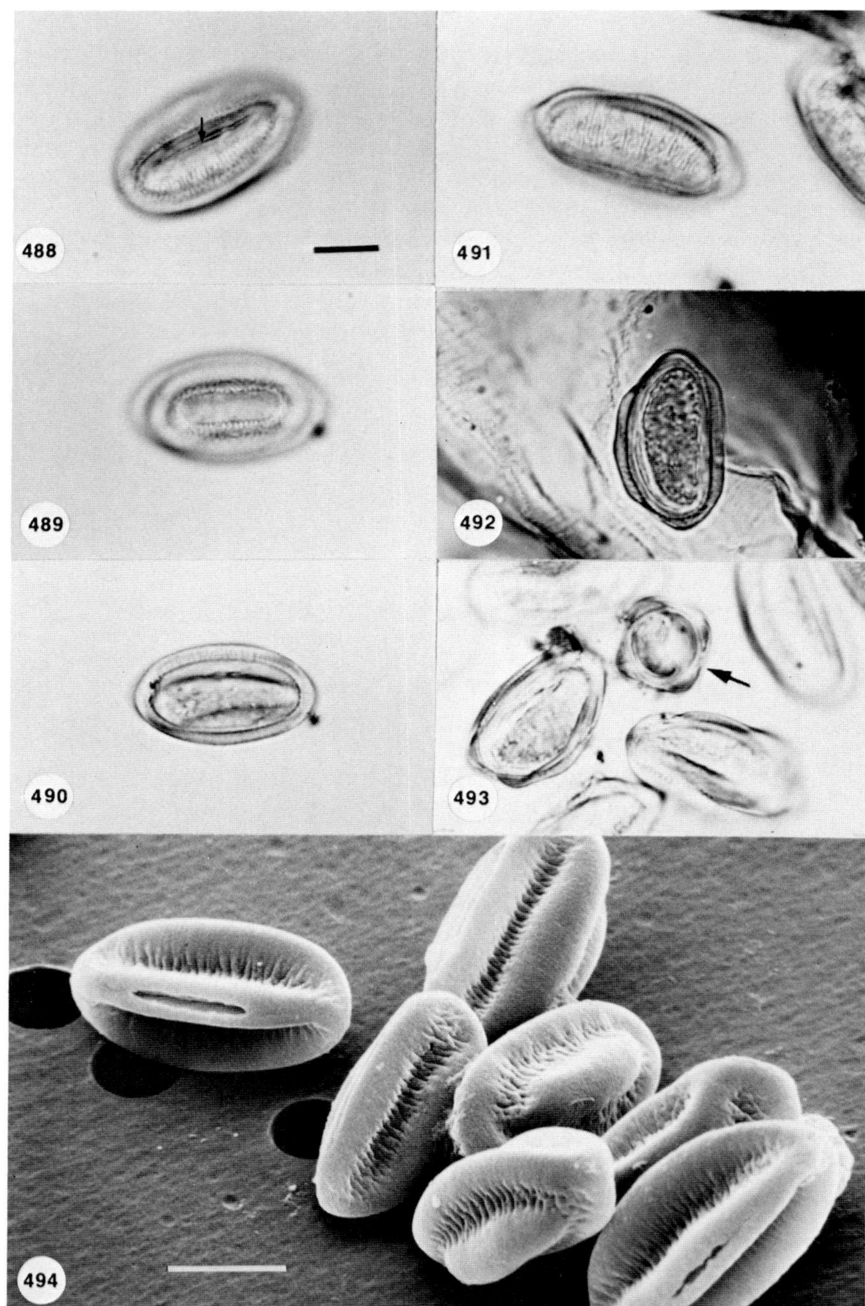
488
489
490
491
492
493
494

gled in end view. Although sporangial ornamentation of *C. carolinianus*, distinction may be made in that ridging in the latter is highly irregular and results in sporangia appearing generally pentangulate in end view. In some collections of *C. quadrangulatus* var. *quadrangulatus*, irregular sporangia have been observed that are quite similar to those of *C. carolinianus*. In all such collections, however, the typical form predominates. Yet, because of this similarity, further collection and study of these forms may necessitate reduction of *C. carolinianus* to synonomy with *C. quadrangulatus* var. *quadrangulatus*.

Coelomomyces quadrangulatus var. ***irregularis*** Couch & Dodge ex Couch, *J. Elisha Mitchell Sci. Soc.* **78,** 135 (1962)

Coelomomyces quadrangulatus var. *irregularis* Couch & Dodge, *J. Elisha Mitchell Sci. Soc.* **63,** 69 (1947)

Figures 495–498

DESCRIPTION: Hyphae as in *C. quadrangulatus* var. *quadrangulatus*. Resting sporangia as in the typical variety, 15–21 × 23–41 μm, but with outer wall surface highly irregular to crenulate. Thin-walled sporangia not observed.

TYPE HOST AND COLLECTION SITES: *Anopheles punctipennis* (Say); Georgia (collected by SMCG).

DEPOSITORY FOR TYPE: BPI. A paratype, collected in Georgia on the same host as the type, is deposited at NCU.

SPECIMENS EXAMINED AND OTHER COLLECTIONS:

Type, paratype

COMMENTS: Although resting sporangia of *C. quadrangulatus* var. *irregularis* appear four-angled in end view, distinction from other varieties of this species may be made on the basis of the highly irregular sporangia found in this variety.

Figs. 488–494. *Coelomomyces quadrangulatus* var. *quadrangulatus.* Resting sporangia from *Anopheles punctipennis* collected in Georgia. Fig. 488. Surface view, dorsal side (dehiscence slit, arrow). Bar = 10 μm, applies for Figs. 488–493. Fig. 489. Surface view, ventral side. Fig. 490. Midoptical section, face view. Fig. 491. Surface view of sporangial side. Fig. 492. Midoptical section, side view. Fig. 493. Midoptical section end view (arrow). Fig. 494. Scanning electron micrograph showing several sporangia from different perspectives. Note the prominent dehiscence slit visible along the dorsal ridge of two of the sporangia. Bar = 10 μm.

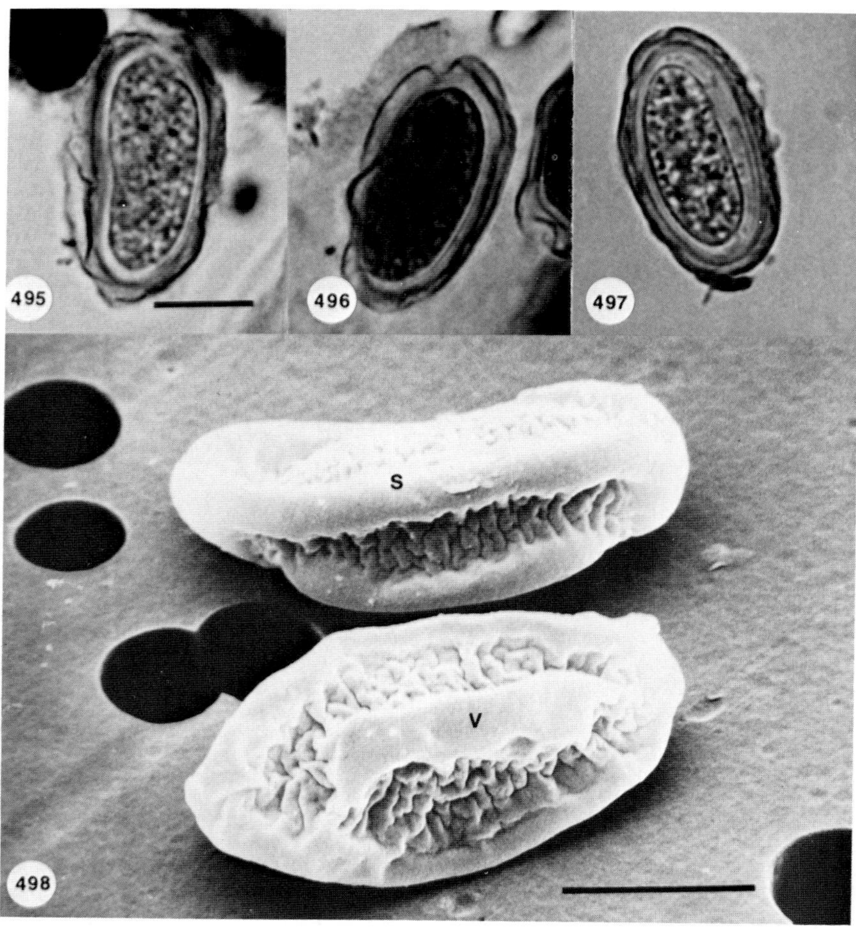

Figs. 495–498. *Coelomomyces quadrangulatus* var. *irregularis.* Resting sporangia of type specimen from *Anopheles punctipennis* collected in Georgia. Figs. 495–497. Midoptical section, side view. Bar = 10 μm. Fig. 498. Scanning electron micrograph showing side view (S) and ventral view (V) of two different sporangia. Bar = 10 μm.

Coelomomyces quadrangulatus var. ***lamborni*** Couch & Dodge ex Couch *J. Elisha Mitchell Sci. Soc.* **78,** 135 (1962)

Coelomomyces quadrangulatus var. *lamborni* Couch & Dodge, *J. Elisha Mitchell Sci. Soc.* **63,** 69 (1947)

(Not illustrated.)

DESCRIPTION: Hyphae not observed. Resting sporangia as in the typical variety except larger, 26–31 × 36–58 μm. Thin-walled sporangia not observed.

TYPE HOST AND COLLECTION SITE: *Aedes* (*Stegomyia*) *scutellaris* (Walker)*; Malay States (collected by Lamborn).

DEPOSITORY FOR TYPE: This specimen was discovered on the type slide of *C. stegomyiae* deposited in the laboratory of Dr. D. Keilin, Cambridge University, Cambridge, England. However, recent attempts to locate this material have been unsuccessful.

SPECIMENS EXAMINED AND OTHER COLLECTIONS:

Type (Although studied by J. N. C. prior to preparation of the original description, this variety has neither been studied nor collected subsequently.)

COMMENTS: *Coelomomyces quadrangulatus* var. *lamborni* is distinguished by its resting sporangia, which are larger than in any other variety of this species. Although included here, this taxon must be considered inadequately known because of the limited material on which the type was based and because of its lack of further collection or study.

Coelomomyces quadrangulatus var. ***parvus*** Laird ex Laird, *J. Elisha Mitchell Sci. Soc.* **78,** 132 (1962)

Coelomomyces quadrangulatus var. *parvus* Laird, *Ecology* **40,** 206 (1959)

Figures 499–500

DESCRIPTION: Hyphae 10–17 μm thick. Resting sporangia measuring 9–16 × 12–21 μm, ornamented with longitudinal ridges giving sporangia a quadrangular appearance in end view. Thin-walled sporangia not observed.

TYPE HOST AND COLLECTION SITE: *Culex* (*Culex*) *tritaeniorhynchus summorosus* Dyar; Pasir Panjang, Singapore, Malaysia.

DEPOSITORY FOR TYPE: BPI.

SPECIMENS EXAMINED AND OTHER COLLECTIONS:

Type

* Thought by Laird (1965a) to have been misidentified and to more likely have been *Aedes albopictus* (Skuse).

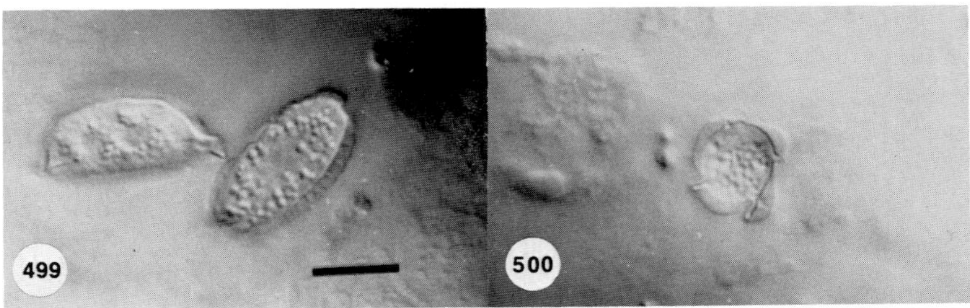

Figs. 499 and 500. *Coelomomyces quadrangulatus* var. *parvus*. Resting sporangia of type specimen from *Culex tritaeniorhynchus summorosus* collected from Singapore, Malaysia. Fig. 499. Mid-optical section, side view. Bar = 10 μm. Fig. 500. Mid-optical section, end view. Scale as in Fig. 499.

COMMENTS: This variety is distinguished by resting sporangia that are smaller than in any other variety of this species. Type specimens sent to Couch by Laird in 1967 (since returned) confirmed the quadrangulate appearance in end view of resting sporangia of this species. The minute pits described by Couch (1945) to occur on sporangia of the typical variety of *C. quadrangulatus* and by Laird (1959a) to occur also on sporangia of *C. quadrangulatus* var. *parvus* have been shown by SEM views of the typical variety to be end views of internal wall lacunae and not actually surface features. Such "pitting" is evident only in optical sections of the wall.

Coelomomyces raffaelei var. ***raffaelei*** Coluzzi & Rioux, *Riv. Malariol.* **41,** 29 (1962)

Figures 501–512

DESCRIPTION: Hyphae not observed. Resting sporangia ellipsoid, 18–26 × 33–49 μm, slightly flattened on dorsal side; surface ornamented with irregular-to-elongate anastomosing ridges (4–6 μm wide × 2–3 μm high) enclosing circular or irregular-to-elongate depressed areas with punctae/striae; dorsal surface with longitudinal ridge containing central dehiscence slit, ridge bordered by elongate depressed areas with punctae/striae aligned perpendicular to the ridge. In midoptical section: end view, ridges as 4–6 lobes alternating with depressed areas around periphery of sporangia; side view, elongate ridge with dehiscence slit as wing along dorsal edge, irregular ridges appearing as domes alternating with depressed areas at poles and along the ventral edge. Thin-walled sporangia not observed.

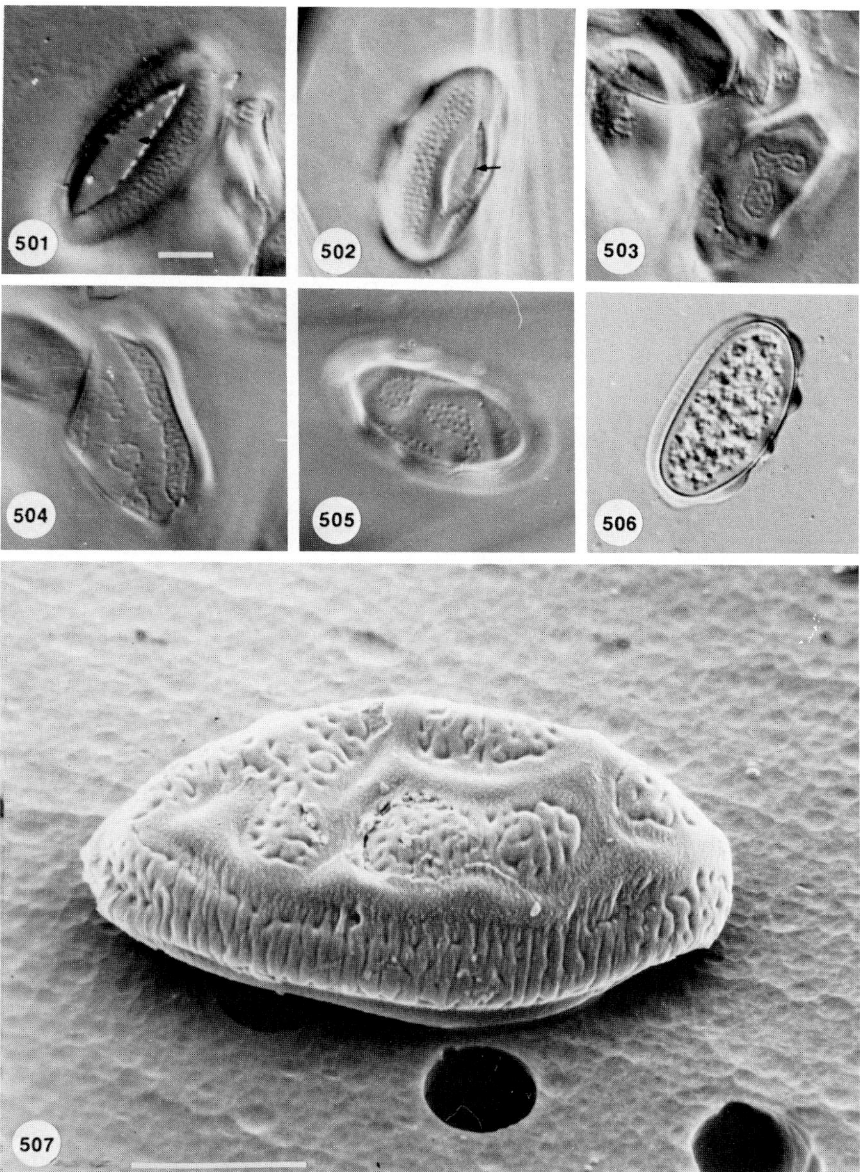

Figs. 501–507. *Coelomomyces raffaelei* var. *raffaelei*. Resting sporangia of paratype from *Anopheles claviger* collected in Monticelli, Italy. Figs. 501 and 502. Surface view of dorsal side showing open dehiscence slit (arrow). Bar = 10 μm. Figs. 503–505. Surface views of sporangial side. Scale as in Fig. 501. Fig. 506. Midoptical section, side view. Scale as in Fig. 501. Fig. 507. Scanning electron micrograph showing side view. Bar = 10 μm.

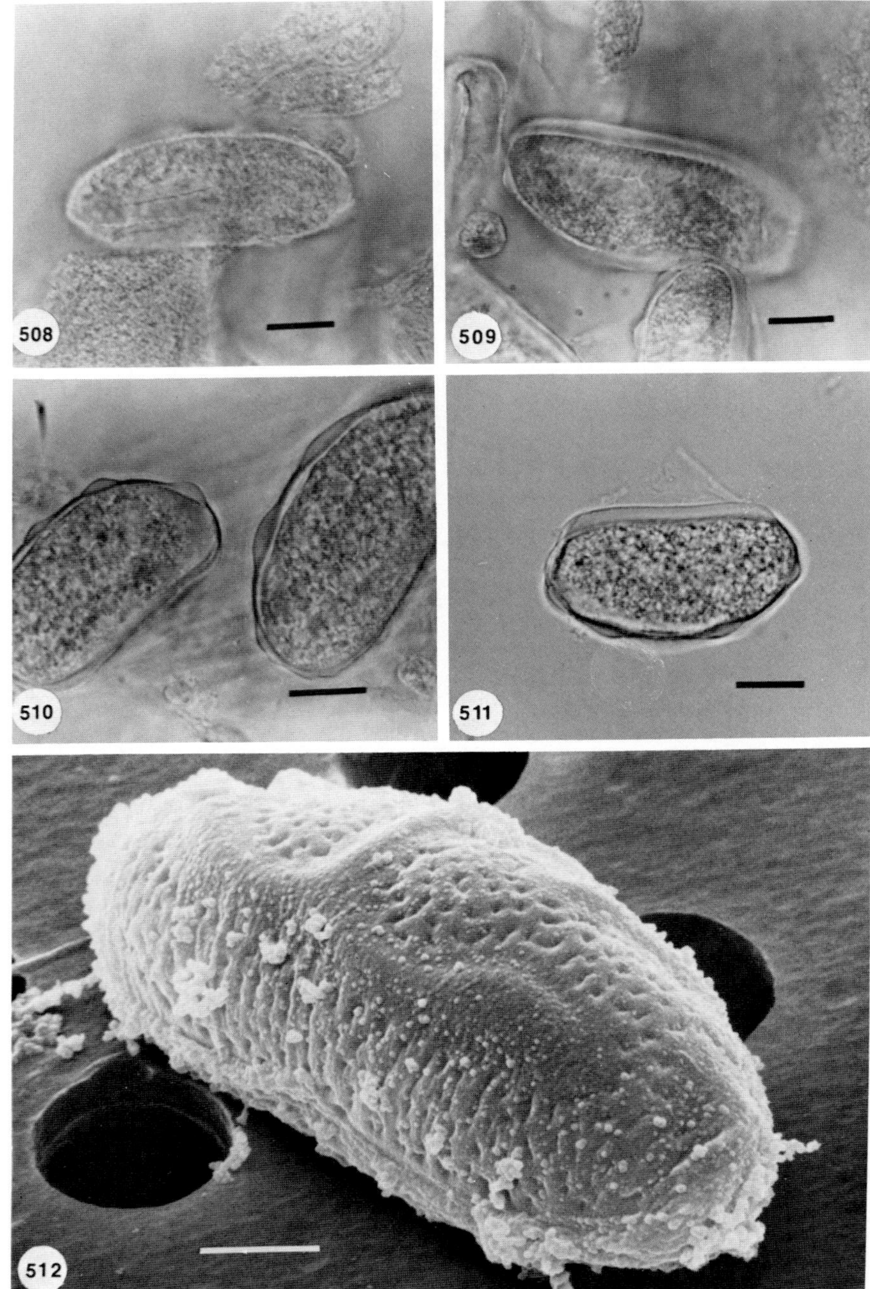
508
509
510
511
512

Type Host and Collection Site: *Anopheles claviger* (Meig.); Valley of Forma Quesa, Monticelli (Frasinone), Italy (collected by M. Coluzzi).

Depository for Type: Stazione Sperimentale di Entomologia Medicae Veterinaria, Monticelli (Frasinone), Italy. Paratypes are deposited also at the Laboratoire de Parasitologie, Universite de Montpellier, Montepellier, France, and at NCU.

Specimens Examined and Other Collections:

Paratype

Corsica

Anopheles claviger; Furiani, Corsica; Rodhain; 1925; NCU.

Comments: Resting sporangia of *C. raffaelei* var. *raffaelei* closely resemble those of *C. africanus;* however, those of the former have larger, more prominent ridges and more prominent bands of punctae/striae bordering the dehiscence slit. Further, depressed areas on the side and ventral face of the sporangial wall of *C. raffaelei* var. *raffaelei* are irregularly shaped, whereas comparable areas in *C. africanus* are circular. While sporangia of *C. raffaelei* var. *parvus* bear ornamentation identical to that of the typical variety, sporangia of the latter are significantly larger than those of *C. raffaelei* var. *parvus.*

The specimen of *C. raffaelei* var. *raffaelei* from Corsica was found by Rodhain among old entomological collections of the Institut Pasteur, Paris. The infected larva of *Anopheles claviger* was collected near Furiani, Corsica on 31 December, 1925, and mounted in a permanent medium (probably balsam). For study, after dissolving the mounting medium in tybel, the specimen was rehydrated through an ethanol series. After this, one portion was prepared for study by SEM and another mounted in lactophenol.

Coelomomyces raffaelei var. ***parvus*** Laird *et al.*, *J. Parasitol.* **61,** 539 (1975)

Coelomomyces raffaelei var. *parvum* Laird *et al.*, *J. Parasitol.* **61,** 539 (1975)

Figures 513–519

Figs. 508–512. *Coelomomyces raffaelei* var. *raffaelei.* Resting sporangia from *Anopheles claviger* collected in Furiani, Corsica. Fig. 508. Surface view of dorsal surface. Bar = 10 μm. Fig. 509. Surface view of sporangial side. Bar = 10 μm. Figs. 510 and 511. Midoptical section, side view. Bars = 10 μm. Fig. 512. Scanning electron micrograph showing ventral surface. Bar = 5 μm.

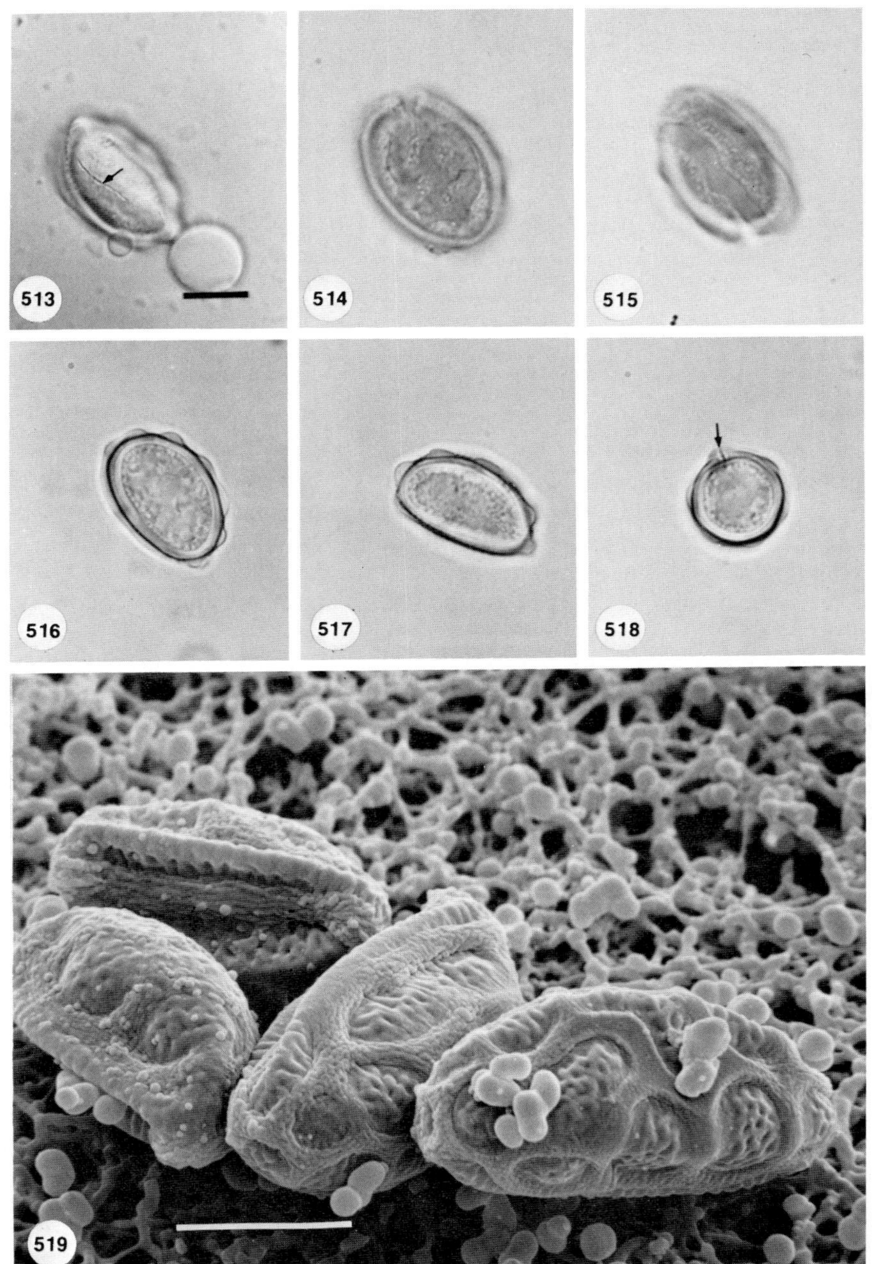
513
514
515
516
517
518
519

DESCRIPTION: Hyphae not observed. Resting sporangia as in the typical variety except smaller, 12–21 × 21–34 μm. Thin-walled sporangia not observed.

TYPE HOST AND COLLECTION SITE: *Anopheles sinensis* Wiede.; Nagasaki, Japan (collected by M. Mogi).

DEPOSITORY FOR TYPE: Holotype not designated. Syntypes (cotypes) deposited at TNS and BPI.

SPECIMENS EXAMINED AND OTHER COLLECTIONS:

Syntype (BPI)

COMMENTS: See Comments for the typical variety.

Coelomomyces reticulatus sp. nov. var. ***reticulatus*** Couch & Walker

Figures 520–524

DESCRIPTION: Hyphae 5–18 μm in diameter. Resting sporangia ellipsoid, slightly flattened on dorsal side, 27–55 × 43–80 μm; surface ornamented with a recticulum of dome-like ridges (2–3 μm wide and 2–3 μm high) enclosing circular to smoothly polygonal, depressed areas (5–12.5 μm across) with scattered punctae (0.5 μm in diameter and 1 μm apart); dehiscence slit central along an elongate longitudinal ridge bordered by circular to polygonal depressed areas. In midoptical section: ridges in end view as truncate domes that are interconnected by a wing of material (ridges in side view) encircling the sporangium. Thin-walled sporangia not observed.

TYPE HOST AND COLLECTION SITE: *Anopheles kochi* Dön.; Tampin, Malaya (collected by Wharton).

DEPOSITORY FOR TYPE: BPI. An isotype is on deposit at NCU.

LATIN DIAGNOSIS: Hyphae 5–18 μm crassae. Sporangia perennantia ellipsoidea, dorso aliquantilum applanato, 27–55 × 43–80 μm; superficie

---

Figs. 513–519. *Coelomomyces raffaelei* var. *parvus.* Resting sporangia of syntype from *Anopheles sinensis* collected in Nagasaki, Japan. Fig. 513. Surface view, dorsal side (dehiscence slit, arrow). Bar = 10 μm. Figs. 514 and 515. Surface view of sporangial side. Bars = 10 μm. Figs. 516 and 517. Midoptical section, side view. Bars = 10 μm. Fig. 518. Midoptical section, end view. Note the dehiscence slit visible in the middorsal ridge (arrow). Scale as in Fig. 517. Fig. 519. Scanning electron micrograph showing side view of several sporangia. Bar = 10 μm.

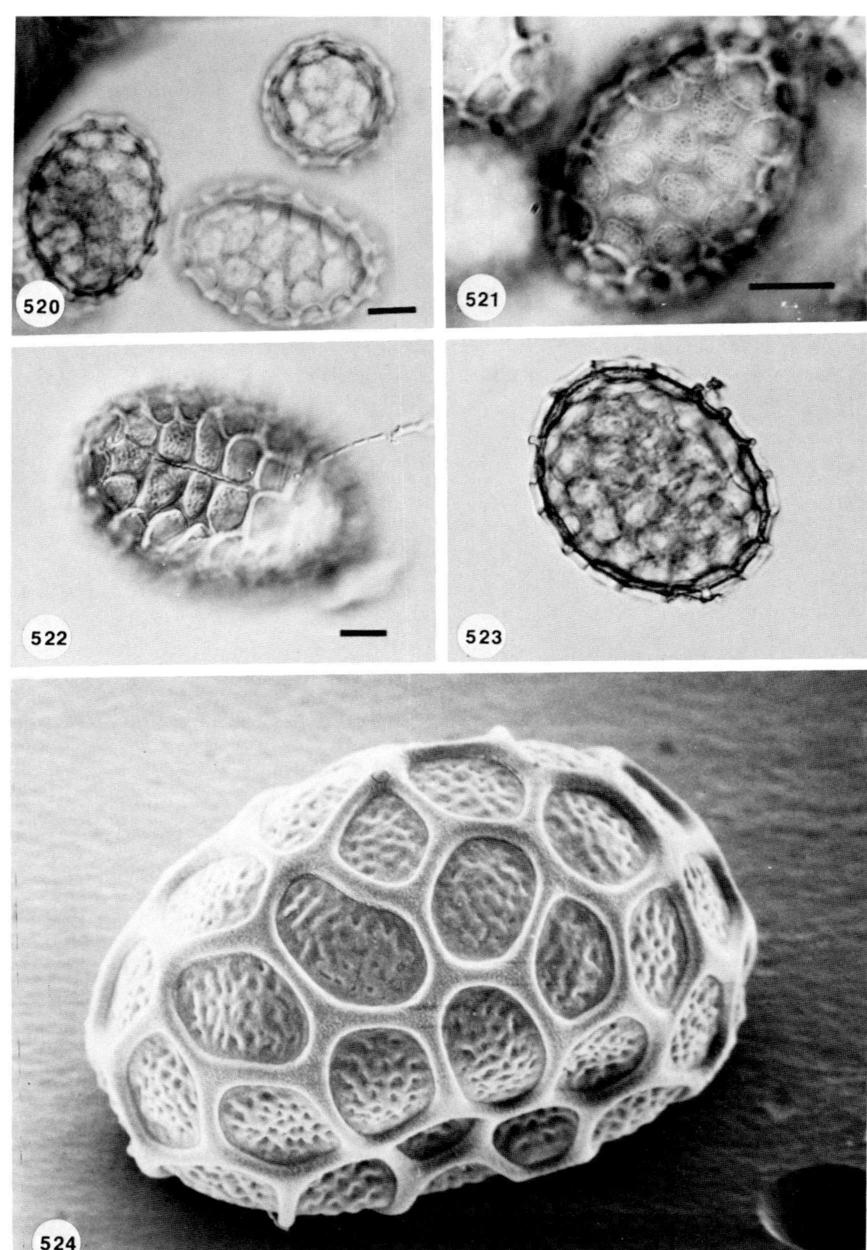
520
521
522
523
524

reticulo e liris tholiformibus 2–3 μm latis et altis composito areolas depressas rotundas vel aeque polygonias 5–12.5 μm latas punctis sparsis 0.5 μm diametro, inter se 1 mm, distantibus notatas includentibus, ornata; incisura dehiscentiae per liram longitudinalem elongatam ab areis depressis rotundis vel polygoniis marginatam media; quasi in sectione media inspecta, a cacumine, liris instar tholorum truncatorum ab ala, scilicet lira ab latere observata, sporangium cingente, conjunctorum. Sporangia tenuiter tunicata non observata.

SPECIMENS EXAMINED OTHER THAN TYPE AND ISOTYPE:

India
Anopheles jeyporiensis James; Wynaad; Iyengar; ?; NCU.
Java
Culex tritaeniorhynchus Giles; Ciranjang; Nelson; 1974; NCU.
Malaya
Anopheles kochi Dön.; Tampin; Wharton; 1953; NCU.

COMMENTS: Sporangia of *C. reticulatus* varieties appear similar to those of both *C. couchii* and *C. lairdi*. Distinction from both may be made in that ridging on sporangia of *C. reticulatus* varieties appears as truncate domes in optical section (edge view), whereas similar views of ridging in *C. lairdi* and *C. couchii* reveals them as low, rounded domes. Further distinction is possible in that ridges on the sporangial surface of *C. reticulatus* varieties are narrower and higher. Also, the dehiscence slit on sporangia of *C. couchii* is bordered on both sides by longitudinal depressed areas with punctae, whereas the same area on sporangia of *C. reticulatus* varieties bears circular to polygonal depressed areas, as over the sporangia. Further distinction from *C. lairdi* may be made in that depressed areas of *C. reticulatus* varieties are prominently punctate, whereas similar areas on sporangia of *C. lairdi* bear only a few central punctae. Sporangia of *C. reticulatus* varieties bear some resemblance also to those of *C. finlayae*, but ridges on the sporangial surface of the latter are toothlike in optical section (edge view) and enclose large polygo-

---

Figs. 520 and 521. *Coelomomyces reticulatus* var. *reticulatus*. Resting sporangia of type specimen from *Anopheles kochi* collected in Tampin, Malaya. Fig. 520. Low-magnification micrograph showing general sporangial morphology. Bar = 10 μm. Fig. 521. Surface view of sporangial side. Scale as in Fig. 520.

Figs. 522–524. *Coelomomyces reticulatus* var. *reticulatus*. Resting sporangia from *Culex tritaeniorhynchus* collected in Ciranjang, Java. Fig. 522. Surface view, dorsal side (dehiscence slit, arrow). Bar = 10 μm. Fig. 523. Midoptical section, side view. Bar = 10 μm. Fig. 524. Scanning electron micrograph showing ventral surface. Bar = 10 μm.

nal areas, whereas ridges in *C. reticulatus* varieties are as truncate domes in optical section (edge view).

Although lack of original material makes positive identification impossible, it is likely that the Type 4 sporangium described by Walker (1938) from larvae of *Anopheles gambiae* was a specimen of *C. reticulatus* var. *reticulatus*.

The specific epithet of *C. reticulatus* is from the diminutive *reticulum* L. (small net), in agreement with the network of ridges covering the surface of resting sporangia.

Coelomomyces reticulatus var. ***parvus*** Couch, Farr & Mora var. nov.

Figures 525–527

Description: Characteristics as for the typical variety, except sporangia measuring 22–35 × 33.5–45 μm.

Type Host and Collection Site: *Aedomyia squamipennis* (L.-A.); Acacias, Meta, Columbia (collected by Farr and Mora).

Depository for Type: BPI.

Latin Diagnosis: Characteres varietatis typicae, sporangiis 22–35 × 33.5–45 μm exceptis.

Specimens Examined Other than Type:

Colombia

Aedomyia squamipennis (L.-A.); Acacias, Meta; Farr and Mora; 1979, 1980; NCU.

Comments: Although sporangial ornamentation in *C. reticulatus* var. *parvus* is identical to that of the typical variety, the smaller sporangial dimensions of the former, which very rarely overlap those of the latter, are considered justification for varietal status.

One should note that, at present, this is the only report of *Coelomomyces* sp. from South America.

The varietal name of *C. reticulatus* var. *parvus* is derived from *parvus* L. (small), in recognition of its resting sporangia being smaller than in the typical variety.

Coelomomyces rugosus Couch & Service sp. nov.

Figures 528–533

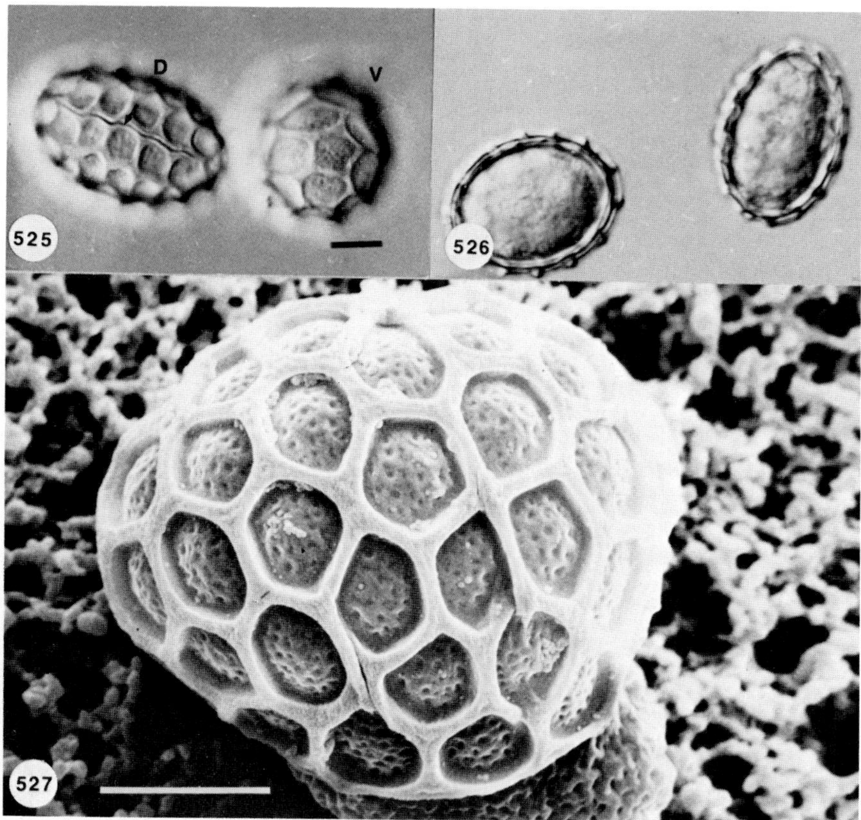

Figs. 525–527. *Coelomomyces reticulatus* var. *parvus.* Resting sporangia of type from *Aedomyia squamipennis* collected in Meta, Colombia. Scale as in Fig. 524. Fig. 525. Surface view of two sporangia, one seen in dorsal view (D), the other seen in ventral/end view (V). Note the dehiscence slit (arrow) visible in the sporangium in dorsal view. Fig. 526. Midoptical section, face view. Fig. 527. Scanning electron micrograph showing sporangial side. Bar = 10 μm.

DESCRIPTION: Hyphae 4–15 μm in diameter. Resting sporangia broadly ellipsoid, 26–39 × 30–62 μm; surface appearing rugose to crenulate as a result of numerous irregular ridges and depressed areas; dehiscence slit central along a prominent longitudinal ridge that is bordered by a 5- to 8-μm wide depressed area with irregular striae oriented perpendicular to the ridge. In midoptical section: side view, ridging along dorsal edge as a vertically striate wing and at poles and along ventral edge as a striate, irregularly scalloped fringe; face view, irregular ridging as an irregularly

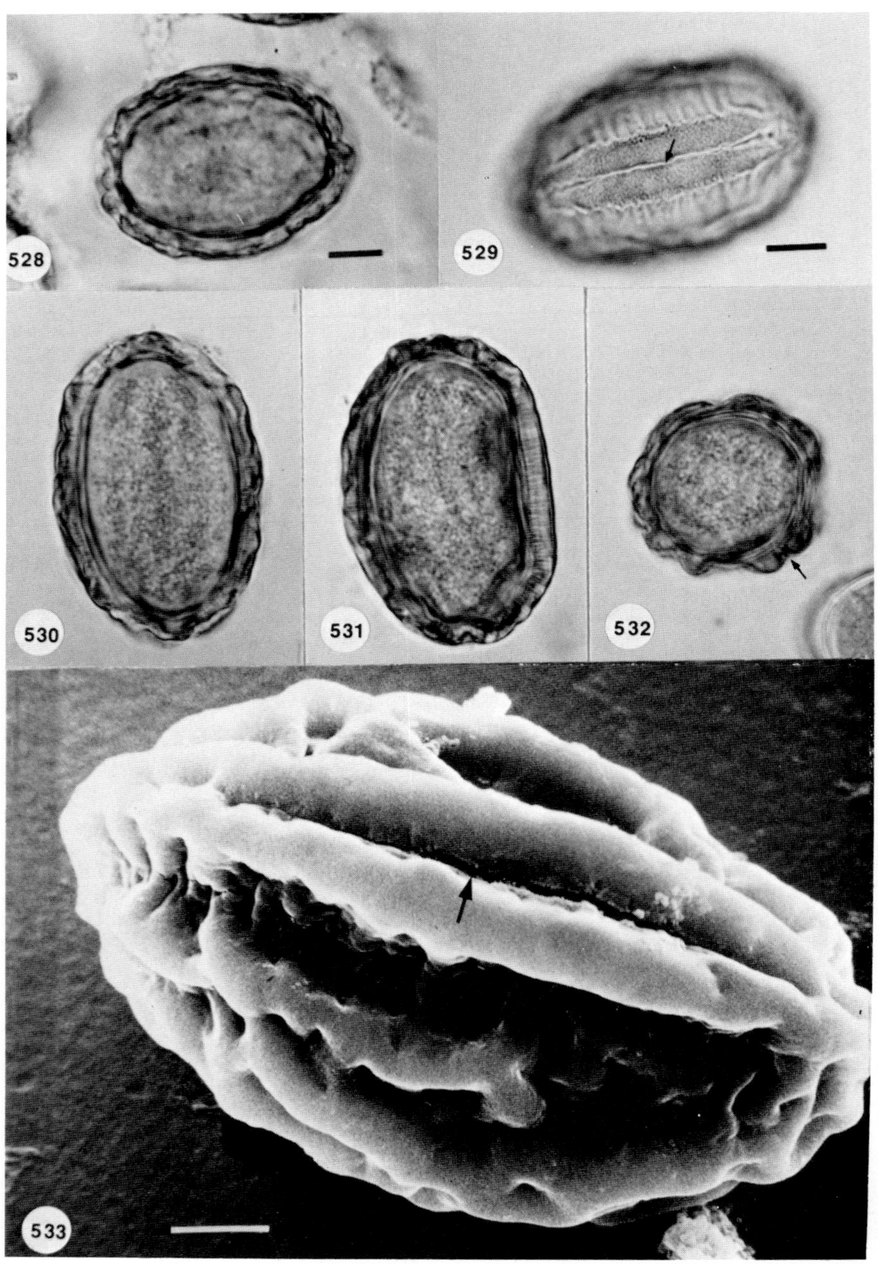
528
529
530
531
532
533

scalloped fringe encircling the sporangium; end view, dehiscence slit prominent in a raised ridge on dorsal surface. Thin-walled sporangia ellipsoid, 23–33 × 24–48 μm.

TYPE HOST AND COLLECTION SITE: *Anopheles gambiae* Giles; Kenya, Africa (collected by Service).

DEPOSITORY FOR TYPE: BPI. Several isotypes are on deposit at NCU.

LATIN DIAGNOSIS: Hyphae 4–15 μm crassae. Sporangia perennantia late ellipsoidea, 26–39 × 30–62 μm; superficie specie rugosa vel crenulata ob liras numerosas enormes atque areolas depressas; incisura dehiscentiae per liram longitudinalem prominentem ab areola depressa 5–8 μm lata, striis enormibus rectis notata marginatam media; quasi in sectione media inspecta a latere, liris secundum marginem dorsalem instar alae verticaliter striatae, per ventralem et ad polos instar fimbriae striatae, inaequaliter sinuosae, adversus lirae enormibus instar fimbriae inaequaliter sinuosae sporangium cingentis, a cacumine incisura in lira e superficie dorsali eminente conspicua. Sporangia tenuiter tunicata elipsoidea, 23–33 × 24–48 μm.

SPECIMENS EXAMINED OTHER THAN TYPE AND ISOTYPES:

India

Anopheles nigerrimus Giles; Trivandrum; Iyengar; 1970; NCU; BPI.*

COMMENTS: *Coelomomyces rugosus* may be distinguished on the basis of its unique, irregularly ornamented resting sporangia. While similar sporangia are occasionally observed as teratologic forms of other species, the collection in Africa of a larva of *Anopheles gambiae* that was completely filled with similar, irregular sporangia supports the establishment of a new species. The specimen cited above from India, although having sporangia resembling those of *C. rugosus*, was in association with *C. iliensis* var. *indicus* and therefore may simply represent irregularly

* Specimen present on type slide of *C. iliensis* var. *indicus* and on other slides of same variety retained at NCU.

Figs. 528–533. *Coelomomyces rugosus.* Resting sporangia of type from *Anopheles gambiae* collected in Kenya, Africa. Fig. 528. Midoptical section, face view. Bar = 10 μm. Fig. 529. Surface view, dorsal face (dehiscence slit, arrow). Bar = 10 μm. Fig. 530. Midoptical section, face view. Scale as for Fig. 530. Fig. 531. Midoptical section, side view. Scale as for Fig. 530. Fig. 532. Midoptical section, end view. Note the dehiscence slit (arrow) visible in the middorsal ridge. Scale as in Fig. 529. Fig. 533. Scanning electron micrograph showing dorsal face of a sporangium (dehiscence slit, arrow). Bar = 5 μm.

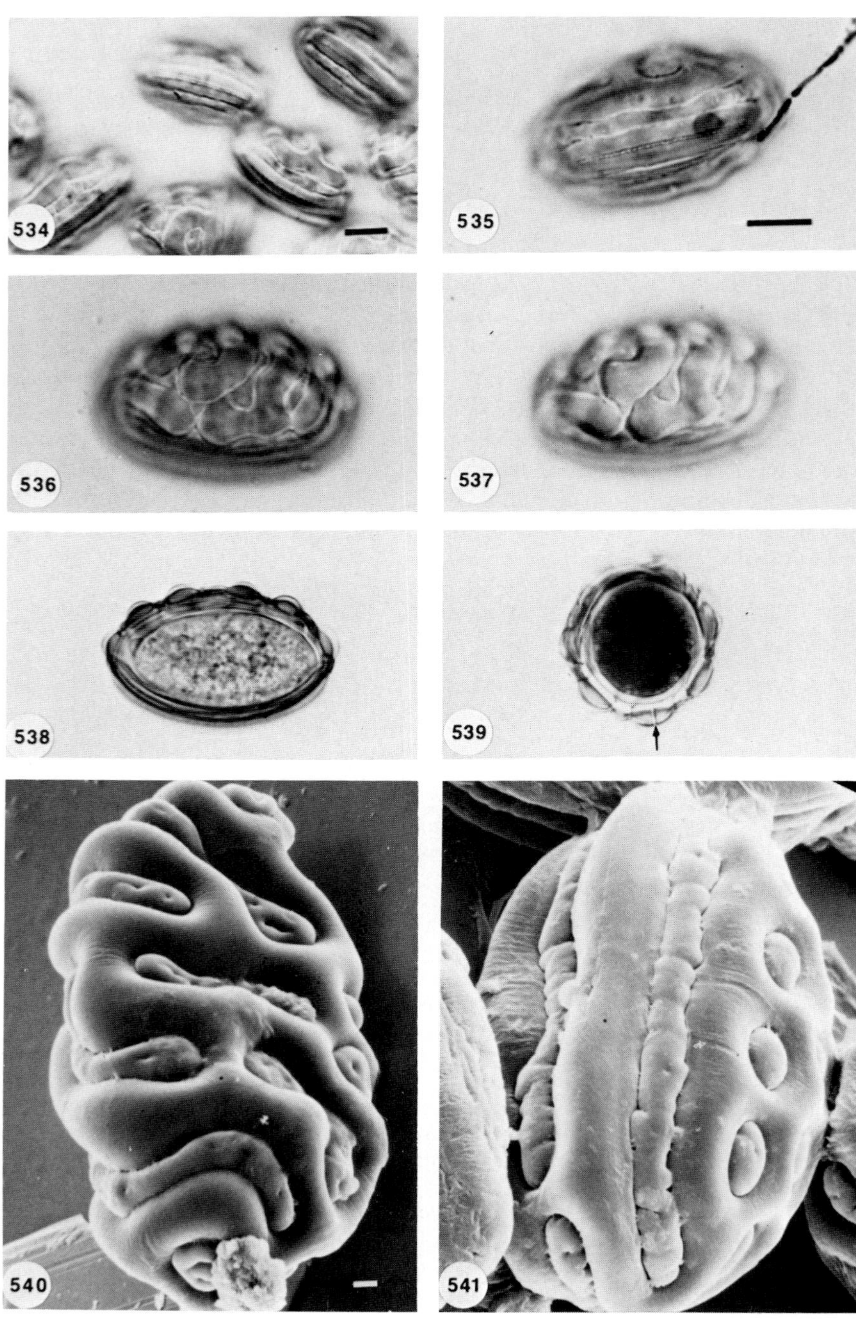
534
535
536
537
538
539
540
541

formed sporangia of this variety. Additional collections from India will be needed for resolution of this problem.

The specific epithet of *C. rugosus* is from *rugosus* L. (wrinkled), in recognition of its wrinkled sporangial wall.

Coelomomyces sculptosporus Couch & Dodge ex Couch, *J. Elisha Mitchell Sci. Soc.* **78,** 135 (1962)

Coelomomyces sculptosporus Couch & Dodge, *J. Elisha Mitchell Sci. Soc.* **63,** 69 (1947)

Figures 534–541

DESCRIPTION: Hyphae 4–17 μm thick. Resting sporangia ellipsoid, 22–31 × 35–58 μm; surface ornamented with irregular, anastomosing high ridges (3–5 μm high × 4–8 μm wide) enclosing and/or alternating with elongate, irregular, to occasionally circular (domelike) lower ridges (1–2.5 μm high × 2–4 μm wide) with infrequent punctae; ridges often originating from a high, elongate ridge, with central dehiscence slit, on dorsal surface, continuing to ventral surface, first as elongate ridges then as irregular to circular ridges. In midoptical section: side view, dorsal ridge with dehiscence slit appearing as wing along dorsal edge, irregular ridges as alternating high and low domes at poles and along ventral edge; end view, ridges as alternating high and low domes surrounding the sporangium, the dehiscence slit often clearly visible in end view of the prominent dorsal ridge. Thin-walled sporangia not observed.

TYPE HOST AND COLLECTION SITE: *Anopheles punctipennis* Say; Macon, Georgia (collected by SMCG). Paratypes on *Anopheles punctipennis* Say and *Anopheles crucians* Wiede.; Georgia (collected by SMCG).

DEPOSITORY FOR TYPE: BPI. Several paratypes are on deposit at NCU.

SPECIMENS EXAMINED AND OTHER COLLECTIONS:

Figs. 534–541. *Coelomomyces sculptosporus.* Resting sporangia of type from *Anopheles punctipennis* collected in Georgia. Fig. 534. Surface view of several sporangia. Bar = 10 μm. Fig. 535. Surface view of dorsal face. Bar = 10 μm. Figs. 536 and 537. Surface view of the same sporangium showing different appearance via bright-field microscropy (Fig. 536) and interference–contrast microscropy (Fig. 537). Scale as in Fig. 535. Fig. 538. Midoptical section, side view. Scale as in Fig. 535. Fig. 539. Midoptical section, end view. Dehiscence slit, arrow. Scale as in Fig. 535. Figs. 540 and 541. Scanning electron micrographs showing ventral side (Fig. 540) and dorsal side (Fig. 541). Bar = 1 μm.

Type, paratypes

United States

Anopheles walkeri Theo.; Minnesota; Barr; 1954; NCU; (Barr, 1958; Laird, 1961).

COMMENTS: Sporangial ornamentation in *C. sculptosporus* is similar to that of both *C. cribrosus* and *C. bisymmetricus*. In fact, some collections of *C. sculptosporus* may contain a few sporangia indistinguishable from either of these two. In general, however, distinction from each may be made in that sporangia of *C. sculptosporus* are irregularly sculptured and lack the dome-like, circular ridges that cover sporangia of *C. cribrosus* and the bisymmetrical ornamentation characteristic of *C. bisymmetricus*.

Coelomomyces seriostriatus Couch & Davies sp. nov.

Figures 542–546

DESCRIPTION: Hyphae not observed. Resting sporangia broadly ellipsoid, 20.5–32 × 39–59.5 μm; surface ornamented with longitudinal rows of transverse striae (0.5 μm wide and 1.5–2 μm long) separated by areas 2.3–3.5 μm wide devoid of striae; transverse striae spaced ~2 μm apart in concentric rings between poles; dehiscence slit central along a longitudinal ridge bordered by transverase striae. In midoptical section: side view, striae giving the sporangial wall a scalloped appearance. Thin-walled sporangia not observed.

TYPE HOST AND COLLECTION SITE: *Culex portesi* S. & A. ♀; Turure Forest, Trinidad (collected by Davies).

DEPOSITORY FOR TYPE: BPI.

LATIN DIAGNOSIS: Hyphae non visae. Sporangia perennantia late ellipsoidea, 20.5–32 × 39–59.5 μm; superficie seriebus longitudinalbus striarum transversarum 0.5 μm latarum, 1.5–2 μm longarum, ab areolis 2.3–3.5 μm latis striis carentibus separatis, ornata; striis transversis inter se ~2 μm distantibus in anulis concentricis inter polos ordinatis; incisura dehiscentiae per liram longitudinalem ab striis transversis marginatam media; quasi in sectione media inspecta e latere, striis in tunicam sporangii speciem introrsus sinuatam imponentibus. Sporangia tenuiter tunicata non visa.

SPECIMENS EXAMINED OTHER THAN TYPE:

Trinidad

Culex portesi S. & A. ♀; Turure Forest; Davies; 1967, 1968; NCU.

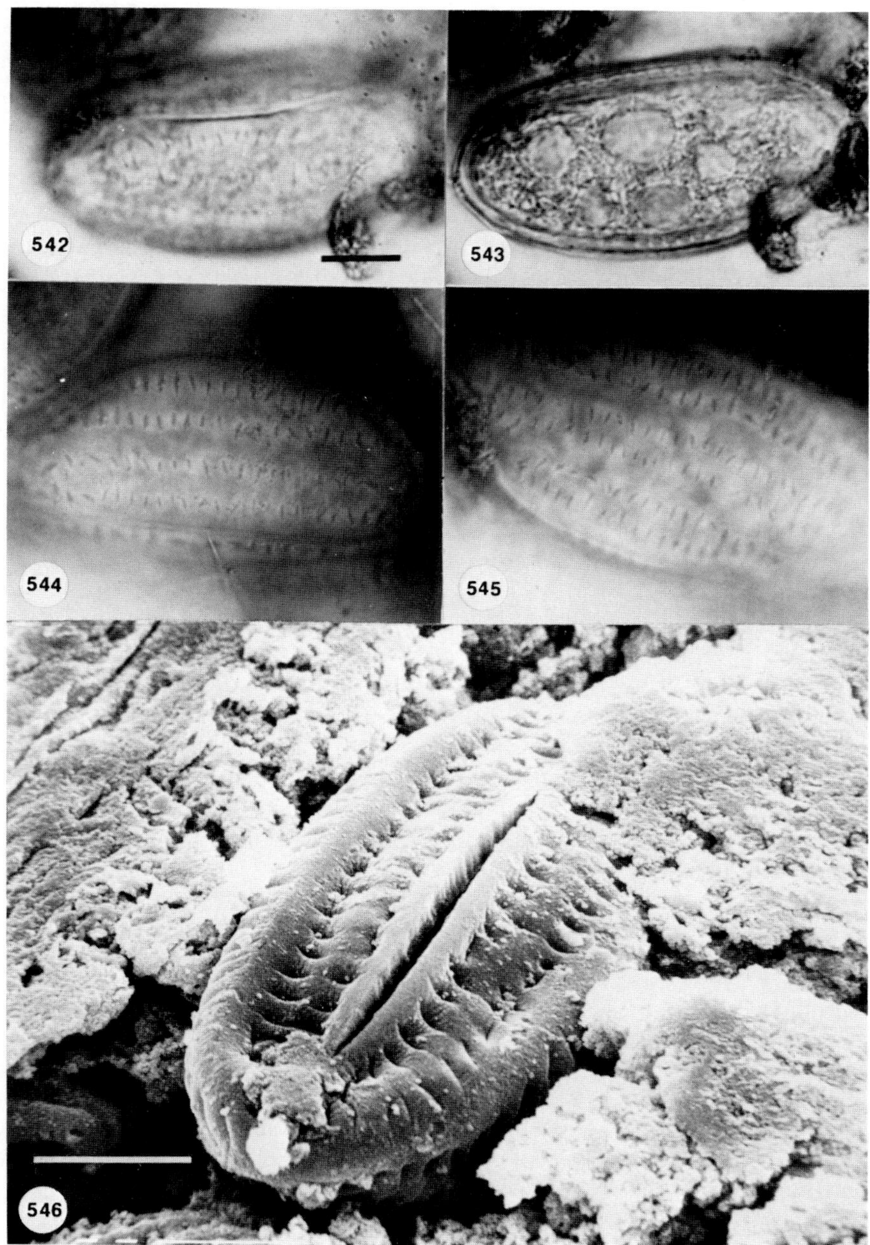

Figs. 542–546. *Coelomomyces seriostriatus.* Resting sporangia of type from *Culex portesi* collected in Trinidad. Fig. 542. Surface view of dorsal side. Dehiscence slit, arrow. Bar = 10 μm. Fig. 543. Midoptical section, side view. Scale as for Fig. 542. Figs. 544 and 545. Surface views of sporangial side. Scale as for Fig. 542. Fig. 546. Scanning electron micrograph showing dorsal face of a sporangium with its prominent, central dehiscence slit. Bar = 10 μm.

COMMENTS: The unique ornamentation of the resting sporangial wall of *C. seriostriatus* distinguishes this species from all other species of *Coelomomyces*. The species was collected only from Turure Forest, Trinidad, and J. B. Davies (personal communication) reports that during 1969–1970, "we found quite a number of adults (*C. portesi*) whose ovaries were parasitized." In such infections, there was no external evidence of parasitism and dissection was necessary for detection. Davies indicates further that, while parasitism of larvae undoubtedly occurs, detection of such infection is likely to be rare due to the infrequency with which larvae of *C. portesi* are collected.

The specific epithet of *C. seriostriatus* is after *series* L. (row) and *striae* L. (line, streak, or groove), in recognition of the rows of striae ornamenting resting sporangia of this species.

Coelomomyces stegomyiae var. *stegomyiae* Keilin, *Parasitology* **13,** 225 (1921)

Coelomomyces stegomyiae var. *rotumae* Laird, *Can. J. Zool.* **37,** 781 (1959)

Figures 547–554

DESCRIPTION: Hyphae 2–6 μm thick. Resting sporangia ellipsoid, 15–45 × 25–100 μm; surface ornamented with scattered punctae (0.5 μm in diameter and 1–5 μm apart) that are bordered by 0.25- to 0.5-μm-wide depressed margins; via SEM, papillae (0.5–1 μm in diameter) are seen interspersed among the punctae. Wall 1.5–4 μm thick. Thin-walled sporangia not observed.*

TYPE HOST AND COLLECTION SITE: *Stegomyia scutellaris* Walter†; Kajang, Malaya (collected by W. A. Lamborn).

NEOTYPE HOST AND COLLECTION SITE: *Aedes albopictus* (Skuse); Ceylon (collected by Rajapaksa, 1964).

DEPOSITORY FOR TYPE: The type was retained in the laboratory of Dr. D. Keilin (deceased), Cambridge University, Cambridge, England, and despite repeated efforts has been unobtainable. The neotype is deposited at the BPI.

* Although Keilin (1921) referred to thin-walled sporangia in the original description of *C. stegomyiae,* they have not been observed in subsequent collections of this species. It is likely that Keilin was instead observing immature resting sporangia.

† A likely misidentification, probably *Aedes albopictus* (Skuse) (Laird, 1956a).

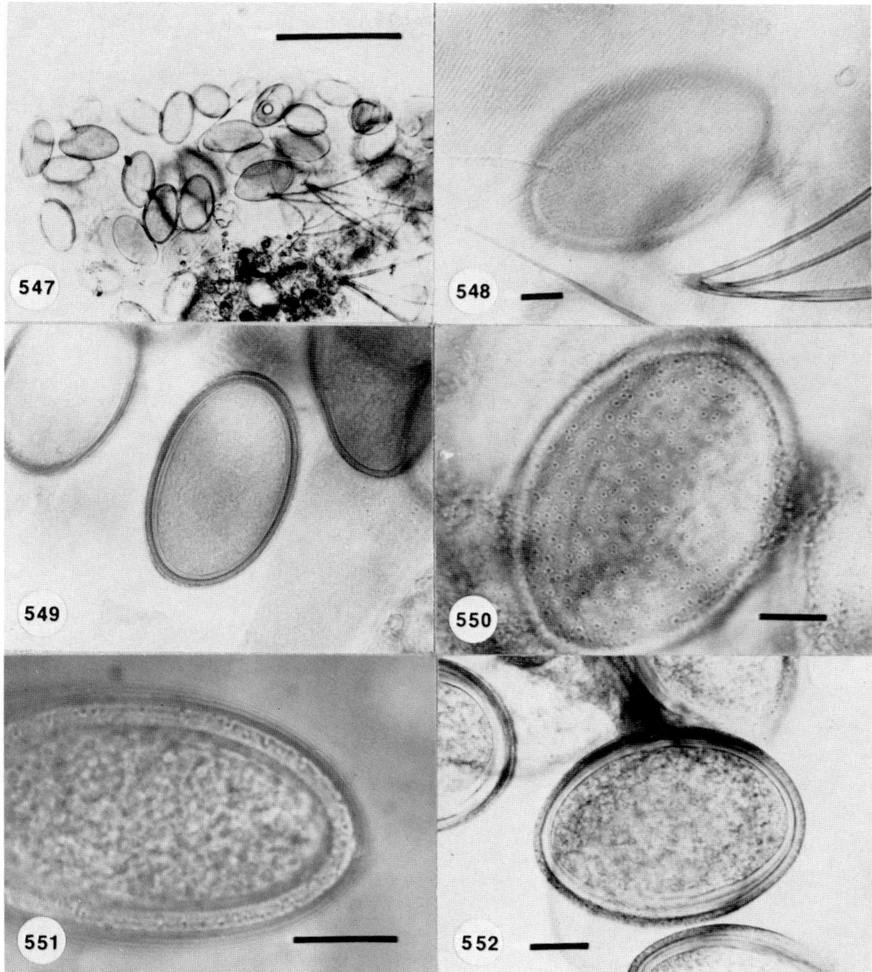

Figs. 547–549. *Coelomomyces stegomyiae* var. *stegomyiae*. Resting sporangia from *Aedes aegypti* collected in the Philippines. Fig. 547. Low-magnification view showing general sporangial shape. Bar = 100 μm. Fig. 548. Surface view of sporangial side. Bar = 10 μm. Fig. 549. Midoptical section, side view. Scale as for Fig. 547.

Fig. 550. *Coelomomyces stegomyiae* var. *stegomyiae*. Resting sporangia from *Aedes multifolium* collected in New Guinea. Surface view of sporangial side. Bar = 10 μm.

Fig. 551. *Coelomomyces stegomyiae* var. *stegomyiae*. Resting sporangium from *Aedes polynesiensis* collected in the Tokelau Islands. Midoptical section, side view. Bar = 10 μm.

Fig. 552. *Coelomomyces stegomyiae* var. *stegomyiae*. Resting sporangia from *Aedes multifolium* collected in New Guinea. Midoptical section, side view. Bar = 10 μm.

553
554

Specimens Examined and Other Collections:
See Table VIII.

Comments: *Coelomomyces stegomyiae* may be easily distinguished from other species of *Coelomomyces* by its resting sporangia that are ornamented with widely spaced punctae having depressed margins. Although sporangial ornamentation in this species is similar to that of *C. psorophorae*, punctae on sporangia of the latter lack depressed margins. While this feature is at times difficult to discern, careful focusing on the sporangial surface at a magnification of ×400 should reveal whether depressed margins are present. Resting sporangia of *C. stegomyiae* varieties have ornamentation similar also to that of *C. ponticulus*. In fact, distinction may be especially difficult since in all collections of *C. ponticulus* sporangia of *C. stegomyiae* have been present also. By careful observation, recognition of the two may be made in that punctae on resting sporangia of *C. ponticulus* are irregular and have a central bridge, whereas those of *C. stegomyiae* are circular and bordered.

It should be noted that all collections of *C. stegomyiae* vars. have been on hosts occurring in discarded receptacles, spent nuts, or tree holes, a feature shared with *C. macleayae* and *C. dentialatus*.

Coelomomyces stegomyiae var. ***chapmani*** Laird *et al., Can. J. Zool.* **17,** 333 (1980)

Figures 555–556

Description: Characteristics as for the typical variety except resting sporangia frequently distinctly fusiform.

Type Host and Collection Site: *Aedes albopictus* (Skuse); Taitung, Taiwan.

Depository for Type: BPI (#71915).

Specimens Examined and Other Collections (indicated by brackets): None.

Figs. 553 and 554. *Coelomomyces stegomyiae* var. *stegomyiae*. Scanning electron micrographs of resting sporangia of type from *Aedes albopictus* collected in Sri Lanka. Fig. 553. Dorsal view of sporangial face. Bar = 10 μm. Fig. 554. Enlargement of area near dehiscence slit (arrow) showing bordered pits. Bar = 1 μm.

TABLE VIII. Specimens Examined and Other Collections (Indicated by Brackets) of *Coelomomyces stegomyiae* var. *stegomyiae* Keilin

Location	Host	Collector	Collection site	Sporangial size (μm)	Reference	Depository
Africa	*Aedes scatophagoides* Theo.	Muspratt	Zambia	33.5–58 × 65–93	Muspratt, 1946a	NCU
Ceylon	*Aedes aegypti* (L.)	Rajapaksa	Various sites	19–40 × 28–58	Rajapaksa, 1964	NCU
	Aedes albopictus (Skuse)	Rajapaksa	Various sites	19–40 × 28–58	Rajapaksa, 1964	NCU
Fiji Islands	*Aedes polynesiensis* Marks	Pillai	Koro Island	26–41 × 39–67		NCU
	Aedes sp.	Laird	Rotuma Island	19.5–30 × 40–60	Laird, 1959b[a]	US
Japan	*Aedes riversi* Boh. and Ing.	Suenaga	unknown	30–45 × 40–70		NCU
	Aedes (Stegomyia) flavopictus Yamada[b]	Mogi	Mikura-jima Island	19–34 × 35–56	Nolan and Mogi, 1980	NCU
Malaya	*Aedes albopictus* (Skuse)	Lamborn	Kajang	20–30 × 37.5–57	Keilin, 1921[c]	
New Guinea	*Aedes multifolium* King and Hoog.	Huang	Port Moresby	43–68 × 58–83[d]	Huang, 1968	NCU
	[*Aedes multifolium* King and Hoog.	?	?	?	Briggs, 1967	?]
	[*Aedes quadrispinatus* King and Hoog.	?	?	?	Briggs, 1967	?]
	[*Aedes variabilis* Huang	?	?	?	Briggs, 1967	?]
Philippine Islands	*Aedes aegypti* (L.)	Villacarlos	College Laguna	27–35 × 47–73	Villacarlos and Gabriel, 1974	NCU
Singapore	[*Aedes albopictus* (Skuse)	Laird	Station I[e]	17–29 × 29–53	Laird, 1959a	?]
	[*Aedes albopictus* (Skuse)	Laird	Station II[e]	17–27 × 31–51	Laird, 1959a	?]
	[*Aedes albopictus* (Skuse)	Laird	Station III[e]	18–30 × 35–58	Laird, 1959a	?]
	[*Aedes aegypti* (L.)	Laird	?	18–28 × 34–56	Laird, 1959a	?]
	[*Armigeres obturbans* (Walker)	Laird	?	16–29 × 28–54	Laird, 1959a	?]
Solomon Islands	[*Aedes (Stegomyia) scutellaris* Walker	Laird	Rennell Island	15–24 × 28–56	Laird. 1956a	]
	Aedes hebrideus Edwards					
	[*Culex (Culex) annulirostris* (Skuse)	Maffi	Vanikoro Island	38–64 × 55–100	Briggs, 1969; Maffi and Genga, 1970	?]
Thailand	*Aedes subalpictus* (Skuse)	?	Kanchanaburi	20–33 × 40–52	Hembree, 1979	NCU
Tokelau Islands	*Aedes polynesiensis* Marks	Introduced	Nokunono Island	18–32 × 25–52	Laird, 1966	?
United States	[*Wyeomyia vanduzeei* Dyar and Knab	Hall and Anthony	Everglades National Park, Fla.	20–34	Hall and Anthony, 1979	?]

[a] Includes description of *C. stegomyiae* var. *rotumae*.
[b] Combined infection with *C. ponticulus*.
[c] Description of type. Examined by Couch and Dodge (1947) and returned.
[d] Dimensions inaccurate due to mounting.
[e] See Laird (1959a) for explanation.

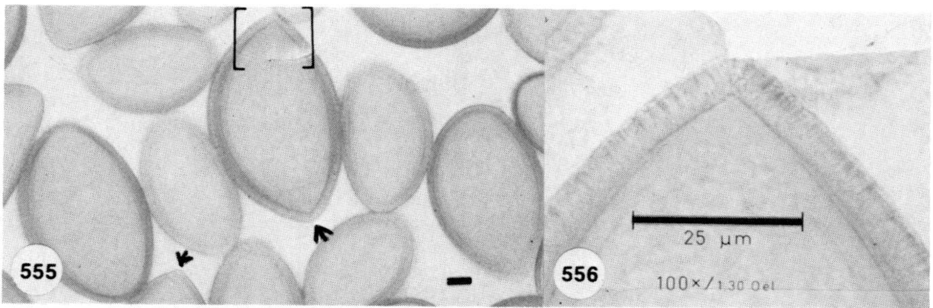

Figs. 555 and 556. *Coelomomyces stegomyiae* var. *chapmani.* Resting sporangia of type specimen from *Aedes albopictus* collected in Taiwan. Fig. 555. Low-magnification micrograph showing general sporangial morphology. Note pointed tip evident on some of the sporangia (arrows and brackets). Bar = 10 μm. Fig. 556. Sporangial tip showing pointing typical of this variety.

Taiwan
[*Armigeres subalbatus* (Coq.); Taipei; ?; ?; ?; (Laird *et al.*, 1980).]
[*Topomyia yanbarensis* Mayagi; Taipei; ?; ?; ?; (Laird *et al.*, 1980).]
[*Tripteroides bambusa* (Yamada); Haulien, Taipei; ?; ?; ?; (Laird *et al.*, 1980).]

COMMENTS: *Coelomomyces stegomyiae* var. *chapmani* is distinguished from the typical variety in that up to 50% of the resting sporangia of the former are characterized by bipolar pointing resulting in a typically fusiform shape. Other sporangia of this variety are, however, shaped as in the typical variety.

Coelomomyces sulcatus Couch & Iyengar sp. nov.

Figures 557–564

DESCRIPTION: Hyphae 6–10 μm in diameter. Resting sporangia ellipsoid, 19–24 × 30–37 μm; surface ornamented with elongate, finely pitted ridges (~4 μm wide up to sporangial length) enclosing elongate depressed areas (furrows) with transverse striae/punctae; ridging as two dorsal ridges bordering a central depressed area with dehiscence slit, a lateral ridge encircling the sporangium and anastomosing with the dorsal ridges at the poles, and a ventral, continuous, elliptical ridge with central depressed area. In midoptical section: face view, lateral ridge as a wing encircling the sporangium; side view, dorsal ridge as wing along dorsal surface and continuing around poles, ventral ridge as a short, insular wing along ventral edge; end view, ridges as usually five domes alternating with

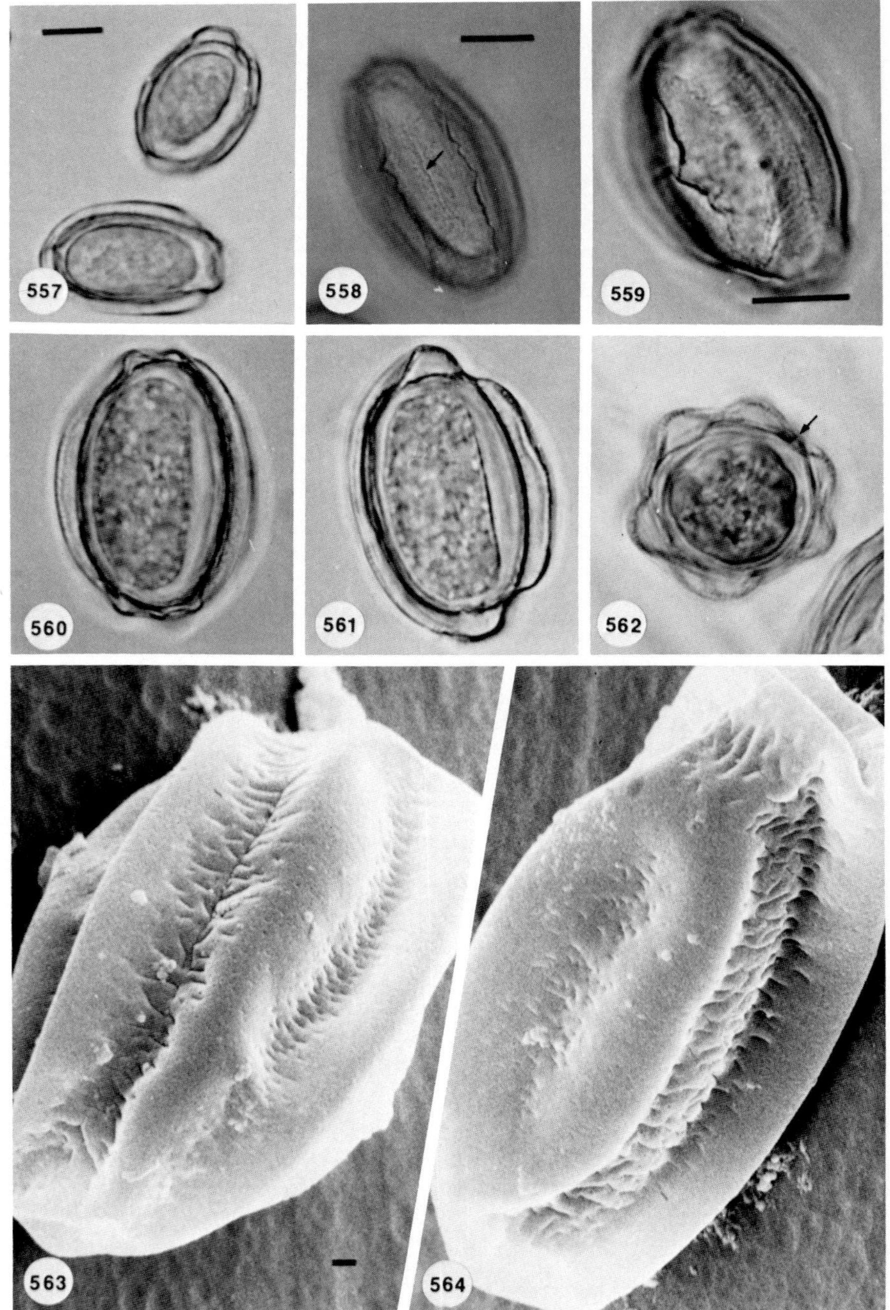

557
558
559
560
561
562
563
564

depressed areas around sporangial periphery; fine punctae on surface of ridges obvious also as punctae at periphery of wings (ridges in optical section). Thin-walled sporangia not observed.

Type Host and Collection Site: *Culex vishnui* (Theo.); Trivandrum, India (collected by Iyengar).

Depository for Type: BPI. An isotype is on deposit at NCU.

Latin Diagnosis: Hyphae 6–10 μm crassae. Sporangia perennantia ellipsoidea, 19–24 × 30–37 μm, superficie liris elongatis, subtiliter foveolatis, ~4 μm latis, ad sporangii longitudinem decorata, areolas depressas elongatus, scilicet sulcas, striis punctisque transversis notatas, includentibus; liris instar lirarum duarum dorsalium aream depressam mediam incisura dehiscentiae praeditam marginantium, atque lirae lateralis sporangium cingentis et ad polos cum liris dorsalibus anastomosantis, atque ventralis, continuae, ellipticae, areolam centralem depressam includentis; quasi in sectione media adversus inspecta, lira laterali instar alae sporangium cingentis; e latera lira dorsali instar alae per superficiem dorsalem et circum polos continuae, ventrali instar alae per superficiem dorsalem et circum polos continuae, ventrali instar alae insularis brevis secundum marginem ventralem; e cacumine, liris instar tholorum vulgo quinque cum areolis depressis alternantium circum sporangii; punctis subtilibus in lirarum superficie etiam instar punctorum ad alarum scilicet lirarum sicut in sectione inspectarum, limitem obviis. Sporangia tenuiter tunicata non observata.

Specimens Examined Other than Type and Isotype: None.

Comments: *Coelomomyces sulcatus* is distinguished from all known species on the basis of the ornamentation and size of its resting sporangia. Particularly distinctive is the characteristic manner in which the dorsal and ventral ridges appear in some views as elongate domes on the opposing faces of sporangia. The specific epithet of *C. sulcatus* is from *sulcatus* L. (furrowed or grooved) in recognition of the distinctive depressed areas that appear as furrows between the ridges on resting sporangia of this species.

Figs. 557–564. *Coelomomyces sulcatus.* Resting sporangia of type specimen from *Culex vishnui* collected in Trivandrum, India. Fig. 557. Low-magnification micrograph showing general sporangial morphology. Bar = 10 μm. Fig. 558. Dorsal side, face view (dehiscence slit, arrow). Bar = 10 μm. Fig. 559. Surface view of sporangial side. Bar = 10 μm. Figs. 560 and 561. Midoptical section, side view. Scale as in Fig. 559. Fig. 562. Midoptical section, end view (dehiscence slit, arrow). Scale as in Fig. 559. Figs. 563 & 564. Scanning electron micrographs of resting sporangia showing dorsal side, Fig. 563, and ventral side, Fig. 564. Bar = 2 μm.

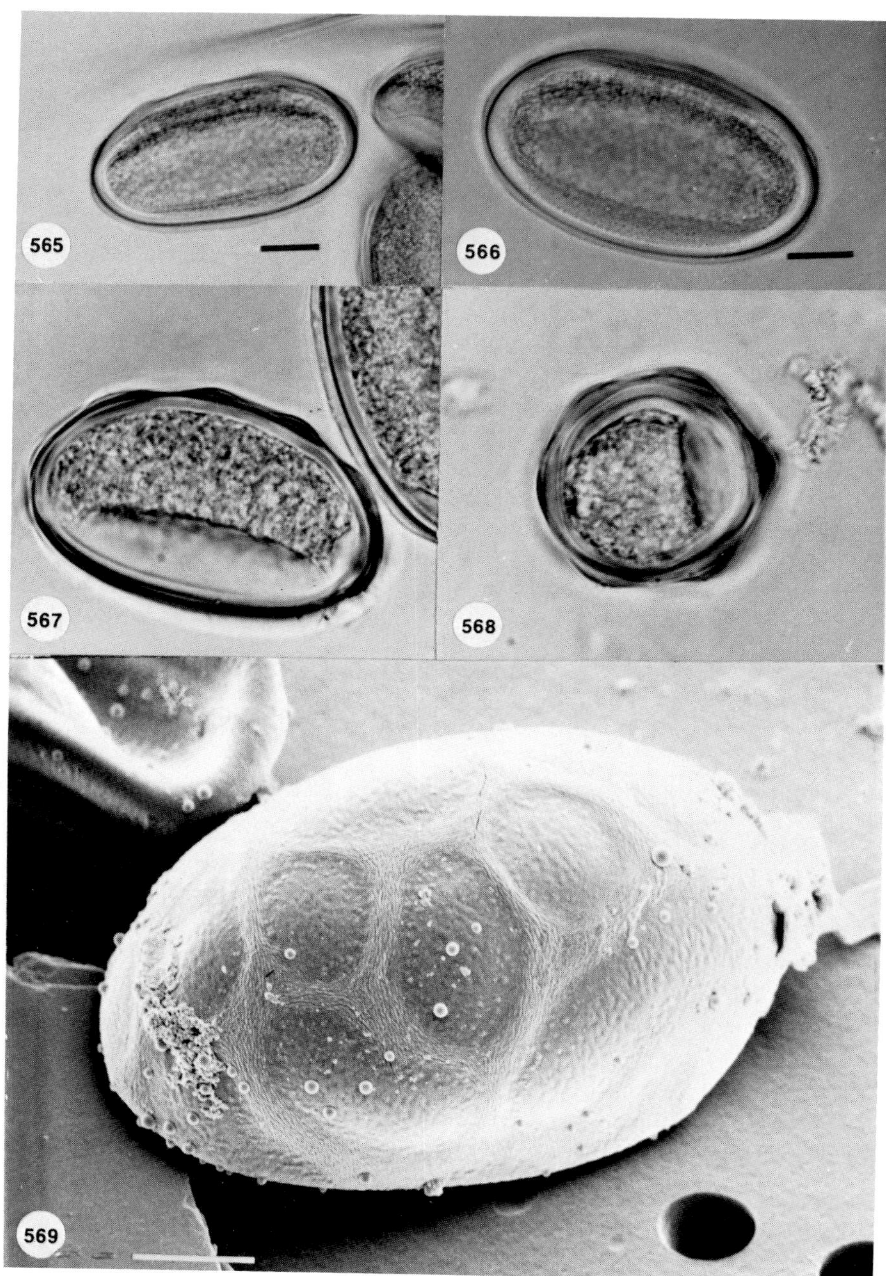
565
566
567
568
569

Coelomomyces thailandensis Couch, Gould, & Hembree sp. nov.

Figures 565–569

DESCRIPTION: Hyphae 10–26 μm in diameter. Resting sporangia broadly ellipsoid, 24–33.5 × 38–71 μm; surface ornamented with broad, low ridges (2–4 μm wide and 1–1.5 μm high) enclosing circular (predominately on ventral surface, 8–10.5 μm in diameter) to elongate (predominately on dorsal surface, 8–10.5 μm wide and up to 65 μm long) depressed areas with central punctae (0.5 μm wide and ~1 μm apart, covering all but a 2- to 4-μm smooth border at periphery of depressed areas); dehiscence slit central along a longitudinal ridge. In midoptical section: side view, ridges as a wing or domes at sporangial periphery; end view, ridges giving sporangia a 5- to 6-lobed appearance. Thin-walled sporangia not observed.

TYPE HOST AND COLLECTION SITE: *Anopheles bengalensis;* Ban Wung Mut, Prachinburi, Thailand (collected by AFRIMS: see Hembree, 1979, host/pathogen association No. 3).

DEPOSITORY FOR TYPE: BPI. Two isotypes are on deposit at NCU.

LATIN DIAGNOSIS: Hyphae 10–26 μm crassae. Sporangia perennantia late ellipsoidea, 24–33.5 × 38–71 μm; superficie liris latis, humilibus 2–4 μm latis, 1–1.5 μm altis, areolas depressas includentibus, praecipue in superficie ventrali rotundas, 8–10.5 μm diametro, praecipue in dorsali elongatas 8–10.5 μm latas, usque ad 65 μm longas, punctis centralibus 0.5 μm latis, inter se ~1 μm distantibus, depressionum totam superficiem margine levi 2–4 μm lato excepto obtinentibus, ornata; incisura dehiscentiae per liram longitudinalem media; quasi in sectione media inspecta e latere, liris instar alae vel tholorum circum sporangium, e cacumine liris speciem 5–6 lobatam sporangiis imponentibus. Sporangia tenuiter tunicata non observata.

SPECIMENS EXAMINED OTHER THAN TYPE:

Figs. 565–569. *Coelomomyces thailandensis.* Resting sporangia of type specimen from *Anopheles bengalensis* collected in Thailand. Fig. 565. Midoptical section, side view. Bar = 10 μm. Fig. 566. Surface view of sporangial side showing pitting. Bar = 10 μm. Fig. 567. Midoptical section, side view. Scale as in Fig. 566. Fig. 568. Midoptical section, end view. Scale as in Fig. 566. Fig. 569. Scanning electron micrograph showing sporangial side and ventral surface. Bar = 10 μm.

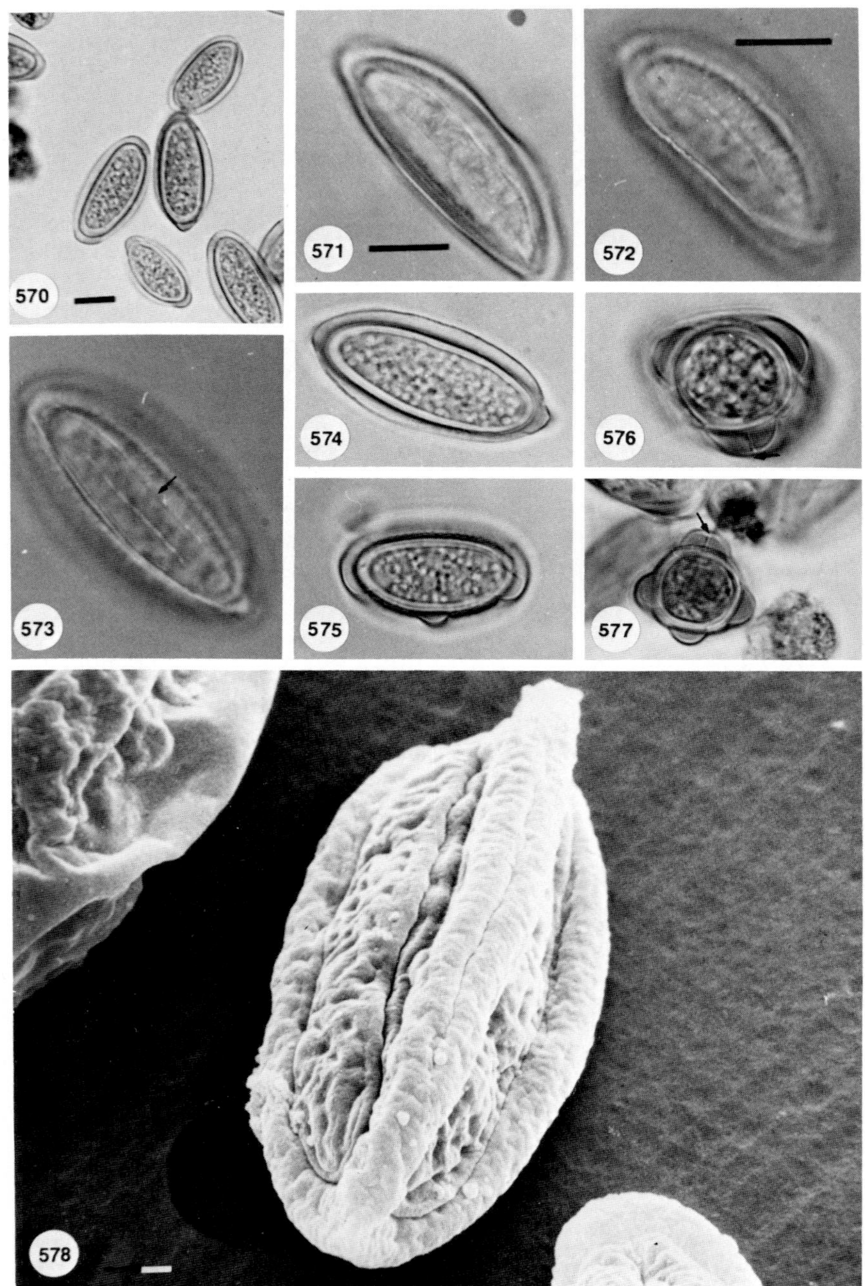
570
571
572
573
574
575
576
577
578

Thailand

Anopheles bengalensis; Ban Wung Mut, Prachinburi; AFRIMS #06137; ?; NCU; (Hembree, 1979).

COMMENTS: Although sporangial ornamentation in *C. thailandensis* resembles somewhat that of *C. africanus,* distinction of the two is easy in that sporangial dimensions of the former are larger and do not overlap those of the latter. While sporangial ornamentation of *C. thailandensis* is difficult to discern via LM observation of sporangial surfaces, the wings and domes evident in optical section at the sporangial periphery are characteristic for this species.

The specific epithet for *C. thailandensis* is in reference to its having been collected only in Thailand. Part of a large collection of mosquito larvae from this area (Hembree, 1979; host/pathogen association No. 3), all hosts for *C. thailandensis* were from a stream margin.

Coelomomyces triangulatus Couch & Martin sp. nov.

Figures 570–578

DESCRIPTION: Hyphae 6–9 μm in diameter. Resting sporangia ellipsoid, oblong–ellipsoid, to fusiform 11–17 × 19–35 μm; surface ornamented with three longitudinal ridges (4–5 μm high and 4–6 μm wide) that anastomose at the poles; elongate depressed areas between ridges with irregular striae, dehiscence slit centrally located along one of the longitudinal ridges. In midoptical section: face view, ridges as a wing that encircles the sporangium, but with end of the dorsal ridge occasionally appearing as a dome at each pole; side view, dorsal ridge as a wing along dorsal edge and continuing around poles, ventral edge lacking wing or with a central dome formed by an extension of one of the lateral ridges into the ventral depressed area; end view, ridges as domes giving the sporangium a triangular appearance. Thin-walled sporangia not observed.

Figs. 570–578. *Coelomomyces triangulatus.* Resting sporangia of type specimen from *Anopheles punctipennis* collected in Georgia. Fig. 570. Low-magnification micrograph showing general sporangial shape. Bar = 10 μm. Figs. 571 and 572. Surface views of sporangial side. Bar = 10 μm. Fig. 573. Surface view of dorsal side showing dehiscence slit (arrow). Scale as in Fig. 572. Fig. 574. Midoptical section, face view. Scale as in Fig. 572. Fig. 575. Midoptical section, side view. Scale as in Fig. 572. Figs. 576 and 577. Midoptical section, end view (dehiscence slit, arrow). Scale as in Fig. 572. Fig. 578. Scanning electron micrograph showing dorsal and end view. Note the dehiscence slit visible along the middorsal ridge. Bar = 1 μm.

Type Host and Collection Site: *Anopheles punctipennis* (Say); Thomasville, Georgia (collected by SMGC).

Depository for Type: BPI.

Latin Diagnosis: Hyphae 6–9 μm crassae. Sporangia perennantia ellipsoidea, oblongo–ellipsoidea, vel fusiformia, 11–17 × 19–35 μm; superficie liris tribus longitudinalibus 4–5 μm altis, 4–6 μm latis, ad polos anastomosantibus, ornata, areolis depressis elongatis inter liras striis enormibus notatis; incisura dehiscentiae per lirae longitudinalis unae medium posita; quasi in sectione media adversus inspecta, lirae instar alae sporangium cingentis, dorsali lira interdum instar tholi ad polum utrumque, e latere, lira dorsali instar alae per marginem dorsalem circum polos productae, margine ventrali ala carente vel tholo centraili ornato ab productione unae lirae lateralis in areolam depressam ventralem efformato; ad cacumine, liris instar tholorum in sporangium speciem triangulam imponentibus. Sporangia tenuiter tunicata non observata.

Specimens Examined Other than Type:

United States

Anopheles crucians Wiede.; Thomasville, Georgia; SMGC; 1945; NCU.

Anopheles punctipennis (Say); Thomasville, Georgia; SMGC; 1945; NCU.

Anopheles punctipennis (Say); Hanover County, Virginia; Martin; 1973; Martin.

Comments: Although sporangia of *C. triangulatus* resemble those of *C. quadrangulatus,* distinction may be made in that sporangia of the former are smaller and are triangular in end view whereas those of the latter are quadrangular. A further distinction may be made in that depressed areas between ridges of *C. quadrangulatus* are continuous with the ridges and bear prominent parallel to anastomosing transverse striae, but those of *C. triangulatus* are delineated from the ridges by a continuous groove and bear irregular, scattered striae.

The specific epithet of *C. triangulatus* is from *tri* L. (three) and *angulatus* L. (angled), in agreement with its resting sporangia, which are three-angled in end view.

Coelomomyces tuberculatus Bland & Rodhain sp. nov.

Figures 579–583

Description: Hyphae not seen. Resting sporangia ellipsoid, 10–12.5 × 17.5–22.5 μm; ventral surface and sides ornamented with anastomosing ridges (1–1.5 μm high and 1.5–2 μm wide) enclosing circular areas containing flattened domes or tubercles measuring 4–6.5 μm in diameter; ridges becoming elongate on dorsal surface, with dehiscence slit central along a prominent longitudinal ridge. In midoptical section: side view, ridges appearing as four to six pointed domes along ventral edge and at poles, and as an elongate wing along the dorsal edge; end view, ridges as four to usually five domes at sporangial periphery, giving the sporangium a stellate appearance; dehiscence slit prominent in the dorsal ridge. Thin-walled sporangia not observed.

Type Host and Collection Site: *Culex antennatus* Beck.; Manakara, Madagascar (collected by Rodhain).

Depository for Type: BPI.

Latin Diagnosis: Hyphae non visae. Sporangia perennantia ellipsoidea, 10–12.5 × 17.5–22.5 μm, superficie ventrali lateribusque liris anastomosantibus 1–1.5 μm altis, 1.5–2 μm latis, areas rotundas habentes tholus applanatos vel tubercula 4–6.5 μm diametro includentis ornata; lirae in superficie dorsali elongatae, incisura dehiscentiae per liram prominentem longitudinalem media; sectione media opticali a latere observata, lirae instar tholorum quater ad sexies apiculatorum per marginem ventralem et ad polos, alae elongatae autem per dorsalem; a cacumine, lirae instar tholorum quattuor vel plerumque quinque ad sporangii marginem, speciem stellatam efficientium; incisura dehiscentiae per liram dorsalem prominens. Sporangia tenuiter tunicata non visa.

Specimens Examined Other than Type: None.

Comments: Description of this new species is based on a single collection from a rice field near Manakara, Madagascar. Although the original specimen, collected 30 April 1976 and mounted in "gomme au chloral," was badly distorted, rehydration of the specimen and study by LM and SEM disclosed features enabling description of a new species. The specific epithet of *C. tuberculatus* is in recognition of the domelike ridges or tubercles, after *tuberculum* L. (a knoblike or wartlike excrescence), covering the ventral surface and sides of resting sporangia.

Coelomomyces tuzetae Manier *et al., Ann. Parasitol. Hum. Comp.* **45,** 119 (1970)

Figures: See Manier *et al.* (1970)

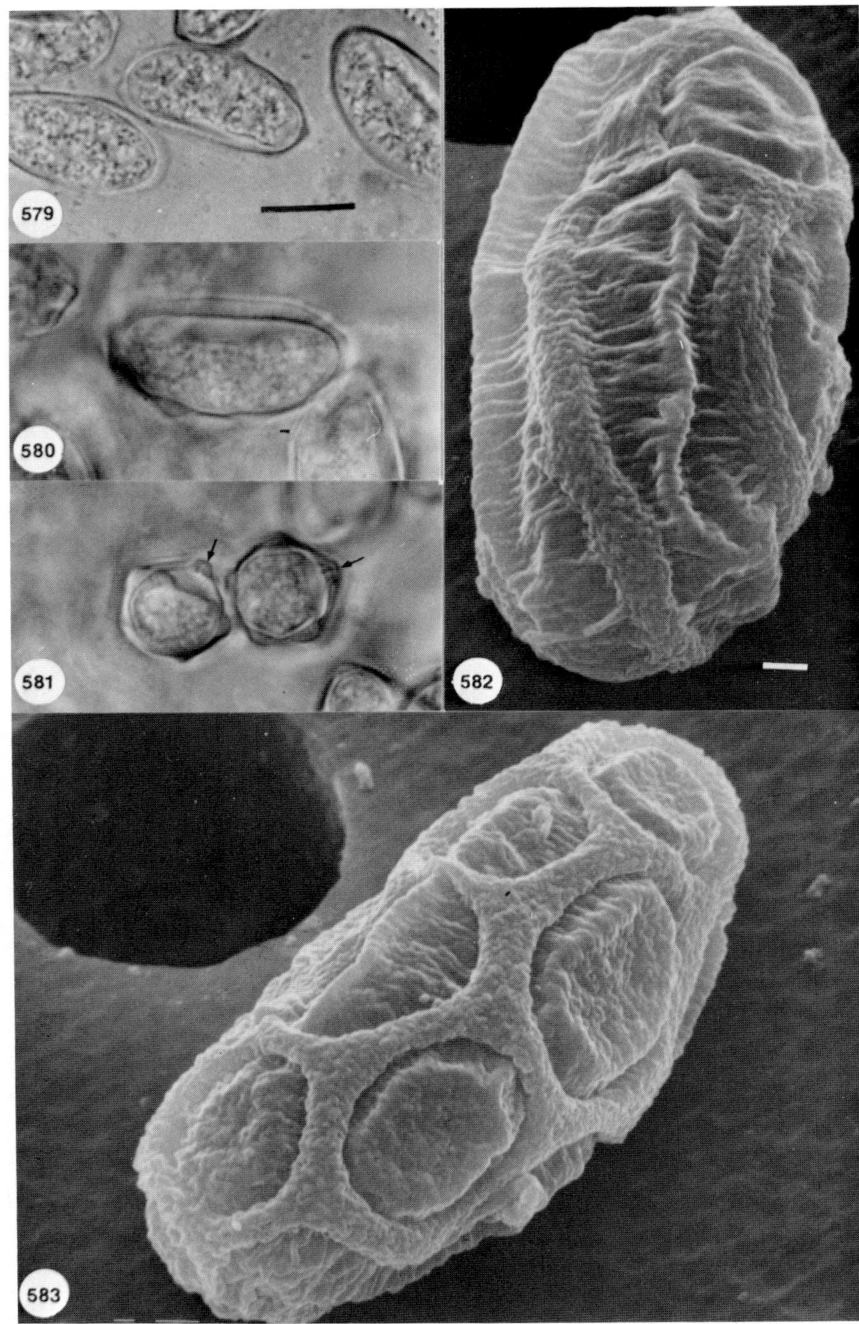
579
580
581
582
583

Description: Hyphae 10–25 μm wide, frequently branched. Resting sporangia broad–ellipsoid to ovoid, slightly flattened on one side, 18–30 × 27–47 μm; surface lacking ornamentation. Thin-walled sporangia not observed.

Type Host and Collection Site: *Orthocladius lignicola* Kieffer; Montpellier, France.

Depository for Type: Laboratoire de Zoologie I, Faculte des Sciences, 34, Montepellier, France (Specimen #LIR-641).

Specimens Examined and Other Collections:

France

Two *C. tuzetae* specimens were provided by Mme. Manier for study; one a prepared slide of a sectioned chironomid host, the other, a specimen dissected in blue-lactophenol. Collection data for these specimens were not available. All reported collections of *C. tuzetae* (Manier *et al.*, 1970) were between 1964 and 1968 from the vicinity of Montpellier, France, on larvae of either *Orthocladius* sp. or *Cricotopus* sp.

Comments: Resting sporangia of *C. tuzetae* are similar to those of *C. walkeri* and *C. pentagulatus* in that wall ornamentation is lacking in all. Distinction from *C. walkeri* may be made in that resting sporangia of *C. walkeri* are larger and occur, so far as known, only on mosquito hosts. Separation from *C. pentangulatus* is possible since resting sporangia of *C. pentagulatus* are five-angled in end view and are narrower than those of *C. tuzetae*.

Coelomomyces uranotaeniae Couch ex Couch, *J. Elisha Mitchell Sci. Soc.* **78,** 135 (1962)

Coelomomyces uranotaeniae Couch, *J. Elisha Mitchell Sci. Soc.* **61,** 124 (1945)

Figures 584–588

Description: Hyphae 4–12 μm thick, elongate, branched. Resting sporangia ellipsoid 21–30 × 29–45 μm; surface ornamented with steep,

---

Figs. 579–583. *Coelomomyces tuberculatus.* Resting sporangia of type specimen from *Culex antennatus* collected in Manakara, Madagascar. Figs. 579 and 580. Midoptical section, side view. Bar = 10 μm. Fig. 581. Midoptical section, end view (dehiscence slit, arrow). Scale as in Fig. 579. Figs. 582 and 583. Scanning electron micrographs showing a side view (Fig. 582), and a ventral view (Fig. 583). Bar = 1 μm.

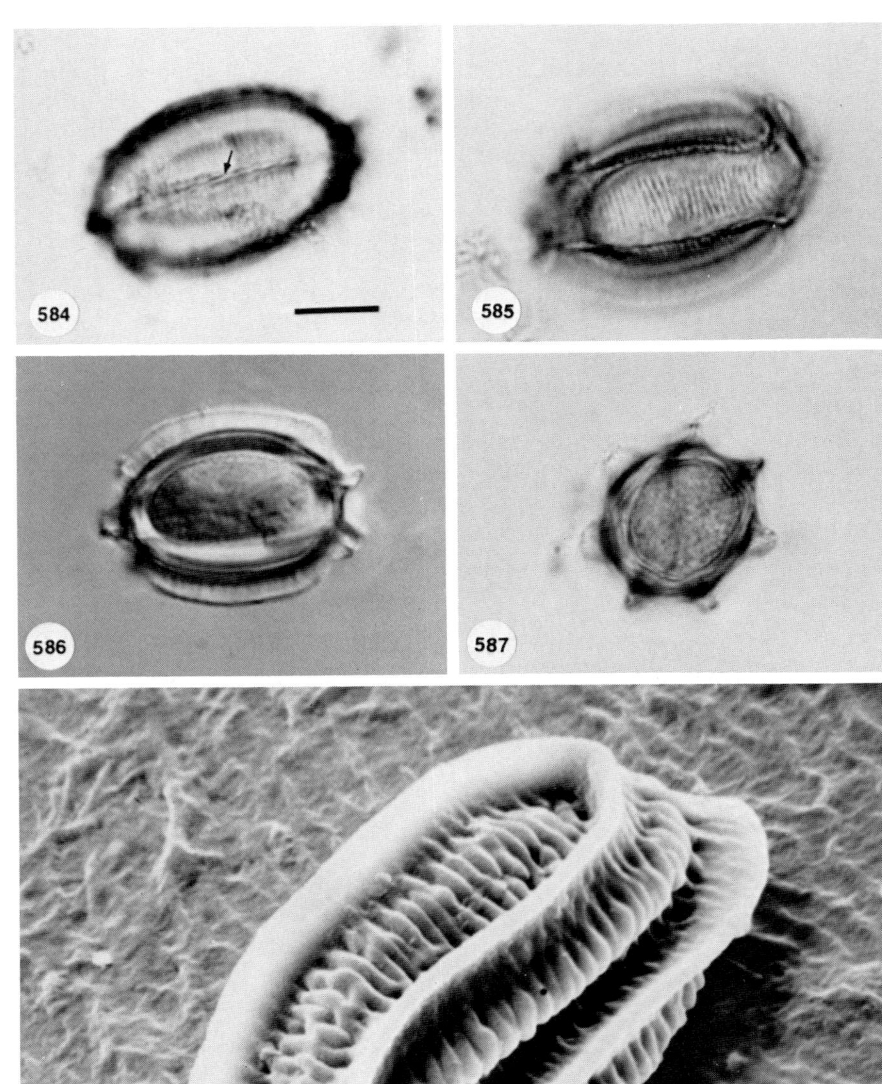

584
585
586
587
588

longitudinal ridges that anastomose at poles; ridges 4–5 μm tall and 7–10 μm apart, area between ridges with fine parallel to anastomosing, transverse striae. In midoptical section: side or face view, ridges as vertically striate wing encircling sporangia; end view, ridges as 7–8 triangular teeth at periphery of sporangia. Depressed areas of wall between ridges 2–3 μm thick. Thin-walled sporangia not observed.

TYPE HOST AND COLLECTION SITE: *Uranotaenia sappharina* Osten-Sacken; Georgia (collected by SMCG).

DEPOSITORY FOR TYPE: BPI.

SPECIMENS EXAMINED AND OTHER COLLECTIONS:

Type

United States

Uranotaenia sappharina Osten-Sacken; Georgia; SMCG; 1945; NCU; (Couch and Dodge, 1947).

Uranotaenia sappharina Osten-Sacken; Georgia; Chapman; 1968; NCU.

COMMENTS: Resting sporangia of *C. uranotaeniae* have ornamentation resembling somewhat that of *C. macleayae*. In the latter, however, ridging is more irregular and more frequently anastomosing. Also, depressed areas between ridges in sporangia of *C. uranotaenia* have prominent transverse striae, whereas similar areas in *C. macleayae* have punctae arranged in rows.

Coelomomyces utahensis Romney, Couch & Nielsen sp. nov.

Figures 589–593

DESCRIPTION: Hyphae 4.5–15 μm in diameter. Resting sporangia ellipsoid, slightly flattened on one side, 39–71 × 54–97 μm; surface ornamented with irregular, scattered, circular to elongate punctae (~1 μm in diameter when circular and 1–3 μm long, 2–5 μm apart), appearing larger and more irregular in optical section immediately below the wall surface;

Figs. 584–588. *Coelomomyces uranotaeniae.* Resting sporangia of type specimen from *Uranotaenia sappharina* collected in Georgia. Fig. 584. Surface view, dorsal side (dehiscence slit, arrow). Bar = 10 μm. Fig. 585. Surface view of sporangial side. Scale as in Fig. 584. Fig. 586. Midoptical section, side view. Scale as in Fig. 584. Fig. 587. Midoptical section, end view. Scale as in Fig. 584. Fig. 588. Scanning electron micrograph showing sporangial side. Bar = 10 μm.

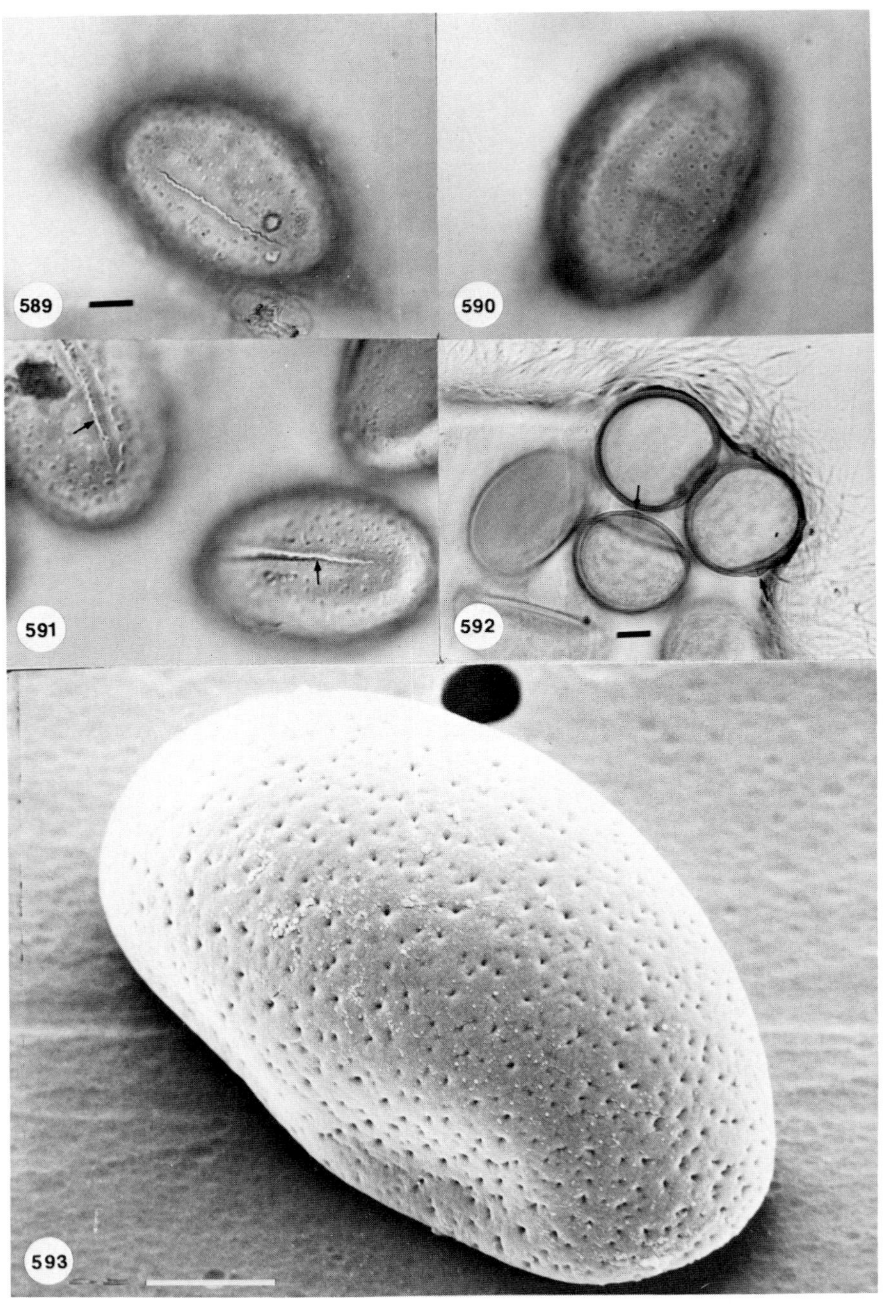
589
590
591
592
593

dehiscence slit in prominent, low dorsal ridge that is devoid of punctae itself, but bordered by punctae. Wall 2–3.5 μm thick. Thin-walled sporangia not observed.

TYPE HOST AND COLLECTION SITE: *Aedes atropalpus nielseni* Craig & O'Meara; Salt Lake City, Utah (collected by Romney).

ALTERNATE HOST: *Potamocypris smaragdina* (Vávra) (H. C. Whisler, personal communication).

DEPOSITORY FOR TYPE: BPI.

LATIN DIAGNOSIS: Hyphae 4.5–15 μm crassae. Sporangia perennantia ellipsoidea, uno latere paulum applanata, 39–71 × 54–97 μm; superficie punctis enormibus, sparsis, rotundis, ~1 μm diametro, vel elongatis, 1–3 μm longis, inter se 2–5 μm distantibus, quasi in sectione media inspecta specie majoribus, enormioribus proxime sub tunicae superficiem, ornata; incisura dehiscentiae per liram prominentem, humilem, dorsalem puncta carentem, punctis autem marginatam inclusa; tunica 2–3.5 μm crassa. Sporangia tenuiter tunicata non observata.

SPECIMENS EXAMINED AND OTHER COLLECTIONS:

United States

Aedes altropalpus epactius; Laboratory Infection; 1973, 1975; NCU.

Aedes atropalpus nielseni Craig and O'Meara; Utah*; Romney; 1970; NCU; (Romney *et al.*, 1971).

Culex tarsalis Coq.; Utah*; Romney; 1970, 1972; NCU; (Romney *et al.*, 1971).

Culiseta incidens (Thompson); Utah*; Romney; 1970, 1972; NCU; (Romney *et al.*, 1971).

COMMENTS: Sporangial ornamentation in *C. utahensis* resembles that of *C. psorophorae*, *C. stegomyiae*, and *C. punctatus*. Distinction from the three may be made in that the dehiscence slit on sporangia of *C. utahensis* lies along a longitudinal, low ridge that is devoid of punctae, a feature

* Infection reported from both field and laboratory.

Figs. 589–593. *Coelomomyces utahensis.* Resting sporangia of type specimen from *Aedes atropalpas nielseni* from Utah. Fig. 589. Surface view, dorsal side (dehiscence slit, arrow). Bar = 10 μm. Fig. 590. Surface view, ventral side. Scale as in Fig. 589. Fig. 591. Interference–contrast optics showing surface view of the dorsal side (dehiscence slit, arrow). Scale as in Fig. 589. Fig. 592. Midoptical section, end view (dehiscence slit, arrow). Bar = 10 μm. Fig. 593. Scanning electron micrograph showing ventral side. Bar = 10 μm.

absent in the other three species. Additionally, the punctae on sporangia of *C. utahensis* are more irregular than in either *C. psorophorae* and *C. stegomyiae,* and shorter and less prominent than in *C. punctatus*. For comparison with *C. lacunosus* vars., see Comments for *C. lacunosus* var. *lacunosus*.

Coelomomyces utahensis has been the subject of extensive laboratory infection studies involving intergeneric transfer of infection (Romney *et al.,* 1971). This species is unique in having been collected only on mosquitoes from rock pools in Utah. Romney *et al.* (1971) gives complete ecological data for these collection sites.

The specific epithet of *C. utahensis* is in recognition of this species having been collected only in Utah.

Coelomomyces walkeri van Thiel ex van Thiel, *J. Elisha Mitchell Sci. Soc.* **78,** 135 (1962)

Coelomomyces walkeri van Thiel, *J. Parsitol.* **40,** 271 (1954)

(Not illustrated.)

DESCRIPTION: Hyphae 1.4–12.6 μm in diameter. Resting sporangia broad–ellipsoid, 30–46.5 × 45–65 μm; surface lacking ornamentation. Thin-walled sporangia not observed.

TYPE HOST AND COLLECTION SITE: *Anopheles tesselatus* Theo. ♀; Marunda, Java (collected by van Thiel).

NEOTYPE HOST AND COLLECTION SITE: *Anopheles gambiae* Giles ♀; Upper Volta, Africa (Rodhain and Brengues, 1974).

DEPOSITORY FOR TYPE: The type, retained in the laboratory of Dr. P. H. van Thiel, has been lost (P. H. van Thiel, personal communication). The neotype is on deposit at IP.

SPECIMENS EXAMINED AND OTHER COLLECTIONS (indicated by brackets):

Neotype

Africa

[*Anopheles funestus* Giles ♀; Sierra Leone; Walker; 1936–1938; ?; (Walker, 1938).]

[*Anopheles gambiae* Giles ♀; Sierra Leone, Walker; 1936–1938; ?; (Walker, 1938).]

[*Anopheles gambiae* Giles ♀; Upper Volta; Coz; 1972; ?; (Coz, 1973).]

Anopheles gambiae Giles ♀; Upper Volta; Rodhain; 1967; ?; (Rodhain and Brengues, 1974).*

Java

[*Anopheles tessalatus* Theo. ♀; Marunda; van Thiel; 1954; lost; (van Thiel, 1954).]†

Philippines

Anopheles annulatus de Rook ♀‡; Tala; Ungureanu; 1966; NCU.

COMMENTS: Van Thiel (1954), in describing *C. walkeri* from specimens of *Anopheles tessalatus* collected in Java, claimed resting sporangia of his specimens to be identical to those of "Type 1" described by Walker (1938) on *A. gambiae* and *A. funestus* from Upper Volta, Africa, in that resting sporangia of both forms were similar in size, lacked ornamentation, and were restricted primarily to the ovaries of hosts. Similar specimens have been collected also by Coz (1973) and Rodhain and Brengues (1974) from Africa and by Ungureanu from the Philippines.‡

Resting sporangia of *C. walkeri* are similar to those of *C. pentangulatus* and *C. tuzetae* in that wall ornamentation is lacking in all. Distinction from both may be made, however, in that resting sporangia of *C. walkeri* are larger than in either. Additionally, sporangia of *C. pentagulatus* appear five-angled in end view whereas those of *C. walkeri* appear circular. Further distinction for *C. tuzetae* is possible in that *C. walkeri* has been collected only on anopheline (mosquito) hosts, whereas *C. tuzetae* occurs on chironomids.

E. Excluded Taxa

Coelomomyces ascariformis van Thiel ex van Thiel, *J. Elisha Mitchell Sci. Soc.* **78,** 135 (1962)

Coelomomyces ascariformis van Thiel, *J. Parasitol.* **40,** 271 (1954)

(Not illustrated.)

The brief description of *C. ascariformis* provided by van Thiel (1954) was based on a specimen described by Manalang (1930) that was thought by van Thiel to be identical to Walker's "Type 4" sporangium (Walker,

* Neotype collection.

† Type collection.

‡ Although resting sporangia of the Philippine specimen are smaller (20.5–33.5 × 39–54 μm) than in the description and are badly distorted, record of this specimen is included here because of similarities in host infection site (ovaries) and lack of sporangial ornamentation.

1938; Plate 4, Fig. 4). Laird (1956a) disagreed, in that the wall of the specimen described by Manalang was ribbed whereas that of Walker's "Type 4" was crenate. Close comparison of Manalang's figures and description with Walker's figures show this to indeed be the case. In his Latin diagnosis for *C. ascariformis,* van Thiel (1962) described Manalang's material but cited Walker's Plate 4, Figure 4, as the type. In the absence of specimens of either, *C. ascariformis* must be deemed a *nomen confusum*.

Other reported collections of *C. ascariformis* have been by Rodhain and Gayral (1971), Coz (1973), and Rodhain and Brengues (1974). While it is possible that the specimens described in these reports are of the same species as that figured by Walker (1938) and described by van Thiel (1954) as *C. ascariformis,* in the absence of specimens of the latter it is impossible to be conclusive. In this regard, the specimens referred to by Rodhain and Gayral (1971) have been studied and are here recognized to be the same as *C. orbiculostriatus* sp. nov.

Coelomomyces beirnei Weiser & McCauley ex Weiser & McCauley, *Can. J. Zool.* **50,** 365 (1972)

Coelomomyces beirnei Weiser & McCauley, *Can. J. Zool.* **49,** 65 (1971)

Figure 594

Coelomomyces beirnei was described originally by Weiser and McCauley (1971) from three species of chironomid larvae (*Tanytarsus* sp. A, *Tanytarsus* sp. B, and *Tanytarsini* sp. A) collected from Marion Lake, British Columbia, Canada between November 1968 and March 1969. Only sporangia in various stages of maturity were observed, described as "rather uniform in size, elongated oval, slightly flattened on one side and measured 21–22 μm by 8–10 μm." While the sporangial wall was smooth or with a protuberance on the flattened side, internally the sporangia contained a "minute resting body located either close to or at one pole" and in some cases two or three vacuoles (see Weiser and McCauley, 1971; Figs. 2C and D and 3A–D). Neither a dehiscence slit nor spore release was observed.

While examination of co-type material of *C. beirnei,* provided by Weiser to C. B. and J. N. C., revealed features identical to those described previously (Weiser and McCauley, 1971), this organism is apparently not of the genus *Coelomomyces* and is probably a protozoan (possibly a sporozoan) rather than a fungus. The failure to observe hyphae, even though numerous specimens were collected, the absence of a dehiscence slit, the small size of the sporangia, and finally, the thin sporangial wall enclosing a highly vacuolate, irregular internal cytoplasm all make this organism highly suspect as a species of *Coelomomyces*.

Coelomomyces ciferrii nomen nudum Arêa Leão & Pedroso, *Mycopathol. Mycol. Appl.* **26,** 305 (1965)

Figures 595–597

The taxon *C. ciferrii,* as described by Arêa Leão and Pedroso (1965), lacked a Latin diagnosis and designation of a type, therefore lacking also valid publication. From specimens provided J. N. C. by Arêa Leão, it would be possible to now validly describe this supposed new species of *Coelomomyces*. However, careful examination of this parasite of eggs of *Lutzomyea* sp. (*Phlebotomus*) raises several questions as to its placement in the genus *Coelomomyces*. Of particular concern is the absence of a dehiscence slit in "resting sporangia" of this organism. Other features atypical for species of *Coelomomyces* are its disc-shaped "sporangia" and its very thick wall. As noted by Dedet and Laird (1981), characteristics of *C. ciferrii* indicate that this organism is more likely to be an entomopathogenic member of the Entomophthorales than a species of *Coelomomyces*.

Coelomomyces dubitskii Shcherbak *et al., Dopov. Akad. Nauk Ukr. RSR, Ser. B: Geol., Khim. Biol. Nauki* **8,** 758 (1977)

(Not illustrated.)

The specimen described by Shcherbak *et al.* (1977) as *C. dubitskii,* a *nomen nudum,* is recognized to be *C. iliensis* var. *iliensis*. However, in keeping with the intention of Shcherbak *et al.* (1977) to honor Dr. Dubitskij, the new species that was mistakenly identified by Shcherbak *et al.* (1977) as *C. iliensis* is here described as *C. dubitskii* sp. nov.

Coelomomyces milkoi Dudka & Koval, *Novit. Syst. Plant. Vasc.* **10,** 88 (1973)

(Not illustrated.)

Described originally (Dudka *et al.,* 1973) as a member of the genus *Coelomomyces, C. milkoi* has since been studied further by Dubitskij *et*

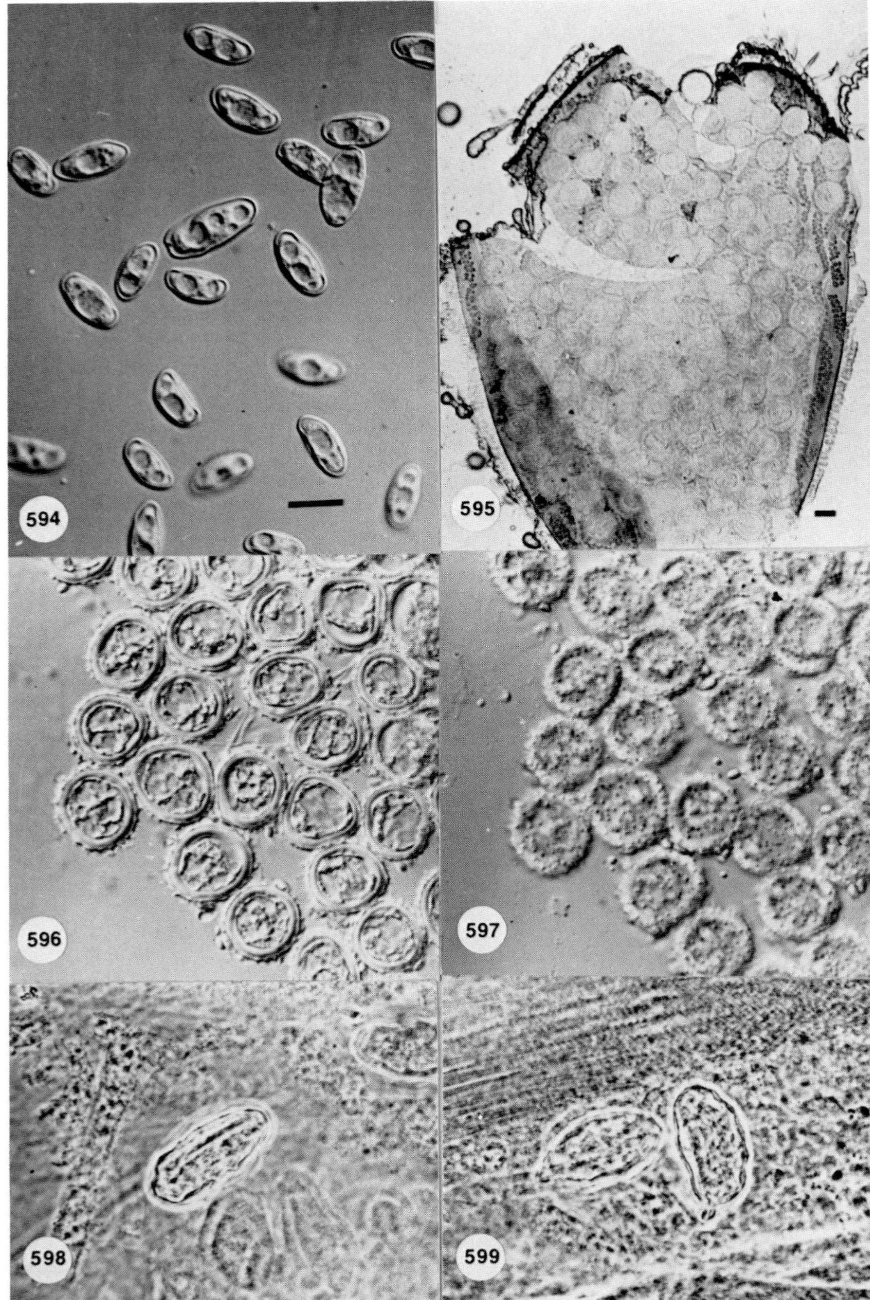

594
595
596
597
598
599

al. (1974) and Nam and Dubitskij (1977), and redescribed by Couch *et al.* (1979) as *Tabanomyces milkoi,* a member of the Entomophthorales.

Coelomomyces notonectae (Bogoyavlensky) Keilin, *Parasitology* **19,** 365 (1927)

Zografia notonectae Bogoyavlensky, *Arch. Soc. Russe. Protistenkd.* **1,** 113 (1922)

(Not illustrated.)

Zografia notonectae, described by Bogoyavlensky (1922) from a species of *Notonecta* (Hemiptera), was determined by Keilin (1927) to be so similar to his previously described species, *Coelomomyces stegomyiae,* that they should be included in the same genus; *Zografia notonectae* therefore became *Coelomomyces notonectae,* since the name *Coelomomyces* had priority over *Zografia.* Although the organism described by Bogoyavlensky is undoubtedly a member of the genus *Coelomomyces,* the limited description provided for its resting sporangia make it impossible to recognize this organism and distinguish it from other species of *Coelomomyces. Coelomomyces notonectae* must therefore be considered an incompletely known species.

Coelomomyces solomonis Laird, *Bull. R. Soc. N. Z.* **6,** 1 (1956a)

Figures 598–599

As described originally (Laird, 1956a) from larvae of *Anopheles punctulatus* (Dön.) collected in the British Solomon Islands, resting sporangia of *C. solomonis* appeared oval in face view but with one surface convex and the other flattened, measured 18.5–24 × 32–46 μm (10–21 μm thick), and lacked ornamentation. Although the type of this species, deposited in the National Museum of New Zealand, Wellington, has been lost (P. J. Brownsey, personal communication to M. Laird), Manier *et al.* (1970) presented observations on this species that were based on the type. In

Figs. 594–599. Excluded taxa. Fig. 594. "Resting sporangia" of "*Coelomomyces beirnei.*" Bar = 10 μm. Figs. 595–597. "Resting sporangia" of "*Coelomomyces ciferrii*". Fig. 595. Low-magnification micrograph showing "sporangia" inside an egg of *Lutzomyea* sp. Bar = 10 μm. Fig. 596. Midoptical section. Scale as in Fig. 595. Fig. 597. Surface view of "sporangia". Scale as in Fig. 595. Figs. 598 and 599. Resting sporangia of *Coelomomyces solomonis.* Scale as in Fig. 595.

contrast to the original description, these authors report the outer wall of resting sporangia to be ornamented with papillae and hyaline ridges, "une paroi externe ornée de papilles et de crêtes hyalines." Similarly, sporangia of a type specimen sent to J. N. C. by Laird in 1967 (since returned) exhibited apparent sculpturing of the outer wall (Figs. 598–599). In view of this contradiction with the original description, and since type specimens are unavailable, *C. solomonis* must be considered an incompletely known species.

F. Incompletely Known Forms

In many reported infections involving species of *Coelomomyces,* species identification is often omitted. The reasons for this are many, but generally involve the investigator's lack of familiarity with the group or the absence of material suitable for species identification or its description as a new species. In such instances, however, continued study of the specimen and/or additional collections of it may enable its identification as either a known or possibly a new species. In this regard, Table IX lists the major published references to specimens identified simply as "*Coelomomyces* sp." and, when known, provides information regarding the disposition of such forms. The following incompletely known forms were considered in the present study, but could not be described or classified on the basis of the material available. (Collection data is given as host collection site, collector, date, and depository.)

Specimen No. 1

Collection Data: *Culex territans* Walker; New Haven, Connecticut; Anderson; 1965, 1967; NCU.

Comments: This sample consists solely of larvae containing thin-walled sporangia (measuring 17–24 × 28–45 μm). Unfortunately, because of the typical shape of the sporangia (ellipsoid to allantoid) and the absence of resting sporangia, species identification is impossible. In living material, sporulation was, however, noted to occur (J. N. Couch, unpublished) within 1 day after sporangia were dissected from larvae. This report constitutes the first record of thin-walled sporangia in the Western Hemisphere.

Specimen No. 2

Collection Data: *Culex territans* Walker; Lake Charles, Louisiana; Chapman; 1969; NCU.

TABLE IX. A Summary of the Major Reports of Unidentified Species of *Coelomomyces*

Reference	Host	Collection site	Species identification or comments
Briggs (1967)	*Simulium ornatum nitidifrons*	Morocco	No *Coelomomyces* sp. present
Briggs (1969)	*Aedes excrucians* ♀	College, Alaska	= *C. borealis* var. *borealis*
	Culex orientalis	Vladivostok, U.S.S.R.	Probably *C. iliensis* var. *iliensis* (Dubitskij *et al.*, 1973)
	Culex peccator	United States	
	Culex portesi	Trinidad	= *C. seriostriatus*
Chapman and Woodard (1966)	*Aedes sollicitans*	Louisiana	= *C. psorophorae* var. *halophilus*
	Aedes taeniorhynchus	Louisiana	= *C. psorophorae* var. *halophilus*
	Aedes vexans	Louisiana	= *C. psorophorae* var. *psorophorae* "complex"
	Culex restuans	Louisiana	= *C. psorophorae* var. *psorophorae* "complex"
	Culex salinarius	Louisiana	= *C. psorophorae* var. *psorophorae* "complex"
	Culiseta inornata	Louisiana	= *C. psorophorae* var. *psorophorae* "complex"
	Psorophora ciliata	Louisiana	= *C. psorophorae* var. *psorophorae* "complex"
	Psorophora howardii	Louisiana	= *C. psorophorae* var. *psorophorae* "complex"
Chapman *et al.* (1967)	*Aedes sollicitans*	Louisiana	= *C. psorophorae* var. *halophilus*
Chapman *et al.* (1969)	*Aedes triseriatus*	Louisiana	= *C. macleayae*
	Anopheles bradleyi	Louisiana	Probably *C. punctatus*
	Culex peccator	Louisiana	—
Fedder *et al.* (1971)	*Aedes togoi* ♀	Vladivostok, U.S.S.R.	= *C. psorophorae* var. *psorophorae* "complex"
Genga and Maffi (1973)	*Aedes hebrideus*	Solomon Islands	Specimen thought similar to *C. solomonis*, but Rodhain and Fauran (1975) report *C. finlayae* on the same host from New Hebrides
Hall and Anthony (1979)	*Wyeomyia vanduzeei*	Florida	Likely *C. stegomyiae* var. *stegomyiae*
Hembree (1979)	*Aedes albopictus*	Thailand	= *C. macleayae* and *C. stegomyiae*
	Aedes mediopunctatus	Thailand	= *C. macleayae*

(*continued*)

TABLE IX. (Continued)

Reference	Host	Collection site	Species identification or comments
	Aedes pseudalbopictus	Thailand	= *C. macleayae*
	Anopheles aconitus	Thailand	= *C. iyengarii*
	Anopheles bengalensis	Thailand	= *C. thailandensis*
	Anopheles navipes	Thailand	= *C. couchii*
	Anopheles vagus	Thailand	= *C. indicus*
	Armigeres longipalpis	Thailand	= *C. macleayae*
	Culex fuscocephalus	Thailand	*C. muspratti*i
	Culex vishnui	Thailand	*C. orbiculostriatus*
	Toxyrhynchites gravelyi	Thailand	*C. macleayae*
	Toxyrhynchites sp.	Thailand	*C. macleayae*
	Uranotaenia campestris	Thailand	*C. celatus*
Kellen *et al.* (1963)	*Aedes melanimon* ♀	California	= *C. psorophorae*
Khaliulin and Ivanov (1973)	*Aedes caspius dorsalis*	Maryiskaya, U.S.S.R.	Species reported by Shcherbac *et al.* (1977) to be infected with *C. psorophorae*
	Aedes cyprius	Maryiskaya, U.S.S.R.	—
	Culex sp.	Maryiskaya, U.S.S.R.	—
Lacour and Rageau (1957)	*Culex pipens fatigans*	New Caledonia	—
Lum (1963)	*Aedes taeniorhynchus*	Florida	*C. psorophorae* var. *halophilus*
Maffi and Genga (1970)	*Anopheles farauti*	British Solomon Islands	= *C. lairdi*
	Culex annulirostris	British Solomon Islands	—
	Culex sp.	British Solomon Islands	—
	Uranotaenia barnesi	British Solomon Islands	—

Manalang (1930)	*Anopheles* spp.	Philippines	Referred to by Manalang as "Coccidium" and described by van Thiel as *C. ascariformis*. Thought by Laird (1956a) to be *C. indicus*.
McCrae (1972)	*Aedes simpsoni*	Uganda, Africa	—
Muspratt (1946a)	*Aedes scatophagoides*	Zambia, Africa	"Type c" = *C. stegomyiae*
	Anopheles squamosus	Zambia, Africa	"Type b" = *C. orbicularis*
	Cyperus sp.	Livingstone, Zambia	"Type d" = not *Coelomomyces*. Probably vesicular *Arbuscular mycorrhiza*
Rajapaksa (1964)	*Aedes aegypti*	Ceylon	*C. stegomyiae* and *C. dentialatus*
	Aedes albopictus	Ceylon	*C. stegomyiae* and *C. dentialatus*
	Aedes sp.	Ceylon	*C. dentialatus*
	Armigeres obturbans	Ceylon	*C. dentialatus*
	Culex gelidus	Ceylon	*C. elegans*
	Culex pipiens fatigans	Ceylon	—
Ramalingam (1966)	*Aedes aegypti*	Tonga	Specimens too distorted for identification
	Aedes oceanicus	American Samoa	Specimens too distorted for identification
	Aedes polynesiensis	American Samoa	= *C. macleayae?*
	Aedes samoanus	Western Samoa	= *C. stegomyiae?*
	Aedes tutuilae	American Samoa	Specimens too distorted for identification
Ribeiro *et al.* (1981)	*Anopheles squamosus*	Southern Angola, Africa	—
	Culex argenteopunctatus kingii	Northern Angola and Cabinda, Africa	—
	Culex guiarti	Northern Angola and Cabinda, Africa	—
Romney *et al.* (1971)	*Aedes atropalus epactius*	Utah	= *C. utahensis*
Service (1974)	*Aedes cantans*	Monks Wood, England	*C. borealis* var. *borealis*
Weiser (1977b)	*Simulium ornatum*	Czechoslovakia	Published photographs (Weiser and Undeen, 1981) show specimens unlike any member of the genus *Coelomomyces*

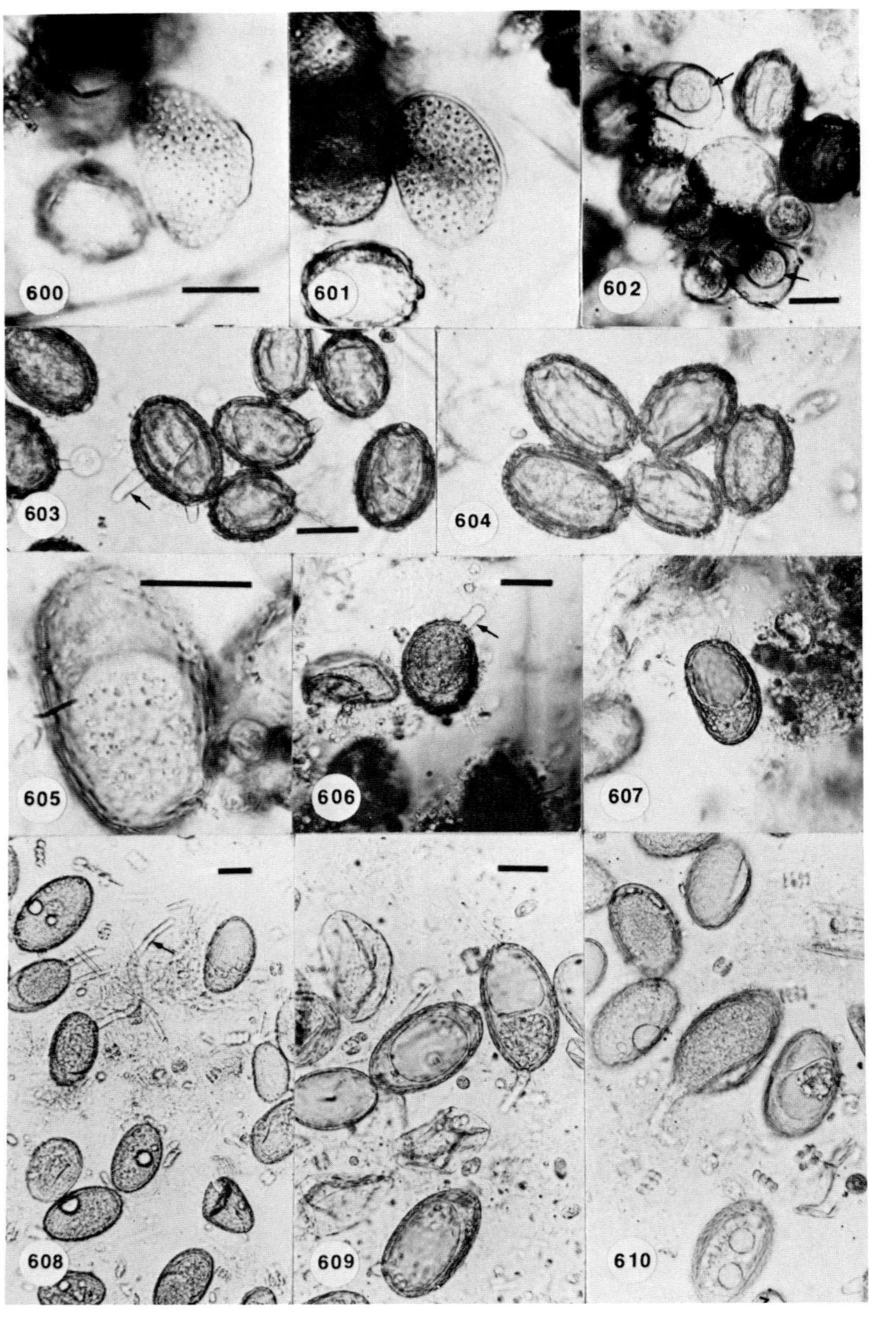
600
601
602
603
604
605
606
607
608
609
610

COMMENTS: As with specimen no. 1, this sample includes larvae containing only thin walled sporangia (measuring 19–26 × 33–58 μm). Although species identification is impossible due to the absence of resting sporangia, similarities in host and sporangial size between this specimen and specimen no. 1 indicate the two may be of the same species.

Specimen No. 3

COLLECTION DATA: *Anopheles funestus* Giles ♀; Kapeta, Liberia, Africa; Giglioli; 1955; NCU.

COLLECTION DATA: *Anopheles gambiae* Giles ♀; Gebedin, Liberia, Africa; Giglioli; 1955; NCU.

COMMENTS: For both specimens, sporangia were observed only in the ovaries of hosts, measured 11–20 × 22–35 μm, and lacked ornamentation. A similar specimen ("Type 2") was described by Walker (1938) from ovaries of the same two hosts collected also in Africa. It is possible, therefore, that all may be of the same taxon. However, this determination must await study of additional material. These forms may represent a small variety of *C. walkeri*.

G. Parasites of Resting Sporangia

Although not of common occurrence, it is not unusual to encounter other fungi growing on or inside sporangia of *Coelomomyces* spp. In most instances little is known of the exact nature of this relationship, and the identity of the supposed parasite is often in question. However, a brief description of observed parasites is included here as a basis for further

Figs. 600–610. Parasites of resting sporangia of various species of *Coelomomyces*. Figs. 600–602. *Rozella*-like internal parasite of thin-walled sporangia of *Coelomomyces anophelesicus*. Bar = 10 μm. Figs. 600 and 601. Zoosporangia of parasite with partially cleaved zoospores. Note that the zoosporangium of the parasite completely fills the thin-walled sporangia. Fig. 602. Smooth-walled, resting sporangia of parasite (arrow) seen inside the thin-walled sporangia of the host. Bar = 10 μm. Figs. 603 and 604. *Olpidium*-like internal parasite of resting sporangia of *Coelomomyces indicus*. Note the discharge tubes that are seen emerging through the sporangial wall of the host (arrow). Bar = 10 μm. Figs. 605–607. *Olpidium*-like internal parasite of resting sporangia of *Coelomomyces punctatus*. Bars = 10 μm. Fig. 605. Young sporangium of parasite inside resting sporangium of host. Fig. 606. Mature sporangium of parasite with discharge tube (arrow). Fig. 607. Empty sporangium of parasite following dehiscence of zoospores. Figs. 608–610. *Olpidium*-like parasite of resting sporangia of *Coelomomyces punctatus*, perhaps *Olpidium longicollum*. Note the long discharge tube characteristic of this parasite (arrow). Bars = 10 μm.

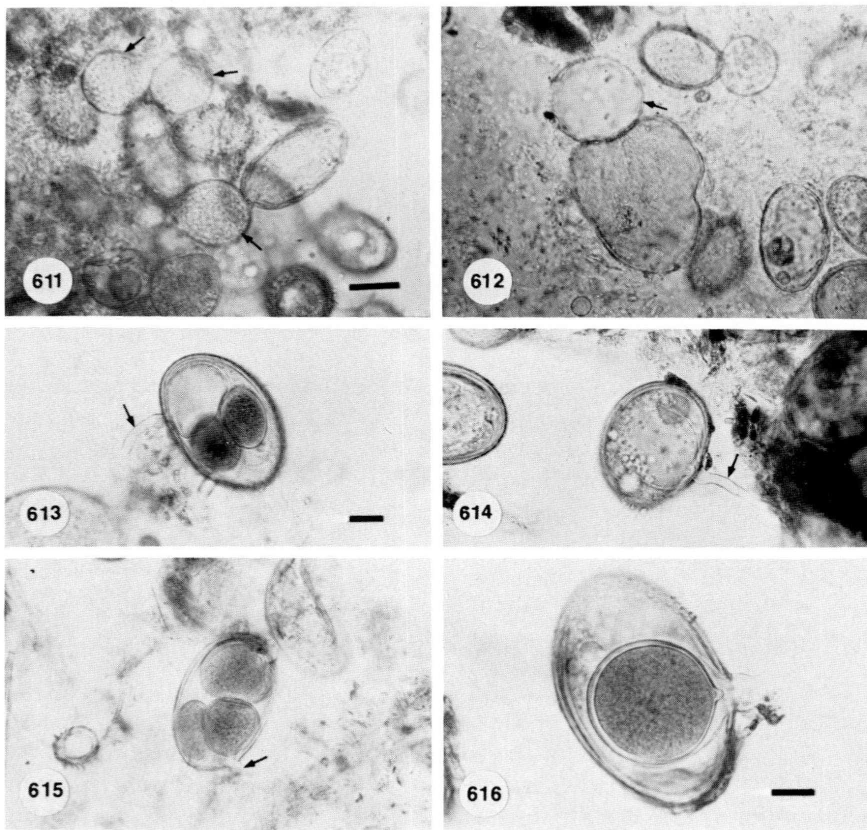

Figs. 611–616. Parasites of resting sporangia of various species of *Coelomomyces*. Figs. 611 and 612. Infection of resting sporangia of *Coelomomyces punctatus* by an epibiotic, eucarpic chytrid, possibly *Rhizophydium globosum* (arrow). Bar = 10 μm. Figs. 613–616. Infection of resting sporangia of *Coelomomyces utahensis* by a holocarpic form of *Catenaria,* possibly *Catenaria anguillulae*. Note the prominent exit tubes (arrow) visible in Figs. 613–615 and the resting spore with a prominent "pip" or projection at its apex seen in Fig. 616. Bars = 10 μm.

work and because the presence of such parasites may cause problems in species identification. In all instances, parasites were observed on sporangia that had been dissected from mosquito larvae and mounted in water on slides for studies of sporulation. In many cases sporangia had been repeatedly allowed to dry and then rewet. Nothing is known of the occurrence of such parasites under natural conditions. Dr. J. S. Karling assisted in the tentative identification of parasites.

1. Infection of thin-walled sporangia of *Coelomomyces anophelesicus* by a *Rozella*-like internal parasite. The parasite was characterized by zoosporangia which completely filled the thin-walled sporangia, and by smooth-walled resting sporangia. No other stages were observed. (Figures 600–602.)

2. *Olpidium*-like internal parasite of resting sporangia of *Coelomomyces indicus*. (Figures 603–604.)

3. *Olpidium*-like internal parasite of resting sporangia of *Coelomomyces punctatus*. (Figures 605–607.)

4. *Olpidium*-like parasite of resting sporangia of *Coelomomyces punctatus*. The long exit tube characteristic of this parasite suggests that this may be *Olpidium longicollum*. (Figures 608–610.)

5. Infection of resting sporangia of *Coelomomyces punctatus* by an epibiotic, eucarpic chytrid, possibly *Rhizophydium globosum*. (Figures 611 and 612.)

6. Infection of sporangia of *Coelomomyces utahensis* by a holocarpic form of *Catenaria*, possibly *Catenaria anguillulae*. This parasite was characterized by zoosporangia with prominent exit tubes and by resting spores bearing a "pip" or projection at the apex, a common feature of such spores in *Catenaria* spp. (Figures 613–616.)

7. Umphlett (1961) reported *Hyphochytrium catenoides* parasites in resting sporangia of *Coelomomyces dodgei*. Various stages in development and sporulation of the parasite were followed in detail. Since the parasite was successfully cultured on corn meal agar, Umphlett concluded *H. catenoides* to be a faculatative parasite.

REFERENCES

Andreadis, T. G., and Magnarelli, L. A. (1984). New variants of the *Coelomomyces psorophorae* "complex" (Chytridiomycetes: Blastocladiales) from the salt marsh mosquitoes *Aedes cantator* and *Aedes sollicitans* (Diptera: Culicidae). *J. Med. Entomol.* **21,** 379–383.

Anthony, D. W., Chapman, H. C., and Hazard, E. I. (1971). Scanning electron microscopy of the sporangia of species of *Coelomomyces* (Blastocladiales: Coelomomycetaceae). *J. Invertebr. Pathol.* **17,** 395–403.

Arêa Leãe, A. E., and Pedroso, M. C. (1965). Novo especie do genero *Coelomomyces* parasito de ovos de *Phlebotomous*. *Mycopathol. Mycol. Appl.* **26,** 305–307.

Barr, A. R. (1958). The mosquitoes of Minnesota (Diptera: Aulicidae: Culicinae). *Minn., Agric. Exp. Stn., Tech. Bull.* **228,** 154.

Bland, C. E., and Couch, J. N. (1973). Scanning electron microscopy of sporangia of *Coelomomyces*. *Can. J. Bot.* **51,** 1325–1330.

Bogoyavlensky, N. (1922). *Zoografia notonectae,* n. g., n. sp. *Arch. Soc. Russe Protistenkd.* **1,** 113–119.

Briggs, J. D. (1967). "1966 Activities of the WHO International Reference Centre for Diagnosis of Diseases of Vectors," WHO/VBC/67.8. World Health Organ., Rome.

Briggs, J. D. (1969). "1968 Activities of the WHO International Reference Center for Diagnosis of Diseases of Vectors," WHO/VBC/69.171. World Health Organ., Rome.

Chandrahas, R. K., and Rajagopalan, P. K. (1979). Mosquito breeding and the natural parasitism of larvae by a fungus, *Coelomomyces* and a mermithid nematode, *Romanomeris,* in paddy fields in Pondicherry. *J. Med. Res.* **69,** 63–70.

Chapman, H. C., and Glenn, F. E., Jr. (1972). Incidence of the fungus *Coelomomyces punctatus* and *C. dodgei* in larval populations of the mosquito *Anopheles crucians* in two Louisiana ponds. *J. Invertebr. Pathol.* **19,** 256–261.

Chapman, H. C., and Woodard, D. B. (1966). *Coelomomyces* (Blastocladiales: Coelomomycetaceae) infections in Louisiana mosquitoes. *Mosq. News* **26**(2), 121–123.

Chapman, H. C., Woodard, D. B., and Petersen, J. J. (1967). Pathogens and Parasites in Louisiana Culicidae and Chaoboridae. *Proc. N. J. Mosq. Exterm. Assoc.* **54,** 54–60.

Chapman, H. C., Clark, T. B., Peterson, J. J., and Woodard, D. B. (1969). A two-year survey of pathogens and parasites of Culicidae, Chaoboridae, and Ceratopogonidae in Louisiana. *Proc. Annu. Meet. N. J. Mosq. Exterm. Assoc.* **56,** 203–212.

Coluzzi, M., and Rioux, J. A. (1962). Primo reperto in Italia di larve di *Anopheles* parassitate de funghi del genere *Coelomomyces* Keilin. Descrizione di *Coelomomyces raffaelei* n. sp. (Blastocladiales: Coelomomycetaceae). *Riv. Malariol.* **41,** 29–37.

Couch, J. N. (1945). Revision of the genus *Coelomomyces* parasitic in insect larvae. *J. Elisha Mitchell Sci. Soc.* **61,** 124–136.

Couch, J. N. (1962). Validation of the family Coelomycetaceae and certain species and varieties of *Coelomomyces. J. Elisha Mitchell Sci. Soc.* **78,** 136–138.

Couch, J. N., and Dodge, H. R. (1947). Further observations on *Coelomomyces,* parasitic on mosquito larvae. *J. Elisha Mitchell Sci. Soc.* **63,** 69–79.

Couch, J. N., Andreeva, R. V., Laird, M., and Nolan, R. A. (1979). *Tabanomyces milkoi* n. gen. n. comb. a fungal pathogen of house flies. *Proc. Natl. Acad. Sci. U.S.A.* **76,** 2299–2302.

Coz, J. (1973). Contribution à l'étude du parasitisme des anapheles Ouest-Africain. Mermithidae et *Coelomomyces. Cah. ORSTOM, Ser. Entomol. Med. Parasitol.* **11,** 237–241.

Dedet, J. P., and Laird, M. (1981). Un cas Algérien de parasitisme de *Phlebotomous* (*Paraphlebotomous*) *sargenti* Parrot 1917 par Entomophthorale. *Can. J. Zool.* **59,** 323–325.

Dubitskij, A. M. (1978). "Biological Control of Bloodsucking Flies in the USSR." Alma Ata "Nauka" of the Kagokh SSR.

Dubitskij, A. M., Danebekov, A. E., and Deshevykh, N. D. (1970). The discovery in mosquito larvae of a fungus of the genus *Coelomomyces* in southeastern Kazakhstan. *Med. Parazitol. Parazit. Bolezni* **39,** 737–738.

Dubitskij, A. M., Dzerzhinskii, V. A., and Danebekov, A. E. (1973). A new species of pathogenic fungus of the genus *Coelomomyces. Mikol. Fitopatol.* **7,** 136–139.

Dubitskij, A. M., Deshevykh, N. D., and Andreeva, R. V. (1974). Activation conditions of resting sporangia of the entomopathogenic fungus *Coelomomyces milkoi* under laboratory conditions. *Parazitologiya* **8,** 175–178.

Dudka, I. A., Koval, E. Z., and Andreeva, R. V. (1973). A new species of *Coelomomyces, C. milkoi,* from the hemocoel of larval horseflies. *Novit. Syst. Plant. Vasc.* **10,** 88.

Ekstein, F. (1922). Beiträge zur Kentniss der Stechmückenparasiten. *Zentralbl. Bakteriol., Parasitenkd., Infektionskr. Hyg., Abt. 1: Orig.* **88,** 128.

Fedder, M. L., Danielevskii, M. L., and Reznik, E. P. (1971). Finding of parasitic fungus of the genus *Coelomomyces* (Phycomycetes, Blastocladiales) in *Aedes togoi* Theobald mosquitoes in the Maritime territory. *Med. Parazitol. Parazit. Bolezni,* **40**(2), 201–204.

Federici, B. A. (1977). Laboratory infection of *Anopheles freeborni* with the parasitic fungi *Coelomomyces dodgei* and *Coelomomyces punctatus. Proc. Pap. Annu. Conf. Calif. Mosq. Vector Control Assoc.* **45,** 107–108.

Federici, B. A. (1979). Experimental hybridization of *Coelomomyces dodgei* and *Coelomomyces punctatus. Proc. Natl. Acad. Sci. U.S.A.* **76,** 4425–4428.

Federici, B. A., and Chapman, H. C. (1977). *Coelomomyces dodgei.* Establishment of an *in vivo* laboratory culture. *J. Invertebr. Pathol.* **30,** 288–297.

Federici, B. A., and Roberts, D. W. (1975). Experimental laboratory infection of mosquito larvae with fungi of the genus *Coelomomyces.* I. Experiments with *Coelomomyces psorophorae* var. in *Aedes taeniorhynchus* and Coelomomyces psorophorae var. in *Culiseta inornata. J. Invertebr. Pathol.* **26,** 21–27.

Federici, B. A., and Roberts, D. W. (1976). Experimental laboratory infection of mosquito larvae with fungi of the genus Coelomomyces. II. Experiments with *Coelomomyces punctatus* in *Anopheles quadrimaculatus. J. Invertebr. Pathol.* **27,** 333–341.

Federici, B. A., Smedley, G., and van Leuken, W. (1975). Mosquito host range tests with *Coelomomyces punctatus. Ann. Entomol. Soc. Am.* **68,** 669–670.

Gad, A. M., and Sadek, S. (1968). Experimental infection of *Anopheles pharoensis* larvae with *Coelomomyces indicus. J. Egypt. Public Health Assoc.* **43,** 387–391.

Gad, A. M., Sadek, S., and Fateen, Y. A. (1967). The occurrence of *Coelomomyces indicus* Iyengar in Egypt. *U. A. R. Mosq. News* **27,** 201–202.

Genga, R., and Maffi, M. (1973). *Coelomomyces* sp. infection in *Aedes* (*Stegomyia*) *herbrideus* Edwards on Nupani Island, Santa Cruz Group, Solomons. *J. Med. Entomol.* **10**(4), 413–414.

Gugnani, H. C., Wattal, B. L., and Kalra, N. L. (1965). A note on *Coelomomyces* infection in mosquito larvae. *Bull—Indian Soc. Malar. Commun. Dis.* **2,** 333–337.

Haddow, A. J. (1942). The mosquito fauna and climate of native huts at Kisumu, Keneya. *Bull. Entomol. Res.* **33,** 91–142.

Hall, D. W., and Anthony, D. W. (1979). *Coelomomyces* sp. (Phycomycetes, Blastocladiales) from the mosquito *Wyeomyia vanduzeei* (Diptera, Culicidae). *J. Med. Entomol.* **16,** 84.

Hembree, S. C. (1979). Preliminary report of some mosquito pathogens from Thailand. *Mosq. News* **39,** 575–678.

Huang, Y. M. (1968). *Aedes* (*Verrallina*) of the Papuan Subregion (Diptera: Culicidae). *Pac. Insects Monog.* **17,** 1–74.

Iyengar, M. O. T. (1935). Two new fungi of the genus *Coelomomyces* parasitic in larvae of *Anopheles. Parasitology* **27,** 440–449.

Iyengar, M. O. T. (1962). Validation of the species *Coelomomyces anophelesicus* and *Coelomomyces indicus. J. Elisha Mitchell Sci. Soc.* **78,** 133–134.

Keilin, D. (1921). On a new type of fungi: *Coelomomyces stegomyiae,* n. g., n. sp., parasitic in the body-cavity of the larva of *Stegomyia scutellaris* Walker (Diptera, Nematocera, Culicidae). *Parasitology* **13,** 225–234.

Keilin, D. (1927). On *Coelomomyces stegomyiae* and *Zografia notonectae,* fungi parasitic in insects. *Parasitology* **19,** 365–367.

Kellen, W. R. (1963). Research on biological control of mosquitoes. *Proc. Pap. Annu. Conf. Calif. Mosq. Control Assoc.* **31,** 23–25.

Kellen, W. R., Clark, T. B., and Lindegren, J. E. (1963). A new host record for *Coelomomyces psorophorae* Couch in California (Blastocladiales: Coelomomycetaceae). *J. Insect Pathol.* **5,** 167–173.

Khaliulin, G. L., and Ivanov, S. L. (1973). Parasitic fungus *Coelomomyces* sp. in mosquito larvae in the Maryiskaya ASSR. *Med. Parazitol. Parazit. Bolezni* **42,** 487.

Kiser, D. R. (1976). Occurrence of the fungus *Coelomomyces* in larval populations of North Carolina salt marsh mosquitoes. M.S. Thesis, East Carolina University, Greenville, North Carolina.

Kuznetzov, V. G., and Mikheeva, A. I. (1970). The occurrence of the fungus *Coelomomyces* in larvae of *Aedes* in the Soviet Far East. *Parazitologiya* **4,** 392–393.

Lacour, M., and Rageau, J. (1957). Equête épideḿiologique et entomologique sur la filariose de Bancroft en Nouvelle Calédonie et Dépendances. *South Pac. Comm., Tech Pap.* **110,** 1–24.

Laird, M. (1956a). Studies of mosquitoes and fresh water ecology in the South Pacific. *Bull—R. Soc. N. Z.* **6,** 1–213.

Laird, M. (1956b). A new species of *Coelomomyces* (fungi) from Tasmanian mosquito larvae. *J. Parsitol.* **42,** 53–55.

Laird, M. (1959a). Parasites of Singapore mosquitoes, with particular reference to the significance of larval epibionts as an index of habit pollution. *Ecology* **40,** 206–221.

Laird, M. (1959b). Fungal parasites of mosquito larvae from the oriental and Australian regions, with a key to the genus *Coelomomyces* (Blastocladiales: Coelomomycetaceae). *Can. J. Zool.* **37,** 781–791.

Laird, M. (1961). New American locality records for four species of *Coelomomyces* (Blastocladiales, Coelomomycetaceae). *J. Insect Pathol.* **3,** 249–253.

Laird, M. (1962). Validation of certain species of *Coelomomyces* (Blastocladiales, Coelomomycetaceae) described from the Australian and Oriental regions. *J. Elisha Mitchell Sci. Soc.* **78,** 132–133.

Laird, M. (1966). "Integrated Control and *Aedes polynesiensis:* An outline of the Tokelau Islands Project, and its Results," WHO/EBL/66.69, WHO/FIL/66.63, WHO/Vector Control/66.204. World Health Organ., Geneva.

Laird, M., Nolan, R. A., and Mogi, M. (1975). *Coelomomyces omorii* sp. n. and *C. raffaelei* Coluzzi and Rioux var. *parvum* var. n. from mosquitoes in Japan. *J. Parasitol.* **61,** 539–544.

Laird, M., Nolan, R. A., and Lien, J. C. (1980). *Coelomomyces skgomyiae* var. *Chapmani* var. nov. with new host records for *Coelomomyces* from mosquitoes of Taiwan. *Can. J. Zool.* **58,** 1836–1844.

Lum, P. T. M. (1963). The infection of *Aedes taeniorhynchus* (Wiedemann) and *Psorophora howardii* Coquillett by the fungus *Coelomomyces. J. Insect Pathol.* **5,** 157–166.

McCrae, A. W. R. (1972). Age-composition of man-biting *Aedes* (*Stegomyia*) *simpsoni* (Theobald) (Diptera: Culicidae) in Bwamba County, Uganda. *J. Med. Entomol.* **9,** 545–550.

Madelin, M. F. (1968). Studies on the infection by *Coelomomyces indicus* of *Anopheles gambiae. J. Elisha Mitchell Sci. Soc.* **84,** 115–124.

Maffi, M., and Genga, R. (1970). Contributo alla conoscenza dell'infestazione da *Coelomomyces* nei Culicidi delle Salomone Britanniche, Oceania. *Parassitologia* (*Rome*) **12,** 171–178.

Maffi, M., and Nolan, R. A. (1977). *Coelomomyces lairdi,* n. sp., a fungal parasite of larvae of the *Anopheles* (*Cellia*) *punctulantus* comples (Diptera: Culicidae) from the highlands of Irian Jaya (Indonesian New Guinea). *J. Med. Entomol.* **14,** 29–32.

Manalang, C. (1930). Coccidiosis in Anopheles mosquitoes. *Philipp. J. Sci.* **42,** 279.

Manier, J. F., Rioux, J. A., Coste, F., and Murand, J. (1970). *Coelomomyces fugetae* n. sp. (Blastocladiales, Coelomomycetaceae) parasite of chironomid larvae (Diptera, chironomidae). *Ann. Parasitol. Hum. Comp.* **45,** 119–128.

Mogi, M., Laird, M., Nolan, R. A., and Kurihara, T. (1976). *Coelomomyces* fungi in rice field mosquito larvae of Nagasaki with notice of a long overlooked record of this genus from Japan. *Trop. Med.* **18,** 71–74.

Morozov, V. A. (1967). The discovery of the parasitic fungus *Coelomomyces* in the larvae of *Aedes* in the vicinity of Krasnodar. *Med. Parazitol. Parazit. Bolezni* **36,** 353.

Muspratt, J. (1946a). On *Coelomomyces* fungi causing high mortality of *Anopheles gambiae* larvae in Rhodesia. *Ann. Trop. Med. Parasitol.* **40,** 10–17.

Muspratt, J. (1946b). Experimental infection of the larvae of *Anopheles gambiae* (Dept., Culicidae) with a *Coelomomyces* fungus. *Nature (London)* **158,** 202.

Muspratt, J. (1962). "Destruction of the Larvae of *Anopheles gambiae* Giles by a *Coelomomyces* Fungus," WHO/EBL/2, WHO/Vector Control/2. World Health Organ., Rome.

Muspratt, J. (1963a). Destruction of the larvae *Anopheles gambiae* Giles by a *Coelomomyces* fungus. *Bull. W. H. O.* **29,** 81–86.

Muspratt, J. (1963b). "Progress Report (May 1963) on Investigations Concerning Three Mosquito Pathogens at Livingstone, Northern Rhodesia," WHO/EBL/12. World Health Organ, Geneva.

Muspratt, J. (1964). "Parasitilogy of Larval Mosquitos, Especially *Culex pipiens fatigans* Wied., at Rangoon, Burma, WHO/EBL/18. World Health Organ., Geneva.

Nam, E. A. (1981). Biology of the entomopathogenic fungus *Coelomomyces iliensis* Dubit. Dzersh. et Daneb. and the possibilities of utilizing it as a biological control measure against bloodsucking mosquitoes. Author's abstract of dissertation for the degree of Candidate of Biological Sciences 1981. Alma Ata Inst. Zool. of the Academy of Sciences of the Kazakh SSR.

Nam, E. A., and Dubitskij, A. M. (1977). "A Review of the Systematic Status of the Fungus *Coelomomyces milkoi* and a Description of a New Genus of the Entomophthoraceous Fungi *Pseudocoelomomyces* (Entomophthorales), VINITI No. 1309–77 Dep., UDK 591. 69–577. I: 576. Akad. Sci. Kazakh. SSR, Inst. Zool. Alma Ata, USSR.

Nolan, R. A. (1978). *Coelomomyces canadense* stat. et. comb. nov. (Blastocladiales: Coelomomycetaceae) from Psechocladius sp. G. (Diptera: Chironomidae). *Can. J. Bot.* **56,** 2303–2306.

Nolan, R. A., and Mogi, M. (1980). *Coelomomyces ponticulus,* new species of fungal parasite of larval *Aedes flavopictus* (Diptera: Culicidae) from Mikura-Jima Island, Japan. *J. Med. Entomol.* **17,** 333–337.

Nolan, R. A., and Taylor, B. (1979). *Coelomomyces couchii* n. sp., a fungal parasite of larval *Anopheles farauti* (Diptera, Culicidae) from Gizo Island, Solomon Islands. *J. Med. Entomol.* **16,** 297–299.

Nolan, R. A., Laird, M., Chapman, H. C., and Glenn, F. E., Jr. (1973). A mosquito parasite from a mosquito predator. *J. Invertebr. Pathol.* **21,** 172–175.

Peterson, A. (1964). "Entomological Techniques; and How to Work with Insects," 10th ed. Edwards, Ann Arbor, Michigan.

Pillai, J. S. (1969). A *Coelomomyces* infection of the *Aedes australis* in New Zealand. *J. Invertebr. Pathol.* **14,** 93–95.

Pillai, J. S. (1971). *Coelomomyces opifexi* (Pillai and Smith) (Coelomomycetaceae: Blastocladiales). I. Its distribution and the ecology of infection pools in New Zealand. *Hydrobiologia* **38,** 425–436.

Pillai, J. S., and Rakai, I. (1970). *Coelomomyces macleayae* (Laird), a parasite of *Aedes polynesiensis* (Marks) in Fiji. *J. Med. Entomol.* **7,** 125–126.

Pillai, J. S., and Smith, J. M. B. (1968). Fungal pathogens of mosquitoes in New Zealand. I. *Coelomomyces opifexi* sp. n., on the mosquito *Opifexi fuscus* Hutton. *J. Invertebr. Pathol.* **11,** 316–320.

Pillai, J. S., Wong, T. L., and Dogshun, T. J. (1976). Copepods are essential hosts for the development of a Coelomomyces parasitizing mosquito larvae. *J. Med. Entomol.* **13,** 49–50.

Rajapaksa, N. (1963). "Records of a Survey of *Coelomomyces* Infections in Mosquito Larvae Carried out in the South-West Coastal Belt of Ceylon," WHO/EBL/8, WHO/Vector Control/31. World Health Organ., Geneva.

Rajapaksa, N. (1964). Survey for *Coelomomyces* infections in mosquito larvae in the south-west coastal belt of Ceylon. *Bull. W. H. O.* **30,** 149–151.

Ramalingam, S. (1966). *Coelomomyces* infections in mosquito larvae in the South Pacific. *Med. J. Malaya* **20,** 334.

Răsín, K. (1928). *Coelomomyces chironomi* n. sp. *Věstn. Sjezdu Ĉesk.* Přírodozp. **3,** 146.

Răsín, K. (1929). *Coelomomyces chironomi* n. sp. houba crzopasici v dutině tělní larev chironoma. *Biol. Spisy Vys. Školy Zvěrolek., Brno* **8,** 13.

Ribeiro, H., Ramos, H. C., and Pires, C. A. (1981). First records of Coelomomyces and mermithids in mosquitoes of Angola, Africa. *Mosq. News* **41,** 381.

Rioux, J. A., and Pech, J. (1960). *Coelomomyces grassei* n. sp. parasite *d'Anopheles gambiae* (note preliminaire). *Acta Trop.* **17,** 179–182.

Rioux, J. A., and Pech, J. (1962). Validation of *Coelomomyces grassei* Rioux and Pech. *J. Elisha Mitchell Sci. Soc.* **78,** 134–135.

Rodhain, F., and Brengues, J. (1974). Présence de champignons du genre Coelomomyces chez des Anopheles en Haute-Volta. *Ann. Parasitol. Hum. Comp.* **49**(2), 241–246.

Rodhain, F., and Fauran, P. (1975). Infection naturelle d'*Aedes* (*S.*) *hebrideus* par un *Coelomomyces* aux Nouvelle-Hébrides. *Ann. Parasitol. Hum. Comp.* **50,** 643–646.

Rodhain, F., and Garyral, P. (1971). Nouveaux cas de parasitisme de larves d'Anopheles par des champiguons du genre *Coelomomyces* en Republique de Haute-Volta. *Ann. Parasitol. Hum. Comp.* **46**(3), 295–300.

Romney, S. V., Boreham, M. M., and Nielsen, L. T. (1971). Intergeneric transmission of *Coelomomyces* infections in the laboratory. *Utah Mosq. Abat. Assoc. Proc.* **24,** 18.

Service, M. W. (1974). Further results of catches of Culicoides (Diptera: Ceratopogonidae) and mosquitoes from suction traps. *J. Med. Entomol.* **11,** 571–579.

Service, W. M. (1976). "Mosquito Ecology: Field Sampling Methods." Wiley, New York.

Shapiro, M., and Roberts, D. W. (1976). Growth of *Coelomomyces psorophorae* mycelium in vitro. *J. Invertebr. Pathol.* **27,** 399.

Shcherbak, V. P., Shacherban, Z. P., Koval, E. Z., and Andreeva, R. V. (1977). Electron microscopic and histological examination of certain entomogenous fungi of the genus *Coelomomyces*. *Dopov. Akad. Nauk Ukr. RSR, Ser. B: Geol., Khim. Biol. Nanki* **8,** 758–762.

Shemanchuk, J. A. (1959). Note on *Coelomomyces psorophorae* Couch, a fungus parasitic on mosquito larvae. *Can. Entomol.* **91,** 743–744.

Taylor, B. W. (1980). *Coelomomyces* infection of the adult female mosquito *Aedes trivittatus* (Coq.) in Manitoba. *Can. J. Zool.* **58,** 1215–1219.

Travland, L. B. (1979). Structures of the motile cells of *Coelomomyces psorophorae* and function of the zygote in encystment on a host. *Can. J. Bot.* **57,** 1021–1035.

Umphlett, C. J. (1961). Comparative studies in the genus *Coelomomyces* Keilin. Ph. D. Dissertation, University of North Carolina, Chapel Hill.

van Thiel, P. H. (1954). Trematode, gregarine and fungus parasites of A *Anopheles* mosquitoes. *J. Parasitol.* **40,** 271–279.

van Thiel, P. H. (1962). Validation of *Coelomomyces ascariformis*. *J. Elisha Mitchell Sci. Soc.* **78,** 135.

Villacarlos, L. T., and Gabriel, B. P. (1974). Some microbial pathogens of 4 species of mosquitoes. *Kalikasan* **3,** 1–12.

Walker, A. J. (1938). Fungal infections of mosquitoes, especially of *Anopheles costalis*. *Ann. Trop. Med. Parasitol.* **32,** 231–244.

Weiser, J. (1976). The intermediary host for the fungus *Coelomomyces chironomi*. *J. Invertebr. Pathol.* **28,** 273–274.

Weiser, J. (1977a). The crustacean intermediary host of the fungus *Coelomomyces chironomi*. *Ceska Mykol.* **31,** 81–90.

Weiser, J. (1977b). "An Atlas of Insect Diseases," 2nd rev. ed. Junk, The Hague.

Weiser, J., and McCauley, V. J. E. (1971). Two *Coelomomyces* infections of Chironomidae (Diptera) larvae in Marion Lake, British Columbia. *Can. J. Zool.* **50,** 365.

Weiser, J., and Undeen, A. H. (1981). Diseases of blackflies. *In* "Blackflies: The Future for Biological Methods In Integrated Control" (M. Laird, ed.), pp. 181–189. Academic Press, London.

Weiser, J., and Vv́ra, J. (1964). Zur Verbrietung der *Coelomomyces*. Pilze in europäischen Insekten. *Z. Tropenmed. Parasitol.* **15,** 38–42.

Whisler, H. C., Shemanchuk, J. A., and Travland, L. B. (1972). Germination of the resistant sporangia of *Coelomomyces psorophorae*. *J. Invertebr. Pathol.* **19,** 139–147.

Whisler, H. C., Zebold, S. L., and Shemanchuk, J. A. (1974). Alternate host for mosquito parasite *Coelomomyces*. *Nature (London)* **251,** 715–716.

Whisler, H. C., Zebold, S. O., and Shemanchuk, J. A. (1975). Life history of *Coelomomyces psorophorae*. *Proc. Natl. Acad. Sci. U.S.A.* **72,** 963–966.

Zebold, S. L., Whisler, H. C., Shemanchuk, J. A., and Travland, L. B. (1979). Host specificity and penetration in the mosquito pathogen *Coelomomyces psorophorae*. *Can. J. Bot.* **57,** 2766–2770.

Zharov, A. A. (1973). Detection of a parasitic fungus *Coelomomyces psorophorae* Couch (Phycomycetes: Blastocladiales) in *Aedes vexans* Meigen mosquitoes in the Astrakhan region. *Med. Parazitol. Parazit. Bolezni* **42,** 485–487.

5 Experimental Systematics

BRIAN A. FEDERICI

Division of Biological Control
Department of Entomology
University of California
Riverside, California

I. INTRODUCTION

A. The Species Concept in *Coelomomyces*

The genus *Coelomomyces* is the largest group of fungi in the order Blastocladiales. Since the original description of the genus by Keilin (1921), more than 60 species have been described, all of which are obligate parasites of aquatic arthropods. Most species have been described from mosquito larvae, within which these fungi produce thousands of resting sporangia prior to larval death. As there are now over 2500 known species of mosquitoes, and the search for species of *Coelomomyces* on a global scale has been rather limited, the genus may eventually be shown to consist of several hundred species.

Taxa in *Coelomomyces* are based on a variety of properties including characteristics of hyphae and sporangia, geographical distribution, habitat, and mosquito host range. The two most important taxonomic characters are sporangial surface structure, because it varies considerably from species to species, and mosquito host range, which in most species of *Coelomomyces* is generally thought to be restricted to one or a few closely related mosquito species. The extensive use of the latter two characters in descriptions of species attests to their taxonomic value. However, where closely related species are concerned, the validity of both of these characters has come into question as a result of recent studies that demonstrated that many species bear strong similarities in sporangial ornamentation, and the mosquito host range, where examined, was broader than realized. For example, in a scanning electron microscope study, Bland and Couch (1973) were able to divide 21 species into eight groups based on similarities in sporangial surface structure, although they noted that these groupings did not necessarily indicate relationships. In some groups, such as the *C. dodgei* complex, sporangial features intergraded from one species to another, whereas in others, in particular the large *C. psorophorae* complex, sporangial surface structure was markedly uniform throughout the group. Anthony *et al.* (1971) also found similarities in the sporangia of species within the *C. dodgei* and *C. psorophorae* groups. With regard to host range, experimental studies of *C. dodgei, C. opifexi, C. psorophorae,* and *C. punctatus* have shown that each of these species can infect and reproduce in mosquitoes other than those from which they were originally isolated (Federici *et al.,* 1975; Federici, 1977a; Federici, 1982; Pillai, 1969; Zebold *et al.,* 1979).

These studies made it apparent that improvements were necessary in the species concept for *Coelomomyces,* particularly for determining the relationship of species thought to be closely related and for establishing the taxonomic status of isolates so similar to known species that they are considered varieties. The discovery of a sexual phase in several species of the genus (Whisler, Chapter 2) provided a mechanism for clarifying the relationships of species and varieties by studying the properties of the sexual stages and progeny produced through hybridization. Domestication of *C. dodgei* and *C. punctatus,* species known to be closely related, provided a model experimental system to study interspecific relationships. The remainder of this chapter will be devoted primarily to a review of previous literature on *C. dodgei* and *C. punctatus,* and recent studies of the gametophytic phase and hybrids of these species. These studies provide substantial evidence that although closely related, *C. dodgei* and *C. punctatus* are distinct and valid species. Furthermore, it is shown that two properties of the gametophytic phase, the time of gametangial dehiscence and viability of hybrid gametophytes are of taxonomic significance for these species and may be of similar value for other species of the genus.

B. General Taxonomy of the *C. dodgei* Complex

Originally, Couch (1945) described *C. dodgei* from infected larvae of *Anopheles crucians, An. punctipennis,* and *An. quadrimaculatus* collected in Georgia. He characterized the sporangial surface as consisting of rounded or elongated pits or, more frequently, long grooves arranged along the length of the sporangium or in variously curved patterns. Later, based on studies of sporangia from additional infected larvae, Couch and Dodge (1947) separated *C. dodgei* into three species: (1) *C. punctatus,* parasitic in *An. quadrimaculatus* with the sporangial surface characterized by minute rounded or elongated pits (Chapter 4, Figs. 483–487); (2) *C. dodgei* emended, with the host range limited to *An. crucians* and the sporangial surface characterized by rounded or elongated pits or narrow grooves, or with pits on one side and grooves on the other (Chapter 4, Figs. 267–273); and (3) *C. lativittatus,* parasitic in *An. crucians* with the sporangial surface exhibiting wide longitudinal bands on one side and irregular or transverse bands on the other, very rarely with a few pits. The banded appearance in *C. lativittatus* is created by the long grooves (Chapter 4, Figs. 373–380).

Although data on the geographic distribution of these species are sparse, the accuracy of the sporangial descriptions is supported by the collection of isolates conforming to each species type from over a broad range of the eastern United States. In addition to the original isolates from Georgia, isolates of *C. punctatus* have been reported from North Carolina (Couch, 1972) and Louisiana (Chapman and Glenn, 1972); isolates of *C. dodgei* have been reported from Louisiana (Chapman and Glenn, 1972) and Ohio (Mead, 1949); and isolates of *C. lativittatus* have been reported from Minnesota (Laird, 1961), Connecticut, Pennsylvania, and Mississippi (Couch and Bland, Chapter 4).

Nevertheless, more recent studies of sporangial surface structure and mosquito host range have provided evidence *C. dodgei* and *C. punctatus* may be varieties of the same species. Using scanning electron microscopy to study sporangial ornamentation in detail, Anthony *et al.* (1971), Bland and Couch (1973), and Federici (1979) found that some sporangia characteristic of *C. dodgei* occurred in larvae infected with *C. punctatus,* and conversely that sporangia characteristic of *C. punctatus* occurred in larvae infected with *C. dodgei.* Bland and Couch (1973) pointed out that there was an intergrading of the predominant surface structure of species similar to *C. dodgei,* ranging from pits only in *C. punctatus,* to pits and furrows in *C. dodgei,* the furrows delimiting bands, to furrows only in *C. lativittatus.* Based on their studies, both Anthony *et al.* (1971) and Bland and Couch (1973) suggested cross-infection studies would have to be undertaken to establish the validity of *C. dodgei* and *C. punctatus.* The status of these species was further confused by the demonstration that *C. dodgei* and *C. punctatus* could be hybridized, and that in some larvae infected with hybrids, most of the sporangia formed were characteristic of the third species of the complex, *C. lativittatus* (Federici, 1979).

Moreover, the establishment of *in vivo* cultures of *C. dodgei* and *C. punctatus* (Couch and Bland, Chapter 9) enabled experimental studies of host range which revealed that each species was capable of infecting a variety of anopheline species, and importantly, that they shared several hosts in common. The known host range of these species based on infected larvae obtained from natural habitats or in laboratory infection trials is shown in Table X. For comparison, the host range of *C. lativitta-*

TABLE X. Known Anopheline Host Range of Species of the *Coelomomyces dodgei* Complex

C. dodgei	*C. lativittatus*	*C. punctatus*
An. crucians[a,d,e]	*An. crucians*[a,e]	*An. crucians*[a,d]
An. freeborni[b,f]	*An. earlei*[a,i]	*An. freeborni*[b,f]
An. quadrimaculatus[b,e,g]	*An. punctipennis*[a,j]	*An. quadrimaculatus*[a,e]
An. punctipennis[a,j]		*An. bradleyi*[a,c]
		An. stephensi[b,h]

[a] Infected larvae collected from natural habitats.
[b] Larvae infected experimentally in the laboratory.
[c] Chapman (1974).
[d] Chapman and Glenn (1972).
[e] Couch and Dodge (1947).
[f] Federici (1977a).
[g] Federici and Chapman (1977).
[h] Federici *et al.* (1975).
[i] Laird (1961).
[j] Mead (1949).

tus is also shown, all records for which were obtained from field-collected larvae.

C. Life History of *C. dodgei* and *C. punctatus*

In his initial revision of the genus, Couch (1945) queried whether species of *Coelomomyces* might have an alternation of hosts, and sporophytic and gametophytic generations. Yet it was not until quite recently that Whisler *et al.* (1974, 1975) demonstrated that the life history of *C. psorophorae* involved an obligate alternation of sporophytic and gametophytic generations between a mosquito and copepod host. Since this important discovery, several other species of *Coelomomyces,* including *C. dodgei* and *C. punctatus,* have been shown to have life cycles of the euallomyces type, with sporophytic and gametophytic generations alternating between a dipteran and crustacean host, respectively (Whisler, Chapter 2).

The life cycles of *C. dodgei* and *C. punctatus,* important to understanding the discussion below, are similar and parallel the life cycle of the free-living blastocladiaceous fungus, *Blastocladiella variabilis* (Harder and Sörgel, 1938). In both *C. dodgei* and *C. punctatus,* the sporophyte produces resting sporangia in the hemocoel of anopheline larvae. Infection normally results in death, after which the larva putrefies, liberating sporangia into the aquatic environment. Meiosis and spore cleavage occur in these sporangia over a period of 48 hr, resulting in the formation of uniflagellate haploid spores referred to as meiospores. After release, these spores seek out and invade a crustacean host. Presently, the only known host for *C. dodgei* and *C. punctatus* is the cyclopoid copepod, *Cyclops vernalis* (Federici and Roberts, 1976; Federici and Chapman, 1977). Within the copepod, the fungus forms a gametophytic thallus that, by branching dichotomously, grows throughout the hemocoel, maturing in 6–12 days. At maturity, the gametophyte in essence becomes a gametangium, and within it uniflagellate gametes are formed during the 16-hr period preceding dehiscence. At a predetermined time, the gametes begin swarming within the gametangium, causing it to rupture, killing the copepod and releasing gametes into its hemocoel. Shortly thereafter, usually within a few minutes, the copepod body ruptures, apparently as a result of enzymatic weakening of the integument at intersegmental membranes, and the gametes escape. Gametes find mates and form biflagellate zygotes, which seek out and invade other anopheline larvae, completing the life cycle.

In the gametophytic phase of *C. dodgei* and *C. punctatus,* gametangia and gametes of opposite mating type can be easily recognized because they are different colors (Federici, 1977). Those designated male are orange due to the presence of β-carotene, whereas those referred to as female are light amber (Federici and Thompson, 1979). As the gametophyte develops within the copepod, the host appears either orange or amber in color, depending on the sex of the meiospore by which it was infected, except in copepods infected by both meiospore types, in which case both colors develop. In addition to allowing infected copepods to be distinguished from uninfected specimens, this pigmentation enables the mating type of the gametophyte developing with the copepod to be easily identified.

II. PERIODICITY AND SYNCHRONIZATION OF GAMETANGIAL DEHISCENCE

A. Formation of Gametangia, Gametogenesis, and Dehiscence

As far as is known, distinct ovoidal gametangia such as those found in species of *Allomyces* and *Blastocladiella* are not formed in *Coelomomyces*. Rather, gametangial development is holocarpic with the entire mass of the gametophyte at maturity cleaving to form gametes. The gametophyte becomes the gametangium, and after dehiscence all that remains is the transparent wall-like sheath that bounded the gametophyte as it developed.

In *C. dodgei* and *C. punctatus,* gametogenesis and most of the synthesis of β-carotene in male structures occurs during the 24-hr period prior to dehiscence. In most cases, gametogenesis appears complete 1–2 hr prior to gamete release, although whether the gametes are actually functional at this time is not known at present. Nevertheless, after formation gametes remain inactive until just prior to a predetermined diurnal dehiscence period of "time window" of 2–3 hr. As this period approaches, the gametes begin to swarm slowly around one another within the gametangium. Typically, movement begins at the edges or tips of gametangia and spreads inwardly. The rate of swarming gradually increases, eventually becoming quite rapid. Then, suddenly, the gametangium ruptures, releasing the gametes into the body cavity of the copepod. The period from the initiation of swarming until the gametangium ruptures usually lasts about 30–45 min. As noted above, rupture of the gametangium kills the copepod and within a few minutes the copepod body itself ruptures, permitting the gametes to escape.

In cases where male and female gametophytes develop within the same copepod, gametogenesis and dehiscence are synchronized between both, with dehiscence occurring at about the same time in each gametangium.

Although gametangial dehiscence in *C. dodgei* and *C. punctatus* is synchronous within each species, the gametes of each species are released during significantly different, nonoverlapping periods of the day (Federici, 1983).

B. Diurnal Periodicity of Gametangial Dehiscence in *C. dodgei*

That the dehiscence of gametangia in *C. dodgei* is synchronous was first indicated by the large number of deaths of patently infected copepods that occurred approximately 9 hr after the onset of the light period in laboratory cultures reared on an 8:16-hr dark:light photoperiod. As will be discussed in more detail below, the most critical event in the determination of the time of dehiscence in *C. dodgei* is not the onset of the light period, but rather the initiation of the dark phase during the final 24-hr period of gametophyte development. Thus, all times for gametangial dehiscence are expressed in terms of hours after the onset of darkness, even though it is now known that in *C. punctatus* the onset of the light period is also critical.

The time and degree of synchrony of gametangial dehiscence in *C. dodgei* was determined by observing infected copepods during the final 24-hr period of gametophyte development, and recording the time of death, which indicates the gametangium had just dehisced, for each copepod.

Studies of infected copepods from laboratory populations reared under an 8:16-hr dark:light photoperiod demonstrated that gametangia always dehisced within a time frame extending from 16 to 19 hr after the onset of darkness (Federici, 1983). In a population where copepods were observed on three consecutive days, dehiscence occurred during the light period, between 16 hr 30 min and 18 hr and 15 min after the initiation of the dark period (Fig. 617). Gametangia that had not cleaved prior to this period did not dehisce until the next day during the same period. In a separate study of 116 infected copepods from three different populations, the mean time of gametangial dehiscence was 17 hr 25 min, with a standard deviation of 30 min, after the onset of darkness (Table XI).

In most of these experiments copepods were infected by exposing them to meiospores several hours after the onset of the light period. To ensure that the high degree of synchrony obtained for gametangial dehiscence was not due to similar rates of gametophyte development, an experiment

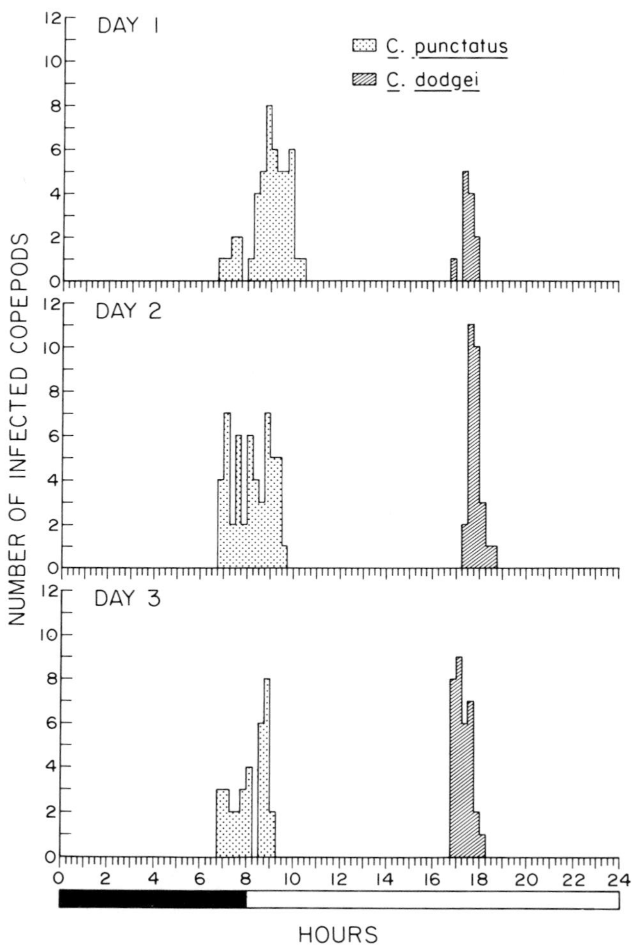

Fig. 617. Diurnal gametangial dehiscence patterns for *C. dodgei* and *C. punctatus* reared on an 8:16-hr dark:light photoperiod. Data are shown for three consecutive days for two separate populations of the copepod host, *Cyclops vernalis,* one infected with *C. dodgei,* the other with *C. punctatus.* Gametangial dehiscence times are expressed as the number of copepods which died as a result of dehiscence in each 15-min interval over the 3-day period.

TABLE XI. Dehiscence Times for Gametangia of *Coelomomyces dodgei* and *C. punctatus* in infected Copepods of *Cyclops vernalis* Reared on an 8 : 16-hr Dark : Light Photoperiod[a]

Species	Orange(♂)			Amber(♀)			Orange(♂) and amber(♀)			Total		
	N	Mean	S	N	Mean	S	N	Mean	S	N	Mean	S
C. punctatus	60	8 : 09	30	17	18 : 07	31	38	8 : 08[b]	26	115	8 : 08	29
C. dodgei	49	17 : 38	26	40	17 : 06	28	27	17 : 28	27	116	17 : 25	30

[a] Times were obtained from three different copepod populations for each species. Values are expressed in terms of hours and minutes after the onset of the dark period. N = number of infected copepods; S = standard deviation.

[b] Gametangia occurring in the same host cleave approximately at the same time.

was conducted in which meiospores were added to one population 5 hr after the onset of the dark period, and to a different population 12 hr later. The mean times of dehiscence for both groups were in the range of 17 hr 10–30 min (Table XII), not significantly different from one another or from the mean dehiscence times for *C. dodgei* obtained in other experiments.

C. Diurnal Periodicity of Gametangial Dehiscence in *C. punctatus*

The determination that gametangial dehiscence in *C. dodgei* followed a diurnal periodicity stimulated studies of gametangial dehiscence in *C. punctatus* to determine whether a diurnal pattern also existed in this closely related species. Using the same techniques described for determination of dehiscence times in *C. dodgei,* it was shown that dehiscence in *C. punctatus* also followed a diurnal periodicity, but, importantly, the period during which dehiscence occurred was significantly different from that of *C. dodgei*. In a population of infected copepods observed for three consecutive days, when reared on an 8:16-hr dark:light photoperiod, dehiscence usually occurred shortly after the onset of the light period, between 6 hr 30 min and 10 hr after the initiation of the dark phase (Fig. 617). The mean time of gametangial dehiscence for *C. punctatus,* determined in a separate study of 115 infected copepods from three different populations, was 8 hr 8 min, with a standard deviation of 29 min, after the onset of the dark period (Table XI).

TABLE XII. Dehiscence Times for Gametangia of *Coelomomyces dodgei* Obtained when Two Separate Populations of Nauplii of *Cyclops vernalis* Were Exposed Once at Two Different Times Separated by 12 Hours

Time of exposure (hours after the onset of the dark period, D)	Replicate	Number of copepods[a]	Time of dehiscence (hours and minutes after the onset of the dark period)[b]
D + 5	1	11	17:30 ± 5
	2	10	17:22 ± 6
	3	21	17:30 ± 5
	Total	42	17:28 ± 3
D + 17	1	10	17:11 ± 5
	2	6	17:02 ± 3
	3	18	17:22 ± 5
	Total	34	17:22 ± 3

[a] For convenience, times for different mating types are not separated here.
[b] Mean ± standard error.

Thus, the dehiscence periods for *C. dodgei* and *C. punctatus* do not overlap, being separated in time by a period of more than 6 hr when gametophytes are reared under an 8:16-hr dark:light regime.

D. Effect of Photoperiod on the Time of Gametangial Dehiscence

Studies have recently been initiated to elucidate the parameters which determine the time of gametangial dehiscence in *C. dodgei* and *C. punctatus*. Although these studies are not complete, they indicate that the photoperiod during the final 24-hr period of gametophyte development is the most critical parameter, but that the actual mechanism by which the time of dehiscence is set differs in these two species.

In *C. dodgei,* experiments demonstrate that it is the onset of the dark phase during the final 24-hr period of gametophyte development that sets the time for dehiscence at approximately 17–19 hr later (Fig. 618). In these experiments the onset and length of the last dark period was varied. In all cases tested, dehiscence occurred between 16–19 hr after the onset of the last dark period. Delaying the onset of the final dark period delayed the time of dehiscence by a period similar to the length of the delay. In contrast to this, the length of the last dark or light period or the onset of the last light period had comparatively little effect on the time of dehiscence. These experiments also demonstrated that the effect of the onset of the dark period could be reversed by insertion of a light period.

In contrast to this, in *C. punctatus* the onset of the final dark period appears to be of relatively little consequence. When cultures were reared in an 8:16-hr dark:light photoperiod for several days and then exposed to continuous light during the final 24-hr period, the gametangia still dehisced approximately 8 hr after the time the dark period would have been initiated, although dehiscence was not as synchronous as normal. Thus it appears that in *C. punctatus* the onset and/or length of the light period during the final 24 hr of gametophyte development determines the time of dehiscence, with the onset and length of the last dark period being comparatively less important.

In summary, the diurnal periodicity of gametangial dehiscence observed in *C. dodgei* and *C. punctatus* is a function of photoperiod, with the final 24-hr period of gametophyte development being the phase during which the time of dehiscence is set. In *C. dodgei,* the onset of the last dark phase sets the time of dehiscence for approximately 17–19 hr later. In *C. punctatus,* the onset of the final dark period appears to have little effect on the time of dehiscence, although it does appear that the gametes are not normally released until it is light. Thus, it is important to note that

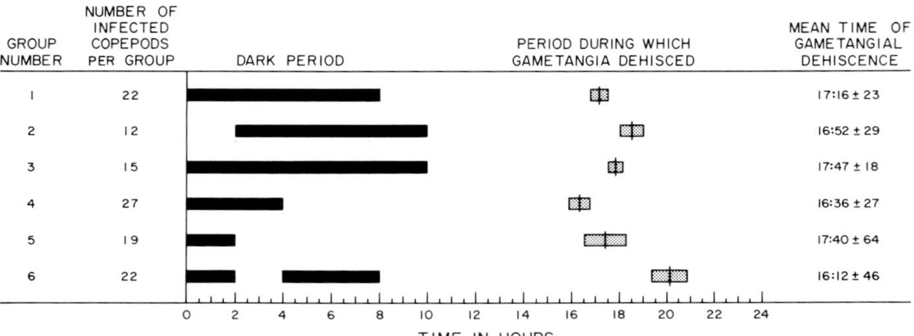

Fig. 618. Effect of the onset and length of the last dark and light periods on gametangial dehiscence in *C. dodgei.* Copepods were infected and reared on an 8:16-hr dark:light photoperiod until the last day of gametophyte development, when the photoperiod was varied for groups 2–6 as indicated in the figure. Group 1 served as the control. Note that in all cases gametangial dehiscence occurred between 16 and 19 hr after the onset of the last dark period, and that delaying the onset of the dark period delayed dehiscence by a period similar to the length of the delay (groups 2 and 6), whereas the length of the light or dark period or onset of the light period had comparatively little effect on the time of dehiscence (groups 3–5). The results obtained with groups 5 and 6 demonstrate that, under the conditions tested, the effect of the onset of darkness is reversible by insertion of a light period. Period of dehiscence indicated as the absolute mean time ± standard deviation of the mean (in minutes) after time zero. Time of dehiscence expressed as the mean time ± standard deviation of the mean (in minutes) relative to the onset of the last dark period.

gametangial dehiscence in both *C. dodgei* and *C. punctatus* occurs during the light phase.

The results of these experiments indicate clearly that the time of gametangial dehiscence is a function of a genetically based, environmentally cued diurnal periodicity. As a specific character that allows *C. dodgei* and *C. punctatus* to be distinguished from one another, the time of gametangial dehiscence is of taxonomic value.

III. TAXONOMIC SIGNIFICANCE OF THE TIME OF GAMETANGIAL DEHISCENCE

The experimental conditions used to determine gametangial dehiscence times in the laboratory are not truly representative of field conditions, in that graded light intensities similar to dawn and dusk periods are not incorporated into the dark:light regimes. Nevertheless, by simulating photoperiod lengths likely to occur under natural conditions in habitats from which species are collected, an indication of the approximate time of dehiscence can be obtained. Even if these values do not accurately reflect dehiscence times in nature, they can still be of taxonomic significance and provide an indication of important biological differences between species, particularly where sympatric species are concerned. For example, by arbitrarily choosing the 8:16-hr dark:light photoperiod, the gametangial dehiscence time for *C. dodgei* can be described as occurring typically 17–19 hr after the onset of darkness, with a mean time of 17 hr 25 min and a standard deviation of 30 min, whereas that for *C. punctatus* is 8–10 hr after the onset of darkness with a mean time of 8 hr 8 min, and a standard deviation of 29 min (Table XI). That the mechanisms controlling the time of dehiscence are different can be noted, but this is not essential to the description as long as the experimental conditions under which the times were determined are carefully defined. Dehiscence times under other or additional photoperiods can also be used, although it is probably more meaningful to use regimes representative of those that occur in habitats of the species under study.

Of course, a character such as the time of gametangial dehiscence would only be necessary for cases where closely related species cannot be separated by more conventional taxonomic methods.

In regard to biological significance, the time of gametangial dehiscence can also indicate potentially important biological information about species. For example, laboratory observations indicate that gametes are not usually motile for more than 20 hr. The synchronous dehiscence of gametangia increases the probability that these relatively short-lived stages will find mates. Additionally, in the case of closely related species that occur sympatrically, dehiscence of each species at different times of the day increases the probability that gametes will encounter mates of the same species. Thus, if two or more species of *Coelomomyces* occur sympatrically and have gametangial dehiscence periods that overlap significantly, it is highly probable they are very distinct species, regardless of similarities in sporangial surface structure or host range.

IV. USE OF EXPERIMENTAL HYBRIDS TO CLARIFY SPECIES RELATIONSHIPS

In organisms with a sexual phase in the life cycle, taxonomic problems can often be resolved through the use of experimental hybrids and study of any progeny produced in the F_1 generation, in subsequent generations, and in back crosses to parental species. Such an approach has been used recently to demonstrate that *C. dodgei* and *C. punctatus* are distinct species rather than varieties of the same species (Federici, 1982). The experimental system employed was relatively simple and will be described briefly here because it is likely that the principles, techniques, and methods of evaluation will be useful in clarifying the relationships of other closely related species and varieties in the genus.

A. Formation of Hybrids

If standard taxonomic characters such as sporangial ornamentation and mosquito host range indicate that one or more varieties or species are closely related, it is probable they will be capable of infecting the same crustacean host, the host from which gametes for hybridization experiments will be obtained. The inability of putatively related species to infect a host known to be susceptible to one of the species is presumptive evidence that the species are not closely related.

In the case of *C. dodgei* and *C. punctatus,* both species infect and reproduce in the copepod *Cyclops vernalis,* with the gametophyte maturing and producing gametes in 6–12 days (Federici, 1979, 1980). For hybridization experiments, infected adult copepods were collected several hours prior to gamete release and separated into groups based on the mating type of the gametangia developing within the hemocoel, amber, orange, or amber and orange. Copepods containing orange or amber gametangia were placed in 1–4 ml of a weak salt solution (mg/l: NaCl, 100; KCl, 4; $CaCl_2$, 6), or in filter-sterilized rearing medium in 35 × 10 mm petri dishes, with 1–6 copepods containing the same type of gametangia per dish. After the gametangia dehisced, reciprocal crosses of *C. dodgei* and *C. punctatus* were made by pooling gametes of opposite mating types from the two species, *C. dodgei* ♀ × *C. punctatus* ♂ and *C. punctatus* ♀ × *C. dodgei* ♂. As controls, intraspecific crosses were made using the same procedures.

Hybrid zygotes were formed in all replicates of both reciprocal crosses, and normal zygotes were formed in all replicates of intraspecific crosses (Federici, 1979, 1982). Zygote formation was initiated as soon as gametes of opposite mating type encountered one another, usually within a few seconds.

In the above experiments, the task of making the hybrids was relatively easy because gametangia and gametes of opposite mating type are differentially pigmented in both species. As noted above, those designated male appear orange due to the presence of β-carotene, whereas female forms, which lack any significant amount of this pigment, appear amber or colorless. However, differential pigmentation of opposite mating types does not occur in all species of *Coelomomyces,* making identification of gamete mating types difficult. For example, no such pigmentation has been reported for *C. psorophorae* or *C. opifexi*. Nevertheless, interspecific mixtures of gametes of opposite mating types can still be made to test the ability of species to hybridize by applying the laws of probability. By setting up a series of infected crustaceans, one per dish, opposite mating types within a species with isogametes can be determined by making a series of crosses using small numbers of gametes—i.e., several thousand of the millions derived from each host—for each cross. The formation of zygotes in any mixture will designate dishes that contain gametes of opposite mating type. Once opposite mating types are determined for two different species, the same procedure can be used for making interspecific gamete mixtures. If no zygotes are formed after the gametes from one dish of one species are pooled in separate mixtures with gametes from 10 hosts of a different species, the odds are $(\frac{1}{2})^{10}$, or 1 in 2048, that the lack of zygote formation is due to mixing gametes of the same mating type. Thus, it can be reasonably concluded that the two test species are incapable of forming hybrid zygotes. If zygotes are formed in any cross, they can be tested for infectivity.

B. Infection of Larvae

In the crosses of *C. dodgei* and *C. punctatus*, larvae of *An. freeborni* or *An. quadrimaculatus* were infected by placing first-instars directly in the petri dishes with the zygotes within 30 min of zygote formation. From 10 to 50 larvae were used, depending on how many gametes had been used for the cross. The larvae were left in the dish with the zygotes for 24 hr and then transferred to a standard larval rearing medium and reared normally.

C. Properties of *C. dodgei–C. punctatus* hybrids

1. Sporophytic Generation

The hybrid zygotes formed in both reciprocal crosses of *C. dodgei* and *C. punctatus* were infective for mosquito larvae and produced sporophytes similar to those of the parental species with regard to growth and production of sporangia, time course of the disease, and gross pathology. The sporophytes proliferated throughout the larval hemocoel over a period of approximately 10 days, yielding as many as 60,000 sporangia in late fourth-instar larvae.

The hybrid resting sporangia were similar in size and shape to those of the parental species, although significant differences in sporangial surface structure were observed (Federici, 1979, 1982). All the hybrid sporangia in any one larva tended to be similar in surface structure, although the surface structure varied considerably from one larva to another. In most larvae, the sporangia bore characteristics of both parental types with the surface structure consisting of a mixture of pits and furrows (Fig. 619). Because *C. dodgei* is described as consisting primarily of pits and furrows, often with pits on one side of the sporangium, furrows on the other, most hybrid sporangia resembled this species. However, larvae were encountered in which the majority of the sporangia were typical of either *C. punctatus* or *C. lativittatus*. Significantly, sporangia representative of both the wide-banded and narrow-banded variants of *C. lativittatus* were formed by the hybrids (Fig. 619, k and o).

With regard to function, hybrid sporangia appeared to undergo normal meiospore cleavage and dehiscence. After being flooded with water, 50–80% of the sporangia dehisced in 48 hr, a rate and amount typical for meiospore cleavage and release in both *C. dodgei* and *C. punctatus*.

2. Gametophytic Generation

Hybrid *C. dodgei–C. punctatus* meiospores appeared as viable as those of the parental species. When combined with groups of 48- to 72-hr-old copepod nauplii, the meiospores were capable of finding and encysting upon these hosts preferentially on intersegmental membranes.

Until this point in the life cycle, no significant biological differences were apparent between the parental species and the hybrids. However, although the meiospores were capable of encysting upon the intersegmental membranes, no mature gametophytes were ever produced in these copepods. The *C. dodgei–C. punctatus* hybrid gametophyte, therefore, is inviable. The comparative viability of different stages of the gametophytic and sporophytic generations of the parental species and hybrids is summarized in Table XIII.

The reasons for the inviability of the hybrid gametophyte are not known at present, although it may be due to a disruption or lack of coordination of gametophytic developmental or regulatory functions. Some of these functions differ between the parental species, such as the timing of gametangial dehiscence, and are obviously complex. In contrast to the hybrid zygote and sporophyte, which are diploid and contain a full haploid complement of genes from each parental species, the gametophyte is haploid with only a single allele present for each gene. Improper pairing of chromosomes at meiosis may result in uneven or random chromosomal segregation, leaving each meiospore with an incomplete genetic complement, or genetic recombination at meiosis may result in a haploid recombinant genome incapable of expressing essential gametophytic functions.

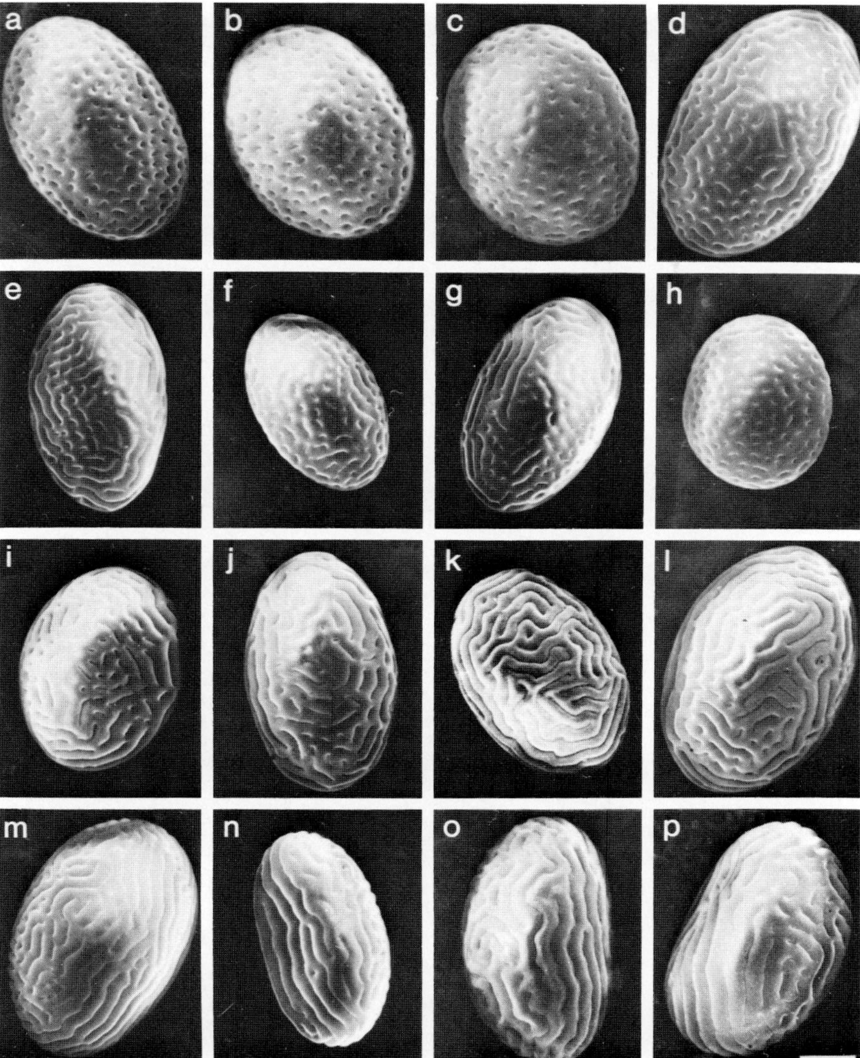

Fig. 619. Scanning electron micrographs illustrating the surface structure of sporangia of *C. punctatus, C. dodgei,* and hybrids of these species produced in reciprocal crosses of gametes obtained from infected copepods. (a–c) Sporangia typical of *C. punctatus.* (d) Variant sporangium of *C. punctatus* exhibiting numerous elongated pits, a trait more characteristic of *C. dodgei.* (e–g) Sporangia typical of *C. dodgei.* (h) Variant sporangium of *C. dodgei,* the surface of which is primarily pitted, a trait characteristic of *C. punctatus.* (i–l) Sporangia formed by hybrids made by crossing amber (♀) gametes of *C. punctatus* with orange (♂) gametes of *C. dodgei.* (m–p) Sporangia formed by hybrids made by crossing amber (♀) gametes of *C. dodgei* with orange (♂) gametes of *C. punctatus.* Note that most of the hybrid sporangia are not typical of those of either parental species, and furthermore, that some are similar to the narrow-banded (k) and wide-banded (o) varieties described for *C. lativittatus.* Bar = 10 μm.

TABLE XIII. Comparative Viability of *Coelomomyces dodgei* and *Coelomomyces punctatus* and F_1 Experimental Hybrids of These Species[a]

	Sporophytic generation				Gametophytic generation			
		Mosquito				copepod		
Organism	Zygote formation	Zygote encystment	Sporophyte development	Sporangia formation	Meiosis and dehiscence	Meiospore encystment	Gametophyte development	Gametangial dehiscence
Parental species	+	+	+	+	+	+	+	+
Hybrids	+	+	+	+	+	+	−	−

[a] Viable (+), nonviable (−).

D. Taxonomic Significance of Hybrid Inviability

The incomplete viability of the F_1 generation of *C. dodgei–C. punctatus* hybrids demonstrates that these species are reproductively isolated and, therefore, most probably are valid species. This finding is relevant for the genus as a whole because the use of classical taxonomic techniques to study species of the *C. dodgei* complex, known to be closely related, did not clearly resolve the status of these species (Couch and Dodge, 1947; Anthony *et al.*, 1971; Bland and Couch, 1973; Federici, 1979). It is likely that this technique can be used to examine the status of other species in this and other groups in the genus. For example, although the validity of *C. lativittatus* as a species is uncertain presently, once established in *in vivo* culture, experiments similar to those just described should clarify its taxonomic status. If meiospores obtained from resting sporangia are incapable of forming a viable gametophyte, then it is probable *C. lativittatus* is a naturally occurring hybrid of *C. dodgei* and *C. punctatus*. However, if a viable gametophytic generation can be formed, with no breakdown in subsequent generations, *C. lativittatus* can be considered a valid species.

The occurrence of *C. lativittatus* over a broad geographic range (Couch and Bland, Chapter 4) would appear to indicate it is a valid species. If shown to be so, it will be interesting to determine whether its time of gametangial dehiscence is significantly different from those of *C. dodgei* and *C. punctatus,* whether it will hybridize with either of these species, and, if hybrids are formed, whether they are viable.

Hybridization experiments may find their greatest value in clarifying relationships of the *C. psorophorae* complex, a group of over 19 varieties with overlapping host ranges and marked similarities in sporangial surface structure. The taxonomic status of these varieties is uncertain. However, if hybrids of these can be made, the viability of the hybrids will provide an indication of the degree of reproductive isolation among the varieties. Varieties that can not be hybridized can be elevated to species status.

V. BIOLOGICAL SIGNIFICANCE OF SPECIES-SPECIFIC DIURNAL PERIODICITY OF GAMETANGIAL DEHISCENCE

The demonstration that hybrids of *C. dodgei* and *C. punctatus* are inviable provides an explanation for the existence of isolating mechanisms that prevent or decrease the probability of their formation. Hybrids are a "dead end" and, therefore, of benefit to neither species. That a temporal diurnal isolating mechanism has evolved among *C. dodgei* and *C. punctatus* is probably the result of constraints imposed on these fungi by the brevity of the sexual phase and their reproduction in the same or congeneric sympatric hosts that occur in the same habitats during the same seasons, thereby apparently precluding isolation through host specificity or a seasonal temporal mechanism.

It would appear that the most critical factor in the evolution of the diurnal nature of the temporal isolation is the relatively short life, typically less than 20 hr, of the gamete. Gametes of both species are released during the light phase, are phototactic, and form zygotes immediately upon encountering a mate, whether it is of the same species or not. Thus, under natural conditions gametes released during any day will tend to congregate near the water surface, forming zygotes quickly upon finding mates. The synchronous release of gametes of the same species, rapid mating, and release of gametes of different species during significantly different nonoverlapping periods of the day increase the probability of the formation of intraspecific rather than interspecific zygotes.

VI. SUMMARY AND CONCLUSIONS

Studies of *C. dodgei* and *C. punctatus,* closely related species whose taxonomic status could not be clearly resolved using classical methods, demonstrate that two developmental properties of the gametophytic generation, the time of gametangial dehiscence and the viability of F_1 hybrid gametophytes, are of taxonomic significance for these and possibly other species of the genus. When both species are reared under the same photoperiod, the mean times of gametangial dehiscence for each species differ significantly, regardless of the length of the photoperiod tested. Additionally, although gametes of these species are capable of forming infective hybrid zygotes, the subsequent F_1 gametophyte is inviable. These studies demonstrate that *C. dodgei* and *C. punctatus* are reproductively isolated and thus can be considered distinct and valid species.

It is suggested that the time of gametangial dehiscence and viability of hybrid gametophytes are properties that may be useful in clarifying the taxonomic status of uncertain species and varieties of *Coelomomyces*.

ACKNOWLEDGMENTS

The research discussed in this paper was supported in part by U. S. Public Health Service grant AI 12772 and a grant from the Biological

Control of Vectors Component of the UNDP/World Bank/WHO Special Program for Research and Training in Tropical Diseases.

REFERENCES

Anthony, D. W., Chapman, H. C., and Hazard, E. I. (1971). Scanning electron microscopy of the sporangia of species of *Coelomomyces* (Blastocladiales Coelomomycetaceae). *J. Invertebr. Pathol.* **17,** 395–403.

Bland, C. E., and Couch, J. N. (1973). Scanning electron microscopy of sporangia of *Coelomomyces. Can. J. Bot.* **51,** 1325–1330.

Chapman, H. C. (1974). Biological control of mosquito larvae. *Annu. Rev. Entomol.* **19,** 33–59.

Chapman, H. C., and Glenn, Jr., F. E. (1972). Incidence of the fungus *Coelomomyces punctatus* and *Coelomomyces dodgei* in larval populations of the mosquito *Anopheles crucians* in two Louisiana ponds. *J. Invertebr. Pathol.* **19,** 256–261.

Couch, J. H. (1945). Revision of the genus *Coelomomyces,* parasitic in insect larvae. *J. Elisha Mitchell Sci. Soc.* **61,** 124–136.

Couch, J. N. (1972). Mass production of *Coelomomyces,* a fungus that kills mosquitoes. *Proc. Natl. Acad. Sci. U.S.A.* **69,** 2043–2047.

Couch, J. N., and Dodge, H. R. (1947). Further observations on *Coelomomyces,* parasitic on mosquito larvae. *J. Elisha Mitchell Sci. Soc.* **63,** 69–79.

Federici, B. A. (1977a). Laboratory infection of *Anopheles freeborni* with the parasitic fungi *Coelomomyces dodgei* and *Coelomomyces punctatus. Proc. Pap. Annu. Conf. Calif. Mosq. Vector Control Assoc.* **45,** 107–108.

Federici, B. A. (1977b). Differential pigmentation in the sexual phase of *Coelomomyces. Nature (London)* **267,** 514–515.

Federici, B. A. (1979). Experimental hybridization of *Coelomomyces dodgei* and *Coelomomyces punctatus. Proc. Natl. Acad. Sci. U.S.A.* **76,** 4425–4428.

Federici, B. A. (1980). Production of the mosquito-parasitic fungus, *Coelomomyces dodgei,* through synchronized infection and growth of the intermediate copepod host, *Cyclops vernalis. Entomophaga* **25,** 209–216.

Federici, B. A. (1982). Inviability of interspecific hybrids in the *Coelomomyces dodgei* complex. *Mycologia* **74,** 55–562.

Federici, B. A. (1983). Species-specific gating of gametangial dehiscence as a temporal reproductive isolating mechanism in *Coelomomyces. Proc. Natl. Acad. Sci. U.S.A.* **80,** 604–607.

Federici, B. A., and Chapman, H. C. (1977). *Coelomomyces dodgei* establishment of an *in vivo* laboratory culture. *J. Invertebr. Pathol.* **30,** 288–297.

Federici, B. A., and Roberts, D. W. (1976). Experimental laboratory infection of mosquito larvae with fungi of the genus *Coelomomyces.* II. Experiments with *Coelomomyces punctatus* in *Anopheles quadrimaculatus. J. Invertebr. Pathol.* **27,** 333–341.

Federici, B. A., and Thompson, S. N. (1979). β-Carotene in the Gametophytic phase of *Coelomomyces dodgei. Exp. Mycol.* **3,** 281–284.

Federici, B. A., Smedley, G., and Van Leuken, W. (1975). Mosquito host range tests with *Coelomomyces punctatus. Ann. Entomol. Soc. Am.* **68,** 669–670.

Harder, R., and Sörgel, G. (1938). Uber einen neuen planoisogamen Phycomyceten mit Generations-Wechsel Und Seine Phylogenetische Bedeutung. *Nachr. Ges. Wiss. Goettingen* **3,** 119–127.

Keilin, D. (1921). On a new type of fungus: *Coelomomyces stegomyiae*, n.g., n.sp., parasitic in the body-cavity of the larva of *Stegomyia scutellaris* Walker. (Diptera Nematocera, Culicidae). *Parasitology* **13,** 225–234.

Laird, M. (1961). New American locality records for four species of *Coelomomyces* (Blastocladiales, Coelomomycetaceae). *J. Insect Pathol.* **3,** 249–253.

Mead, F. W. (1949). Ecology of central Ohio mosquitoes. M.S. Thesis, Ohio State University, Columbus.

Pillai, J. S. (1969). A *Coelomomyces* infection of *Aedes australis* in New Zealand. *J. Invertebr. Pathol.* **14,** 93–95.

Whisler, H. C., Zebold, S. L., and Shemanchuk, J. A. (1974). Alternate host for the mosquito parasite *Coelomomyces*. *Nature* (*London*) **251,** 715–716.

Whisler, H. C., Zebold, S. L., and Shemanchuk, J. A. (1975). Life History of *Coelomomyces psorophorae*. *Proc. Natl. Acad. Sci. U.S.A.* **72,** 693–696.

Zebold, S. L., Whisler, H. C., Shemanchuk, J. A., and Travland, L. B. (1979). Host specificity and penetration in the mosquito pathogen *Coelomomyces psorophorae*. *Can. J. Bot.* **57,** 2766–2770.

6 Physiology and Biochemistry

RICHARD A. NOLAN

Department of Biology
Memorial University of Newfoundland
St. John's, Newfoundland, Canada

I. INTRODUCTION

The purpose of the present chapter is severalfold. First, the author wanted to bring together information gathered either in the laboratory or in the field in studies with representatives of the genus *Coelomomyces* that had physiological or biochemical implications. Second, it was hoped that to some degree information from these two sources could be compared. Third, information from the large body of literature dealing with other members of the order Blastocladiales could be used as a background against which to view current knowledge and future needs in research with *Coelomomyces*.

II. PHYSICAL FACTORS

A. pH

Laird (1959) found that successful infection of *Aedes aegypti* larvae occurred in pans containing sediment from the original larval containers, water adjusted to pH 6.6, and dried sporangia of *Coelomomyces stegomyiae* from *Aedes albopictus*. The first study in which an attempt was made to establish some physicochemical parameters in pools in which successful *Coelomomyces* infections of mosquito larvae were occurring was by Muspratt (1962). Muspratt found that larvae of *Anopheles gambiae* were naturally infected in pools in a sandy clay soil where the range for water pH was mostly between 8.0 and 8.5, with 9.2 being the highest value recorded. Madelin (1964, 1968) carried out laboratory infection studies with *A. gambiae* in which a layer of a local clay soil was generally used in the various containers. This soil adjusted the pH of added distilled or rainwater to near 8, thereby providing hydrogen ion conditions, that approximated those found in the field (Muspratt, 1962). Umphlett (1968) examined the prevalence of *Coelomomyces punctatus* infections in *Anopheles quadrimaculatus* in two regions (coves A and B) of a North Carolina lake and found that for cove A (larval infections recorded) the pH ranged from 6.8 to 8.4 (most readings taken were near the lower end of the range), and for cove B (no larval infections recorded) the pH ranged from values approaching 10 to a low just under 8. Umphlett (1968), however, commented that laboratory studies on *C. punctatus* indicated an abundant release of zoospores over the range of pH 3 to 9. Pillai (1971) reported the range in pH values for supralittoral pools in which *Opifex fusca* and *Aedes australis* were infected by *Coelomomyces opifexi* and found for 20 pools (54 samples over a 3 year period) that the values ranged from 6.0 to 8.5 (mean 7.3). The corresponding values for uninfected pools (133 samples) were from 6.3 to 8.8 (mean 7.2). Roberts *et al.* (1973), in a study of the effects of a wide variety of chemicals on the dehiscence of nonsterile resistant sporangia of *Coelomomyces psorophorae* from *Aedes taeniorhynchus,* found that the optimal initial pH for dehiscence in periods of 1 week or less was in the range of 8–9 for Tris, the most active compound tested. However, for glycine (intermediate level of activity) and sarcosine (level of activity not given), the levels of dehiscence after 14 days were approximately twice as high at pH 4.5 as those obtained at pH

9. Nam (1975) studied the conditions that favored germination of sporangia of *Coelomomyces iliensis* from larvae of *Culex modestus* using buffered solutions adjusted to pH 7, 8, 9, and 10. He found that maximum percentages of germination for both thin-walled (80%) and thick-walled (44–45%) sporangia occurred at pH 9 if previous storage had been in 50% glycerin. Whisler *et al.* (1975) carried out successful infection trials of *Culiseta inornata* using *Coelomomyces psorophorae* in a complex admixture that had a pH of 8.2. Federici and Roberts (1976) were able to obtain successful infections of *Anopheles quadrimaculatus* using *Coelomomyces punctatus* within the broad pH range of 4.5 to 10.

Cantino and Horenstein (1959), in studies with *Blastocladiella emersonii,* found that the development of ordinary colorless (OC) plants was approximately the same over the pH range of 6–8 in the dark, whereas in the light the stimulatory effect of light increased with increasing pH. Horenstein and Cantino (1961), in studies with *Blastocladiella britannica,* found that optimum growth occurred over a broad pH plateau of 6–8.

In studies with *B. emersonii,* Cantino (1969) found that the processes leading up to the release of motile cells from resistant sporangia (RS) (cracking of the RS wall and germination—the subsequent processes leading up to motile cell release) responded differently to pH and did not have the same pH optimum. The optimum pH for cracking was near 7.5, whereas the optimum pH for germination was near 6.0. Evidence from studies with *B. britannica* also indicates a two-step process leading up to zoospore release from RS (Cantino, 1970).

Soll and Sonneborn (1969) found that a pH range of approximately 6.0–7.5 was conducive to a reasonably synchronous development of *B. emersonii* zoospores, whereas, pH values above and below this range led to asynchronous zoospore development.

Nolan (1970a) found that the optimum pH range for growth of *Catenaria anguillulae* was 8.2–8.7.

Thus, the generally high optimum pH values for successful infection trials and sporangial germination in *Coelomomyces* are not extreme when viewed in the context of other studies within the order. These other studies also emphasize the need for a more detailed understanding of the role of pH in the germination of both thick- and thin-walled sporangia and in the growth and morphogenesis of both the gametophytic and sporophytic thalli.

B. Temperature

In studies with *Coelomomyces chironomi,* Rašín (1929) determined that approximately 25–30 days at 15–18°C were required for development from an early mycelial stage through sporangium formation. Rašín (1929) studied the effects of freezing sporangia on their viability and found that several sporangia germinated after subsequent thawing at 25°C for 24 hr and an additional incubation for 4 days at room temperature.

Umphlett (1961) and Lum (1963) obtained germination of *Coelomomyces psorophorae* sporangia by initially storing them on moist filter paper for approximately 3–14 days or more at 15°C and then by incubating them at 25–28°C for 3 days followed by flooding of the sporangia which germinated readily within 15–30 min.

In studies with *Coelomomyces punctatus,* Couch (1968, 1972) found that sporangial germination did not occur either below 10°C or above 35°C; Couch adopted a procedure in which sporangia were stored at 10°C on damp filter paper and found that 23–27°C was the optimum range for sporangial germination.

These findings for optimum sporangium germination and sporangial storage procedures were confirmed and adopted or slightly modified by subsequent workers using other species of *Coelomomyces*. Whisler *et al.* (1972, 1975) used 20 ± 2°C for sporangial germination in *Coelomomyces psorophorae* and stored them at a slightly lower temperature (5°C).

Roberts *et al.* (1973) found that the dehiscence level of resistant sporangia of *C. psorophorae* dropped from 23% to 3% in 1.0–1.5 months when stored in water at 10°C, and they used 26°C in laboratory studies.

Pillai and Woo (1973) used either 26 or 28°C in their studies with *Coelomomyces opifexi*. Federici and Roberts (1976) used 23–26°C in studies with *Coelomomyces punctatus*. Federici and Chapman (1977) used 27 ± 2°C for experimental studies with *Coelomomyces dodgei* and 23°C for obtaining sporangial dehiscence. In later studies, Federici (1980) used 26 ± 2°C for *C. dodgei*.

Very little has been done to correlate temperatures in the field with temperatures used in the laboratory. Muspratt (1962) noted that larvae of *Anopheles gambiae* become infected in pools in which the maximum recorded daytime temperature might be in excess of 37.8°C. Madelin (1964, 1968) carried out infection studies using *Coelomomyces indicus* and *Anopheles gambiae* in basins in which daytime water temperature was 28–32°C and night temperature was approximately 25°C. Umphlett (1968) monitored both air and water temperatures at University Lake, near Chapel Hill, North Carolina, and found that infestations of *Anopheles quadrimaculatus* by *Coelomomyces punctatus* were lowest (two collections) following periods when the air temperature was at 38°C. Umphlett (1968) speculated that this might be due to inhibition of germination of resistant sporangia of *C. punctatus* at this temperature, as previously noted by Couch (1968). Pillai (1971) in studies with *C. opifexi* found that the "infected" and "uninfected" pools examined had similar seasonal temperatures. Chapman and Glenn (1972) analyzed data from two Louisiana ponds to see if there was a relationship between the monthly average air temperature and incidence of parasitism by *Coelomomyces dodgei* and *C. punctatus* in *Anopheles crucians*. They divided the monthly temperatures into three categories—(1) <15.6°C, (2) 15.6–26.1°C, and (3) >26.1°C—and they found that for one pond the infection level in the lower temperature range was 6–7 times the level at the highest range.

Thus, except for the pioneering studies by Rašín (1929), most of our knowledge on the effects of temperature regimes and, therefore, the commonly used laboratory storage and testing conditions have evolved from the studies of Couch (1968, 1972).

Hatch (1936) found that gametangium and sporangium formation in *Allomyces arbuscula* could be triggered by temperature shifts of 1.5°C (up or down) provided the temperature change lasted for an undetermined period of between 5–10 min and 1 hr.

Jones (1946) found that a relatively long duration of a low temperature (21.7°C) tended to favor the production of *Allomyces arbuscula*-resistant sporangia and that a relatively long duration of a high culture temperature (26.1°C) tended to limit RS production. Thus, aggregate temperature was an important factor in RS production with an inverse relationship between temperature and RS production.

Machlis and Ossia (1953a) found that RS of *A. arbuscula* matured more rapidly at 25°C than at 21°C as based upon the rate of chromosphere digestion, with RS produced at 28°C behaving in an erratic manner.

Horenstein and Cantino (1961) found that the optimum temperature range for growth of *Blastocladiella britannica* was 16–25°C; however, RS plant formation was favored at 22 and 25°C. Nolan (1970a) found that the optimum temperature for growth of *C. anguillulae* was 25°C.

Youatt *et al.* (1971) found that the optimum temperature for sporangium formation by *A. macrogynus* was near 32°C in standing suspensions. A far better understanding of the role of temperature in all aspects of the biology of *Coelomomyces* is required.

C. Light

Muspratt (1946a) felt that, in the laboratory, a light stimulus played a role in the late stages of germination of sporangia with zoospore liberation beginning after about 15 min of stimulation from a microscope lamp. The possible importance of light was recognized by Madelin (1964, 1968), who used a combination of a 40-watt tubular "darklight" (maximum radiation between 3200 and 3900 Å) and two 250-watt clear-bulb infrared lamps on a

12-hr photoperiod in an attempt to approach conditions in natural breeding pools of *Anopheles gambiae* in Northern Rhodesia in tests conducted in England and found that the darklight allowed vigorous and normal development of the larvae, but enhanced fungal infection was not proven. Thus, the use of photoperiod in infection studies was established early.

Martin (1970) studied zoospore development in resting sporangia of *Coelomomyces punctatus* incubated under a photoperiod, noted positive phototaxis of the zoospores, and took advantage of this phenomenon to obtain massed zoospores. Robertson (1972) reported that the zoospores of *Allomyces reticulatus* were positively phototactic and that the action spectrum extended from 420 to 610 nm with a broad peak from 470 to 525 nm. Robertson (1972) could not correlate this finding with the presence of any previously unreported cellular structure. Martin (1970) also discovered the presence of a paracrystalline body in the zoospores, and Whisler *et al.* (1972) verified its presence in zoospores of *Coelomomyces psorophorae*. The paracrystalline body found in the zoospores of *C. punctatus* (Martin, 1971), *C. indicus* (Madelin and Beckett, 1972), and *C. psorophorae* (Travland, 1979b) but not in the gametes or zygotes of *C. psorophorae* (Travland, 1979) may be viral in nature, as it resembles the variously packed crystalline arrays formed by viruses in some fungi (Lemke, 1979). An RNA virus that can form paracrystalline aggregates has been found in both sporophytic and gametophytic hyphae of *Allomyces arbuscula* (strain Bali 1) (Khandjian *et al.*, 1974; Roos *et al.*, 1976), with the viruses being incorporated into the ribosomal nuclear cap of the mitospores, gametes, and encysted zygotes.

The use of photoperiods in infection studies was also employed by Pillai and Woo (1973), who used fluorescent light sources supplemented by infrared lamps in studies with *Coelomomyces opifexi* and *Aedes australis*. A 16-hr photoperiod was used by Federici and Roberts (1975) in laboratory infection studies with *C. psorophorae* and by Federici and Chapman (1977) in infection studies using *Coelomomyces dodgei* and *Anopheles quadrimaculatus*. However, the induction of sporangial dehiscence was carried out under conditions of natural lighting; Federici (1980) carried out all rearing and experiments under a 16-hr photoperiod. He also found that *C. dodgei* normally killed the copepods (*Cyclops vernalis*) and released the male and female gametes 8–9 hr after initiation of the light period. Whisler *et al.* (1975) used continuous illumination (cool-white fluorescent light at 699.7 lux) in preparing *C. psorophorae* resistant sporangia for zoospore release, whereas infection trials were conducted in diffuse light.

The gametophytic thalli produced by *C. dodgei* in *Cyclops vernalis* are either amber or orange-colored (Federici and Chapman, 1977; Federici,

1977) and are converted into gametangia that release either amber or bright orange gametes, respectively. An orange pigment has also been reported in gametophytic thalli of *C. punctatus* (Federici, 1977). The orange pigment occurring in *C. dodgei* has been identified as β-carotene (Federici and Thompson, 1979), in contrast to the situation in *Blastocladiella emersonii* (Cantino and Hyatt, 1953) and in representatives of the genus *Allomyces* (Emerson and Fox, 1940; Turian and Cantino, 1959), in which γ-carotene predominates and only trace levels, if any, of β-carotene are found.

The carotenoid pigment(s) that occur in the male gametes (and hence zygotes) of at least one representative of *Coelomomyces* may play a role in phototactic responses. Turian and Haxo (1954) have found that reduced levels of carotene do not appear to impair the sexual function of male gametes in *Allomyces javanicus;* however, these gametes do have a reduced rate of motility.

The dark pigment melanin, which is considered to serve a protective function against the harmful effects of ultraviolet radiation, is presumably present in the walls of the thick-walled resistant sporangia of most, if not all, species. However, the presence of melanin has not been definitely established in *Coelomomyces* as it has been in resistant sporangia of *Allomyces neomoniliformis* (Skucas, 1967) and in resistant sporangia of *A. arbuscula* × *A. macrogynus* hybrids (Emerson and Fox, 1940).

Couch (1968) found that sporangial germination in *C. punctatus* occurred equally well in alternating daylight and darkness, and Madelin and Beckett (1972) have found that dehiscence of the thin-walled sporangia of *C. indicus* was unaffected by illumination.

A series of studies (for example, Cantino and Horenstein, 1956) documented CO_2 fixation in the dark by *Blastocladiella emersonii* with enhanced CO_2 fixation and growth in the light, especially during the early stages of growth. The presence of bicarbonate also decreased the rate of lactic acid production in the light, and the studies indicated a relationship between the Krebs cycle and morphogenesis (Cantino and Horenstein, 1956). Cantino and Horenstein (1959) found that ordinary colorless (OC) plants of *Blastocladiella emersonii* exposed for their full generation time to various wavelengths of light were most effectively stimulated by light in the blue portion of the spectrum (400–500 nm) and that OC plants exposed to various light intensities (white fluorescent light) were stimulated in a roughly linear fashion in the range of 0–100 foot-candles (fc) with no further stimulation or inhibition occurring from 100 to 400 fc. Cantino and Horenstein (1959) also investigated the effect of light duration (390 fc, white fluorescent light) and found a linear increase in the stimulation of development for continuous exposures of up to 22 hr (the maximum exposure time tested and also the full generation time at 22°C). Horenstein and Cantino (1961) found that, in the case of *B. britannica*, light inhibited the formation of RS plants, with 27 hr being the maximum exposure period tolerated (total growth period was 50 hr), and bicarbonate either delayed RS formation (4×10^{-3} *M*) or completely inhibited (6×10^{-3} *M* to 2×10^{-2} *M*) growth. Light is also known to affect glucose uptake in *B. britannica* (Horenstein and Cantino, 1964), and in *B. emersonii* light causes a sharp decrease in total glucose-6-phosphate dehydrogenase activity per cell during the last stages of growth, when the light-induced formation of soluble polysaccharide per cell was greatest (Goldstein and Cantino, 1962).

The role of light, both qualitatively and quantitatively, in the diverse structures and functions that essentially comprise the total biology of *Coelomomyces* requires intense laboratory investigation in order to provide information that will allow well-designed biological control programs.

Of special interest would be studies on (1) the possible role of light in that portion of the spectrum from 420 to 525 nm, (2) the function of β-carotene in the gametophytic thalli, gametes and zygotes, and (3) the chemical nature and role of the paracrystalline body.

D. Gases

Rašín (1929) succeeded in germinating sporangia of *C. chironomi* suspended in distilled water in a chamber sealed with lanolin. Couch (1968) found that zoospore discharge in *C. punctatus* sporangia in the "go" condition could be achieved after an additional 15 min of incubation if a cover glass was added. Discharge was achieved in a similar time if the sporangia were instead transferred to a chamber containing sodium pyrogallate; however, the reduced oxygen tension only hastened zoospore release and was not required for it. Umphlett (1968) measured the dissolved oxygen levels in University Lake (Chapel Hill, North Carolina) from late June through early October, a period including most of the summer season for *Anopheles quadrimaculatus* larvae, a host for *C. punctatus*. Larvae (presumably uninfected) were collected as early as June 7. For one site (cove B), Umphlett (1968) postulated that the high levels of oxygen inhibited zoospore release.

Martin (1970) observed that when zoospores of *C. punctatus* were released in water droplets on slides, the zoospores concentrated at the periphery of these droplets; when the zoospores were transferred to a tube and centrifuged and then allowed to remain undisturbed for about 45 min postcentrifugation, the zoospores aggregated in an area just below the meniscus. Both observations indicate the aerobic nature of *C. punctatus* zoospores. Madelin and Beckett (1972) noted that when thin-walled sporangia of *C. indicus* dehisced in water under a cover glass, only those very near the cover glass edge were successful, indicating that this process required free oxygen; this contrasts with the results obtained by Couch (1968) with thick-walled sporangia of *C. punctatus* and by Madelin and Beckett (1972) with thick-walled sporangia of *C. indicus*.

Thus, based on these initial studies, the thick-walled sporangium and the thin-walled sporangium appear to have divergent metabolic requirements, which may relate to the inherent abilities of resistant sporangia to remain viable for long periods of time during which they would probably settle into the depths of ponds where oxygen levels would be expected to be low. The eventual release of zoospores that are both positively phototactic and highly aerobic would again bring the infective units into the upper levels of a body of water.

Roberts *et al.* (1973) exposed resistant sporangia of *C. psorophorae* in 10 ml of liquid in 60-mm petri dishes (approximately 300 sporangia per dish) to nitrogen- and oxygen-saturated atmospheres for 3 and 2 days, respectively, and found, for sporangia incubated in water, that the dehiscence level was 1% in air, 2% in nitrogen, and 0% in oxygen; for sporangia incubated in 5 m*M* Tris (pH 8.9), the dehiscence level was 41% in air, 42% in nitrogen, and 59% in oxygen. A major problem in experimentation with *C. psorophorae* is the difficulty in obtaining sporangia that are free of microbial contamination as a result of the pitted nature of the spore wall or the presence of the membrane within which the sporangium is formed (Nolan and Freake, 1974); indeed, Roberts *et al.* (1973) noted that microbial growth (bacterial species not determined) was present in all treatments in which activation occurred and frequently in treatments in which activation did not occur, and that the situation was not resolved by the addition of antibiotics. Thus, the actual oxygen status could have varied from regime to regime in this and other studies. In studies with *C. psorophorae*-resistant sporangia, Whisler *et al.* (1975) utilized a presumed reduction in oxygen tension in order to initiate zoospore release. Castillo and Roberts (1980) studied the effects of carbon dioxide on *Coelomomyces punctatus* and found that increased levels of CO_2 (up to 10%) neither enhanced mycelial growth nor induced sporangium formation; however, in the absence of information on the status of the oxygen levels in these regimes, it is difficult to arrive at any conclusions.

In light of previous physiological studies with insect pathogens that occur in the hemocoel (for example, Dunphy and Nolan, 1981a), it would seem to be reasonable to anticipate a degree of adaptation of both the gametophytic and sporophytic thalli to conditions of reduced oxygen tension and enhanced CO_2 levels.

E. Salinity

Few experimental studies have been carried out with *Coelomomyces* representatives that occur in brackish water. Pillai (1971) investigated the physical parameters of pools in which larvae of *Aedes australis* were infected by *Coelomomyces opifexi* and found that over a 3-year period the salinity values for infected pools ranged from 0.6 to 20.8‰, and the values for uninfected pools ranged from 8.0 to 42.5‰. The values were subject to wide fluctuation; however, the overall trend indicated that infection occurred in lower salinity pools. This has been reinforced by laboratory studies that indicate that a low salinity value (10.0‰) is required for experimental infection of *A. australis* by *C. opifexi* to be initiated but not maintained (Pillai, 1971). Most of the infected pools are in the range of 3–6‰ (Pillai and Woo, 1973), and during laboratory infection studies salinity values of 4–8‰ were used. Pillai and O'Loughlin (1972) examined the effects of salinity on both the level of sporangial germination and the degree of zoospore activity. At temperatures of 5, 10, 23, and 28°C, the highest level of sporangium germination (75–100%) occurred in distilled water, tap water, brackish pond water, and seawater diluted to salinities of 2.1‰ and 4.2‰. On the other hand, salinities of 8.5‰, 17.0‰, and 35‰ yielded decreasing germination levels; the highest level, 25–50%, was achieved only at 8.5‰ and no germination occurred at 35‰. Maximum zoospore motility was achieved upon release at both 23 and 28°C in distilled water, tap water, brackish pond water (4.2‰), and salinities of 2.1‰, 4.2‰, and 8.5‰. Of those salinities tested, 17‰ or less yielded both sporangial germination and zoospore activity.

Pillai *et al.* (1976) used an artificial pool with a salinity of 3.5‰ for infection studies using the copepod *Tigriopus* sp. near *angulatus, Aedes australis,* and *C. opifexi* and reported that the optimum salinity range was then considered to be between 3 and 4‰. Only one other investigator (Walker, 1938) has dealt with the effects of salinity, in this case with *Coelomomyces africanus* from *Anopheles costalis,* by making dissections into either tap water or normal saline (0.85% NaCl), and no details were given.

Further studies are required in order to determine if, for relevant species, the role of salinity is (1) osmotic in nature, (2) related to a sodium requirement (with possible replacement by other monovalent cations), (3) linked to uptake/carrier systems for either other ions [see Belsky *et al.*

(1970) for studies on the role of sodium in phosphate uptake] or compounds, and (4) any combination of the above.

F. Desiccation

Rašín (1929) succeeded in germinating sporangia of *C. chironomi* that, while in larvae, had frozen in water and then later dried up. Muspratt (1946a) found that when there was intermittent rainfall, the first stages that he could detect as being young fungal thalli in *A. gambiae* occurred in the hemocoel about a week after the pool basins had been filled by rain—provided that this had been preceded by a dry spell and that the rain consisted of one or more heavy showers. Muspratt (1946a) found that thin-walled sporangia of his "types a and c" germinated if they were left with the larval mosquito remains in the water from the breeding site and that zoospores were liberated within 3–6 days. This was in contrast to the use of or requirement for desiccation by thick-walled sporangia of his "types a and c" (Muspratt 1946a, b), with the requirement for desiccation being more evident in his "type a." Umphlett (1961) found that *C. dodgei* resting sporangia from *Anopheles crucians* did not require drying before germination and, in addition, were not killed by desiccation in contrast to previous speculation by Couch (1945). Couch (1945), however, also pointed out that much variation in relation to drying requirements could be expected in *Coelomomyces* RS. Umphlett (1961) found that resting sporangia of *C. psorophorae* from *Aedes taeniorhynchus* were not viable following drying. Couch (1968), however, reported that it was only by using repeated drying and wetting that a succession of germinating RS of *C. psorophorae* from *Culiseta inornata* could be obtained. Pillai and O'Loughlin (1972) felt that the drying up of a pool could lead to the elimination of *Coelomomyces*, whereas if the soil remained moist the sporangia could survive in intact larvae.

Desiccation per se is not required in *Allomyces*, as Emerson (1941) and Hatch (1944) observed the germination of RS *in situ* on sporophytes maintained in water cultures. Hatch and Jones (1944) found that desiccation of resting sporangia influenced the subsequent development of RS zoospores of *Allomyces arbuscula*. Desiccation does not appear to play a role in *Allomyces* in reducing the time required for maturation, as RS produced in nutrient broth initially mature at least 6 days sooner than those produced on agar (a reduction in the length of the maturation period by approximately 70%). A further reduction of the period required for RS maturation to 48–72 hr was obtained by growing the fungus in a defined medium followed by incubation in a dilute salts solution (Machlis and

Ossia, 1953a). Desiccation may play an important role in RS wall conditioning and subsequent germination; however, as Machlis and Ossia (1953a) found, osmotic desiccation (2.5–9.5 atm) not only decreased germination levels but also interfered with normal maturation.

Previous studies on the desiccation of resting sporangia of *Coelomomyces* have been lacking in several respects: (1) there has not been an effort to provide a constant and/or reproducible rate of desiccation, (2) the final extent of desiccation has not been determined, (3) the effects of air versus chemical desiccation have not been determined, and (4) there is a critical need to establish biochemical parameters for use in determining stages of either sporangium activation or dormancy. The studies by Machlis and Ossia (1953a,b) can serve as a useful initial guide; however, until the synchronous production of resistant sporangia in *Coelomomyces* has been achieved under defined conditions, the results will be open to question (see Cantino, 1969, 1970).

G. Osmotic Pressure

Castillo and Roberts (1980) reported that the mycelium of *Coelomomyces punctatus* from *Anopheles quadrimaculatus* tolerated osmotic pressures in the range of 240–440 mOs/kg in different media and in addition could survive exposure to osmotic pressure up to 700 mOs/kg without distortion of hyphal morphology. This study was conducted with stationary cultures and used a wide variety of media to achieve the test osmotic pressures; therefore, the results can serve as general indicators at best, because the wide variety of ionic species that contributed to these osmotic pressures would have their own effects individually and in combination. Additional study is needed in which an osmotic agent that is not metabolized during the period of exposure is used at various concentrations and in which exposure to the various concentrations (osmotic pressures) is followed by exposure to conditions under which subsequent growth and morphogenesis can be assessed (see Dunphy and Nolan, 1979).

III. CHEMICAL FACTORS

A. Auxins

In studies with RS of *Coelomomyces psorphorae* from larvae of *Culiseta inornata,* Nolan (1973) found that low concentrations of indole-3-butyric acid (IBA, 10 μg/ml) stimulated germination, especially if sporangia were incubated with the plant hormone for an initial period (17 days) followed by incubation in dilute salts. The effect was less dramatic and the optimum concentration lower (1 μg/ml) when incubation occurred for the full time in the presence of the test chemical. Roberts *et al.* (1973), in studies with RS of *Coelomomyces psorophorae* from larvae of *Aedes taeniorhynchus*, did not detect any stimulation by IBA (0.01 *M*); however, the controls also had a very low level (0–10%) of dehiscence when incubated continuously in the test solutions.

Machlis and Ossia (1953b) found with *Allomyces arbuscula* (strain Ceylon 1) that the incorporation of low levels of indoleacetic acid (IAA, 1.0 and 2.0 ppm) in the solidified growth medium decreased the time required for meiosporangia to reach maturity and also produced a higher level of sporangium germination. When *A. arbuscula* (strain Burma 1Db) is grown in broth containing 2 ppm IAA, the sporangia are produced sooner (3–4 days); the sporangia also attain a markedly greater level of germination. When *A. arbuscula* (strain Burma 1Db) was grown in broth containing either IAA, IBA, tryptophan or indole, indole failed to induce a higher level of meiosporangium germination. When strain Burma 1Db is grown on a defined medium (tryptophan-free), both the medium after growth and the fungal mycelium contain a substance(s) active in the *Avena* coleoptile curvature test for auxins. Thus, at least one other representative of the order Blastocladiales has been shown to produce an auxinlike substance and to be influenced in its development by auxins and an auxin precursor.

Turian (1957, 1962) found that IAA when supplied externally (or possibly produced internally) induced phosphatase activity in zygotes of *A. macrogynus*.

When studies with *Coelomomyces* have become sufficiently advanced and the chemical nature of RS maturation better understood, then a search for naturally occurring growth hormones should be undertaken and the role of auxins reassessed using synchronously produced RS populations.

B. Amines and Amino Acids

Roberts *et al.* (1973) found that the dehiscence of resistant sporangia of *Coelomomyces psorophorae* from *Aedes taeniorhynchus* was stimulated by compounds with an amino group and either a carboxyl or a methanol group attached to the alpha carbon. The most active compound was Tris, and the response to Tris differed considerably in accordance with the level of bacterial contamination of the sporangia tested: the more contaminated the sporangia, the higher the level of dehiscence. Thus, rigid conclusions are difficult, as the addition of either bacteria-free homogenates of healthy fourth-instar larvae of *Aedes taeniorhynchus* or heat-treated dissected healthy larvae to sporangia induced, after 2 weeks of incubation, levels of dehiscence that were very close, if not equivalent, to those obtained in the same incubation period with Tris. In addition to Tris, a second group of compounds: methionine, glycine, glutamine, diaminopropionic acid, citrulline, L-2-amino-4-ureidobutyric acid, L-2-amino-6-ureidohexanoic acid and Tricine yielded dehiscence levels in the 10–50% range, with Tris producing dehiscence levels in excess of 50% (the controls were presumably in the 0–10% range). It would appear that either bacterial metabolic conversion of the test chemicals or normal metabolic products of larval cells are involved in this study. A variety of amino acids and amines were not active under the conditions tested. One of the active amino acids, glycine, has been reported to stimulate hatching of floodwater mosquito eggs (Clements, 1963). Madelin (1964) did not observe any stimulation of germination of *C. indicus*-resistant sporangia from *Anopheles gambiae* by L-cysteine and L-glutamic acid [0.001, 0.01, and 0.1% (w/v)].

Machlis (1957a,b) found in studies with *Allomyces macrogynus* that, when compounds related to the ornithine–urea cycle (L-glutamic acid, L-aspartic acid, L-arginine, L-proline, L-citrulline, and L-ornithine) were added to a defined medium, (1) growth was greatly stimulated by lag-phase reduction, (2) mannose, which is normally an unsatisfactory carbon and energy source, was used as readily as glucose, and (3) the inhibition of growth by thioctic acid was prevented to varying degrees (see Machlis, 1957c).

Nolan (1970a) in studies with *Catenaria anguillulae* also found that L-aspartic acid, L-glutamic acid, L-arginine, DL-citrulline, and L-ornithine were utilized as sole nitrogen sources for growth. However, L-asparagine was the best single source. When nitrogen was supplied in the form of vitamin-free casamino acids, only lysine and arginine were utilized as early nitrogen sources.

Youatt *et al.* (1971) found that the chief factor controlling differentiation to produce sporangia in *Allomyces* was the amino acid status of the environment and that by eliminating amino acids from an incubation medium into which 16-24-hr-old thalli were transferred, zoosporangia were formed and zoospores released in 3–4 hr at 32°C. The presence of the amino acids L-proline, L-arginine, L-aspartic acid, L-cysteine, L-serine, and L-glutamic acid, individually, did not delay sporangium formation, whereas the individual amino acids L-valine, L-histidine, L-isoleucine, L-methionine, L-phenylalanine, L-threonine, and L-tryptophan delayed sporangium development. Youatt (1973) reported that, in attempts to obtain growth of *Allomyces* on defined media equivalent to that obtained on a glucose/yeast extract or a glucose/vitamin-free casein hydrolysate/thiamine medium, the amino acids L-methionine and L-histidine proved adequate as nitrogen and sulfur sources for *A. macrogynus;* however, L-aspartic acid was added because it permitted the formation of reproductive structures and did not increase oxygen consumption (Youatt *et al.*, 1971). *Allomyces arbuscula* also grew well on the three-amino-acid medium; however, in the case of two hybrids (one predominantly male, one predominantly female), normal growth was obtained with the three amino acids L-proline, L-lysine and L-arginine. Youatt (1973) also found that when the molar ratio of glucose to total added amino acids was 1 : 1 only mitosporangia were formed on the sporophytes and only male gametangia were formed on gametophytes of *A. macrogynus*. When the proportion of amino acids was decreased, there was a shift towards the production of meiosporangia and female gametangia on the appropriate thalli. In contrast to the findings of Machlis and Ossia (1953b), Youatt (1973) found that low levels of tryptophan either had no noticeable effect on or partially inhibited meiosporangium maturation.

The chemotactic response of the mitospores, meiospores, and zygotes of *A. macrogynus* and *A. arbuscula* to casein hydrolysate (Carlile and Machlis, 1965) was caused by the combined action of L-leucine and L-lysine, and the activity of these amino acids was synergistic and specific for the L forms (Machlis, 1969a,b). L-Proline enhanced the response to a mixture of leucine and lysine.

The leucine–lysine immobilization technique of Dill and Fuller (1971) for *A. neomoniliformis* zoospores induced subsequent synchronous development in germlings and offers the possibility of a correlation of biochemical with morphological events (Olson and Fuller, 1971), which could be exploited to advantage in future studies with *Coelomomyces*. This could, for example, be used to produce synchronous infections of mosquitoes or the appropriate copepod/ostracod host.

C. Carbohydrates

Madelin (1964) tested the effects of glucose, fructose, sucrose and fructose 1,6-diphosphate [at 0.001, 0.01, and 0.1% (w/v)] on RS of *C. indicus* from *Anopheles gambiae,* and Roberts *et al.* (1973) tested the effects of sugar alcohols (0.01 *M*) on resistant sporangia of *Coelomomyces psorophorae* from *Aedes taeniorhynchus;* in both cases, stimulation of dehiscence was not noted. In studies with *Coelomomyces psorophorae* from *Aedes trivittatus,* Taylor *et al.* (1980) found that RS did not develop in males and females fed only carbohydrates, but the infection level was up to 56% in blood-fed females.

The importance of media simplification in *Coelomomyces* studies is emphasized by the following studies. Sistrom and Machlis (1955) discovered that in *Allomyces* the presence of a readily utilized sugar such as glucose in low levels (0.05%), added intentionally or the presence of an unidentified compound(s) as a carryover in the zoospores or mycelial fragments used as the inoculum (despite attempts to remove it through starvation), could drastically affect mannose and fructose lag. However, the presence of low levels of glucose (0.05%) did not enhance utilization of sucrose, lactose, galactose, arabinose, xylose, sorbose, raffinose, rhamnose, dulcitol, sorbitol, and mannitol, which had previously been found to be unsuitable as carbon and energy sources for *Allomyces* (Machlis, 1953c; Ingraham and Emerson, 1954).

Machlis (1957b) found that low levels of yeast extract (10 ppm) and casein hydrolysate (10 ppm) enhanced L-mannose but not L-fructose utilization by *A. macrogynus*.

Olson (1973) found that high concentrations of glucose and mannitol were effective in blocking mitosis in the early stages of the sequence in *Allomyces neomoniliformis*.

D. Other Organic Chemicals and Complex Media

Madelin (1964) tested the effects of a wide variety of organic chemicals (e.g., organic acids, such as succinic acid and fumaric acid) and complex components (e.g., 0.001% and 0.1% beef extract and yeast extract) on resistant sporangia of *C. indicus* from *Anopheles gambiae*. Roberts *et al.* (1973) tested the effects of various alcohols, fatty acids, purines and pyrimidines, and related compounds on the dehiscence of resistant sporangia of *Coelomomyces psorophorae* from *Aedes taeniorhynchus*. In both instances these compounds at the concentrations tested did not enhance germination.

Shapiro and Roberts (1976) attempted to culture *C. dodgei* from *Anopheles crucians, C. pentangulatus* from *Culex erraticus,* and *C. psorophorae* from *Psorophora howardii, Culiseta inornata, Aedes taeniorhynchus,* and *A. sollicitans* in regular and supplemented tissue culture and mycoplasma media with and without added mosquito cell lines. Mycelia of *C. psorophorae* from *A. sollicitans* grew in mycoplasma broth supplemented with 25% Grace's medium with 1% yeastolate, 1% lactalbumin hydrolysate, 0.005% myristic acid, and 0.005% palmitic acid added. The culture was maintained and subcultured for 2 years as single, free-floating hyphal fragments, and only partial differentiation into sporangia was noted. Rapid initial growth, which ceased after 34–43 hr, was observed for *C. psorophorae* from *P. howardii* when incubated in Mitsuhashi and Maramorosch medium supplemented with 20% fetal calf serum. Dehiscence of sporangia of *C. psorophorae* from two hosts was noted; however, the zoospores soon rounded up, withdrew their flagella, and failed to develop further. When mycelia of *C. psorophorae* from *A. taeniorhynchus* were incubated in a chamber that separated the fungal cells from cells of *Aedes aegypti* by means of a glass ring and a 0.4-μm Nucleopore filter, abnormal sporangia were produced. They concluded that the mycoplasma broth medium appeared to be the most promising as a basal minimal medium.

Castillo and Roberts (1980) tested the ability of over 50 combinations of vertebrate and invertebrate tissue culture media and microbiological media to support growth of mycelia of *Coelomomyces punctatus* from *Anopheles quadrimaculatus*. The culture conditions used were (1) incubation of mycelia in liquid media, (2) incubation of mycelia in liquid media over an agar base, (3) incubation of mycelia in media with growing mosquito cells, (4) incubation of mycelia in a liquid medium but separated from the insect cells by a 0.22-μm filter, and (5) injection of hyphagens into the hemocoel of larvae of *Musca domestica, Prodenia eridania, Estigmene acrea, Galleria mellonella,* and *Toxorynchites brevipalpis*. The most promising media used were (1) the Mitsuhashi–Maramorosch medium (MM) supplemented with 20% heat-inactivated fetal calf serum and conditioned for 3 weeks by growth of Varma's *Anopheles stephensi* cells, and (2) a modified brain–heart infusion medium {BHM; 2% BHI, + 0.2% glucose, + 0.1% neopeptone, + 0.1% yeastolate, + 20% low-speed supernatant *Anopheles quadrimaculatus* pupal homogenate [2 g fresh weight per 10 ml Grace's (1962) medium], and 15% fetal calf serum}.

Because of the overwhelming complexity of the media which gave the most promising results, especially BHM, it is now (as it was for the investigators at that time) difficult to draw any definitive conclusions. For example, conditioning the MM medium for more than or less than 3 weeks did not stimulate growth of the fungus; also, growth was not observed in media conditioned with cells of either *Anopheles gambiae* or *Aedes aegypti*. However, no attempt at a comparative chemical analysis of the successful and unsuccessful conditioned media is indicated. Castillo and Roberts (1980) found that hyphal bodies incubated in the conditioned MM with various supplements grew to about 20–50 μm in length. A rate of growth for small hyphal fragments comparable to that observed in host larvae was found in one culture using conditioned, supplemented MM to which the tripeptide glycyl-L-histidyl-L-lysine acetate had been added. In this latter medium, small hyphal fragments (100–200 μm long) developed within 24 hr into branched mycelia bearing sporangia in various stages of development; however, this culture contained at least one motile, rod-shaped bacterial contaminant. Hyphae in uncontaminated conditioned, supplemented MM containing the tripeptide did not have a similar rapid rate of growth. Castillo and Roberts (1980) felt that this indicated that the bacteria might have produced a growth-stimulating factor; however, it is also possible that the initial bacterial growth may have either utilized one or more component(s) of this highly complex medium, which then favored more efficient utilization of one or more medium components (as, for example in the case of reduced competition for amino acid carrier systems), or reduced oxygen tension to more acceptable levels. These workers did not isolate the bacterium (or bacteria) and attempt to identify it and its metabolic products when grown on this medium. Fungal growth in the conditioned, supplemented MM medium containing the tripeptide stopped after 24 hr in the absence of the contamination, and resting sporangium development proceeded for roughly 60 hr. However, the sporangia did not produce zoospores.

The BHM medium supported either the elongation of lateral hyphal branches and subsequent sporangium production or the production of sporangia in the main mycelium. These sporangia also did not produce zoospores. The BHM medium was subsequently modified by deleting the individual complex components, and the results indicated that brain–heart infusion was essential to *Coelomomyces* growth and differentiation, with the other components being less important.

Fungal mycelia dissected from the copepod *Cyclops vernalis* remained alive 3–5 days in the media tested; however, no growth or development was observed. One exception occurred when several mycelia differentiated into gametes, beginning with cytoplasmic cleavage 30 min after host dissection and continuing through initial gamete release 30 min later, with observation of free gametes and gamete release continuing for 10 hr in the BHM medium. The activity of the gametes released into BHM was noted to be sluggish compared to previously observed behavior in water or within the copepod hemocoel. The gametes remained active for more than 10 hr in BHM, after which they retracted their flagella and rounded up.

With the exception of acetate, compounds related to the Krebs cycle–citrate, malate, succinate, fumarate, α-ketoglutarate, isocitrate, *cis*-aconitate, and pyruvate—were ineffective in reducing the lag phase of *A. macrogynus* with mannose (Machlis, 1957b). These compounds, with the exception of acetate, cannot function as sole sources of carbon for *A. macrogynus* (Machlis, 1953c). Cantino and Horenstein (1959) found that equimolar amounts of succinate and glyoxylate could substitute for light in stimulatory effects on the development of *B. emersonii*.

Detailed studies of nutritional requirements have been carried out for representatives of only two genera of the order Blastocladiales: *Allomyces arbuscula* (strain Burma 1Db) (Ingraham and Emerson, 1954), *A. macrogynus* (strain Burma 1Da) (Machlis, 1953a,b,c), and later extended outside the subgenus *Euallomyces* with *A. moniliformis* (strain Mexico 46) and *A. cystogenus* (strain Burma 1B) (Nolan, 1969), and *Catenaria anguillulae* (Nolan, 1970a,b,c).

These studies with saprophytic (*Allomyces* spp.) and facultatively parasitic (*Catenaria anguillulae*) representatives indicate that members of the group can be characterized by (1) a requirement for a reduced sulfur source, with methionine readily utilized by both genera and cystathionine by *Catenaria*; (2) the ability to utilize many individual organic nitrogen sources but a variable ability to utilize inorganic nitrogen sources [*Allomyces* species can utilize inorganic nitrogen as a sole nitrogen source, whereas *Catenaria* and *Blastocladiella emersonii* (Barner and Cantino, 1952) cannot utilize inorganic nitrogen as a sole nitrogen source]; and (3) a variable requirement for external sources of vitamins. *Allomyces* requires thiamine, whereas *Catenaria* lacks a requirement for any external vitamin source, one isolate of *Blastocladia pringsheimii* requires thiamine, biotin, and nicotinamide (Cantino, 1948), and another requires thiamine, nicotinamide, and *p*-aminobenzoic acid (Crasemann, 1957). *Blastocladiella emersonii* requires thiamine (Barner and Cantino, 1952), and *Blastocladia ramosa* requires thiamine and nicotinamide (Crasemann, 1957).

There is little indication that previous attempts to culture *Coelomomyces* have taken advantage of the wealth of information available from studies with other representatives of the order Blastocladiales. Instead, a "shotgun" approach has been adopted using extremely complex media and yielding results that are difficult, if not impossible, to evaluate. Media should be devised that provide a diversity of components without resorting to complex, ill-defined mixtures that almost defy complete analysis.

E. Ions

Madelin (1964, 1968) determined the concentration of some ions in basins containing a local (British) clay and Mopane clay from Africa in which successful infections involving *C. indicus* and *Anopheles gambiae* had been established. The results indicated Ca^{2+} 45 ppm, K^+ 25 ppm, and Na^+ 123 ppm. Roberts *et al.* (1973) tested the effects of several inorganic salts (providing the ions Cl^- + Na^+, K^+, Ca^{2+}, Cu^{2+}, Zn^{2+}, Mn^{2+}, or Mg^{2+}; SO_4^{2-} + Ca^{2+}, Fe, Mg^{2+}, Co, Cu, or Zn^{2+}; NCO_3^- + Na^+ or NH_4^+) at concentrations of 0.2 and 20 μg/ml on the level of dehiscence of *C. psorophorae* sporangia from *Aedes taeniorhynchus*; no enhancement of dehiscence was noted. Nolan and Freake (1974) used X-ray analysis of thin sections obtained by the use of low-temperature and anhydrous techniques to determine the elemental composition of sporangial walls of *C. psorophorae* from *Culiseta inornata*. Their data indicated, by percentage (w/w): Al 0.23, Ca 0.12, Fe 0.52, K 1.27, Mg 0.03, Na 1.86, P 2.27, and Si 0.09. Manganese and titanium were not detected. The procedures used were very sensitive and could detect the presence of the membrane within which the sporangia were formed by detecting the K and Na in this membrane. Of the elements detected in the sporangium wall, potassium, sodium, and phosphorus were present above the 1% (w/w) level.

Zebold *et al.* (1979) carried out infection trials using first- or second-instar larvae of a total of 12 mosquito species. These larvae were placed in a salt solution [containing 0.25 g $NaHCO_3$, 0.125 g $MgSO_4 \cdot 7H_2O$, 0.05 g KCl, 0.25 g $Ca(NO_3)_2 \cdot 4H_2O$] to which *Coelomomyces psorophorae* zygotes were added; because larvae were successfully infected in this solution, it can be concluded that none of the components, at the concentrations used, inhibit zygote activity. The rationale for the use of this particular salt solution is, however, not clear.

Suberkropp and Cantino (1972) have found that calcium ions appear to enhance encystment with increasing zoospore population densities above 5×10^6 spores/ml in *B. emersonii.*

Olson (1973) found that Na^+, K^+, Ca^{2+}, and Mg^{2+} ions, expecially at low concentrations, were effective in helping to synchronize mitotic divisions in *Allomyces neomoniliformis* with $CaCl_2$ ($10^{-3}M$).

Soll and Sonneborn (1969, 1972), in studies with *B. emersonii,* found that KCl (5×10^{-2} M) was the key component in a germination solution in which good synchrony of zoospore development was indicated and that KI, KBr, NaCl, CsCl, RbCl, and choline chloride (all at 5×10^{-2} *M*) were equally effective in enhancing synchrony.

The above studies, coupled with those discussed in Section III,B, provide tools with which to explore the possibility of obtaining synchronous infections and development in *Coelomomyces*.

F. Host Hemocyte and Fungal Cell Receptor Sites

Martin (1969, 1970) was the first investigator to examine the ultrastructure of the hyphagen (= hyphal body) and mycelium of a representative of *Coelomomyces*. He studied *C. punctatus* from *Anopheles quadrimaculatus* and found that an early spherical infection stage, the hyphagen (5.4–20.3 μm diameter), gave rise to the mycelium by the elongation of growing points on the hyphagen surface. He found that the hyphagen was surrounded by a single plasma membrane convoluted into many microvilli (0.09–0.33 μm diameter by 0.11– 0.32 μm in length) and postulated not only that the microvilli increased the surface area available for absorption but that it was also possible that they assisted in attachment to mosquito tissues. Although Martin (1969, 1970) observed hemocytes, Powell (1976) in a study of *C. punctatus* from *A. quadrimaculatus* found that the mycelial stage was surrounded by a granular and fibrillar extracellular coat that reacted positively with the periodic acid–silver methenamine procedure for the localization of polysaccharides and that there was no indication of a recognition of this structure by host hemocytes as being nonself. However, Powell (1976) noted the presence of unattached vesicles embedded in granular material, particularly during the latter stages of infection when sporangia were already present, that were apparently engulfed by the mosquito hemocytes. These vesicles appear to be derived from cytoplasmic protuberances; however, once the vesicles are released they are separated from the hyphae by a mass of granular material. The vacuoles of host hemocytes were observed to contain material that resembled the granular material and the vesicles (Powell, 1976). Powell (1976) discussed three features of the cell surface of *C. punctatus* that might enable the fungus to resist mosquito hemocyte and humoral defense mechanisms: (1) the hyphal coat may possess chemical features that are naturally produced or acquired from the host hemolymph, (2) the vesicles may serve to confuse the hemocytes by offering alternate surfaces with which to react, and (3) there are both smooth and irregularly surfaced thalli in the same larva, and the irregularly surfaced thalli may very efficiently absorb nutrients from the hemolymph and as a result leave the cellular defenses of the host weakened.

In studies with *C. psorophorae,* Travland (1979b) found that the gamete, zygote, and zoospore possessed vesicles containing an undetermined substance which was believed to assist in adhesion to hosts for the latter two cell types. In the case of the zygote, these vesicles disappeared within 5 min of cyst initiation. Pseudopodia-like structures were formed by the zygote at points of contact with the mosquito cuticle, and Travland (1977) speculated that a contact response involving specific areas of the host cuticle was involved.

Travland (1979a) and Zebold *et al.* (1979) have studied at the ultrastructure level the physical events that occur during infection of both hosts by *C. psorophorae*. In both cases an appressorium and a narrow penetration tube are formed. The presence of a cuticular collar around the base of the penetration tube has been correlated with successful cuticular penetration in the mosquito, and this process is also accompanied by sporadic melanization (Travland, 1979a). Travland (1979a) speculated that the cuticular collar was a host response to the fungal invasion, with the evidence with regard to melanization being less clear.

As suggested by Travland (1979a), much more work is needed on the *Coelomomyces* infection process for the mosquito and, especially, for the copepod. Information on the enzymatic nature of this process is totally lacking. Also of very great interest is the sequence of events (both physical and chemical) that accompanies the injection of the fungal protoplast into the mosquito epidermal cell (Travland, 1979a). This may, in fact, be a critical stage during which parasite modification occurs, allowing it to enter the hemocoel disguised as self.

The entire area of *Coelomomyces*–host hemocyte recognition from a chemical standpoint needs investigation (see Dunphy and Nolan, 1981b), and the dynamic nature of the cellular interactions in the hemocoel should be studied in detail (see Dunphy and Nolan, 1980a, 1982a). Also important is the need to determine if the host(s) produces any antimicrobial substances (see Dunphy and Nolan, 1980b) and if any mycotoxins are produced (see Dunphy and Nolan, 1982b).

IV. Nutrition and the Overall Life Cycle

In the complex media tested, Castillo and Roberts (1980) observed no growth response for mycelia dissected from copepods in contrast to the response for hyphal bodies from mosquitoes. In contrast, Machlis and Crasemann (1956) found that both the gametophytic and sporophytic thalli of *A. macrogynus* and *A. arbuscula* could grow on the same minimal medium and also responded in a similar fashion when certain amino acids were used as either the nitrogen or carbon source or both. This would seem to indicate, as far as the studies of Castillo and Roberts (1980) can be used, either that the gametophytic and sporophytic generations of *Coelomomyces* may not have the same basic nutritional requirements or that the sporophytic generation is more tolerant of high osmotic pressures and/or other unfavorable growth conditions. One observation of interest is the report by Gugnani *et al.* (1965) of *C. indicus* infections in *Daphnia*. This would appear to indicate, because identifications are based primarily upon the nature of the resistant sporangium, that either the sporophyte had the ability to develop within the nonmosquito host with the obvious physiological and biochemical implications or, in all four specimens observed, the gametophytic thalli produced resistant sporangia, as occasionally occurs in *Allomyces* (see Hatch and Jones, 1944).

The comments by Emerson and Cantino (1948) on the need for pure cultures of aquatic fungi in nearly any type of investigation are certainly appropriate today for *Coelomomyces*.

V. Summary

Once a reliable method of obtaining a clean inoculum is devised, one of the primary goals in studies with *Coelomomyces* spp. should be to simplify the complex media used by Castillo and Roberts (1980). This should be approached from a position of knowledge of the composition of the complex components (such as brain–heart infusion and neopeptone), as has been done for vitamin-free casamino acids (Nolan, 1971; Nolan and Nolan, 1972) and from information on the uptake of components from media. This latter approach has been used successfully by Nolan (1970a). Until defined growth media are available for both the gametophytic and sporophytic generations of at least one isolate of *Coelomomyces,* it will be difficult to assess the effects of individual chemical, and to some extent physical, parameters on growth and morphogenesis.

Another major line of weakness in studies with *Coelomomyces* is the lack of an understanding of (1) the chemical and physical factors important in infection of either host (for example, the role of chemotaxis, phototaxis, and electrotaxis in zoospore or zygote activity), (2) the important physical and chemical changes that occur and permit attachment to and penetration of the appropriate host, (3) the chemical and physical interactions that occur within the host hemocoel, including the fungal–hemocyte interplay, and an understanding of those factors which promote and control fungal morphogenesis, and (4) the factors that are important in the maturation and germination of the male and female gametangia and the thin-walled and resistant sporangia.

These areas constitute a reasonable level of knowledge required for future understanding of the factors involved in host specificity (including nontarget organisms in safety testing programs), mass fermentation production of resistant sporangia, and future genetic engineering studies.

Needless to say, studies with representatives of the genus *Coelomomyces* by workers competent in such areas as physiology, protein chemistry, immunology, genetics, and mass fermentation technology are critical if projected biocontrol programs are to proceed and be successful.

REFERENCES

Barner, H. D., and Cantino, E. C. (1952). Nutritional relationships in a new species of *Blastocladiella*. *Am. J. Bot.* **39,** 746–751.

Belsky, M. M., Goldstein, S., and Menna, M. (1970). Factors affecting phosphate uptake in the marine fungus *Dermocystidium* sp. *J. Gen. Microbiol.* **62,** 399–402.

Cantino, E. C. (1948). The vitamin nutrition of an isolate of *Blastocladia pringsheimii. Am. J. Bot.* **35,** 238–242.

Cantino, E. C. (1969). Physiological age and germinability of resistant sporangia of *Blastocladiella emersonii. Trans. Br. Mycol. Soc.* **53,** 463–467.

Cantino, E. C. (1970). Germination of resistant sporangia of *Blastocladiella britannica:* Bearing on its taxonomic status. *Trans. Br. Mycol. Soc.* **54,** 303–307.

Cantino, E. C., and Horenstein, E. A. (1956). The stimulatory effect of light upon growth and CO_2 fixation in *Blastocladiella.* I. The S.K.I. cycle. *Mycologia* **48,** 777–799.

Cantino, E. C., and Horenstein, E. A. (1959). The stimulatory effect of light upon growth and carbon dioxide fixation in *Blastocladiella.* III. Further studies, *in vivo* and *in vitro. Physiol. Plant.* **12,** 251–263.

Cantino, E. C., and Hyatt, M. T. (1953). Carotenoids and oxidative enzymes in the aquatic phycomycetes *Blastocladiella* and *Rhizophlyctis. Am. J. Bot.* **40,** 688–694.

Carlile, M. J., and Machlis, L. (1965). A comparative study of the chemotaxis of the motile phases of *Allomyces. Am. J. Bot.* **52,** 484–486.

Castillo, J. M., and Roberts, D. W. (1980). *In vitro* studies of *Coelomomyces punctatus* from *Anopheles quadrimaculatus* and *Cylops vernalis. J. Invertebr. Pathol.* **35,** 144–157.

Chapman, H. C., and Glenn, F. E. (1972). Incidence of the fungus *Coelomomyces punctatus* and *C. dodgei* in larval populations of the mosquito *Anopheles crucians* in two Louisiana ponds. *J. Invertebr. Pathol.* **19,** 256–261.

Clements, A. N. (1963). "The Physiology of Mosquitoes." Macmillan, New York.

Couch, J. N. (1945). Revision of the genus *Coelomomyces,* parasitic in insect larvae. *J. Elisha Mitchell Sci. Soc.* **61,** 124–136.

Couch, J. N. (1968). Sporangial germination of *Coelomomyces punctatus* and the conditions favoring the infection of *Anopheles quadrimaculatus* under laboratory conditions. *Proc. J. U.S.-Jpn. Semin. Microb. Control Insect Pests, 1967* pp. 93–105.

Couch, J. N. (1972). Mass production of *Coelomomyces,* a fungus that kills mosquitoes. *Proc. Natl. Acad. Sci. U.S. A.* **69,** 2043–2047.

Crasemann, J. M. (1957). Comparative nutrition of two species of *Blastocladia. Am. J. Bot.* **44,** 218–224.

Dill, B. C., and Fuller, M. S. (1971). Amino acid immobilization of fungal motile cells. *Arch. Mikrobiol.* **78,** 92–98.

Dunphy, G. B., and Nolan, R. A. (1979). Effects of physical factors on protoplasts of *Entomophthora egressa. Mycologia* **71,** 589–602.

Dunphy, G. B., and Nolan, R. A. (1980a). Response of eastern hemlock looper hemocytes to selected stages of *Entomophthora egressa* and other foreign particles. *J. Invertebr. Pathol.* **36,** 71–84.

Dunphy, G. B., and Nolan, R. A. (1980b). Protozoan lysins in the larval eastern spruce budworm hemolymph. *J. Invertebr. Pathol.* **36,** 433–437.

Dunphy, G. B., and Nolan, R. A. (1981a). Comparative physiology of two isolates of *Entomophthora egressa. Mycologia* **73,** 887–903.

Dunphy, G. B., and Nolan, R. A. (1981b). A study of the surface proteins of *Entomophthora egressa* protoplasts and of larval spruce budworm hemocytes. *J. Invertebr. Pathol.* **38,** 352–361.

Dunphy, G. B., and Nolan, R. A. (1982a). Cellular immune responses of spruce budworm larvae to *Entomophthora egressa* protoplasts and other test particles. *J. Invertebr. Pathol.* **39,** 81–92.

Dunphy, G. B., and Nolan, R. A. (1982b). Mycotoxin production by the protoplast stage of *Entomophthora egressa*. *J. Invertebr. Pathol.* **39,** 261–263.

Emerson, R. (1941). An experimental study of the life cycles and taxonomy of *Allomyces*. *Lloydia* **4,** 77–144.

Emerson, R., and Cantino, E. C. (1948). The isolation, growth, and metabolism of *Blastocladia* in pure culture. *Am. J. Bot.* **35,** 157–171.

Emerson, R., and Fox, D. L. (1940). γ-Carotene in the sexual phase of the aquatic fungus *Allomyces*. *Proc. R. Soc. London, Ser. B.* **128,** 275–293.

Federici, B. A. (1977). Differential pigmentation in the sexual phase of *Coelomomyces*. *Nature (London)* **267,** 514–515.

Federici, B. A. (1980). Production of the mosquito-parasitic fungus, *Coelomomyces dodgei,* through synchronized infection and growth of the intermediate copepod host, *Cyclops vernalis*. *Entomophaga* **25,** 209–217.

Federici, B. A., and Chapman, H. C. (1977). *Coelomomyces dodgei*: Establishment of an *in vivo* laboratory culture. *J. Invertebr. Pathol.* **30,** 288–297.

Federici, B. A., and Roberts, D. W. (1975). Experimental laboratory infection of mosquito larvae with fungi of the genus *Coelomomyces*. I. Experiments with *Coelomomyces psorophorae* var. in *Aedes taeniorhynchus* and *Coelomomyces psorophorae* var. in *Culiseta inornata*. *J. Invertebr. Pathol.* **26,** 21–27.

Federici, B. A., and Roberts, D. W. (1976). Experimental laboratory infection of mosquito larvae with fungi of the genus *Coelomomyces*. II. Experiments with *Coelomomyces punctatus* in *Anopheles quadrimaculatus*. *J. Invertebr. Pathol.* **27,** 333–341.

Federici, B. A., and Thompson, S. N. (1979). β-Carotene in the gametophytic phase of *Coelomomyces dodgei*. *Exp. Mycol.* **3,** 281–284.

Goldstein, A., and Cantino, E. C. (1962). Light-stimulated polysaccharide and protein synthesis by synchronized, single generations of *Blastocladiella emersonii*. *J. Gen. Microbiol.* **28,** 689–699.

Gugnani, H. C., Wattal, B. L., and Kalra, N. L. (1965). A note on *Coelomomyces* infection in mosquito larvae. *Bull. Indian Soc. Malaria Commun. Dis.* **2,** 333–337.

Hatch, W. R. (1936). Zonation in *Allomyces arbuscula*. *Mycologia* **28,** 439–444.

Hatch, W. R. (1944). Zoösporogenesis in the resistant sporangia of *Allomyces arbusculus*. *Mycologia* **36,** 650–663.

Hatch, W. R., and Jones, R. C. (1944). An experimental study of alternation of generations in *Allomyces arbusculus*. *Mycologia* **36,** 369–381.

Horenstein, E. A., and Cantino, E. C. (1961). Morphogenesis in and the effect of light on *Blastocladiella britannica* sp. nov. *Trans. Br. Mycol. Soc.* **44,** 185–198.

Horenstein, E. A., and Cantino, E. C. (1964). An effect of light on glucose uptake by the fungus *Blastocladiella britannica*. *J. Gen. Microbiol.* **37,** 59–65.

Ingraham, J. L., and Emerson, R. (1954). Studies of the nutrition and metabolism of the aquatic phycomycete, *Allomyces*. *Am. J. Bot.* **41,** 146–152.

Jones, R. C. (1946). Factors affecting the production of resistant sporangia of *Allomyces arbuscula*. *Mycologia* **38,** 91–102.

Khandjian, E. W., Roos, U.-P., Timberlake, W. E., Eder, L., and Turian, G. (1974). RNA virus-like particles in the Chytridiomycete *Allomyces arbuscula*. *Arch. Microbiol.* **101,** 351–356.

Laird, M. (1959). Parasites of Singapore mosquitoes, with particular reference to the significance of larval epibionts as an index of habitat pollution. *Ecology* **40,** 206–221.

Lemke, P. A., ed. (1979). "Viruses and Plasmids in Fungi." Dekker, New York.

Lum, P. T. M. (1963). The infection of *Aedes taeniorhynchus* (Wiedemann) and *Psorophora howardii* Coquillett by the fungus *Coelomomyces*. *J. Insect Pathol.* **5,** 157–166.

Machlis, L. (1953a). Growth and nutrition of water molds in the subgenus *Euallomyces*. I. Growth factor requirements. *Am. J. Bot.* **40,** 189–195.

Machlis, L. (1953b). Growth and nutrition of water molds in the subgenus *Euallomyces*. II. Optimal composition of the minimal medium. *Am. J. Bot.* **40,** 450–460.

Machlis, L. (1953c). Growth and nutrition of water molds in the subgenus *Euallomyces*. III. Carbon sources. *Am. J. Bot.* **40,** 460–464.

Machlis, L. (1957a). Factors affecting the lag phase of growth of the filamentous fungus, *Allomyces macrogynus*. *Am. J. Bot.* **44,** 113–119.

Machlis, L. (1957b). Effect of certain organic acids on the utilization of mannose and fructose by the filamentous watermold, *Allomyces macrogynus*. *J. Bacteriol.* **73,** 627–631.

Machlis, L. (1957c). Inhibition of the growth of the filamentous fungus, *Allomyces macrogynus,* by lipoic acid. *Arch. Biochem. Biophys.* **70,** 413–418.

Machlis, L. (1969a). Zoospore chemotaxis in the watermold *Allomyces*. *Physiol. Plant.* **22,** 126–139.

Machlis, L. (1969b). Fertilization-induced chemotaxis in the zygotes of the watermold *Allomyces*. *Physiol. Plant.* **22,** 392–400.

Machlis, L., and Crasemann, J. M. (1956). Physiological variation between the generations and among the strains of watermolds in the subgenus *Euallomyces*. *Am. J. Bot.* **43,** 601–611.

Machlis, L., and Ossia, E. (1953a). Maturation on the meiosporangia of *Euallomyces*. I. The effect of cultural conditions. *Am. J. Bot.* **40,** 358–365.

Machlis, L., and Ossia, E. (1953b). Maturation of the meiosporangia of *Euallomyces*. II. Preliminary observations on the effect of auxins. *Am. J. Bot.* **40,** 465–468.

Madelin, M. F. (1964). Laboratory studies on the infection of *Anopheles gambiae* Giles by a species of *Coelomomyces*. *WHO/EBL/***17,** *WHO/Mal/***438,** *WHO/Vector Control/***64,** 1–23.

Madelin, M. F. (1968). Studies on the infection by *Coelomomyces indicus* of *Anopheles gambiae*. *J. Elisha Mitchell Sci. Soc.* **84,** 115–124.

Madelin, M. F., and Beckett, A. (1972). The production of planonts by thin-walled sporangia of the fungus *Coelomomyces indicus,* a parasite of mosquitoes. *J. Gen. Microbiol.* **72,** 185–200.

Martin, W. W. (1969). A morphological and cytological study of the development of *Coelomomyces punctatus* parasitic in *Anopheles quadrimaculatus*. *J. Elisha Mitchell Sci. Soc.* **85,** 59–72.

Martin, W. W. (1970). A morphological and cytological study of *Coelomomyces punctatus*. Ph.D. Thesis, University of North Carolina, Chapel Hill.

Martin, W. W. (1971). The ultrastructure of *Coelomomyces punctatus* zoospores. *J. Elisha Mitchell Sci. Soc.* **87,** 209–221.

Muspratt, J. (1946a). On *Coelomomyces* fungi causing high mortality of *Anopheles gambiae* larvae in Rhodesia. *Ann. Trop. Med. Parasitol.* **40,** 10–17.

Muspratt, J. (1946b). Experimental infection of the larvae of *Anopheles gambiae* (Dipt., Culicidae) with a *Coelomomyces* fungus. *Nature* (*London*) **158,** 202.

Muspratt, J. (1962). Destruction of the larvae of *Anopheles gambiae* Giles by a *Coelomomyces* fungus. *WHO/EBL/***2**, *WHO/Vector Control/***2, 1–11.**

Nam, E. A. (1975). Optimal storage conditions and effect of pH of the medium on germination of sporangia of *Coelomomyces iliensis* Dubit., Dzerzh. and Daneb. *Izv. Akad. Nauk Kaz. SSR, Ser. Biol.* **13,** 46–48 (original in Russian).

Nolan, R. A. (1969). Nutritional requirements for species of *Allomyces*. *Mycologia* **61,** 641–644.

Nolan, R. A. (1970a). The Phycomycete *Catenaria anguillulae*: Growth requirements. *J. Gen. Microbiol.* **60,** 167–180.

Nolan, R. A. (1970b). Sulfur source and vitamin requirements of the aquatic phycomycete, *Catenaria anguillulae*. *Mycologia* **62,** 568–577.

Nolan, R. A. (1970c). Carbon source and micronutrient requirements of the aquatic phycomycete, *Catenaria anguillulae* Sorokin. *Ann. Bot.* (*London*) [N.S.] **34,** 927–939.

Nolan, R. A. (1971). Amino acids and growth factors in vitamin-free casamino acids. *Mycologia* **63,** 1231–1234.

Nolan, R. A. (1973). Effects of plant hormones on germination of *Coelomomyces psorophorae* resistant sporangia. *J. Invertebr. Pathol.* **21,** 26–30.

Nolan, R. A., and Freake, G. W. (1974). An electron microprobe analysis of the resistant sporangium wall of *Coelomomyces psorophorae*. *J. Invertebr. Pathol.* **23,** 121–122.

Nolan, R. A., and Nolan, W. G. (1972). Elemental analysis of vitamin-free casamino acids. *Appl. Microbiol.* **24,** 290–291.

Olson, L. W. (1973). Factors effecting the initiation of mitosis in *Allomyces neo-moniliformis*. *Arch. Mikrobiol.* **91,** 305–311.

Olson, L. W., and Fuller, M. S. (1971). Leucine-lysine synchronization of *Allomyces* germlings. *Arch. Mikrobiol.* **78,** 76–91.

Pillai, J. A. (1971). *Coelomomyces opifexi* (Pillai & Smith). Coelomomycetaceae: Blastocladiales. 1. Its distribution and the ecology of infection pools in New Zealand. *Hydrobiologia* **38,** 425–436.

Pillai, J. S., and O'Loughlin, I. H. (1972). *Coelomomyces opifexi* (Pillai & Smith) Coelomomycetaceae: Blastocladiales II. Experiments in sporangial germination. *Hydrobiologia* **40,** 77–86.

Pillai, J. S., and Woo, A. (1973). *Coelomomyces opifexi* (Pillai & Smith) Coelomomycetaceae: Blastocladiales. III. The laboratory infection of *Aedes australis* (Erichson) larvae. *Hydrobiologia* **41,** 169–181.

Pillai, J. S., Wong, T. L., and Dodgshun, T. J. (1976). Copepods as essential hosts for the development of a *Coelomomyces* parasitizing mosquito larvae. *J. Med. Entomol.* **13,** 49–50.

Powell, M. J. (1976). Ultrastructural changes in the cell surface of *Coelomomyces punctatus* infecting mosquito larvae. *Can. J. Bot.* **54,** 1419–1437.

Rašín, K. (1929). *Coelomomyces chironomi* n.sp., fungus parasitic in the body cavity of chironomid larvae. I. Morphology and biology. *Biol. Spisy, Aka. Vet.* **8,** 1–13. (original in Czechoslovakian).

Roberts, D. W., Shapiro, M., and Rodin, R. L. (1973). Dehiscence of *Coelomomyces psorophorae* sporangia from *Aedes taeniorhynchus:* Induction by amines and amino acids. *J. Invertebr. Pathol.* **22,** 175–181.

Robertson, J. A. (1972). Phototaxis in a new *Allomyces*. *Arch. Mikrobiol.* **85,** 259–266.

Roos, U.-P., Khandjian, E. W., and Turian, G. (1976). RNA virus in *Allomyces arbuscula:* Ultrastructural localization during the life-cycle. *J. Gen. Microbiol.* **95,** 87–95.

Shapiro, M., and Roberts, D. W. (1976). Growth of *Coelomomyces psorophorae* mycelium *in vitro*. *J. Invertebr. Pathol.* **27,** 399–402.

Sistrom, D. E., and Machlis, L. (1955). The effect of D-glucose on the utilization of D-mannose and D-fructose by a filamentous fungus. *J. Bacteriol.* **70,** 50–55.

Skucas, G. P. (1967). Structure and composition of the resistant sporangial wall in the fungus *Allomyces*. *Am. J. Bot.* **54,** 1152–1158.

Soll, D. R., and Sonneborn, D. R. (1969). Zoospore germination in the water mold, *Blastocladiella emersonii*. II. Influence of cellular and environmental variables on germination. *Dev. Biol.* **20,** 218–235.

Soll, D. R., and Sonneborn, D. R. (1972). Zoospore germination in *Blastocladiella emersonii.* IV. Ion control over cell differentiation. *J. Cell Sci.* **10,** 315–333.

Suberkropp, K. F., and Cantino, E. C. (1972). Environmental control of motility and encystment in *Blastocladiella emersonii* zoospores at high population densities. *Trans. Br. Mycol. Soc.* **59,** 463–475.

Taylor, B. W., Harlos, J. A., and Brust, R. A. (1980). *Coelomomyces* infection of the adult female mosquito *Aedes trivittatus* (Coquillett) in Manitoba. *Can. J. Zool.* **58,** 1215–1219.

Travland, L. B. B. (1977). Cytology of *Coelomomyces psorophorae:* Meiospore, gamete, zygote, and initiation of infection. Ph.D. Thesis, University of Washington, Seattle.

Travland, L. B. (1979a). Initiation of infection of mosquito larvae (*Culiseta inornata*) by *Coelomomyces psorophorae. J. Invertebr. Pathol.* **33,** 95–105.

Travland, L. B. (1979b). Structures of the motile cells of *Coelomomyces psorophorae* and function of the zygote in encystment on a host. *Can. J. Bot.* **57,** 1021–1035.

Turian, G. (1957). Activation de la phosphatase acide par l'acide β-indolylacétique et ses répercussions sur la phosphorolyse de l'amidon. *Experientia* **13,** 368–370.

Turian, G. (1962). Cytoplasmic differentiation and dedifferentiation in the fungus *Allomyces. Protoplasma* **54,** 323–327.

Turian, G., and Cantino, E. (1959). Identification du γ-carotène dans les sporanges de résistance du Champignon *Allomyces macrogynus. C. R. Hebd. Seances Acad. Sci.* **249,** 1788–1789.

Turian, G., and Haxo, F. T. (1954). Minor polyene components in the sexual phase of *Allomyces javanicus. Bot. Gaz.* (*Chicago*) **115,** 254–260.

Umphlett, C. J. (1961). Comparative studies in the genus *Coelomomyces* Keilin. Ph.D. Thesis, University of North Carolina, Chapel Hill.

Umphlett, C. J. (1968). Ecology of *Coelomomyces* infections of mosquito larvae. *J. Elisha Mitchell Sci. Soc.* **84,** 108–114.

Walker, A. J. (1938). Fungal infections of mosquitoes, especially of *Anopheles costalis. Ann. Trop. Med. Parasitol.* **32,** 231–244.

Whisler, H. C., Shemanchuk, J. A., and Travland, L. B. (1972). Germination of the resistant sporangia of *Coelomomyces psorophorae. J. Invertebr. Pathol.* **19,** 139–147.

Whisler, H. C., Zebold, S. L., and Shemanchuk, J. A. (1975). Life history of *Coelomomyces psorophorae. Proc. Natl. Acad. Sci. U.S.A.* **72,** 693–696.

Youatt, J. (1973). Sporangium production by *Allomyces* in new chemically defined media. *Trans. Br. Mycol. Soc.* **61,** 257–263.

Youatt, J., Fleming, R., and Jobling, B. (1971). Differentiation in species of *Allomyces:* The production of sporangia. *Aust. J. Biol. Sci.* **24,** 1163–1167.

Zebold, S. L., Whisler, H. C., Shemanchuk, J. A., and Travland, L. B. (1979). Host specificity and penetration in the mosquito pathogen *Coelomomyces psorophorae. Can. J. Bot.* **57,** 2766–2770.

7 Culture

CHARLES E. BLAND

Department of Biology
East Carolina University
Greenville, North Carolina

I. *IN VIVO*

Prior to discovery of the copepod *Cyclops vernalis* as an alternate host in the life cycle of some species of *Coelomomyces* (Whisler *et al.*, 1975), all successful efforts to culture members of this genus involved placing sporulating resting sporangia and mosquito larvae together in water containing a variety of biotic (algae, arthropods, etc.) and abiotic (soil) components (Couch, 1968, 1972; Madelin, 1968; Pillai and Woo, 1973). In the most successful of these efforts, Couch (1968, 1972) obtained continuous culture of *Coelomomyces punctatus* in *Anopheles quadrimaculatus* for a span of approximately 10 years. Although the significance of the copepod was not known, high rates of infection (often up to 100%) were obtained by inoculation and culture techniques that must have inadvertently introduced copepods into the life cycle at exactly the stage necessary for perpetuation of the fungus. In the most successful of these techniques, the fungal inoculum was prepared by macerating approximately 50 heavily infected larvae (collected from successful experiments and stored in 1-qt containers of "algal water"[1] at 10°C) in 500 ml of algal water, which was subsequently diluted with additional water to a final volume of 2000 ml. For infection, approximately 200–800 mosquito eggs were placed with about 200 ml inoculum in containers ("sinks or dishpans or small pans")

[1] Containing primarily the genera *Ankistrodesmus*, *Scenedesmus*, and *Closterium*.

filled to a depth of 5 cm with lake or aerated tap water. By adding inoculum only initially or possibly twice during larval development, a pattern of infection resulting in normal fungal development occurred. However, introduction of large amounts of inoculum during the first three larval ecdyses resulted in 100% infection and death of larvae, but failure of fungal hyphae to differentiate into resting sporangia. Using the technique yielding normal fungal development, Couch (1972) was able to "mass produce" resting sporangia (inoculum) of *C. punctatus*.

Based primarily on the culture techniques developed by Couch (1968, 1972), Whisler *et al.* (1974, 1975) and others (Federici, 1975; Pillai *et al.*, 1976; Weiser, 1977) were able to determine the role of copepods or ostracods (see Chapter 2, Life History) as alternate hosts with mosquitoes during the life cycle of *Coelomomyces* spp. Since this discovery, continuous *in vivo* culture of several species has been established, including *C. psorophorae* (Federici and Roberts, 1975), *C. opifexi* = *C. psorophorae* var. *tasmaniensis* (Wong and Pillai, 1978), *C. dodgei* (Federici and Chapman, 1977), and *C. punctatus* (Federici and Roberts, 1976). Recently, Federici (1980) described improved techniques for culture of *C. dodgei* through synchronized growth and infection of the intermediate copepod host, *Cyclops vernalis*. By this technique, exposure of 2000 48- to 72-hr-old copepod nauplii to 6000 resting sporangia at the time of sporulation yielded approximately 1500 infected copepods; the rate of infection of copepods older than 72 hr (i.e., copepodids and adults) declined sharply (Fig. 620). Continuation of the life cycle via infection of mosquito larvae was as described by Federici and Chapman (1977). A summary (Federici, 1981) of revised techniques necessary for *in vivo* culture of not only *C. dodgei,* but also *C. punctatus,* and *C. psorophorae* follows:

> [*C. dodgei* and *C. punctatus*.] To obtain meiospores, a heavily infected fourth-instar larva (10 larvae for *C. punctatus*) is dissected, thereby liberating sporangia, in 4 ml of distilled water in a 15-mm petri dish. Forty-eight hours later, while meiospores are being released, the contents of the dish are combined with 8,000–10,000 two-day-old copepod nauplii in 1 L of tapwater.[2] This mixture is held at 25°C for 2 days to allow optimum infection of the copepods and then transferred to 12 L of distilled water.[3] Copepods are fed on hardboiled egg yolk and dried brewer's yeast. These cultures produce a total of approximately 1,500 infected copepods from day 7–15. Infected copepods, easily differentiated from the others because of the amber or orange coloration imparted to them by the gametophytes, are picked from the surface when needed for mosquito infection. Mosquitoes are infected by combining one hundred second-instar larvae of *A. quadrimaculatus* in 50 ml of medium from a larval culture along with 12 infected copepods, 6 of each color type, for 48 h. The fungus kills the copepods, forms zygotes, and infects larvae during this period. Approximately eight days later 25% of the larvae will develop patent infections during the fourth-instar.

[2,3] Or preferably, the medium rich in microorganisms described by Federici (1980).

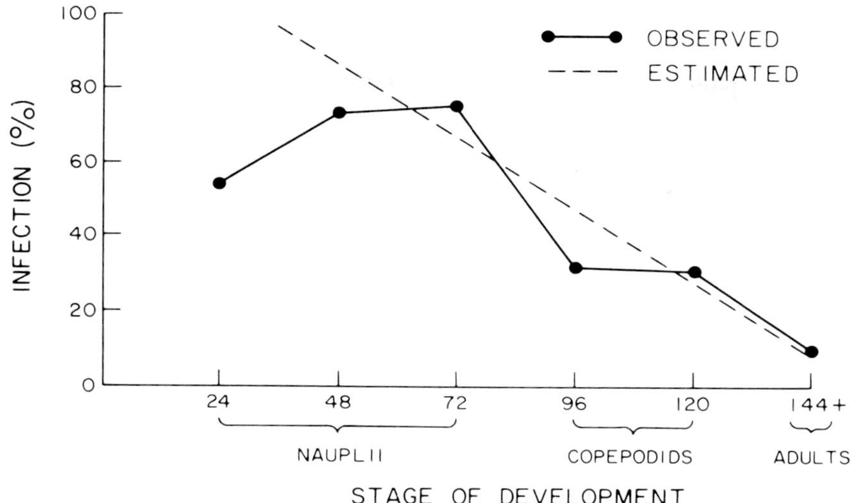

Fig. 620. Observed rates of infection for copepods exposed to meiospores at 24, 48, 72, 96, 120, or 144+ hr old compared with estimated actual infection rates. The observed rate is based on the number of copepods in which gametophytes matured, as an actual infection rate is inaccurate for the 24-hr, and possibly 48-hr, group where many copepods died before the gametophytes matured. The broken line is an estimate of the actual infection rate for these ages, indicative of relative susceptibility, determined by extrapolation of data from the progressively more resistant 72-hr-old nauplii, copepodids, and adults. [From Federici (1980).]

[*C. psorophorae*.] After larval death, sporangia from 10 larvae are washed in distilled water, and collected and stored on Millipore filters at 5°C until use. Copepod infection is obtained by transferring a filter to 100 ml of a 25% dilution of salt solution (Whisler, 1966) along with ca. 200 copepodides of *Cyclops vernalis*. This mixture is held in a petri dish for seven days at 20°C in diffuse light. Meiospore release and copepod infection occur within the first two days, with development and maturation of the gametophytes occurring days three through seven. Copepods are fed nematodes. Approximately 50% of the copepods will develop infections. As the infected copepods begin releasing gametes and zygotes they are combined with 200 second-instar larvae of *C. inornata* in 100 ml of the above salt solution at 50% concentration. Normally, 100% of the larvae develop patent infections by the time they reach the fourth-instar.

In both the procedures just described, Federici (1981) notes that the infection of either copepods or mosquitoes is most easily accomplished when hosts are exposed to fungal inoculum during their early stages of development, when they are most susceptible (Fig. 620, Table XIV). Federici (1980) and H. C. Whisler (personal communication, cited in Federici, 1981) also both report that optimal diets and rearing conditions for hosts are essential to optimal fungal development.

Table XIV. Susceptibility of Different Instar Larvae of *Anopheles quadrimaculatus* Exposed to *Coelomomyces dodgei*[a]

	Larval instar			
Replicate[b]	L_1	L_2	L_3	L_4[c]
1	24	71	25	2
2	59	10	12	3
3	32	23	17	0
4	37	17	16	5
5	35	14	19	2
Mean	37.4	27.0	17.8	2.4

[a] From Federici and Chapman (1977).

[b] In each replicate, 100 larvae of each instar were exposed to 4 amber, 4 orange, and 4 bright orange copepods.

[c] Mosquitoes exposed to L_4 did not develop patent infections until the pupal stage. These figures represent the number of patently infected pupae.

II. *IN VITRO*

In vitro culture of *Coelomomyces* spp., a vital step toward their use in biological control of mosquitoes, has been accomplished on a limited basis for only two species, *C. psorophorae* and *C. punctatus* (Shapiro and Roberts, 1976; Castillo and Roberts, 1980). For *C. psorophorae*,[4] Shapiro and Roberts (1976) obtained growth from hyphal bodies dissected from a larva of *Aedes sollicitans* into a modified mycoplasma broth medium. With subcultures at 2- to 4-week intervals, growth was maintained for 2 years. During this period, partial differentiation of hyphae into sporangia occurred, but mature sporangia were not observed. However, when hyphae were incubated in association with cultured cells of *Aedes aegypti* in a modified Schneider's *Drosophila* medium, pigmented, smooth-walled spheres slightly larger than sporangia were produced, probably representing partially formed sporangia. In the same study, hyphae of *C. psorophorae* from *Psorophora howardii* were noted to grow rapidly for 34–43 hr in Mitsuhashi and Maramorosh (1964) medium supplemented with 20% fetal bovine serum. Attempts to grow *C. psorophorae, C. dodgei,* and *C. pentangulatus* on a variety of other media (approximately 30) were unsuccessful.

Castillo and Roberts (1980) utilized over 50 combinations of known

[4] Likely to be *C. psorophorae* var. *halophilus*.

vertebrate and invertebrate tissue culture media and microbiological media in an attempt to culture hyphae of *C. punctatus* from *Anopheles quadrimaculatus*. In addition, approximately eight media were used in an effort to culture hyphae (gametangia?) of the same species from *Cyclops vernalis*. In this study, hyphae, obtained by careful dissection under aseptic conditions of infected larvae and copepods that had been surface-sterilized, were placed directly in culture media and then incubated under a variety of conditions, or were, in some instances, injected into the hemocoel of other potential hosts (larvae of the housefly, *Musca domestica;* southern armyworm, *Prodenia eridania;* salt marsh caterpillar, *Estigmene acrea;* greater wax moth, *Galleria mellonella;* or the large predatory mosquito, *Toxorhynchites brevipalpis*). Media promoting growth and differentiation of hyphae into resting sporangia are listed in Table XV.[5] Sporangia produced on these media were ornamented and pigmented as in normally produced sporangia (Fig. 621). However, such sporangia lacked a dehiscence slit and failed to sporulate under conditions normally induc-

[5] The abbreviations used by Castillo and Roberts (1980) for the culture media ingredients listed in Tables XV–XVIII are as follows: **A** = amphibian culture medium (Wolf and Quimby, 1964)(GIBCO); **AMP** = adenosine-3′;5′-cyclic monophosphoric acid, sodium salt; **BA** = 7.5% bovine albumin, fraction V (Miles); **BHI** = brain–heart infusion (Difco); **BHM** = 2% BHI + 0.2% glucose + 0.1% neopeptone (Difco) + 0.1% yeastolate (GIBCO) + 20% MP + 15% FBS (modified from Humber, 1975); C-3 = corn stunt spiroplasma medium (Chen and Liao, 1975); **CBHM** = BHM conditioned 3 weeks by growth by Varma's *A. stephensi* cell line; **CEE** = chick embryo extract (GIBCO); **Ch** = cholesterol; **CM** = MM + 20% FBS conditioned 3 weeks by growth of *A. stephensi* cells; **CM199** = medium 199 (Hank's salts) + 20% FBS conditioned 1–3 weeks by growth of FHM cells; **CMg** = MM + 20% FBS, conditioned 3 weeks with *A. gambiae* cells; **CMRL** = medium 1066 (Parker *et al.*, 1957)(GIBCO); CSch = Sch + BA (3 : 1) + 20% FBS conditioned 3 weeks by growth of *A. aegypti* cells; **Er** = β-ecdysterone (Schwarz-Mann); **EU** = whole egg ultrafiltrate (GIBCO); **EY** = hard-boiled egg yolk solution (1 yolk/500 ml H_2O); **FA** = fatty acids (oleic, palmitic, and myristic acids); **FBS** = fetal bovine serum, heat-treated 60°C, 30 min (Microbiological Associates); **FHM** = fathead minnow tissue culture cells; **FrH** = fresh *Estigmene acrea* (Lepidoptera) hemolymph; **FxH** = heat-fixed *E. acrea* hemolymph; **G** = Grace's (Chen *et al.*, 1962) insect tissue culture medium (GIBCO); **GHL** = glycyl-L-histidyl-L-lysine acetate (Calbiochem); **GM** = G as modified by Yunker *et al.* (1967)(GIBCO); **HMB** = G as modified by Hink (1970); **IE** = low-speed supernatant of larval of pupal homogenate of *Culex pipiens pipiens, E. acrea, G. mellonella,* and *M. domestica;* **JH** = synthetic juvenile hormones (Calbiochem); **LH** = lactalbumin hydrolyzate; **Mc** = McCoy's (McCoy *et al.*, 1959) 5A modified medium (GIBCO); **MH** = heat-fixed (60°C, 10 min) *A. quadrimaculatus* larval homogenate; **MHI** = low-speed supernatant *C. punctatus*-infected *A. quadrimaculatus* larval homogenate, heat-fixed 60°C, 10 min; **MM** = Mitsuhashi–Maramorosch (1964) medium; **MP** = low-speed supernatant of *A. quadrimaculatus* pupal homogenate, 2 g fresh wt/10 ml G medium; **PPLO** = mycoplasma broth (GIBCO); **$PPLO_3$** = PPLO + SMI (3 : 1); **Pr** = Ponasterone; **Sch** = Schneider's (1964) *Drosophila* medium; **SMI** = Gm with 1% yeastolate (Difco) and LH and 0.005% myristic and palmitic acids; **Tc** = DL-α-tocopherol (ICN pharmaceutical).

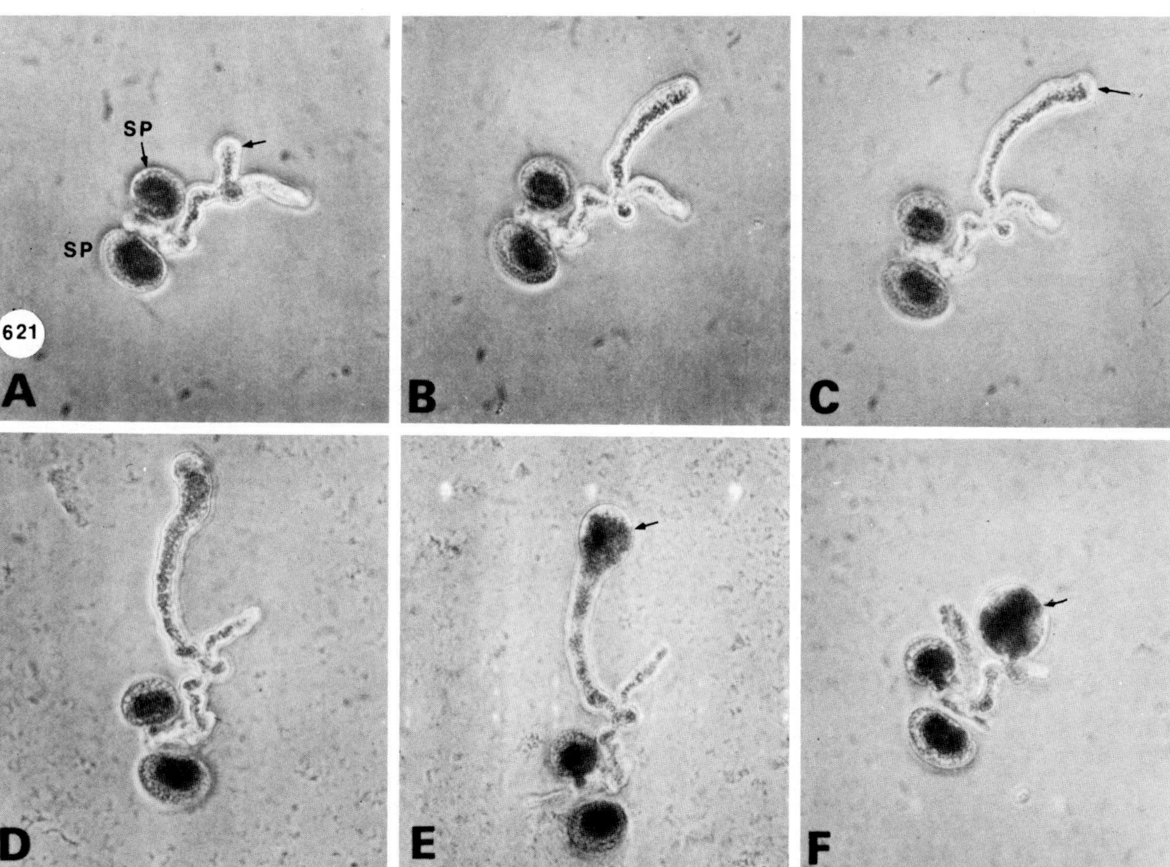
SP
SP
621
A
B
C
D
E
F

Table XV. Media Promoting Growth and Differentiation of Hyphae into Resting Sporangia[a,b]

BHM
BHM + GHL, 20 ng/ml
CBHM
CM
CM + GHL, 10 μg and 20 ng/ml

[a] From Castillo and Roberts (1980).
[b] See p. 353 for explanation of abbreviations.

Table XVI. Media Promoting Growth and Differentiation of Hyphae into Hyphal Bodies[a,b]

BHM + CM (1:1; 2:1; 3:1; 1:2)
BHM + 20% IE
CBHM
CM + PPLO:PPLO$_3$ (1:1) + GHL, 20 ng/ml
CM + PPLO or PPLO$_3$ + EU or EY (2:1:1) + GHL, 20 ng/ml
CM + PPLO$_3$ (1:1) + 3% LH
CM + G (1:7) + 3% LH
CM + MH or MHI (9:1)
CM + FrH (4:1)
G + FBS, 10, 15, or 20%
GM
Mc
Mc + PPLO + EU or EY (2:1:1) + GHL, 20 ng/ml
MM + 20% FBS
MM + EU (8:1) + 20% FBS
PPLO + SMI (3:1)

[a] From Castillo and Roberts (1980).
[b] See p. 353 for explanation of abbreviations.

Fig. 621. Growth and development of *C. punctatus* mycelia from *A. quadrimaculatus*, incubated in BHM (see Table XV for constituents). (A) At 48 hr after dissection, showing two newly developed young resting sporangia (sp). One hyphal branch (arrow) has started to grow. (B) Hyphal branch continuing to elongate after 72 hr. (C) After 96 hr the hyphal tip has begun to swell (arrow), an indication of sporangial initiation. (D) Same hypha showing increased swelling and enlargement of the tip after 127 hr. (E) After 136 hr, sporangium initial (arrow) has become almost spherical in shape, with a thin envelope thicker than the supporting hyphal base. (F) Spherical sporangium initial formed after 192 hr of development in the medium (arrow). × 180. [From Castillo and Roberts (1980).]

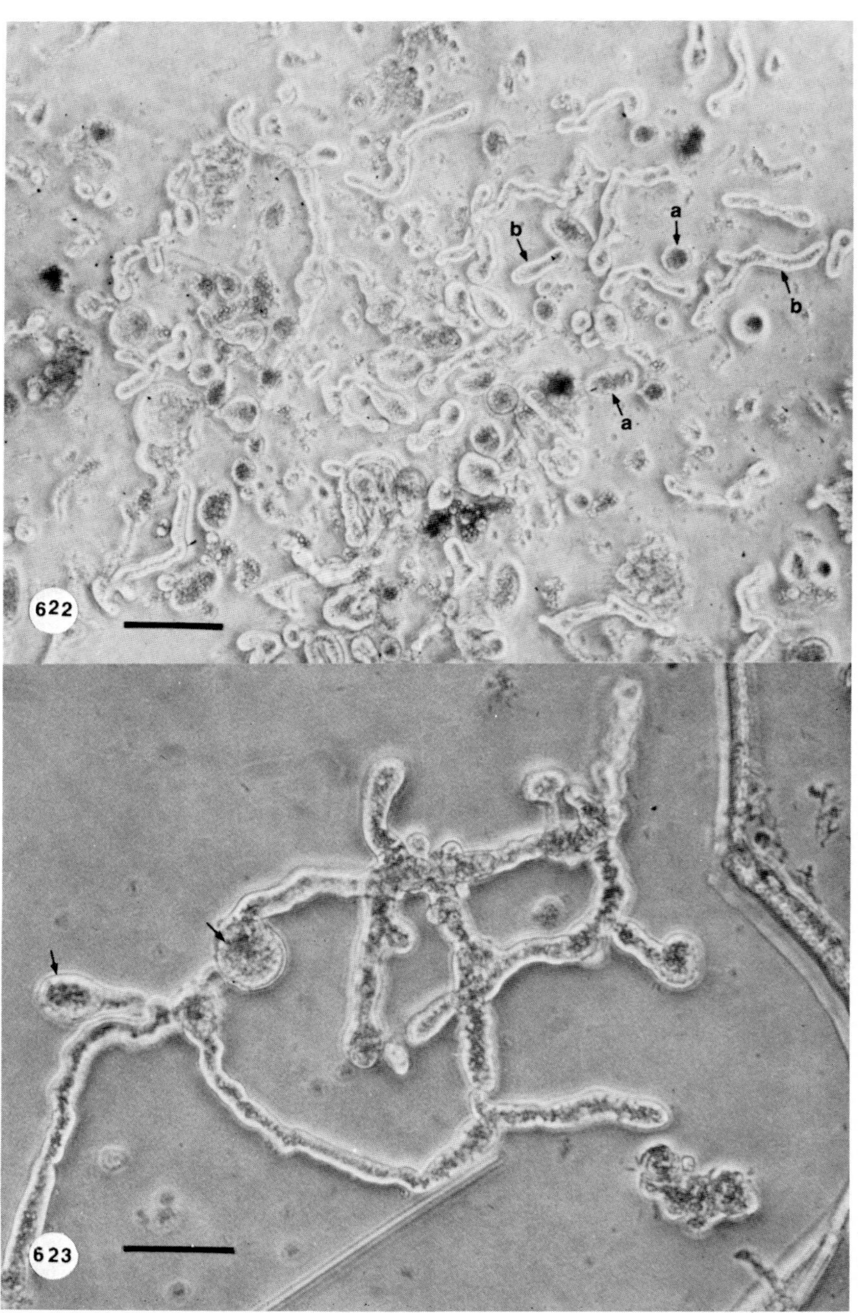
a
b
b
a
622
623

Table XVII. Media Found Unsatisfactory for Growth and Development of Hyphae from *Anopheles quadrimaculatus*[a,b]

Media
A + PPLO + EY (2:1:1) + GHL, 20 ng/ml
BHM + CMRL or Mc or M199 (9:1)
C-3
C-3 + CM (1:1; 2:1)
CM + 10% CEE
CM + PPLO or $PPLO_3$ (5:2) + 10% CEE
CM with ER, PR and/or JH, 50, 100 μg/ml
CM with FA and/or Ch, 100 μg/ml
CM with FHM
CM 199
CMg
CMRL + 20% FBS
CMRL + CM (1:1)
CMRL + CM + PPLO + EU or EY (2:1:1:1) + GHL, 20 ng/ml
CMRL + EU + 1.5% BA + 10% FBS (9:1)
CMRL + PPLO + EU (5:2:2) + 10% FBS + GHL, 20 ng/ml
G + 5% LH + GHL, 20 ng/ml
G + PPLO (8:1) + 5% LH + GHL, 20 ng/ml
GM + CM (2:1) + 5% LH
GM + 20% FBS with FHM
HMG
Mc + CM + PPLO (2:1:1) + GHL, 20 ng/ml
Mc + EU + BA (9:1:1)
MM + EU (8:1) + 20% FBS
MM + 20% FBS + GHL, 10, 20, 50 ng/ml
MM + 20% FBS + 5% FxH + GHL, 20 ng/ml
MM + PPLO (4:1) + 15% FBS + 5% FxH + GHL, 20 ng/ml
MM + $PPLO_3$ (4:1) + 15% FBS + 5% FxH + GHL, 20 ng/ml

[a] From Castillo and Roberts (1980).
[b] See page 353 for explanation of abbreviations.

ing this process. On some media (Table XVI) hyphae failed to produce sporangia, but did differentiate into smooth, spherical hyphal bodies of size and appearance characteristic of sporangial initials (Figs. 622 and 623). On most media (Table XVII) hyphae appeared to remain alive, but there was no growth and/or further development. It was noted that neither increased CO_2 (up to 10%) nor physical manipulation of culture conditions

Figs. 622 and 623. Mycelia of *Coelomomyces punctatus* from *Anopheles quadrimaculatus* larva dissected into CM + GHL, 20 ng/ml (see Table XV for constituents). Fig. 622. At 1 hr after dissection, note spherical or ovoid-shaped hyphagens (a) and irregularly shaped, elongated, simple or branched hyphae (b). Bar = 100 μm. Fig. 623. Mycelial piece from the same culture at 24 hr postdissection. This culture was lightly contaminated by large, motile, rod-shaped bacteria. Arrows indicate different stages of development of sporangial initials. Bar = 50 μm. [From Castillo and Roberts (1980).]

Table XVIII. Media Utilized for Hyphae Dissected from *Cyclops vernalis*[a,b]

BHM
BHM + GHL, 20 ng/ml
CM
CM + GHL, 10 μg and 20 ng/ml
CSch
GM
Mc
MM + 20% FBS

[a] From Castillo and Roberts (1980).
[b] See page 353 for explanation of abbreviations.

enhanced growth or development on any media. Also, no growth occurred from hyphae injected into other organisms. Similarly, hyphae from *C. vernalis* failed to grow or develop on any media tested (Table XVIII). In one instance, however, such hyphae did differentiate gametes shortly after dissection. All gametes formed in this manner encysted within 48 hr, but failed to show further signs of development.

Although success in culturing *Coelomomyces* spp. has been limited, the accomplishments thus far are encouraging and will hopefully lead to further efforts in this area. Of special interest would be successful culture of the gametophyte, as the gametes arising from this stage unite to form zygotes, the units responsible for infection of mosquitoes.

REFERENCES

Castillo, J. M., and Roberts, D. W. (1980). *In vitro* studies of *Coelomomyces punctatus* from *Anopheles quadrimaculatus* and *Cyclops-vernalis*. *J. Invertebr. Pathol.* **35,** 144–157.

Chen, T. A., and Liao, C. H. (1975). Corn stunt spiroplasma: Isolation cultivation, and proof of pathogenicity. *Science* **188,** 1015–1017.

Chen, T. A., Liao, C. H., and Grace, T. D. C. (1962). Establishment of four strains of cells from insect tissues grown *in vitro*. *Nature* (*London*) **195,** 788–789.

Couch, J. N. (1968). Sporangial germination of *Coelomomyces punctatus* and the conditions favoring the infection of *Anopheles quadrimaculatus* under laboratory conditions. *Proc. Jt. U.S.-Jpn. Semin. Microb. Control Insect Pests, 1967* pp. 93–105.

Couch, J. N. (1972). Mass production of *Coelomomyces,* a fungus that kills mosquitoes. *Proc. Natl. Acad. Sci. U.S.A.* **69,** 2043–2047.

Federici, B. A. (1975). *Cyclops vernalis* (Copepoda: Cyclopoida): An alternate host for the fungus, *Coelomomyces punctatus*. *Proc. Pap. Annu. Conf. Calif. Mosq. Control Assoc.* **43,** 172–174.

Federici, B. A. (1980). Production of the mosquito-parasitic fungus, *Coelomomyces dodgei,* through synchronized infection and growth of the intermediate copepod host, *Cyclops vernalis*. *Entomophaga* **25,** 209–217.

Federici, B. A. (1981). Mosquito control by *Culicinomyces, Lagenidium,* and *Coelomomyces. In* "Microbial Control of Pests and Plant Diseases 1970–1980" (H. D. Burges, ed.), pp. 555–572. Academic Press, New York.

Federici, B. A., and Chapman, H. C. (1977). *Coelomomyces dodgei:* Establishment of an *in vivo* laboratory culture. *J. Invertebr. Pathol.* **30**(3), 288–297.

Federici, B. A., and Roberts, D. W. (1975). Experimental laboratory infection of mosquito larvae with fungi of the genus *Coelomomyces.* I. Experiments with *Coelomomyces psorophorae* var. in *Aedes taeniorhynchus* and *Coelomomyces psorophorae* var. in *Culiseta inornata. J. Invertebr. Pathol.* **26,** 21–27.

Federici, B. A., and Roberts, D. W. (1976). Experimental laboratory infection of mosquito larvae with fungi of the genus *Coelomomyces.* II. Experiments with *Coelomomyces punctatus* in *Anopheles quadrimaculatus. J. Invertebr. Pathol.* **27**(3), 333–341.

Hink, W. F. (1970). Established insect cell line from the cabbage looper, *Trichoplusia ni. Nature (London)* **226,** 466–467.

Humber, R. A. (1975). The *in vitro* culture and development of an obligately parasitic fungus of flies. *Abstr., 8th Annu. Meet. Soc. Invertebr. Pathol.* p. 33.

McCoy, T. A., Maxwell, M., and Kruse, P. F., Jr. (1959). Amino acid requirements of the Novikoff Hepatoma *in vitro. Proc. Soc. Exp. Biol. Med.* **100,** 115–118.

Madelin, M. F. (1968). Studies on the infection by *Coelomomyces indicus* of *Anopheles gambiae. J. Elisha Mitchell Sci. Soc.* **84**(1), 115–124.

Mitsuhashi, J., and Maramorosh, K. (1964). Leafhopper tissue culture: Embryonic, nymphal, and imaginal tissues from aseptic insects. *Contrib. Boyce Thompson Inst.* **22,** 435–460.

Parker, R. C., Castor, L. N., and McCulloch, E. A. (1957). Altered cell strains in continuous culture: A general survey. *Spec. Publ. N.Y. Acad. Sci.* **5,** 303.

Pillai, J. S., and Woo, A. (1973). *Coelomomyces opifexi* (Pillai and Smith) Coelomomycetaceae: Blastocladiales. III. The laboratory infection of *Aedes australis* (Erichson) larvae. *Hydrobiologia* **41**(2), 169–181.

Pillai, J. S., Wong, T. L., and Dodgshun, T. J. (1976). Copepods as essential hosts for the development of a *Coelomomyces* parasitizing mosquito larvae. *J. Med. Entomol.* **13,** 49–50.

Schneider, I. (1964). Differentiation of larval dropsophila eye-antennal discs *in vitro. J. Exp. Zool.* **156,** 91–104.

Shapiro, M., and Roberts, D. W. (1976). Growth of *Coelomomyces psorophorae* mycelium *in vitro. J. Invertebr. Pathol.* **27,** 399–402.

Weiser, J. (1977). The crustacean intermediary host of the fungus *Coelomomyces chironomi* Rasin. *Ceska Mykol.* **31**(2), 81–90.

Whisler, H. C. (1966). Host-integrated development in the Amoebidiales. *J. Protozool.* **13,** 183–188.

Whisler, H. C., Zebold, S. L., and Shemanchuk, J. A. (1974). Alternate host for mosquito parasite *Coelomomyces. Nature (London)* **251,** 715–716.

Whisler, H. C., Zebold, S. L., and Shemanchuk, J. A. (1975). Life history of *Coelomomyces psorophorae. Proc. Natl. Acad. Sci. U.S.A.* **72,** 963–966.

Wolf, K., and Quimby, M. C. (1964). Amphibian cell culture: Permanent cell line from the bull frog (*Rana catesbeiana*). *Science* **144,** 1578–1580.

Wong, T. L., and Pillai, J. S. (1978). *Coelomomyces opifexi* Pillai and Smith (Coelomomycetaceae: Blastocladiales). IV. Host range and relative susceptibility of *Aedes australis* and *Opifex fuscus* larvae. *N. Z. J. Zool.* **5,** 807–810.

Yunker, C. E., Vaughn, J. L., and Cory, G. (1967). Adaptation of an insect cell line (Grace's antheraea cells) to medium free of insect hemolymph. *Science* **155,** 1565–1566.

8 Ecology and Use of *Coelomomyces* Species in Biological Control: A Review

H. C. CHAPMAN

Gulf Coast Mosquito Research
Agricultural Research Service
U. S. Department of Agriculture
Lake Charles, Louisiana

I. ECOLOGY (HOST–PARASITE RELATIONSHIPS)

The ecology of many mosquitoes is well known and understood. However, the ecology of ostracods and copepods, as intermediate hosts of *Coelomomyces,* is not well known, and practically nothing is known on the favorable or unfavorable ecological parameters involving the fungus itself. Since *Coelomomyces* species are living entities, it is reasonable to assume that they cannot be forced into an unfavorable habitat and be expected to survive, much less multiply. There are limited numbers of studies involving the collection of environmental or ecological factors in the sites producing naturally occurring *Coelomomyces* infections in mosquitoes, all of which were conducted prior to the discovery of microcrustaceans as intermediate hosts in the life cycle of *Coelomomyces.*

In one of the earliest studies involving the collection of such data on ecological factors, Muspratt (1963) found no correlation in water temperature and pH between infected and uninfected pools containing larvae of *Anopheles gambiae* in Rhodesia. Umphlett (1968) collected data for two seasons on pH, dissolved oxygen, air temperature, relative humidity, and aquatic vegetation from two lake coves in North Carolina (one with *C. punctatus*-infected *An. quadrimaculatus* larvae and the other negative for the fungus). While he did observe some differences, none of these factors appeared to be significant enough to be responsible for the presence or absence of the fungus.

In a 3-year study by Pillai (1971), data on water temperature, pool size, bottom sedimentation, pool biota, and salinity were collected from 400 pools in New Zealand, 21 of which produced *C. opifexi* infections in *Opifex fuscus* or *Aedes australis*. Only significant differences in salinity levels were noted between the infected and uninfected pools. A positive correlation was observed between lower ranges of salinity and the appearance in pools of infected larvae. Such pools served as a source of infection, which then spread to other pools in which favorable conditions occurred.

Chapman and Glenn (1972) showed no important difference between infection levels of *C. dodgei* in *An. crucians* and air temperature in a permanent pond in Louisiana. The infection levels of *C. punctatus* in *An. crucians* in a more temporary pond were 6–7 times lower at the warmest temperatures, but this is merely a reflection of the drought conditions that plagued this pond during the hot summers. Infection levels in the *C. dodgei* pond tended to progressively increase as the larval populations increased (except at the highest population level).

Deshevykh (1973) in the U.S.S.R. reported good survival and high infectivity of *C. iliensis* in larval populations of *Culex modestus* in basins of shallow water with low salt concentrations (1–5%) and temperatures up to 39°C. Deeper water, higher salinities, and lower water temperatures did not seem to favor epizootics.

As mentioned previously, all of these studies were done before the complete life cycle of *Coelomomyces* was known, and thus do not consider the effects of the ecological factors on the intermediate host. One can only conclude that our knowledge on the bioecology of the fungus in nature is about nil. It seems evident that, at least in some instances, the bioecology of the fungal sporangia mimics that of the mosquito species. *Culiseta inornata* larvae occur in southwestern Louisiana normally during the colder months (October to April) in fresh-water situations but also in the salt marshes, even adjacent to the Gulf of Mexico. These salt marsh areas are periodically flooded throughout the year, both by rains and high tides. Thus, the sporangia of *C. psorophorae* complex are often flooded for a 5- to 6-month period when the mosquito host is absent. Past Gulf

Coast Mosquito Research Laboratory observations in the laboratory have shown that only limited numbers of sporangia from a larva of *Cs. inornata* dehisce at any given flooding, a trait that undoubtedly aids in the survival of the fungus in nature. Conversely, a high percentage of the sporangia of *C. psorophorae* complex from larvae of *Ae. sollicitans* (which breeds throughout the year in Louisiana salt marshes) dehisce during the first flooding in the laboratory. These members of *C. psorophorae* complex are not dimorphic and so do not possess the very evident thin- and thick-walled sporangia like *C. indicus, C. africanus,* etc. It is very evident that a great research emphasis is needed on the ecology of the fungal species before the success of their large-scale releases can be positively predicted.

II. NATURAL CONTROL (INFECTION LEVELS AND PERSISTENCE)

Although *Coelomomyces*-infected mosquitoes have been collected from all major continents, many of these reports represent single records and therefore present no evidence on the average infection levels or persistence over time. Other reports encompass continuing collections of infected mosquitoes and present data on the highest infection levels observed such as 96% by *C. psorophorae* in *Psorophora howardii* and *Cs. inornata,* 97% by *C. maclaeyae* in *Ae. triseriatus,* 92% by *C. pentangulatus* in *Cx. peccator,* and 100% in *An. crucians* by *C. punctatus* (Chapman, 1973). Also Pillai (1969) and Pillai and Smith (1968) reported infection levels up to 65% by *C. opifexi* in *Ae. australis* and *Op. fuscus.*

To accurately portray the average level of infection from mosquito populations in a particular site, it is necessary to check such populations at weekly or monthly intervals for at least one or two breeding seasons. Relatively few studies of this type have been done. Weekly surveillance over 3 years of a permanent lake in North Carolina by Umphlett (1970) showed that average yearly infection levels of *C. punctatus* in larvae of *An. quadrimaculatus* decreased from 37 to 26 to 10%. The average infection level for the 3 years was 24%.

In Louisiana, Chapman and Glenn (1972) surveyed weekly larval populations of *An. crucians* in a pond for $4\frac{1}{2}$ years and found that infection levels of *C. punctatus* decreased from 67 to 12%, with an average of 33% over the long study period. Larval populations of *An. crucians* in another pond were dipped weekly for $2\frac{1}{4}$ years and the incidence of *C. dodgei* increased from 24 to 59% with an average infection level of 48% during the time period. The site producing infections of *C. punctatus* was a temporary pond that frequently dried up, whereas the other site possessing *C. dodgei* dried only once during its surveillance.

In the earliest studies with *Coelomomyces,* Muspratt (1946a) estimated mortality of *An. gambiae* by *Coelomomyces,* probably *indicus,* to range as high as 95% over a 4-year period in pools in Rhodesia. Almost 20 years later, pools in the same general area produced almost a 100% infection by *C. indicus* in *An. gambiae* (Muspratt, 1963).

In other, usually less intensive studies, Service (1977a) in Kenya reported infections in collections over several months of principally *C. indicus* in *An. gambiae* (species B of complex) at a average level of 2.1 and 15.9% in ponds and rice fields, respectively. Service (1977b), in a 6-year study of *Ae. cantans* in southern England, found less than 1% of their larvae infected with *Coelomomyces,* probably *psorophorae.*

In India, Gugnani *et al.* (1965) reported an average infection level of 45% in *An. subpictus* larvae by *C. indicus* over a 4-month period.

Shemanchuk (1959) noted a 12% level of infection of *C. psorophorae* in eight weekly collections of *Cs. inornata* in Canada.

Deshevykh (1973) in a two season survey in U.S.S.R. reported that *Cx. modestus* populations showed similar yearly infection levels ranging from 9 to 100% by *Coelomomyces,* probably *iliensis.*

While the Muspratt studies denote persistence of this fungus, very few reports show a yearly persistence from a precise site over long time periods. Umphlett (1970) and Chapman and Glenn (1972) denote persistence of *C. punctatus* for 3 and 4½ years, respectively, from the same habitats.

The Gulf Coast Mosquito Research Laboratory (GCMRL), Lake Charles, Louisiana, has unpublished data that exemplify the excellent persistence of several species of *Coelomomyces* from the same sites. Larval collections were often made weekly or several times a month when time permitted and when water was present in the specific sites. Levels of infection were not always computed, as sometimes only the number of infected larvae was noted. Larvae of *An. crucians* infected with *C. punctatus* were first noticed in two different ponds in early 1966. These two ponds were periodically checked through the years (50 and 182 positive collections) and were still producing infected anopheline larvae in early 1979, 13 years later. The highest levels of infection were 94 and 100%. Another pond produced *C. dodgei* infections in larvae of *An. crucians* from 1968 through 1980 (226 positive collections with the highest infection of 86%). A tree stump hole produced larvae of *Cx. peccator* infected with *C. pentangulatus* from 1968 to 1980 (53 positive collections with the highest infection level of 92%). It must be mentioned that numerous times throughout these years of collections, the specific sites were dry, or flooded but devoid of the mosquito host, or had mosquito hosts that were apparently uninfected. Since *Cx. peccator* breeds only from April through September in Louisiana, the fungus must survive the cooler weather in the absence of the primary host. Lastly, Chapman and Glenn (1972) further stressed persistance when they reported that one site that was checked weekly was either dry or negative for *C. punctatus* for 29 consecutive weeks, after which the fungus reappeared in the mosquito populations.

III. FIELD STUDIES (RELEASES)

Field releases have been very limited even if one considers fungal releases into semifield situations. All releases to date have involved the introduction of infected host larvae and not the intermediate host. The first successful attempt to infect larvae outside of the laboratory was reported by Walker (1938) when he infected larvae of *An. gambiae* by placing them in a cement tank with *C. africanus*. Muspratt (1946a) introduced uninfected larvae of *An. gambiae* into small pot holes lined with mopane clay soil from a *Coelomomyces* infected area, and the introduced larvae developed infections after 48 hr.

Muspratt (1946b) obtained infections in *An. gambiae* when he seeded a concrete trough lined with mopane clay with dried infected cadavers of *An. gambiae*. Laird's 1967 experiment in the Tokelou Islands is covered in the next chapter. Also, Muspratt (1963) successfully introduced infected larvae of *An. gambiae* into a pool, and subsequently healthy larvae of *An. gambiae* developed infections of *Coelomomyces,* probably *indiana*. Chapman (1974) seeded two outdoor artificial ponds, one with *Cs. inornata* infected with *C. psorophorae,* and the other with *An. crucians* infected with *C. punctatus*. Almost 40% of the ensuing exposed larvae of *An. quadrimaculatus* developed infections of *C. punctatus* over a 15-week period. Average weekly infection levels of *C. psorophorae* in introduced larvae of *Cs. inornata* never exceeded 29% over a 1-month period.

Nolan *et al.* (1973) reported the successful infection of *Ae. triseriatus* and *Toxorhynchites rutilus septentrionalis* in two treehole sites seeded with *Ae. triseriatus* cadavers infected with *C. maclaeyae*.

Couch (1972) in North Carolina seeded ditches with inoculum of *C. punctatus* and eggs of *An. quadrimaculatus* through four summers and reported infection levels ranged from near 0% to 100%, with an average of 60%.

Several field trials were conducted by Gad and Sadek (1968) in Egypt when *C. indicus*-infected cadavers of *An. pharoensis* were released on the edges of rice fields and rice-field basins. A rice field seeded with *C. indicus* in 1966 produced infection levels for 3 months in 1967, ranging from 6 to 94% in larvae of *An. pharoensis*. Two of three rice-field basins produced weekly infection levels of 2 and 90% in *An. pharoensis* after it was treated with cadavers infected with *C. indicus*.

Dzerzhinskii *et al.* (1975), in the U.S.S.R., reported infection levels of 9–70%, 13–32%, and 5–17% in larvae of *Cx. modestus,* when a suspension, biomass, or substratum mud, respectively, containing infective units of *C. iliensis* were applied in small-scale field treatments.

Many large areas in the U.S.S.R. have been successfully seeded by releases of larvae infected with *C. iliensis* (personal communications from A. M. Dubitskij).

It is worth noting that all of these field releases were done before the demonstration that the presence of an intermediate host (a microcrustacean) was necessary to obtain mosquito infections.

It is also interesting to note that in the releases just discussed, the inoculative or recycling effect does not seem nearly as persistent as that observed in many natural infections. Either most releases seem to terminate after several weeks or months or there was no follow up on the treated sites in the ensuing years. Since persistence of 12–13 years or more for natural infections by some *Coelomomyces* species occurs, the lack of such persistence from field releases may well indicate problems involving intermediate host populations.

IV. POTENTIAL

In its present state of the art, in which both the mosquito host and microcrustacean intermediate host must be reared *in vivo,* there are essentially three ways that *Coelomomyces* can be used against mosquitoes.

A. Sporangial Releases

Large numbers of infected larvae can be produced in the laboratory, and these dead larvae or their sporangia can be disseminated in the field. No infections will occur in mosquitoes unless adequate numbers of a suitable intermediate host are naturally present in the environment or are released into it with the sporangia. This inoculative release could result in recycling of the fungus and partial to almost complete reduction of future mosquito populations. All field releases to date have been with sporangia.

B. Mosquito-Infective Zygote Releases

The mass production and release of zygotes from infected microcrustaceans would cause an instant reduction and inundative control of the vector or target species. Since the release is applied as a chemical or microbial pesticide, repeated applications must be made to combat new mosquito populations. Therefore, economical mass production of the infected host must be possible. If adequate numbers of the intermediate host are naturally present in the treated area, applications of the zygote could eventually result in some establishment and recyling of the fungus. To date, no zygote field releases per se have been attempted.

C. Combined Releases

Such envisioned field treatments combine releases of both sporangia and zygotes, thus probably insuring a better and quicker establishment of the fungus in the release area, provided that suitable numbers of an adequate intermediate host are present. If the intermediate host population is low or absent, releases of it must be made to insure fungal establishment.

A natural balance between the fungus and hosts may eventually occur, thus producing inadequate reductions of the pest or vector mosquito populations. Hence, periodical reintroductions of the fungus will be necessary.

The potential for the eventual use of various species of *Coelomomyces* by mosquito control districts and vector control groups is promising, but many obstacles remain to be overcome. With the possible exception of some developing countries where labor costs may be low, the application of this fungus to large areas will never occur until both sporangia and mosquito infective zygotes can be produced by *in vitro* means. Also, our very inadequate knowledge of the ecological parameters of both the intermediate host and the fungus (sporangia and zygotes) must be completed. Only then can we successfully duplicate the epizootics that cause continued mortalities of over 90% in nature.

REFERENCES

Chapman, H. C. (1973). Assessment of the potential of some pathogens and parasites of biting flies. *Proc. Symp. Biting Fly Control Environ. Qual., 1972,* pp. 71–77.

Chapman, H. C. (1974). Biological control of mosquito larvae. *Annu. Rev. Entomol.* **19,** 33–59.

Chapman, H. C., and Glenn, F. E., Jr. (1972). Incidence of the fungus *Coelomomyces punctatus* and *C. dodgei* in larval populations of the mosquito *Anopheles crucians* in two Louisiana ponds. *J. Invertebr. Pathol.* **19,** 256–261.

Couch, J. N. (1972). Mass production of *Coelomomyces,* a fungus that kills mosquitoes. *Proc. Natl. Acad. Sci. U.S.A.* **69,** 2043–2047.

Deshevykh, N. D. (1973). Seasonal infection of larval *Culex modestus* by *Coelomomyces* fungi in the Ili River basin. *In* "Regulators of the Numbers of Blood-Sucking Flies in South-Eastern Kazakhstan" (A. M. Dubitskii, ed.), pp. 22–28. Kazka, Acad. Sci., USSR.

Dzerzhinskii, V. A., Dubitskii, A. M., and Deshevykh, N. D. (1975). Determination of the optimum doses for infection of *Culex modestus* larvae with the entomopathogenic fungus *Coelomomyces iliensis. Izv. Akad. Nauk Kaz. SSR, Ser. Biol.* **5,** 52–57.

Gad, A. M., and Sadek, S. (1968). Experimental infection of *Anopheles pharoensis* larvae with *Coelomomyces indicus. J. Egypt. Public Health Assoc.* **43,** 387–391.

Gugnani, H. C., Wattal, B. L., and Kalra, N. L. (1965). A note on *Coelomomyces* infections in mosquito larvae. *Bull.—Indian Soc. Malaria Commun. Dis.* **2,** 333–337.

Muspratt, J. (1946a). On *Coelomomyces* fungi causing high mortality of *Anopheles gambiae* larvae in Rhodesia. *Ann. Trop. Med. Parasitol.* **40,** 10–17.

Muspratt, J. (1946b). Experimental infection of the larvae of *Anopheles gambiae* with a *Coelomomyces* fungus. *Nature* (*London*) **158,** 202.

Muspratt, J. (1963). Destruction of the larvae of *Anopheles gambiae* Giles by a *Coelomomyces* fungus. *Bull. W.H.O.* **29,** 81–86.

Nolan, R. A., Laird, M., Chapman, H. C., and Glenn, F. E., Jr. (1973). A mosquito parasite from a mosquito predator. *J. Invertebr. Pathol.* **21,** 172–175.

Pillai, J. S. (1969). A *Coelomomyces* infection of *Aedes australis* in New Zealand. *J. Invertebr. Pathol.* **14,** 93–95.

Pillai, J. S. (1971). *Coelomomyces opifexi* (Pillai and Smith) Coelomomycetaceae: Blastocladiales. 1. Its distribution and the ecology of infection pools in New Zealand. *Hydrobiologia* **38,** 425–436.

Pillai, J. S., and Smith, J. M. B. (1968). Fungal pathogens of mosquitoes in New Zealand. I. *Coelomomyces opifexi* sp. n. on the mosquito *Opifex fuscus* Hutton. *J. Invertebr. Pathol.* **11,** 316–320.

Service, M. W. (1977a). Mortalities of the immature stages of species B of the *Anopheles gambiae* complex in Kenya: Comparison between rice fields and temporary pools, identification of predators, and effects of insecticidal spraying. *J. Med. Entomol.* **13,** 535–545.

Service, M. W. (1977b). Ecological and biological studies on *Aedes cantans* (Meig.) (Diptera: Culicidae) in southern England. *J. Appl. Ecol.* **14,** 159–196.

Shemanchuk, J. A. (1959). Note on *Coelomomyces psorophorae* Couch, a fungus parasite on mosquito larvae. *Can. Entomol.* **91,** 734–744.

Umphlett, C. J. (1968). Ecology of *Coelomomyces* infections of mosquito larvae. *J. Elisha Mitchell Sci. Soc.* **84,** 108–114.

Umphlett, C. J. (1970). Infection levels of *Coelomomyces punctatus,* an aquatic fungus parasite, in a natural population of the common malaria mosquito, *Anopheles quadrimaculatus. J. Invertebr. Pathol.* **15,** 299–305.

Walker, A. J. (1938). Fungal infections of mosquitoes, especially of *Anopheles costalis. Ann. Trop. Med. Parasitol.* **32,** 231–244.

9 Use of *Coelomomyces* in Biological Control: Introduction of *Coelomomyces stegomyiae* into Nukunono, Tokelau Islands

MARSHALL LAIRD[1]

Research Unit on Vector Pathology
Memorial University of Newfoundland
St. John's, Newfoundland, Canada

[1] Present address: Whangaripo Valley Road, Rural Delivery No. 2, Wellsford, Northland, New Zealand.

I. INTRODUCTION

The eighth session of the Expert Committee on Insecticides, World Health Organization (WHO), Geneva, 1957 (Anonymous, 1958) recognized that arthropod vectors of diseases were developing resistance to synthetic organic pesticides faster than our capacity to solve vector control problems. Possible solutions were felt to include biocontrol through pathogens. In exploring ways to implement pertinent research, WHO noted a suggestion that had been made in Working Paper number 2 for this Expert Committee's seventh session, Geneva, 1956 (Anonymous, 1957). In this document, Laird (1956a) had argued that field trials of such pathogens against mosquitoes might lead to their establishment in new areas where, in the words of Steinhaus (1949), they could "maintain themselves and take a small but significant toll of insects year after year." Laird further suggested that "supported by carefully selected predators, and by appropriate chemical measures (particularly peridomestic anti-adult ones) as well, mosquito parasites might prove an economical and effective means of reducing susceptible populations of dangerous mosquitoes below the level necessary for the continued transmission of human pathogens."

Submitting that the necessary field data could be attained "in the minimum time by taking advantage of the special conditions prevailing in the tropical Pacific, where optimum conditions for the necessary investigations are provided by the isolation and very small area of some of the islands, their limited range and number of larval habitats and their limited mosquito fauna," he pointed out that *Aedes* (*Stegomyia*) *polynesiensis*, the vector of the subperiodic filariasis parasite *Wuchereria bancrofti* throughout tropical Polynesia, was the only known species of mosquito on Nukunono (Laird, 1955). It was recommended that one of the two other Tokelau atolls, Atafu and Fakaofo, could be left untouched as an experimental control, "the remaining island and Nukunono itself being used as introduction sites for *Coelomomyces stegomyiae* and any of the other potential biological control agents agreed upon."

In arranging with the New Zealand authorities that New Zealand's tropical Tokelau Islands dependency could indeed serve as the experimental site, it was decided that the third island would be used not for a second biocontrol experiment but for parallel testing of a chemical larvicide. Atafu was chosen for this purpose. The results of trials there with slow-release dieldrin-cement briquettes duly proved highly encouraging—persistence of field effectiveness in natural and domestic container situations was demonstrated over the next 5 years, after which serum samples from 20 islanders disclosed an average level of only 0.0010 ppm dieldrin (little more than half that then prevailing in the general American population through consumption of foodstuffs earlier treated with this pesticide) (Laird, 1967). Fakaofo was selected for experimental control purposes and Nukunono for the introduction of *C. stegomyiae*.

II. KNOWLEDGE OF *COELOMOMYCES* INFECTIVITY IN 1958

Little was understood and much was assumed about the manner of infection of mosquito larvae by *Coelomomyces* when this project was planned. However, even less was then known of precisely how to achieve large-scale infections in larval Culicidae with, for example, microsporidan protozoa or mermithid nematodes.

In 1958 it was taken for granted that direct exposure of unparasitized larval mosquitoes to infective material derived from dead parasitized mosquitoes would result in new infections: either via the ingestion of the sporangia (followed by dehiscence and release of flagellated zoospores in the gut, these then penetrating the intestinal wall to gain access to the hemocoel) or by dehiscence in the water followed by zoospore invasion of the host through its external cuticle.

Occasional laboratory infections had been achieved among unparasitized mosquito larvae exposed to dried *Coelomomyces* sporangia. One example was Laird's (1959) infection of *Aedes* (*Stegomyia*) *aegypti* larvae with *C. stegomyiae* sporangia derived from *Aedes* (*Stegomyia*) *albopictus* in Singapore in 1956. Because of the primary assumption that infection was direct, no efforts were made at that time to exclude, for example, copepods from the rearing containers. Microcrustaceans were certainly among the concentrates of small aquatic organisms from natural container-type larval habitats added to the rearing pans as food for the experimental mosquitoes. They could also have gained access to the cultures in fresh bottom debris introduced from larval habitats. The same must have applied in the case of earlier laboratory infections.

Thus, Walker (1938), who parasitized some *Anopheles gambiae* larvae with *Coelomomyces africanus* in a concrete tank in Sierra Leone, had added water, vegetation, and sediment to the tank from a natural *An. gambiae* larval habitat. Muspratt (1946) achieved his infections of *An. gambiae* larvae by *Coelomomyces indicus* in pot-holes dug in the mopane-clay soil of an "infected area" after heavy rain, commenting perceptively that "the possibility cannot entirely be excluded of transmission of the infection by some other inhabitants of the soil, or even by a winged insect." These routes nevertheless seemed unlikely to Muspratt (1946) because, in his examinations of associated organisms, none had "ever been found to contain the fungus."

III. ACTUAL PROCESS OF INVASION AND INFECTION OF MOSQUITO LARVAE BY *COELOMOMYCES*

In describing conditions favoring laboratory infections of *Anopheles quadrimaculatus,* Couch (1968) stressed that no previous workers "had been able to get infection of the larvae without using soil and detritus from the location in which infected larvae had been found," but went on to state that although he had achieved a few infections without soil "our best results have come when soil was used." However, Couch's first successful infection experiments in a sink lacking soil were undertaken in the presence of small algal mats (*Spirogyra, Oedogonium*). In discussing features that cause failure of infection, he included the eating of sporangia "by an ostracod or by other small aquatic animals," noting that microscopic examination revealed sporangia in the gut contents of such ostracods.

Madelin (1968) succeeded in infecting *An. gambiae* by *Coelomomyces indicus* in the laboratory in Bristol, U.K., but once again soil (mopane clay imported from Africa with WHO assistance) was among the features shared by his successful tests, as was "the presence of water fleas which probably originated from the mopane clay."

Couch (1972) then described successful *in vivo* mass production of *C. punctatus* in *An. quadrimaculatus* in the laboratory. His experimental pans contained a dense growth of unicellular green algae and Protozoa and, presumably, other aquatic microorganisms as well.

Next, Whisler *et al.* (1974, 1975) revealed the role of a microcrustacean intermediate host (*Cyclops vernalis*) in the life cycle of *Coelomomyces psorophorae*. Federici and Roberts (1976) reported that planonts (= zoospores) of *C. punctatus* released from sporangia developed in *An. quadrimaculatus* are not infective for mosquito larvae, their data suggesting that the intermediate host from which uniflagellate planonts infective for mosquitoes were derived was again the copepod, *Cyclops vernalis*. Other cyclopids have since been reported to play the same role for additional species of *Coelomomyces* elsewhere. Ostracods were first shown to also be capable of serving as intermediate hosts for *Coelomomyces* by Weiser (1976). Federici (1977, 1979) has confirmed that the life cycle is of the *Euallomyces* type, with gametophyte and sporophyte generations alternating. The latter (in the mosquito host) leads to the formation of meiosporangia (= sporangia), which release haploid meiospores after the host's death. They parasitize the microcrustacean host, each developing into a gametophyte that cleaves at maturation to form thousands of haploid gametes. The death of the microcrustacean host releases these. Opposite mating types fuse to form biflagellate zygotes (= zoospores), which infect mosquito larvae to initiate sporophyte formation. It remained for Travland (1979) to demonstrate that the transcuticular route applies for the entry of microcrustacean-derived planonts of *Coelomomyces* into mosquito larvae.

All these events lay far in the future when the 1958 introduction of *C. stegomyiae* into Nukunono was planned.

IV. COLLECTION OF INFECTIVE *COELOMOMYCES STEGOMYIAE* MATERIAL AT SINGAPORE, MARCH TO AUGUST, 1958

Dr. D. H. Colless of the Department of Parasitology, University of Malaya at Singapore (now University of Singapore), was to accompany me to Nukunono as a WHO Consultant. He supervised the preparation of *Coelomomyces stegomyiae* sporangial (= meiosporangial) concentrates from 48 (2.0%) of 2454 *Aedes albopictus* larval habitats sampled at Singapore. A further 16 habitats of several hundred randomly sampled had sporangia free in the detritus, although none of the larvae present were infected (Laird and Colless, 1959). The true percentage of positive habitats, which appeared to show a nonrandom distribution in both time and space, was thus probably distinctly higher than 2.0%. Monthly percentages of infection varied from 0.6 to 3.1% (Table XIX). These fluctuations presumably reflected rainfall conditions, among others. Interestingly, positive habitats tended to occur in groups in the same general locality. This suggested localized dispersal of the pathogen following an inoculation of infective material—perhaps via positive females dying after oviposition.

The following four paragraphs (and Tables XIX–XXI) are reproduced verbatim from Laird and Colless (1959) except for emendment of the original Table numbers 2 and 3 to XX and XXI. They summarize studies conducted by Dr. D. H. Colless and technical staff under his direction.

> In 37 habitats positive for *Coelomomyces*, the percentage infection rates, at the time of collection, were almost exactly the same in 2nd, 3rd, and 4th instar larvae (Table XX). The time spent in the 1st instar is probably rarely long enough for infections to appear, though they may be acquired and appear in the 2nd instar. A single infection was, in fact, seen in a 1st instar larva in a very slow-growing laboratory culture. Generally, it seems that susceptibility to infection is of the same order in all instars.

TABLE XIX. Monthly Totals of Habitats Yielding Parasitized Larvae (Singapore)

Month	Number of collections	Habitats with parasitized larvae	
		Number	% Of total
March	144	4	2.7
April	163	5	3.1
May	314	2	0.6
June	288	4	1.4
July	706	24	2.9
Total	2454	48	2.0

Larval parasitaemia in naturally seeded habitats was usually heavy, and the larvae failed to develop any further. In several instances, however, mainly in laboratory-infected larvae, with rather light infections, normal pupation and eclosion of the adult was observed. Of such adults, 4 out of 9 females and 4 out of 10 males showed numerous sporangia circulating in the body cavity, though none were found in the ovaries or alimentary tract. Two which died in the pupal stage proved to be packed with sporangia. Where pupation and eclosion occurred, not all sporangia carried over to the next stage, since many were found to remain in the anal papillae of the cast larval skin, and, on one occasion, in the pupal paddles. This probably explains why some 60% of the adults no longer had sporangia, since in lightly infected larvae most of the sporangia tend to wash back into the anal papillae.

Most *Coelomomyces*-positive containers found in the field were brought to the laboratory, where their uninfected larvae were returned to them, and a watch was kept for further infections. The debris always contained many apparently viable sporangia, but, in most batches, few further infections were obtained, although kept for long periods. Many attempts at laboratory infection were made, using sporangia from parasitized larvae, either fresh, or after various periods of desiccation. Most of these met with failure. Infections resulted in two cases, however, in one after 13 days and in the other after 24 days. Dried sporangia were employed on both occasions, but as the larvae were maintained in water and debris from the field, it is possible that undried

TABLE XX. Parasite Rates According to Instar in 37 *Coelomomyces*-Positive Habitats (Singapore)

Instar	Number of larvae examined	Number parasitized	% Parasitized
4	765	103	13.4
3	372	49	13.2
2	196	27	13.8
1	120	0	0.0
Total	1453	179	12.3

TABLE XXI. Proportions of *Coelomomyces*-Positive Habitats According to Permanence of Container (Singapore)

Container	Number examined	Number positive	% Positive
Permanent[a]			
Intact bottles	37	0	0.0
Other	781	19	2.4
Total	818	19	2.3
Impermanent[b]	1660	30	1.8

[a] Made of glass, stone, rubber, heavy metal, etc.
[b] Made of tinplate, canvas, etc.

sporangia were also introduced. Another success was obtained when 9 apparently uninfected treeholes were seeded with dried sporangia; infected larvae were found in one only, after $2\frac{1}{2}$ months. . . . [1]

It is submitted that natural seeding of habitats might well occur through the medium of sporangia from dried infective debris; e.g., from tins which have rusted away. It is possible that infected adults occasionally die on the water in unseeded habitats, but this must be extraordinarily rare. Most natural infections are too heavy for further development of the concerned larvae. Only those which acquire infections shortly before pupation are likely to survive. Moreover, the probability that an infected adult will actually die in a larval habitat is very low. It seems far more likely that sporangia are blown by the wind, or carried on the bodies of scavenging insects or crabs [see p. 380]. This hypothesis is in accord with the tendency, noted above, for seeded habitats to occur in groups. The figures in Table XXI may also be of significance. As might be expected, there were relatively more positive records from habitats in containers made of the more permanent materials (glass, stone, heavy metal, etc.) than from those in tins, etc., which rust away fairly rapidly. However, no infections were ever found in larvae from unbroken bottles. The number examined is admittedly very small, but the fact does agree with the hypothesis of air-borne dispersal of sporangia, since the small orifice in an intact bottle would offer little access to such sporangia.

On two occasions the dehiscence of mosquito-derived (meio)sporangia was observed under the microscope, from 50–100 zoospores (= haploid meiospores) being released into the water. One particular batch of (meio)sporangia-laden debris from a treehole continued to give regular infections in the laboratory over a period of 4 months (presumably reflecting the conjoint presence of gametophyte-bearing microcrustaceans and consequently of mosquito-parasitizing biflagellate zygotes). While one aliquot from this container did not produce its first infection until 2 months

[1] With hindsight, suitable microcrustacean intermediate hosts for *C. stegomyiae* were presumably present in the fresh field-derived debris introduced into the experimental container in the laboratory and in the apparently uninfected treeholes.

had passed, a second (apparently identical) aliquot regularly produced infections from the start. At the time it was assumed that the explanation might lie in "some essential feature of pre-conditioning of the sporangia, or the physico-chemical condition of the water" (Laird and Colless, 1959). With current information, it seems probable that the explanation really lay in the scarcity of gametophyte-bearing microcrustaceans in one culture and their abundance in the other.

Bottom debris from laboratory cultures in which large numbers of dead, parasitized *A. albopictus* larvae were allowed to decompose in sediment from natural habitats formed a large part of the inoculum for the Nukunono trial. This intermittently dried and rewetted[2] material was divided into two batches with high and medium concentrations of *C. stegomyiae* (meio)sporangia, numbered 1 and 2, respectively. Inoculum 3 consisted of intermittently dried and rewetted sediment from a treehole from which parasitized larvae had been collected for several months. In the week prior to departure, the author (who was then on McGill University's faculty, at the Institute of Parasitology, Macdonald College, Quebec, Canada) worked with Dr. Colless at Singapore. All (58) infected larvae collected during this time were held in distilled water. Thirty of these larvae died, inoculum 4 comprising filter paper onto which their bodies were individually dried. Twenty-eight larvae remained alive, and, with the approval of authorities in countries where it was necessary to change flights on the way to the Tokelaus (i.e., Australia, Fiji, and Western Samoa), these were transported alive. Only two of them survived the 8-day journey to Nukunono. They were killed upon arrival and their bodies were added to the rich concentration of fresh sporangia at the bottom of the jar, constituting inoculum 5.

V. INTRODUCTION OF *COELOMOMYCES STEGOMYIAE* AT NUKUNONO, SEPTEMBER 3–22, 1958

Dr. Colless and I reached Nukunono aboard a Royal New Zealand Air Force (RNZAF) Sunderland flying boat on the morning of 3 September. Our field laboratory was set up in quarters kindly provided by the Roman Catholic Mission. That afternoon, 25 second- and third-instar larvae of *A. polynesiensis* from a household water container (44-gal oil drum) and a

[2] This routine was adopted to ensure the hatching of any *A. albopictus* or other insect eggs present. Frequent checks of the rewetted cultures at Singapore (and later at Nukunono, too) revealed neither insects nor any other macroscopic aquatic organisms.

nearby "tungu" (a local type of reservoir formerly of much importance in the Tokelaus, hollowed out into the lower part of the bole of a standing coconut palm; see Figure 624 and Laird *et al.*, 1982), were placed in a bowl with water from their habitats. The freshest inoculum in our possession, inoculum 5, was now examined microscopically. It proved to contain numerous (meio)sporangia in which the central protoplasm was undergoing division with motile meiospores evident in some cases. There were also some already dehisced empty (meio)sporangia.

No other aquatic organisms except some protozoans and bacteria were seen. There were certainly no animals as large as microcrustaceans present. As already mentioned, the larvae in question had been held in distilled water at Singapore, and during the journey, and the only food supplied was powdered yeast. The whole of inoculum 5 was now poured into the above bowl, to make a total of 250 ml of liquid, powdered yeast again being provided as food (Laird and Colless, 1959). By the following day, some of the *A. polynesiensis* larvae had become sluggish, apparently through some unexplained environmental circumstance. On 5 September (day 2) mortality began. By 11 September (day 8) the last five larvae had died. Each of them exhibited an abundant mycelium, some of the hyphae being terminally swollen at the onset of (meio)sporangial formation in the hemocoel. This indication of the susceptibility of local *A. polynesiensis* to infection by *C. stegomyiae* was the only such evidence obtained during the 1958 stay at Nukunono.

At the time, it was assumed that direct infection had taken place through zoospore (= meiospore) invasion from the dehisced free (meio)sporangia. With hindsight, one can only postulate that a suitable microcrustacean host had been present in the water poured into the bowl from the tungu and/or oil drum sampled. Collections made at Nukunono in 1953 (Laird, 1955) had already disclosed the presence of ostracods in a Nukunono (Motuhanga) tungu (Laird, 1956b); subsequent intensive 1958 collecting on all "motu" (= islets) of the atoll was to reveal the abundance of both ostracods and copepods in this habitat, and at least one of the latter is a confirmed host for *Coelomomyces*.

It is now known that the infective biflagellate zygotes, formed by fusion of gametes of opposite mating types escaping from dead copepods, parasitize mosquito larvae rapidly enough for these to display "typical *Coelomomyces* hyphal bodies and sporangia within a few days of inoculation" (Whisler *et al.*, 1974). Before field use, inocula 1–3 were twice flooded with water from a tungu, left overnight in basins tightly covered by polyethylene sheeting, and redried, as a final check against any insect eggs in them remaining viable. While (meio)sporangia remained plentiful, and especially so in inoculum 1, only a few bacteria, and rotifers and protozoa

Fig. 624. A characteristic tungu. Note the rainwater grooves (in one of which is a microscope slide with lines drawn 1 in apart for scale) running into the apex of the aperture.

of cosmopolitan species already present in the Tokelaus, were otherwise seen.

Over the next 3 weeks, inoculum 1 was pipetted into natural-container larval habitats of *A. polynesiensis* from Tokelau motu at the northeastern corner of Nukunono's lagoon and into other habitats on smaller motu southwards to Na Motu Ha. From here southwards to Fakahuapolo, inoculum 3 was used alone. Inoculum 2 was used alone on motu of the western reef (Te Kamu, Te Fakanava, Te Puka) north of the village motu, which, like the atoll itself, bears the name Nukunono. A mixture of inoculum 2 and 3 was employed everywhere else, total "seedings" as indicated by Laird and Colless (1959) being as follows:

Tungu	486
Other treeholes	181
Coconuts	88
Bracts	6
	761

In the light of the above, the earlier mortalities of Nukunono *A. polynesiensis* exposed to inoculum 5 on 3 September are more likely to be ascribable to the "unexplained environmental circumstance" mentioned by Laird and Colless (1959). Nevertheless, this particular experiment demonstrated the onset of *C. stegomyiae* (meio)sporangial formation in *A. polynesiensis* within what are now regarded as the usual time constraints. Couch (1972) and Federici and Roberts (1976) did not observe their first hyphal bodies in previously "clean" *C. punctatus* until 7 days after exposure to dehiscing zoospores (= meiospores), about 5 of the intervening days later being shown to be passed in the microcrustacean host (Federici, 1980).

VI. MONITORING SURVEYS AT NUKUNONO, 1959–1980

A. November 10–11, 1959

Toward the end of 1959, the New Zealand authorities arranged for an interim progress report. An RNZAF flying boat briefly visited all three atolls, enabling Wing Commander J. W. G. McDougall (Deputy Director of Medical Services, RNZAF) and G. W. Gibbs (Victoria University of Wellington) to monitor the results of the dieldrin-cement briquette experiment at Atafu and the *C. stegomyiae* experiment at Nukunono.

In an hour's collecting at Motuhanga, they found 11 tungu with larval *A. polynesiensis* present. Twenty-seven larvae, from two tungu, exhibited numerous (meio)sporangia of *C. stegomyiae*.

B. April 12–24, 1960

Again with New Zealand transportation assistance via RNZAF flying boats and a chartered copra boat, I visited all three atolls as a WHO Consultant (this time without Dr. Colless). Having only 12 days at Nukunono and working without the assistance of another scientist, there was insufficient time for a thorough survey of the approximately 45 motu. However, 118 of 667 tungu and natural treeholes "seeded" with *C. stegomyiae* inoculum in 1958 were relocated (Laird, 1960). (Meio)sporangia-packed larvae were collected from 11 (9.3%) of these, and viable resting sporangia were identified in the bottom debris of three more. The overall percentage of parasitized larvae per habitat was twice that derived from the preliminary field studies at Singapore, the fungal establishment achieved representing from three to seven times the natural incidence of the pathogen in the latter locality. Furthermore, evidence of natural dispersal from hand-seeded habitats was obtained, for parasitized larvae were found in two half-coconut shells (which from their condition obviously postdated the 1958 visit) a short distance from three *Coelomomyces*-positive tungu. A crab (*Sesarma gardineri,* Figure 625), common in tungu, was actually seen to dart to the ground from a tungu and within seconds splash through the contents of a nearby half-coconut shell, pointing to a probable manner of spread of infective material, whether (meio)-sporangia or gametophyte-bearing microcrustaceans.

All 1960 records of *C. stegomyiae* were obtained from only three of the motu of Nukunono—Nukunono village motu (Motuhanga and Vao districts), Tokelau motu, and Taulagapapa motu—where either inoculum 5 (Motuhanga), inoculum 1 (Tokelau), or a mixture of inocula 1 and 2 had originally been introduced. Biting collections yielded appreciably fewer female *A. polynesiensis* than had been the case 18 months previously. Nineteen 15-min adult catches (one man protected by repellent collecting from the back of another not so protected, who counted mosquitoes attempting to bite the front parts of his body) were repeated at the same sites and same times of day, as in 1958. The 1960 mean was only half (30.5) that of the former one (61.8). While too much significance cannot be attached to this reduction in mosquito biting incidence in view of the relatively small number of catches on which the figures are based and because fluctuations of this order could simply reflect differing rainfall conditions, it should be noted that the 1958 and 1960 adult catches at

Fig. 625. *Sesarma gardineri,* a crab commonly found in tungu, illustrated resting in a rainwater entry groove.

Fakaofo (the control atoll) compared very closely with one another:

1958	Mean (14 catches)	65.9
1960	Mean (14 catches)	82.4

C. July 1963

A New Zealand medical expedition to the Tokelaus (Dr. Ian Prior *et al.*) kindly facilitated a further progress report, collecting *A. polynesiensis* larvae from 35 tungu and other container habitats on Nukunono. By then,

I was without laboratory facilities as a member of the WHO Secretariat in Geneva, Switzerland, and Professor J. N. Couch (University of North Carolina, Chapel Hill, North Carolina) undertook the examination of the collections. Thirteen (37.1%) of the 35 larval habitats sampled proved to include parasitized larvae, a fourfold increase over the 1960 figures (Laird, 1967).

Professor Couch suspected that a second species of *Coelomomyces* was present in the material, together with possible hybrids between it and the dominant *C. stegomyiae*. This led to the suggestion (Laird, 1967) that two species of the genus instead of one might have been introduced into Nukunono in 1958 (double infections having been rather commonly reported with respect to *Coelomomyces*), the predominance of the typical (meio)sporangia of *C. stegomyiae* having masked the presence of far smaller numbers of the second species.

Again with hindsight, such an eventuality is now considered highly unlikely. As is evident from other chapters of this volume, most of the infections formerly thought to be due to two species of *Coelomomyces* in a single host have since turned out to be single-species infections characterized by the association of relatively thick-walled (sometimes smooth, but more commonly variously ornamented) resting sporangia with thin-walled sporangia—the latter being smoothly ovoid in fresh state, but subject to shrinkage and crinkling of the wall by fixatives, preservatives, and mounting media. This can confuse even experienced mycologists, finding themselves examining (for example) balsam mounts of what appear to be variously ridged sporangia accompanying thicker-walled resting sporangia much less subject to such loss of shape through the action of reagents.

D. January 18 to February 23, 1967

Dr. A. D. Hinckley's visit to Nukunono was primarily for the purpose of surveying the coconut rhinoceros beetle problem, *Oryctes rhinoceros* having become established after a presumably ship-borne introduction during 1964 (Hinckley, 1969). Together with M. Williams of the Tokelau rat control project, he examined approximately 100 *A. polynesiensis* larvae in the course of a broader insect survey. Only two of these proved positive for *C. stegomyiae,* while a few bore vorticellids (A. D. Hinckley, personal communication, 1967). This disappointing result arrived too late for inclusion in the summary of results published that year (Laird, 1967).

E. May 7–29, 1968

In the course of his own rat control investigations, Dr. K. Wodzicki collected 190 *A. polynesiensis* larvae for me at Nukunono. I examined them at the Memorial University of Newfoundland later that year and found only one from the Vao district, Nukunono motu, to be positive for resting sporangia of *C. stegomyiae*. Dr. Wodzicki's samples originated from six of the outer reef islets as well. However, none of them were from the second location where positives were found in 1960, Tokelau motu; further, his only material from the northern end of Taulagapapa motu was from Avalu—15 larvae from there were negative, as were 52 larvae in 1980.

F. November 2–4, 1980

Once more in the capacity of a WHO Consultant, I visited the Tokelaus aboard the Trans-Globe Expedition ship "Benjamin Bowring" in early November. We called at Fakaofo and Nukunono only, with two working days at the latter atoll.

Because of the limited time available, all collecting was concentrated upon the three motu where *C. stegomyiae*-positive *A. polynesiensis* larvae were found in 1960.

Thirteen containers (half-coconut shells, coconut-frond spathes, fallen rat-gnawed coconuts) were found in the Vao district of Nukunono motu and sampled. The 105 third- and fourth-instar larvae collected from them were stored in 7% formalin for later examination. Vao tungu had been filled in with sand and coral fragments in the mosquito control efforts of recent years.

A first visit to Tokelau motu was marred by extremely heavy rain, which made it impossible to dip larvae from tungu with any precision because of the total disturbance of the water. However, none of these tungu had been blocked up, and physicochemical readings were taken from eight large ones. The water temperature averaged 20°C, the pH varied from 6.9 to 7.2, and the conductivity readings were 280, 310, 320, 340, 490, 570, 820, and 840 μS (mean, 496 μS). Several of the obviously very old palms with tungu still showed clear traces of the blaze marks made by bushknife on the trunk above the tungu following the original 1958 seeding. Only seven third- and fourth-instar *A. polynesiensis* larvae were dipped fortuitously from five of these tungu. They were taken back alive to Nukunono motu, where microscopic examination that evening proved all to be negative for *Coelomomyces*.

Tokelau was revisited on 3 November, a calm, sunny day, providing ideal collecting conditions. Fourteen tungu sampled yielded 122 third- and fourth-instar *A. polynesiensis* larvae. Collecting in the northern sector of Taulagapapa motu furnished 52 more larvae from Avalu and another 215 larvae from two tungu at Te Kava Kava/Hiolonga Tautai. One of the latter tungu was very old and large, holding about 100 liters of water. It yielded some 20 third- and fourth-instar larvae per dip, and 215 larvae were collected from it.

Following the return to St. John's, all 501 larvae collected were individually examined by high-power microscopy. All 52 from Avalu proved negative for fungi, but two (3.8%) of them exhibited a few *Vorticella microstoma* on the sides of the thorax and abdomen. This observation was of some interest in that during the 1958–1960 surveys in Nukunono, when a total of 5 weeks were spent there, it was noted that larvae from tungu were remarkably free from stalked ciliates. As these are particularly characteristic of water of relatively high organic content, their absence and that of such other epibionts as filamentous bacteria serves as a good biological indication of potable water.

A single larva collected from Tokelau motu was positive for *Coelomomyces,* its hemocoel exhibiting mycelium only. Of the 215 larvae from Te Kava Kava/Hiolonga Tautai, seven proved positive for *C. stegomyiae* (meio)sporangia, with five of them derived from a single tungu. My field notes contrasted the rarity of larvae in this one with their abundance in the other from which 210 larvae were dipped, and noted that the few collected exhibited yellowish-white discoloration of the abdomen and thorax. None of the 105 Vao larvae were parasitized either.

Thus, only 8 (1.6%) of 501 third- and fourth-instars of *A. polynesiensis,* from all types of habitats sampled, proved positive for *Coelomomyces.* All mature (meio)sporangia were in every respect comparable with those of *C. stegomyiae* originally brought from Singapore, typical examples from 1958 and 1980 material being illustrated in Figs. 626 and 627, respectively.

As in 1958–1960, there were abundant copepods and ostracods in the tungu sampled. Some of these were preserved in 7% formalin. The late Dr. J. P. Harding of the British Museum of Natural History identified as *Cypridopsis* sp. some ostracods that I had collected from Motuhanga district, Nukunono motu, in 1953 (Laird, 1956b). This earlier material was subsequently reexamined by Victor and Fernando (1978), who described the ostracod as *Cypridopsis lairdi.* Dr. R. Victor (personal communication, 1981) reported that all specimens collected in 1980 were referable to this species. The copepods proved to be *Bryocyclops fidjiensis* and *Elaphoidella taroi.* Dr. H. Yeatman (personal communication, 1981), who kindly provided the identifications, wrote that "these were to be expected because the Tokelau group . . . is not very far from the Samoan and Fiji Islands, where these species are common."

Elaphoidella taroi has only recently (Toohey *et al.*, 1982) been demonstrated to be a good intermediate host of *Coelomomyces* sp. in Fiji. These workers' experiments there showed that *B. fidjiensis* is only marginally significant in this respect.

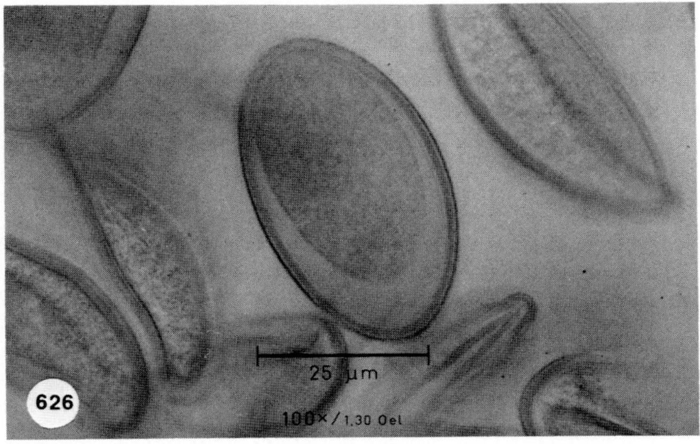

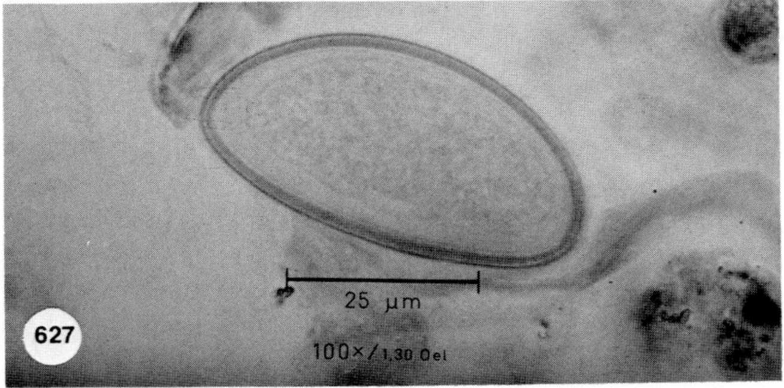

Figs. 626 and 627. Meiosporangium of *Coelomomyces stegomyiae*. Fig. 626. Zeiss Axiomat photomicrograph from a Singapore *Aedes albopictus* preparation, 1958. Fig. 627. Zeiss Axiomat photomicrograph from *Aedes polynesiensis,* Nukunono, 1980.

VII. DISCUSSION AND CONCLUSIONS

In 1958, as already indicated, Colless and I had accepted the prevailing assumption that zoospores discharged from dehiscing *Coelomomyces stegomyiae* sporangia derived from parasitized aedine larvae, would infect larval *Aedes polynesiensis* directly. The subsequent demonstration of establishment of the fungus from 1959 to the high peak of 37.1% habitats positive in 1963 was duly interpreted as evidence that such infections had taken place as expected. It is now obvious that our project had benefited from serendipity, suitable intermediate hosts already being present as part of the original aquatic fauna. Whether or not the ostracod *C. lairdi* may also be a suitable intermediate host remains to be established; of course, it would be desirable to repeat at Nukunono the life cycle achieved by Toohey *et al.* (1982) in Fiji.

In view of what was already known in 1958 of the high incidence of certain *Coelomomyces* populations in nature—especially at the *C. indicus* site that J. Muspratt (1963) had for some 20 years kept under intermittent examination in an *Anopheles gambiae* population near the Victoria Falls, in what is now Zambia—it had been hoped that the inoculative introduction of *C. stegomyiae* into Nukunono might lead to high levels of infection in an isolated nonimmune host.

There was certainly encouraging evidence for this up to 1963. However, a disappointingly low level of infection was already evident by 1967/1968. The level now reported, 22 years after the 1958 introduction, indicates that a host–parasite balance has been attained, with fungal infections persisting, but at low incidence, not significantly higher than the original source of the material, Singapore. There, as already mentioned, while only 48 (2.0%) of 2454 *Aedes albopictus* larval habitats sampled in 1958 contained infected larvae, the presence of viable sporangia in bottom debris of many habitats had led to the suspicion that perhaps as many as 5% of *A. albopictus* habitats were really *Coelomomyces*-positive (Laird and Colless, 1959).

Of the 40 larval habitats of *A. polynesiensis* sampled in 1980 on the once (1963) heavily infected motu of Nukunono atoll, at least three tungu (7.5% of the sample) from the motu of Tokelau and Taulagapapa were positive for *C. stegomyiae*. However, of the 501 larvae collected from all habitats, only 8 (1.6%) proved to be parasitized. The original Singapore figures thus bear close comparison with the 1980 ones from Nukunono.

Especially in view of the presumption of the post-1966 hurricane arrival of the microsporidan *Vavraia culicis* and the gregarine *Lankesteria culicis* at Nukunono, through the introduction and establishment of parasitized *A. aegypti* (Laird, 1982), it may be asked whether a reintroduction of the fungus might have taken place in the same manner. It is submitted that this is most unlikely. Although *A. aegypti* is susceptible to *C. stegomyiae,* and *Coelomomyces* sp. has been reported from its larvae in Samoa and Tonga (Ramalingam, 1966), natural incidence tends to be low. Also, as is generally the case with *Coelomomyces* spp., parasitized examples almost always die in the larval stage. The microsporidan and gregarine just mentioned both normally pass through into adult mosquitoes. However, the survival of *Coelomomyces* into adult mosquitoes with the production of mature sporangia is rare enough to be noteworthy. In 1958 at Singapore, for example, Colless found that while, in several instances studied in the laboratory, light infections of *A. albopictus* larvae by *C. stegomyiae* did not preclude normal pupation and eclosion (4 of 9 resultant females and 9 of 10 males showing numerous sporangia in the body cavity), "larval parasitaemia in naturally seeded habitats was usually heavy, and the larvae failed to develop any further" (Laird and Colless, 1959).

Again, both *V. culicis* and *L. culicis* appear to be still confined to the village motu of Nukunono, while seven of the eight of the *C. stegomyiae*-positive larvae collected in 1980 originated from the motu of Tokelau (some 11 km across the lagoon) and Taulagapapa (some 5 km across the lagoon). It is therefore presumed that the few parasitized *A. polynesiensis* larvae found during the 1980 Nukunono survey represented persistence of the establishment achieved 22 years previously.

Because the level of infection now existing is quite close to that evident from the 1958 survey of the parent *C. stegomyiae* stock in *A. albopictus* in Singapore, it is felt that the inoculative approach followed at Nukunono simply achieved what may be typical of accidental range-extensions of *Coelomomyces* spp. in nature: an initial establishment, with a temporary build-up of parasites in a nonimmune host population (to 37.1% in the 1963 sample examined by Couch), followed by the achievement of a balance between host and parasite at a relatively low level of infection. From the similarity between the findings in 1967 and 1968 and those now reported for 1980, this balance may already have been achieved rather more than a decade ago.

It is thus considered that future attempts to use *Coelomomyces* spp. against mosquitoes would best be planned on a repetitive, inundative basis. At the present time, such progress as has been made towards *in vivo* mass production cannot assure the quantities of infective material necessary for such ventures. There is perhaps a possibility that mass production of infective copepods, as outlined by Federici (1980), might help to some extent. As some suitable freshwater copepods are already virtually cosmopolitan anyway, there would not seem to be any significant environmental hazard in this approach. Nevertheless, relevant progress remains slow. Even when large-scale production of infective material becomes practical, it will still be necessary to undertake further safety-related studies, as in the case of all fungal material designed for handling in bulk quantities.

It might also be borne in mind that it has long been known that microcrustaceans compete for food with mosquito larvae (Weed, 1924) and sometimes prey upon the smallest stages of the latter. It should be noted that Platzer and MacKenzie-Graham (1980) have shown that copepods [the species concerned, *Cyclops vernalis*, being a good host for *Coelomomyces* spp. (Federici, 1980)] attack the preparasites of another biocontrol agent with anti-*Stegomyia* potential: *Romanomermis culicivorax*. However, Laird *et al.* (1982) have demonstrated that *R. culicivorax* has survived in at least one Fakaofo tungu for almost 3 years, despite the fact that on that atoll, too, copepods and ostracods are universally abundant in tungu. Another negative quality of certain copepods (although hardly one of potential significance in South Pacific atolls!) is that in Africa and Asia cyclopids (e.g., *C. vernalis*!) are the hosts of the causal agent of dracunculiasis.

Furthermore, two other fungal agents pathogenic for mosquitoes, *Lagenidium giganteum* (Washino, 1981) and *Culicinomyces clavisporus* (Sweeney, 1981), have now become prominent as candidate biocontrol agents. Being mass-producible by known *in vitro* technology (Sweeney, 1981), the latter fungus is at present closer than *Coelomomyces* spp. to practicality.

It may of course be questioned whether *Coelomomyces* fungi, now that they are known to require two hosts for completion of the life cycle, should continue to be ranked as leading biocontrol candidates. Perhaps a decision should be shelved pending the outcome of current work towards the development of practical *in vitro* culture procedures, recent advances toward which are considered in detail in Chapter 7.

To end on a less pessimistic note with respect to the future viability of practical biocontrol application of *Coelomomyces* spp. against mosquitoes, and with specific regard to the 25-year Tokelau project, it is reflected that the main achievement of the 1958–1960 investigations was probably the fact that they concentrated wide interest upon biocontrol in medical entomology two decades ago. Sustained interest on the part of the many invertebrate pathologists who have concentrated their efforts in this field since then has led to a vast expansion of knowledge (much of this work also supported by WHO), which is progressively broadening the choice of promising biocontrol candidates suited for intensive exploitation as elements in future integrated mosquito control methodologies.

ACKNOWLEDGMENTS

Appreciation is expressed to the World Health Organization for the financial support that made possible the 1958, 1960, and 1980 investigations in the Tokelaus. None of the visits to the Tokelaus would have been possible without transportation assistance from the Royal New Zealand Air Force (1953, 1958, 1960), the Royal New Zealand Navy (1958) and the Office for Tokelau Affairs (1960, 1980). The efforts of Dr. D. H. Colless at Singapore were fundamental to the collection of *C. stegomyiae* infective material in 1958, and appreciation is also expressed to others who were then members of his group at the University's Department of Parasitology: Cheong Chee Hock, the late Rev. C. A. Caldwell, W. T. Chellappah, and Enche' Rosnari bin H. Omar. The then-chairman of the Department of Parasitology, Professor A. A. Sandosham, and the late Professor T. W. M. Cameron, Director of McGill University's Institute of Parasitology while I was on the Faculty there from 1958–1961, were both generous in facilitating Dr. Colless's and my participation in the project, the 1958 field work for which had a heavy contribution from Dr. Colless. On each visit to Nukunono, The Faipule and Pulenu'u arranged for small boat transportation and field assistance, and the work could not have been completed in the time available without the enthusiastic assistance of many Tokelauans who must remain nameless for reasons of space. Special thanks are due to Father Camille Desrosiers, S. M., and Kalolo Perez for hospitality at Nukunono in 1958–1960 and 1980 respectively. Finally, warm appreciation is expressed to Dr. K. Wodzicki, OBE, for his kindness in collecting the 1968 specimens referred to herein, to members of the Office of Tokelau Affairs, Apia (not forgetting Opeta, who accompanied me as Field Assistant on the 1960 visit), and to others named in the body of this chapter who helped in various ways.

REFERENCES

Anonymous (1957). Insecticides. Seventh report of the Expert Committee on Insecticides. *W. H. O. Tech. Rep. Ser.* **125,** 1–31.

Anonymous (1958). Insect resistance and vector control. Eighth report of the Expert Committee on Insecticides. *W. H. O. Tech. Rep. Ser.* **153,** 1–67.

Couch, J. N. (1968). Sporangial germination of *Coelomomyces punctatus* and the conditions favoring the infection of *Anopheles quadrimaculatus* under laboratory conditions. *Proc. Jt. U.S.-Jpn. Semin. Microb. Control Insect Pests, 1967* pp. 93–105.

Couch, J. N. (1972). Mass production of *Coelomomyces* a fungus that kills mosquitoes. *Proc. Natl. Acad. Sci. U.S.A.* **69,** 2043–2047.

Federici, B. A. (1977). Laboratory infection of *Anopheles freeborni* with the parasitic fungi *Coelomomyces dodgei* and *Coelomomyces punctatus*. *Proc. Pap. Annu. Conf. Calif. Mosq., Vector Control Assoc.* **45,** 107–108.

Federici, B. A. (1979). Experimental hybridization of *Coelomomyces dodgei* and *Coelomomyces punctatus*. *Proc. Natl. Acad. Sci. U.S.A.* **76,** 4425–4428.

Federici, B. A. (1980). Production of the mosquito-parasitic fungus, *Coelomomyces dodgei*, through synchronized infection and growth of the intermediate copepod host, *Cyclops vernalis*. *Entomophaga* **25,** 209–217.

Federici, B. A., and Roberts, D. W. (1976). Experimental laboratory infection of mosquito larvae with fungi of the genus *Coelomomyces*. II. Experiments with *Coelomomyces punctatus* in *Anopheles quadrimaculatus*. *J. Invertebr. Pathol.* **27,** 333–341.

Hinckley, A. D. (1969). Ecology of terrestrial arthropods on the Tokelau Islands. *Atoll Res. Bull.* **124,** 1–18.

Laird, M. (1955). Notes on the mosquitoes of the Gilbert, Ellice and Tokelau Islands, and on filariasis in the latter group. *Bull. Entomol. Res.* **46,** 291–300.

Laird, M. (1956a). A place for parasites and invertebrate predators in mosquito control. 7th Session of the Expert Committee on Insecticides, Geneva, 17 April, 1956. *W. H. O. Work. Pap.* No. 2.

Laird, M. (1956b). Studies of mosquitoes and freshwater ecology in the South Pacific. *Bull. R. Soc. N. Z.* **6,** 1–213.

Laird, M. (1959). Parasites of Singapore mosquitoes, with particular reference to the significance of larval epibionts as an index of habitat pollution. *Ecology* **40,** 206–221.

Laird, M. (1960). "Experiments Towards the Biological Control of Mosquitos in the Tokelau Islands," 2nd rep. Summarized result of assessment surveys made in November 1959 and April 1960. Mimeogr. Doc., *W. H. O. MHO/PA/173.60.* WHO, Geneva.

Laird, M. (1967). A coral island experiment: A new approach to mosquito control. *WHO Chron.* **21,** 18–26.

Laird, M. (1982). Gregarine and microsporidan protozoa in *Aedes polynesiensis,* Tokelau Islands. Recent accidental importations? *Can. J. Zool.* **60,** 1922–1929.

Laird, M., and Colless, D. H. (1959). "Experiments towards the Biological Control of Mosquitos in the Tokelau Islands," 1st rep. August-October, 1958. Mimeogr. Doc., *W. H. O. MHO/PA/93.59.* WHO, Geneva.

Laird, M., Urdang, J., and Tinielu, I. (1982). Establishment and long-term survival of *Romanomermis culicivorax* in mosquito habitats, Tokelau Islands. *Mosq. News* **42,** 86–92.

Madelin, F. (1968). Studies on the infection by *Coelomomyces indicus* of *Anopheles gambiae. J. Elisha Mitchell Sci. Soc.* **84,** 115–124.

Muspratt, J. (1946). On *Coelomomyces* fungi causing high mortality of *Anopheles gambiae* larvae in Rhodesia. *Ann. Trop. Med. Parasitol.* **40,** 10–17.

Muspratt, J. (1963). Destruction of the larvae of *Anopheles gambiae* Giles by a *Coelomomyces* fungus. *Bull. W. H. O.* **29,** 81–86.

Platzer, E. G., and MacKenzie-Graham, L. L. (1980). *Cyclops vernalis* as a predator of the preparasitic stages of *Romanomermis culicivorax. Mosq. News* **40,** 252–257.

Ramalingam, S. (1966). *Coelomomyces* infections in mosquito larvae in the South Pacific. *Med. J. Malaya* **20,** 334.

Steinhaus, E. A. (1949). "Principles of Insect Pathology." McGraw-Hill, New York.

Sweeney, A. W. (1981). Prospects for the use of *Culicinomyces* fungi for biocontrol of mosquitoes. *In* "Biocontrol of Medical and Veterinary Pests" (M. Laird, ed.), pp. 105–121. Praeger, New York.

Toohey, M. K., Prakash, G., Goettel, M. S., and Pillai, J. S. (1982). *Elaphoidella taroi*: the intermediate copepod host in Fiji for the mosquito pathogenic fungus *Coelomomyces. J. Invertebr. Pathol.* **40,** 378–382.

Travland, L. B. (1979). Initiation of infection of mosquito larvae (*Culiseta inornata*) by *Coelomomyces psorophorae. J. Invertebr. Pathol.* **33,** 95–105.

Victor, R., and Fernando, C. H. (1978). Systematics and ecological notes on Ostracoda from container habitats of some South Pacific islands. *Can. J. Zool.* **56,** 414–422.

Walker, A. J. (1938). Fungal infections of mosquitoes, especially of *Anopheles costalis. Ann. Trop. Med. Parasitol.* **32,** 231–244.

Washino, R. K. (1981). Biocontrol of mosquitoes associated with California rice fields with special reference to the recycling of *Lagenidium giganteum* Couch and other microbial agents. *In* "biocontrol of Medical and Veterinary Pests" (M. Laird, ed.), pp. 122–139. Praeger, New York.

Weed, A. C. (1924). Another factor in mosquito control. *Ecology* **5,** 110–111.

Weiser, J. (1976). The intermediary host for the fungus *Coelomomyces chironomi. J. Invertebr. Pathol.* **28,** 273–274.

Whisler, H. C., Zeobold, S. L., and Shemanchuk, J. A. (1974). Alternate host for mosquito parasite *Coelomomyces. Nature (London)* **251,** 715–716.

Whisler, H. C., Zeobold, S. L., and Shemanchuk, J. A. (1975). Life history of *Coelomomyces psorophorae. Proc. Natl. Acad. Sci. U.S.A.* **72,** 693–696.

Index

T

U

V

W

X

Y

Z